Firefighter Safety and Survival

Firefighter Safety and Survival

Don Zimmerman

DELMAR
CENGAGE Learning

Australia • Brazil • Japan • Korea • Mexico • Singapore • Spain • United Kingdom • United States

Firefighter Safety and Survival
Don Zimmerman

Vice President, Editorial: Dave Garza

Director of Learning Solutions: Sandy Clark

Senior Acquisitions Editor: Janet Maker

Managing Editor: Larry Main

Development Editor: Julie Scardiglia

Vice President, Marketing: Jennifer Baker

Marketing Director: Deborah Yarnell

Marketing Manager: Mark Linton

Associate Marketing Manager: Erica Ropitzky

Senior Production Director: Wendy Troeger

Production Manager: Mark Bernard

Senior Content Project Manager: Jennifer Hanley

Senior Art Director: Casey Kirchmayer

Cover image photography: Brain Fowler

Cover image pilot: Mark Makee, Makee insurance

© 2012 Delmar, Cengage Learning

ALL RIGHTS RESERVED. No part of this work covered by the copyright herein may be reproduced, transmitted, stored, or used in any form or by any means graphic, electronic, or mechanical, including but not limited to photocopying, recording, scanning, digitizing, taping, Web distribution, information networks, or information storage and retrieval systems, except as permitted under Section 107 or 108 of the 1976 United States Copyright Act, without the prior written permission of the publisher.

> For product information and technology assistance, contact us at
> **Cengage Learning Customer & Sales Support,
> 1-800-354-9706**
>
> For permission to use material from this text or product, submit all requests online at **www.cengage.com/permissions.**
>
> Further permissions questions can be e-mailed to
> **permissionrequest@cengage.com**

Library of Congress Control Number: 2011934947

ISBN-13: 978-1-111-30660-1

ISBN-10: 1-111-30660-5

Delmar
5 Maxwell Drive
Clifton Park, NY 12065-2919
USA

Cengage Learning is a leading provider of customized learning solutions with office locations around the globe, including Singapore, the United Kingdom, Australia, Mexico, Brazil, and Japan. Locate your local office at: **international.cengage.com/region**

Cengage Learning products are represented in Canada by Nelson Education, Ltd.

To learn more about Delmar, visit **www.cengage.com/delmar**

Purchase any of our products at your local college store or at our preferred online store **www.cengagebrain.com**

Notice to the Reader
Publisher does not warrant or guarantee any of the products described herein or perform any independent analysis in connection with any of the product information contained herein. Publisher does not assume, and expressly disclaims, any obligation to obtain and include information other than that provided to it by the manufacturer. The reader is expressly warned to consider and adopt all safety precautions that might be indicated by the activities described herein and to avoid all potential hazards. By following the instructions contained herein, the reader willingly assumes all risks in connection with such instructions. The publisher makes no representations or warranties of any kind, including but not limited to, the warranties of fitness for particular purpose or merchantability, nor are any such representations implied with respect to the material set forth herein, and the publisher takes no responsibility with respect to such material. The publisher shall not be liable for any special, consequential, or exemplary damages resulting, in whole or part, from the readers' use of, or reliance upon, this material.

Printed in the United States of America
1 2 3 4 5 6 7 15 14 13 12 11

CONTENTS

FOREWORD / IX

PREFACE / XI

ABOUT THE AUTHOR / XVIII

ACKNOWLEDGMENTS / XIX

FESHE / XXII

INTRODUCTION / XXVI

Chapter 1 DEFINING A CULTURAL CHANGE / 1

Change / 2
Leadership During Change / 7
Safety Culture / 15
Cultural Compliance / 29
Summary / 29
Key Terms / 30
Review Questions / 31
Firefighting Website Resources / 31
Notes / 31

Chapter 2 ENHANCING ACCOUNTABILITY / 32

Accountability / 33
Blame / 35
Personal Accountability / 38
Improving Personal Accountability / 41
Organizational Responsibility / 47
Health and Safety Accountability / 49
Summary / 56
Key Terms / 56
Review Questions / 57
Firefighting Website Resources / 58
Notes / 58

Chapter 3 APPLYING RISK MANAGEMENT TECHNIQUES / 59

Risk-Benefit Analysis / 60
Risk Management: Overview / 63
Decision Making / 64
Incident Management / 69
Risk Management at the Strategic Level / 77
Risk Management at the Tactical Level / 81

Risk Management at the Task Level / 83
Improving the Safety of the Fireground / 93
Summary / 93
Key Terms / 94
Review Questions / 95
Firefighting Website Resources / 95
Notes / 96

Chapter 4 ELIMINATING UNSAFE ACTS / 97

The Foundation of Unsafe Practices / 98
Firefighter Empowerment / 100
Accidents / 103
Recognition of Unsafe Acts / 108
Preventing Unsafe Acts Before They Occur / 112
Stopping Unsafe Acts in Progress / 115
Crew Resource Management / 121
Summary / 130
Key Terms / 131
Review Questions / 132
Student Activity / 132
Firefighting Website Resources / 133
Notes / 133

Chapter 5 IMPLEMENTING TRAINING AND CERTIFICATION STANDARDS / 134

Opposing Concepts / 135
How We Got Here / 138
Evaluating Where We Are / 139
Establishing Where We Should Be / 151
Summary / 156
Key Terms / 156
Review Questions / 157
Firefighting Website Resources / 158
Notes / 158

Chapter 6 DEVELOPING MEDICAL AND FITNESS STANDARDS / 159

A Common Push for Health and Fitness / 162
Body Metabolism / 164
Physical Agility Testing / 176
Maintenance Versus Repairs / 182
Efficiency Factors of Firefighting / 184
Summary / 187
Key Terms / 187
Review Questions / 189
Firefighter Website Resources / 190
Notes / 190

Chapter 7 CREATING A RESEARCH AGENDA / 191

Research as a Tool / 194
National Research Agenda / 197
Data Collection / 215
Summary / 219
Key Terms / 220
Review Questions / 221
Student Activity / 221
Firefighter Website Resources / 222
Notes / 222

Chapter 8 UTILIZING AVAILABLE TECHNOLOGY / 223

Overreliance on Technology / 224
Utilization of Technology / 226
Types of Line-of-Duty Deaths / 228
Summary / 245

Key Terms / 245
Review Questions / 246
Firefighter Website Resources / 247

Additional Resources / 247
Notes / 247

Chapter 9 INVESTIGATING FATALITIES, INJURIES, AND NEAR-MISSES / 249

Investigating Fatalities / 251
Near-Miss Reporting / 253
Building a Successful System / 259
Summary / 268
Key Terms / 269

Review Questions / 270
Student Activity / 271
Firefighter Website Resources / 271
Additional Resources / 272
Notes / 272

Chapter 10 BLENDING GRANTS AND SAFETY / 273

Who Offers Grants? The Grantors / 276
Classifications of Grants / 279
Choosing a Grant / 285
Components of a Grant / 287
Awarding of a Grant / 295
Summary / 296

Key Terms / 297
Review Questions / 298
Student Activity / 298
Firefighter Website Resources / 299
Additional Resources / 299
Notes / 299

Chapter 11 ESTABLISHING RESPONSE STANDARDS / 300

Terminology / 302
Levels of Standard Policies and Procedures / 309
Implementing the Standards / 321
Summary / 322

Key Terms / 323
Review Questions / 324
Firefighter Website Resources / 324
Additional Resources / 324
Notes / 325

Chapter 12 EXAMINING RESPONSE TO VIOLENT INCIDENTS / 326

Terrorism / 327
Job Description / 330
Standard Procedures / 332
Type of Incident / 337
Summary / 344

Key Terms / 345
Review Questions / 345
Firefighting Website Resources / 346
Additional Resources / 346
Notes / 346

Chapter 13 PROVIDING EMOTIONAL SUPPORT / 347

Stress: Overview / 348
Post-Traumatic Stress Disorder / 352
The Future / 363
Summary / 366

Key Terms / 367
Review Questions / 368
Additional Resources / 368
Notes / 368

Chapter 14 ENABLING PUBLIC EDUCATION / 369

Prevention Versus Preparation / 370
Progressive Message / 372
Developing a Program / 375
Partnerships / 386
Life Safety Education for Firefighter Safety / 391
Summary / 394
Key Terms / 394
Review Questions / 395
Notes / 395

Chapter 15 ADVOCATING RESIDENTIAL FIRE SPRINKLERS / 396

Residential Sprinklers / 397
The Opposition / 401
Different Paths / 410
Summary / 420
Key Terms / 420
Review Questions / 421
Additional Resources / 421
Notes / 422

Chapter 16 ENGINEERING SAFETY INTO EQUIPMENT / 423

The Safety Curve / 425
The Origination of Safety / 428
Designing Safety / 435
Retrofitting Safety / 445
Summary / 449
Key Terms / 450
Review Questions / 451
Student Activity / 451
Additional Resources / 451
Notes / 451

APPENDICES

Appendix A   SOP-1 No-Fault Management / 453

Appendix B   SOP-2 Crew Resource Management / 457

Appendix C   SOP-3 Residential Sprinkler Retrofits / 460

Appendix D   SOP-4 Purchasing Policy / 463

Appendix E   SOP-5 Secured Equipment / 466

Appendix F   Firefighter Injuries and Fatalities Report / 469

Appendix G   Fire Department Self-Survey: Analyzing Your Profile / 472

GLOSSARY / 475

INDEX / 487

FOREWORD

By now, everyone has heard of terms such as *Everyone Goes Home*, the *16 Firefighter Life Safety Initiatives, Courage to Be Safe*, the *LACK program, line-of-duty death*, and *PSOB*. The 16 Firefighter Life Safety Initiatives (16 FLSI) were developed from the Life Safety Summit in 2004 and were put together using data from line-of-duty death investigations, the reports of firefighter injuries, and data collected from fire officials (private and public) and survivors. Six domains were established: (1) Prevention; (2) Structural Firefighting; (3) Wildland Operation; (4) Health, Wellness, and Fitness; (5) Vehicles; and (6) Training. When we focus on these domains within our scope of work, we will reduce line-of-duty deaths and injuries.

    We are now in a position to create a culture change within our own fire departments to reduce line-of-duty deaths and injuries and support the crusade that the National Fallen Firefighters Foundation has created under the leadership of Ron Sarniki. Several program directors, regional advocates, and support staff assist in this endeavor. In addition, in each state, there are advocates and support staff. On the average, we still lose 100 firefighters in the line of duty and approximately 80,000 firefighters are injured each year. Heart attacks and vehicle crashes are still the leading cause of firefighter fatalities. A change in the culture can help. A change in attitude can help. A realization that you are responsible can help. Our leaders have to take responsibility and lead by example and empower firefighters and officers to help make decisions to provide a better safety environment for our firefighters.

    One way we can really help with a culture change is to start with our new recruits in their fire cadet school and have resources available to us older/senior firefighters. You may even call us blue hair dinosaurs! Most firefighters believe in the 16 FLSI. Many fire officers realize that you just can't drop the 16 FLSI on the laps of the firefighters and expect them to understand how to implement them. It is one thing to read them and say, "Yeah, we got it," but do you really understand them and work within the initiatives from your senior staff members to the probationary firefighter on the department? An example would be Initiative 4, *All firefighters must be empowered to stop unsafe acts.* Easily said, but how does a new firefighter tell a senior firefighter that he or she is doing something unsafe and needs to stop? No one likes to be criticized as a human being in general and when you are a firefighter, it becomes worse. So we have to start this training and mind-set in cadet school with our new firefighters (culture change), and we have to retrain some of our officers and senior firefighters how to accept new ideas and constructive criticism. This book uses some real-world situations to demonstrate how to use and receive "empowerment."

Don Zimmerman has taken the 16 FLSI and broken them down so we all can relate to them and better understand them in a real-world environment. As you read each chapter, all 16 of them, you will relate to the examples this text provides. Don has a talent in communications that many of us wish we had in the fire service. His imagination for demonstration delivers an outstanding product in *Firefighter Safety and Survival.* He has taken on the challenge to help create and support the culture change within the fire service and provide ways to assist fire departments around the world to implement the 16 FLSI. Don carries on the mission of the Everyone Goes Home® Firefighter Life Safety Initiatives Program with all of us in mind.

Sometimes the unpredictable happens on the fireground and our firefighters get hurt or die. It is up to all of us to make sure we are working within standard operating guidelines in a safe manner. No one is saying not to fight fires, but we need to do it in a calculated and safe manner within the incident action plan. With the help of all fire officers, we can change the culture in the fire service to reduce firefighter line-of-duty deaths and injuries. Please use this book to your advantage and understand that we all have a responsibility in firefighter safety. As officers, we must empower others to feed us information to make wise decisions. As chief officers, we must provide the leadership that is expected of us. It is our duty!

Ron Terriaco
Deputy Fire Chief Concord Twp. Fire Department
Lake County, Ohio
Ohio Advocate, Everyone Goes Home 16 FLSI Program

PREFACE

INTENT OF THIS BOOK

This text is written to the Fire and Emergency Services Higher Education (FESHE) course outcomes, and includes components of the course outline for Firefighter Safety and Survival. Its primary use will be for postsecondary education. It also mirrors the 16 Firefighter Life Safety Initiatives, with one chapter devoted to implementing each initiative. This makes it suitable for promotional testing as well.

Three of the latest trends with the fire service (as it relates to college education) are following the FESHE guidelines, a greater emphasis on the application of risk management, and online courses. This textbook is different than any safety book ever written. Unlike other textbooks that merely touch the surface of safety, this book delves into the 16 initiatives and gives realistic examples from fire and emergency services that each student can apply. Every chapter is full of situations, which could each be used as a case study on its own. Instructors can start discussions or assign projects based on the specific examples. By writing a textbook from this vantage point, it actually teaches the readers skills in risk management as they progress through the book. The text also has several new methods of increasing safety and implementing the initiatives never introduced to the fire service before. But possibly the best feature of the book is its application for online classes. Because of the amount of new information provided, distance learners can participate in online discussions and projects which would in turn encourage them to apply the changes and raise the level of safety in their geographical area.

WHY I WROTE THIS BOOK

This book is the culmination of years of studying, listening, teaching, experimenting, hard work, and asking questions. When the Fallen Firefighters Foundation released the 16 initiatives, I looked forward to getting a glimpse of our future: Finally, a blueprint we can use to build a safer fire service, complete with a supply list and instructions. I attended lectures and classes, and listened to advocates speak. I watched every Life Safety Resource Kit and read online white papers looking for solutions, searching for a way to make my fire department safer. As I researched, I realized the initiatives were a brilliant list of conceptual drawings and cutting-edge ideas, and it was up to instructors to fill in the details and teach proper and safe practices. But where to start? For instance, what is empowerment, and how does a rookie firefighter stop an unsafe practice? What's a safety culture and how would we adopt it? Sure we should purchase safe equipment, but what is *safe equipment*?

This is where my mind-set was when I started on this project. What I have to offer, and what I strive to offer in this book, is ambition and creativity. I enjoy taking a problem and tearing it apart. I like to research it and look at it from different angles. I live for developing unique solutions, which I believe is sometimes a blessing and other times a curse. As for the initiatives, I decided early on that the only way to implement them was by comparing other industries to fire and emergency services. Many of them have already solved problems we are up against and have solutions that work. Although we can't carry over every aspect, they provide great resources for making our own solutions.

I had three ground rules I tried to stick to while writing. First off, no firefighters would die in this book. Rather than use real fatalities as case examples, the fictitious stories bring the reader into the situations. Emergency responders are thrown into uncomfortable situations all the time, so the reader should have no problem fitting into the scenarios and being forced to make decisions. When appropriate, actual line-of-duty death reports can easily be researched by the student online. The second rule was that the book would be written like we were sitting on the front bumper of the engine solving the world's problems. The first initiative looks to embrace a safety culture, which means safety is built in to what we do. Instead of preaching safety from a pulpit in a green vest with eye protection in place, safety needs to be reintroduced as *normal*. There's no reason why safety and operations can't get along. The final rule was the book would give concrete examples. When I asked Deputy Chief Ron Teriacco to proofread the manuscript, I told him to not let me wimp out on my examples. There are far too many textbooks that touch on an issue, but don't provide real examples of difficult situations.

I believe I have accomplished my three goals, and hope that it helps emergency responders of all ages and experience levels to bring up the level of their own safety. For example, I hope a fire department that has questions about Initiative 9 can pull the book off the shelf, turn to Chapter 9, and find some real ways to fix a problem. I hope students who read it reluctantly because they have to will come away with a feeling that safety is feasible, and walk away with at least one lesson that changes their behavior. But most of all, I hope this book helps the National Fallen Firefighters Foundation achieve their goal of reducing the number of line-of-duty deaths. There are far too many needless injuries and fatalities in our line of work. Thank you for the opportunity.

HOW THIS BOOK IS ORGANIZED

The chapters in this book align with the 16 Initiatives, as well as the FESHE course objectives. The initiatives are supported by solutions identified from other industries that could be applied to fire and emergency services. Each chapter is designed to encourage implementation of the initiatives. Even the chapter titles start with an action verb to set the tone.

CHAPTER 1, DEFINING A CULTURAL CHANGE *Initiative 1: Define and advocate the need for a cultural change within the fire service relating to safety; incorporating leadership, management, supervision, accountability and personal responsibility.* This chapter examines the culture of fire and emergency services, and what cultural change is necessary to increase safety. The nuclear industry sheds light on what the definition of a safety culture is, and how it successfully implemented changes.

CHAPTER 2, ENHANCING ACCOUNTABILITY *Initiative 2: Enhance the personal and organizational accountability for health and safety throughout the fire service.* Chapter 2 looks at what it means to be accountable, and how to enhance accountability in your fire department or agency. Examples of both personal and organizational accountability for safety are identified, including the introduction to a safety system that can be designed to measure progress.

CHAPTER 3, APPLYING RISK MANAGEMENT TECHNIQUES *Initiative 3: Focus greater attention on the integration of risk management with incident management at all levels, including strategic, tactical, and planning responsibilities.* Getting safety into the operational component begins in Chapter 3. By using the U.S. Coast Guard's SPE model, students learn how to assign a number to a situation to better understand the risk involved. After establishing risk, the benefit can be weighed to establish if the action has a desired effect. OSHA's two-in, two-out rule is examined and applied to rapid intervention. Communications is covered in depth, with comparisons to other stressful communications.

CHAPTER 4, ELIMINATING UNSAFE ACTS *Initiative 4: All firefighters must be empowered to stop unsafe practices.* In this chapter, empowerment is viewed from the emergency responder's eyes. Numerous situations involve the application of crew resource management (CRM) techniques, specifically the challenge and response model that is extremely effective in applying empowerment.

CHAPTER 5, IMPLEMENTING TRAINING AND CERTIFICATION STANDARDS *Initiative 5: Develop and implement national standards for training, qualifications, and certification (including regular recertification) that are equally applicable to all firefighters based on the duties they are expected to perform.* Chapter 5 looks into the benefits of national certifications and recertification, and explores organizations such as Fire and Emergency Services

Higher Education (FESHE). NFPA professional qualifications are discussed, as is the benefits of credentialing emergency responders.

CHAPTER 6, DEVELOPING MEDICAL AND FITNESS STANDARDS *Initiative 6: Develop and implement national medical and physical fitness standards that are equally applicable to all firefighters, based on the duties they are expected to perform.* This chapter takes medical and fitness information, and packages into a comparison that most firefighters can understand. It examines programs by the National Volunteer Fire Council, the International Associations of Fire Chiefs and Fire Fighters, and the NFPA.

CHAPTER 7, CREATING A RESEARCH AGENDA *Initiative 7: Create a national research agenda and data collection system that relates to the initiatives.* Although Chapter 7 defines research and data collection, it gives specific examples of research to each of the initiatives.

CHAPTER 8, UTILIZING AVAILABLE TECHNOLOGY *Initiative 8: Utilize available technology wherever it can produce higher levels of health and safety.* Technology can be used to increase safety in a variety of instances. This chapter compares the leading causes of injury and death to technological advances that may be used to produce higher levels of health and safety in each category.

CHAPTER 9, INVESTIGATING FATALITIES, INJURIES, AND NEAR-MISSES *Initiative 9: Thoroughly investigate all firefighter fatalities, injuries, and near misses.* This chapter teaches how to conduct investigations, in an effort to identify the root cause and contributing factors of an event. Injury patterns are explained, and a system of leverage points is identified as a way of organizing the information gained from the investigation. The final component is to assemble the findings and get them out and available for use to prevent the next occurrence.

CHAPTER 10, BLENDING GRANTS AND SAFETY *Initiative 10: Grant programs should support the implementation of safe practices and/or mandate safe practices as an eligibility requirement.* Grants are becoming very popular ways to increase funding for emergency agencies. This chapter identifies the different types of grants available, along with several different funding sources. Common components of grants are explained along with suggestions by experts for writing a successful grant. This base of knowledge is used to identify how grants can be used to increase safety, and to create an FLSI-16 registry of agencies that meet the criteria for application.

CHAPTER 11, ESTABLISHING RESPONSE STANDARDS *Initiative 11: National standards for emergency response policies and procedures should be developed and championed.* Two of the difficulties of applying national standards to emergency response are the vast differences between agencies and identifying what an emergency response includes. Chapter 11 begins by explaining where standards come from, and defining policies, procedures, and guidelines. It then takes common policies and procedures of emergency response agencies that can be improved for health and safety, and categorizes them into three different series of procedures that can be molded to fit any agency.

CHAPTER 12, EXAMINING RESPONSE TO VIOLENT INCIDENTS *Initiative 12: National protocols for response to violent incidents should be developed and championed.* This initiative addresses a concern that is likely to be more of a safety issue in the future. Terrorism has already proven to be catastrophic to firefighters and other emergency workers. Building off Chapter 11, this chapter uses the same system of common procedures to meet the needs of the initiative. Additionally, it addresses safety concerns seldom mentioned before regarding uniforms and procedures if responders find themselves in the middle of a violent event.

CHAPTER 13, PROVIDING EMOTIONAL SUPPORT *Initiative 13: Firefighters and their families must have access to counseling and psychological support.* Chapter 13 looks at the effects of psychological trauma and the treatment options used for post-traumatic stress disorder. This information is bridged to emergency responders with options such as employee assistance programs, critical incident stress management, and chaplaincies. It also takes a look at the influence of modern technology and how it may increase the use of debriefing systems through relative or total anonymity.

CHAPTER 14, ENABLING PUBLIC EDUCATION *Initiative 14: Public education must receive more resources and be championed as a critical fire and life safety program.* Fire and life safety education has been proven to help fire and emergency services reach their goals of protecting lives and property. By reducing the number and severity of incidents, education also increases our safety. Chapter 14 identifies ways public education can be improved, especially with the use of technology. It also walks the student through the development of a program using the champion model from NFPA.

CHAPTER 15, ADVOCATING RESIDENTIAL SPRINKLERS *Initiative 15: Advocacy must be strengthened for the enforcement of codes and the installation of home fire sprinklers.* This chapter starts by giving some history on home sprinklers, then breaks down NFPA 13D, the standard for residential fire sprinklers. It explains the arguments for and against mandatory sprinklers, and gives facts gained from studies. It also provides information on the benefits of code enforcement. Partial systems are discussed, and ways to promote them for firefighter safety are established.

CHAPTER 16, ENGINEERING SAFETY INTO EQUIPMENT *Initiative 16: Safety must be a primary consideration in the design of apparatus and equipment.* The last chapter identifies the reasons why safety should be designed into equipment, and some of the ways that it can occur. It explains the importance of manufacturers being involved in the process, as well as end users that push the increase in safety. Finally, firefighter inventions and suggestions are addressed, proposing a clearinghouse to ensure that they will actually increase safety.

The textbook is written in an easy-to-read format and includes volunteer, part-time paid, EMS, fire police, career departments, and wildland firefighting.

Features of This Book

- **CORRELATES TO FESHE AND 16 FIREFIGHTER LIFE SAFETY INITIATIVES:** The information presented in this book not only correlates to the FESHE Firefighter Safety and Survival course and the 16 Firefighter Life Safety Initiatives, but also seeks to provide guidance for these safety concepts.
- **EXAMPLES, AND MORE EXAMPLES:** Each chapter is full of realistic situations to help reinforce the importance of safety and how to apply safe practices in training, at the firehouse, and during response.
- **TEACHING AND IMPROVING RISK MANAGEMENT SKILLS:** This book immerses readers into situations where choices must be made, and how the right choice can lead to a positive outcome.
- **LEARNING FROM THE SUCCESS OF OTHERS:** This book not only explains the 16 Firefighter Life Safety Initiatives, but also examines new ideas from other industries for their applicability to specific needs of the fire service.
- **ADDITIONAL RESOURCES:** Sources for additional information, including fire-related websites, offer readers an opportunity to conduct further research on topics of interest.
- **ROBUST APPENDICES:** The appendices include a number of helpful tools, including examples of SOPs for new ideas introduced throughout the book, a look at the sobering firefighter fatality statistics as reported in 2011, and a form to analyze your profile which helps to answer the all-important question—how safe is your organization?

SUPPLEMENTS TO THIS BOOK

Instructor Resources

The *Instructor Resources CD* includes tools to help the instructor prepare for classroom teaching and student evaluation:

- **LESSON PLANS** are based on chapter learning objectives and correlate to the corresponding PPT presentations.
- **ANSWERS TO QUESTIONS** provide a quick reference to validate learning.
- **POWERPOINT® PRESENTATIONS** align with the Lesson Plans and include photos and graphics to enhance classroom presentations.

- **TEST BANKS** are available for each chapter and are editable to allow instructors to revise questions or create new tests based on the existing questions.
- **APPENDICES** are included in electronic form for instruction purposes, as well as to help users improve current practices within their own departments.
- **IMAGE GALLERY** offers additional photos and graphics from the book to enable instructors to create their own presentations, or add to existing presentations.

WebTutor

WebTutor Advantage on WebCT and Blackboard includes a content-rich and web-based alternative to the classroom. These web-based platforms enhance learning and facilitate teaching, and include features such as class notes, quizzes, a discussion board, interactive games, and more!

ABOUT THE AUTHOR

Don Zimmerman has been active in the fire service since high school. He worked his way through the ranks of the Hambden (Ohio) Volunteer Fire Department, with the last 15 years as the assistant chief. He's currently serving as a career lieutenant and paramedic with the City of Mentor (Ohio) Fire Department. He teaches basic recruit training for firefighters at Auburn Career Center, teaches fire service degree and paramedic classes at Lakeland Community College, and is a disaster coordinator and EMS instructor for the EMSI at the University Hospitals Health System in Cleveland, Ohio.

ACKNOWLEDGMENTS

I would like to thank some of the people who made this possible. First to my wife, Charlie, and my children, Allison and TJ, thank you for allowing me the time to devote to this project.

Along with the publisher, I would like to thank the reviewers who participated in this project and offered valuable feedback as we developed the manuscript:

Robert Colameta
Owner, Public Safety Education Network
National Fallen Firefighters Foundation Courage to Be Safe National Program Manager
Everett Massachusetts Fire Department
Everett, MA

Christopher L. Gilbert
Enforcement Officer/Training Officer/Captain
Alachua County Fire Rescue
Alachua County Environmental Protection Department
Gainesville, FL

Richard P. Kasko
Fire Chief, GE Global Research
Schenectady County Community College
Schenectady, NY

Dave Martinichio
Deputy Chief
NYS Office of Fire Prevention & Control
NYS Academy of Fire Science
Montour Falls, NY

David McEntire
Program Coordinator, Emergency Administration and Planning
University of North Texas
Corinth, TX

Dr. Debra Mertz
FESHE Associate—Professional Development Committee
St. Clairsville, OH

Paul E. Nelson, MS
Fire Chief
North Fond du Lac Fire and EMS
Fond du Lac, WI

Craig Schwinge
Captain (ret) San Jose Fire Department
Adjunct Instructor, Fire Technology
Cabrillo College
Aptos, CA

Kevin J. Sehlmeyer
Chief of Training, Grand Rapids Fire Regional Training Center
Grand Rapids, MI

David Smyth
Fire Service Education Specialist
Office of the State Fire Academy
PA State Fire Academy
Lewistown, PA

Faron Taylor
Office of the Maryland State Fire Marshal (retired)
Whiteford, MD

To the leadership and members of the Hambden and Mentor Fire Departments, thank you for putting up with my new ideas and tactics over the years. You were clearly good sports. To Deputy Chief Ron Teriacco, thanks for reading every word in every chapter several times and giving me the guidance to help you achieve your goals with the EGH program. Thank you Chief Scott Hildenbrand for serving as my law enforcement expert, a guy who proves cops aren't that bad! Thanks to the great leadership and membership of FESHE and TRADE, I learned so much from you in the past two years. Lee Silvi and Rich Hall, thanks for the help on Chapter 5 (and thanks for not killing me which I know was on your minds). Thank you Paul Mannion for the airline connections used throughout the book, and to Doctor Tim Ziegenfuss, thanks for the latest on the medical world for Chapter 6. It was a great honor to learn from Chief Billy Goldfeder, B/C Mike Alder, D/C Mat Fratus, Chief Don Barnes, B/C Mark Emery, Chief Dennis Compton, D/C Nelson Bryner, and Lt. Jeffrey Neil. Thanks to Charles R. Jones and especially to Dan Madrzykowski, you are by far the most valuable asset to the fire service when it comes to the facts. Indianapolis Fire Department and Maryland Fire and Rescue Institute, phenomenal job on your firefighter health studies. Thanks to Lt. Rob Gandee, Ken Winter, Jeff Davis, B/C Bob Lloyd, Jeremy Szydlowski, Matt Henk, A/C Rich Vandevander, Brian Fowler, Lt. Tim Tobin, Captain John Blaugh, D/C John Kloski, Erica Ziegler, Bob Gahr, Scott Beverege, and the Mentor Fire Museum for providing great photographs. Thank you Ralph Boomer and Suzanne Jackson, RN for your insight on CISM, and Reverend Gary Russell for your leadership and guidance. Marsha Geisler, thanks for the help (pub-ed rocks)! Big thanks to Julie Scardiglia for clarifying the use of *affects* and *effects*, it was truly a pleasure, and Janet Maker for your vision. And finally thanks to the great authors and speakers who gave me the base of knowledge that got me where I am. Avillo, Bowman, Brennan, Brunacini, Carter, Coleman, Dodson, Dunn, Layman, Mittendorf, and Norman, I truly am not worthy.

FIRE AND EMERGENCY SERVICES HIGHER EDUCATION (FESHE) CORRELATION GUIDE

In June 2001, The U.S. Fire Administration hosted the third annual Fire and Emergency Services Higher Education Conference, at the National Fire Academy campus, in Emmitsburg, Maryland. Attendees from state and local fire service training agencies, as well as colleges and universities with fire-related degree programs attended the conference and participated in work groups. Among the significant outcomes of the working groups was the development of standard titles, outcomes, and descriptions for six core associate-level courses for the model fire science curriculum that had been developed by the group the previous year.

The six core courses are Building Construction for Fire Protection, Fire Behavior and Combustion, Fire Prevention, Fire Protection Systems, Principles of Emergency Services, and Principles of Fire and Emergency Services Safety and Survival.

FESHE CONTENT AREA COMPARISON

The following table correlates the Model Curriculum Course Outcomes for the Principles of Fire and Emergency Services Safety and Survival requirements to this textbook's chapters.

FIRE AND EMERGENCY SERVICES HIGHER EDUCATION (FESHE)

COURSE CORRELATION GRID

Principles of Fire and Emergency Services Safety and Survival

Course Description:	This course introduces the basic principles and history related to the national firefighter life safety initiatives, focusing on the need for cultural and behavior change throughout the emergency services.	
Text Description:	This book is organized by chapter based on the 16 fire life safety initiatives, and closely follows the FESHE outcomes for Principles of Fire and Emergency Services Safety and Survival.	***Firefighter Safety and Survival***
FESHE Outcomes:		**Chapter Reference**
1	Define and describe the need for a cultural and behavioral change within the emergency services relating to safety, incorporating leadership, supervision, accountability, and personal responsibility.	1, 2
2	Explain the need for enhancements of personal and organizational accountability for health and safety.	2
3	Define how the concepts of risk management affect strategic and tactical decision making.	3
4	Describe and evaluate circumstances that might constitute an unsafe act.	4
5	Explain the concept of empowering all emergency services personnel to stop unsafe acts.	4
6	Validate the need for national training standards as they correlate to professional development inclusive of qualifications, certifications, and re-certifications.	5
7	Defend the need for annual medical evaluations and the establishment of physical fitness criteria for emergency services personnel throughout their careers.	6
8	Explain the vital role of local departments in national research and data collection systems.	7
9	Illustrate how technological advancements can produce higher levels of emergency services safety and survival.	8
10	Explain the importance of investigating all near-misses, injuries, and fatalities.	9

(Continues)

Principles of Fire and Emergency Services Safety and Survival

FESHE Outcomes:		Chapter Reference
11	Discuss how incorporating the lessons learned from investigations can support cultural change throughout the emergency services.	9
12	Describe how obtaining grants can support safety and survival initiatives.	10
13	Formulate an awareness of how adopting standardized policies for responding to emergency scenes can minimize near-misses, injuries, and deaths.	11
14	Explain how the increase in violent incidents impacts safety for emergency services personnel, when responding to emergency scenes.	12
15	Recognize the need for counseling and psychological support for emergency services personnel and their families, as well as identify access to local resources and services.	13
16	Describe the importance of public education as a critical component of life safety programs.	14
17	Discuss the importance of fire sprinklers and code enforcement.	15
18	Explain the importance of safety in the design of apparatus and equipment.	16

FESHE Outline:		Chapter Reference
	I. Introduction	
	A. History of fire service culture	1
	B. Organizational culture	1
	C. Individual role in culture / behavior	1, 2
	D. History of line of duty deaths and injuries statistics	Appendix
	E. Defining the nature of the problem	1, 2
	II. The national context, health and safety	
	A. NFPA, OSHA	3, 7
	B. Medical and Fitness Standards	6
	C. Data Collection (NFIRS)	7
	D. Research / Investigation NIST, NIOSH	7

FIRE AND EMERGENCY SERVICES HIGHER EDUCATION (FESHE)

Principles of Fire and Emergency Services Safety and Survival

FESHE Outline:	Chapter Reference
III. Training, equipment, response	
A. Training, certification, and credentialing	5
B. Apparatus and equipment	8, 16
C. Emergency response—response to emergency scenes	11
D. Violent Incidents	12
E. Emerging Technologies	8
IV. Organizational health and safety profile	
A. Personal and organizational accountability	2
B. Present condition / culture	1, 9
C. Investigations- internal	9
D. Analyzing your profile	Appendix
E. Utilizing grants to meet needs	10
V. Risk Management	
A. Risk management concepts and practices	3
B. Unsafe acts	4
C. Empowerment definition	4
VI. Prevention	
A. Home fire sprinklers	15
B. Code enforcement	15
C. Public education / fire and life safety	14
D. Counseling and psychological support	13

INTRODUCTION

They say that the only thing constant is change, with fire and emergency services being no exception. The intent of this book is not only to prove that firefighter safety and survival can be improved, but that using the 16 Life Safety Initiatives as a guide is a creative and effective way to do it.

Created by the Everyone Goes Home campaign and the National Fallen Firefighters Foundation, the initiatives target specific goals to reduce the number of line-of-duty deaths in the fire service. Safety is not a very fascinating subject, but this book attempts to keep the reader's interest and spark some creative juices to make a change for safety.

So how do we get past the *"Safety First* sticker on a hard hat" phase and into a fire service culture that actually weaves safety and risk management throughout it? In addition, how do we meld the best aspects of the rich history and raw tradition of firefighting with the ergonomic, aerodynamic, bicycle helmet injury prevention fire service we might fear? The first thing working against us is that safety has always been a "bolt-on" component to the fire service (both literally and figuratively).

In a literal sense, a department could order the safety component from a catalog, and simply bolt it on; for example, the first PASS alarms clipped on to existing SCBA straps, making them safer. Of course, someone might forget to turn it on, or it might even fall off during an intense search and rescue. Still, it was an improvement. Eventually, thanks to NFPA standards and manufacturers, they became integrated and second nature. You can't take the safety component off, and new firefighters can't imagine an SCBA without an integrated PASS. It's even becoming more and more difficult to lay it on the table and point out specifically "this is the SCBA and this is the PASS."

On the figurative side, we also use bolt-on safety components. Recently, accessories such as rapid intervention crews (RIC, RIT, FAST), personnel accountability reports (PARs), and air management procedures (AMPs) became hot topics and departments grabbed them up. It became common place to attach one of these to trainings and emergency scenes. Every one of them started as a brilliant idea, and evolved into a system that still has the potential to save firefighters' lives. Unfortunately, some departments still have these hanging on the side of an airpack. For example, survey fire departments around the country, and ask how many have ever assembled a RIT for a fire scene. Now ask how many have one at every structure fire scene. What about how many have a policy or procedure of RIT job responsibilities and radio channels? Ask about what specific jobs on the fire scene they can accomplish without permission from command. Can they throw a ladder, pull a backup line, or ventilate a window? Can they assist getting a victim out a window? Some believe RIT stands on a tarp next to command until being deployed. Others argue RIT should do what is possible to prevent a firefighter from getting trapped, to avoid rescue. All of a sudden the answers get a little cloudy.

It should be obvious to fire and emergency services that a bolt-on approach is a great way to institute new ideas and equipment. If it doesn't work, simply unbolt it and throw it in a drawer. But if it has potential, it eventually has to be built in to the equipment. Safety must be weaved seamlessly into the emergency culture. Each chapter is full of examples from inside and outside the fire service of how the initiatives can be implemented. These ideas will hopefully unlock specific opportunities for fire departments to expand on the initiatives without actually giving up history and tradition.

The problem with integrating safety into our culture is that safety has always been a "soft" issue. Not soft as in fluffy or flexible, but soft as in offering no concrete road to travel. We know water weighs 8.34 pounds per gallon, and is relatively incompressible. You can multiply 29.7 times the squared diameter of an open orifice times the square root of the flowing pressure and calculate exactly how many gallons you are flowing every minute. This is because hydraulics is a "hard" subject. You can wish for less BTUs and more GPMs in the middle of the night, but facts are facts. If the BTUs surpass the GPMs, you're going to miss breakfast. Every facet of safety on the other hand, from rapid intervention teams to mayday communications is soft. When it comes to health and fitness, what is fit? Which homes should be sprinklered and how do we convince firefighters, homeowners, and builders it's a good idea?

Because of this relative lack of hard data in regard to safety, we must expand our view and see how similar situations are handled in other professions. Sometimes the comparisons aren't apples to apples, but we have to at least agree that they are similar in size, shape, texture, color, and even come from similar trees. In fact, it is also fair to assume that disease, insects, nutrients, and weather will similarly affect them. Wholesale costs for growers are relatively the same, and most of the equipment used for the production, harvest, shipping, and storage is the same for both. Even end-market users, grocery stores, and farmers markets would buy the product. In reality, there is no other practical way to reduce injuries and death in the fire service without comparing apples to pears. There are differences between them, but in order to actually make changes to reduce line-of-duty deaths in the fire service, we need to spend the next 16 chapters concentrating on the similarities.

We can't just go out and compare firefighting to tax preparation. The common denominator has to be life preservation for us as well as for the public, and frankly we're not talking paper cuts. In an effort to pick a fruit that is close to us, we must use examples, theories, and practices from industries that share our unique features:

1. High risk—The potential for disaster is significant.
2. Life safety—The potential for multiple fatalities is present.
3. Human response—The potential for outcomes is based primarily on decisions that are made, sometimes with limited information.

Any profession that meets these three conditions deserves a ticket to our debate. We can objectively view how their past disasters occurred, how similar problems were handled, and what they did to ensure it would never happen again. We can look at the policies and procedures they have developed based on their experiences. Finally, we can use some of their knowledge to apply to our problems.

Nowhere in this book does it say we can't do our job. We still have to respond to the new emergencies that society continues to throw at us, bandage the injured, extinguish the fires, and mitigate the situations. We must do it promptly, professionally, and safely. There is absolutely no reason why safety and operations can't work together. Even though they are represented by two distinctly different boxes on a command flowchart, it is possible for them to coexist and even encourage each other. What's at issue here is creating a safety culture in the fire service. The goal is to first do our job—just do it safer. Firefighting is dangerous work. We will continue to lose firefighters every year in the line of duty. However, there's little justification to lose one or two every week.

The idea behind this illustration is that the 16 initiatives build off each other. However, without a solid safety culture (root system), it's like a tree leaning over a river. Everything looks healthy but all it takes is one good rain to topple the tree. A fire department without a strong safety culture is doomed even if it looks healthy above ground.

CHAPTER 1

Defining a Cultural Change

LEARNING OBJECTIVES

- Define culture.
- Discuss the need for cultural change in the fire service relating to safety.
- Discuss the aspects of a safety culture within fire and emergency services.
- Discuss the gaps between a safety culture and the existing culture of emergency services.
- Discuss why change occurs in an organization.
- List examples of how an organization can use change to its advantage.
- List some of the reasons people resist change, and give examples in fire and emergency services.
- List examples of other industries that have used leadership, management, and supervision to develop components of a safety culture.
- Discuss ways to use the successes of other industries as a catalyst for "advocating" a safety culture within fire and emergency services.

1. Define and advocate the need for a cultural change within the fire service relating to safety; incorporating leadership, management, supervision, accountability, and personal responsibility.

CHANGE

For the past 200 years, the United States has gone through several significant changes. Time has proven that adapting to circumstances is both normal and vital to our existence. From the Boston Tea Party to the Great Depression, uncertainty has always given way to progress. The present generation is no exception. On September 11, 2001, we watched an attack that would change life forever. It moved terrorism from the six o'clock news to our front window. It put emergency responders on the front line, and security ahead of freedom. We watched a progression from metal detectors at airports to removing shoes and the use of full-body scanners. In 2006, a plot was uncovered that involved disguising explosives in drink bottles, which led to an immediate ban on liquids over a certain amount. Heightened security has also spread to other forms of mass transportation as well as to recreational and sporting events. It affects the way emergency responders do their jobs and the dangers they face. The security changes were drastic, like nothing our generation has ever seen. But as with all changes, they eventually become normal to us.

Emergency Services Change

The big change that is currently under way in fire and emergency services is the change to a **safety culture.** Partially due to our involvement in terrorism response and the fact that we have continually failed to decrease our **line-of-duty deaths (LODDs),** a change is imminent. For the purposes of this book, **culture** can be described as the values, customs, and traditions of a group of people or an organization. A change to a safety culture doesn't necessarily *abandon* those values, but instead *incorporates* safety into our customs. Other industries have proven that advocating a change for safety has not disrupted their operations or put them out of business. Rather, a safety culture allows them to do their job more safely. The change involves weaving the theories of **risk management** into everything we do. As with all change, there is resistance. Fire and emergency services have long attempted to maintain a balance between both being progressive and remaining traditional. We are a passionate profession with a rich history and generally don't embrace change. When change does come, it's usually a slow and arduous process. Although it's human nature to resist change, some firefighters tend to excel in resisting change. This stress is apparent as soon as a new theory on firefighting emerges or a piece of equipment is introduced.

Take simultaneous **positive-pressure ventilation (PPV)** during a fire attack, for example. Some experts believe that a well-orchestrated structural fire attack can take place using fans in tandem with hoselines to reduce the loss of life and property both safely and efficiently. By creating an increased pressure of fresh air at the point of the attack, interior floor–level temperatures and poisonous gas concentrations can be reduced, giving trapped victims a better chance of survival. Smoke and heat are cleared to the seat of the fire, allowing for a quicker search and attack on the fire, while forcing a vent on the other side. Besides increasing safety, the odds of a successful attack can be improved by fighting fires in which natural ventilation disrupts firefighting efforts, such as in basement or wind-driven fires.

Other firefighters feel PPV on attack is a dangerous practice because it creates unsafe situations that threaten the building, the safety of occupants, and firefighters. PPV not only adds oxygen to the fire, increasing its intensity, but it also could force the fire into concealed spaces or adjacent areas that it might not normally spread to, such as closets or soffits. Natural fire spread patterns are accelerated into areas such as attics or other exposures. With construction materials and methods used today already speeding up the fire, our goal should be to slow the spread of fire rather than to increase it. It's hard to believe that creating a "wind-driven fire" is a useful tactic. Unfortunate victims in remote areas located between the fire and vent point are exposed to lethal concentrations of poisonous gasses and increased heat conditions that make a rescue highly unlikely.

This debate has excellent points on both sides, made by very intelligent firefighters and officers. It's also fair to say that because gas-powered fans weren't readily available 30 years ago, it's a fairly new debate and one that is sure to continue for years to come. Is one side right and one wrong? As in most debates, the answer lies somewhere in the middle. Many firefighters believe that under the right conditions,

- when the location of the fire is known,
- and an attack team is ready,
- and a water supply is established,
- and a safe external vent point is adjacent to the seat of the fire,

then positive pressure on attack can be incredibly effective. But if you miss one component, you may burn it to the ground.

This example is one of the most common change models in a fire department. It doesn't necessarily begin with the fire chief, but often originates lower in the ranks. It starts with a belief in a new concept and a desire to add it to the firefighter's arsenal. It involves research, debate, and, ultimately, a decision by administration to change. Some embrace it as forward thinking, whereas others continue to resist. Hopefully, the department continues to train hard, make sound decisions on the subject, and adjust their **standard operating procedures (SOPs)** to make it successful. As victories mount with the new tactic, dissention dissolves. Time eventually heals wounds, and retirements take care of the rest. As new firefighters join, it's as though the tactic was there forever. Change has taken place.

Types of Change

Change can be either **reactive** or **proactive.** Society is generally reactive regarding safety issues. For instance, take a small town that consists of two state routes intersecting in the center. Although the speed limit on both roads is 55 mph, one road has stop signs. Over the years, traffic has increased and so have serious crashes. Local officials complain to the state about the safety of the intersection, and the speed limit is reduced to 40 mph in the town center. After two teenagers are killed one spring, the state adds flashing red lights to help draw attention to the stop signs. People continue to complain and ask for a stoplight. Several years later and after another fatality, a stop light is finally installed. It is not uncommon for safety improvements to come at a cost, especially when tax dollars are at work. **Department of Transportation (DOT)** officials simply can't afford to install stoplights at every intersection in the state. They are forced to use **data** in their ranking of which intersections most need the improvements. In this case, "data," unfortunately, is lives. Although we might jump to the conclusion that it's just a governmental tactic, this reactive approach is more of a human trait.

Let's say you assist at a motorcycle crash in a remote area. A 15-year-old boy was involved in a dirt bike crash and received a serious head injury while riding without a helmet. Would you expect the parents to order his younger brother to wear a helmet when riding in the future, or maybe even sell the dirt bike? It's obvious that the parents would institute *reactive* rules and force change on his siblings. This in turn could reduce the severity of any future injuries to members of the family. The question is, why don't all families have *proactive* helmet rules before a debilitating injury causes a change? One big reason is because human nature says "it won't happen to me." It could be that the general public isn't convinced that wearing a helmet really makes that much of a difference. Some might feel it's a free country and parents need to decide what's best for their children. Depending on media coverage and the political climate following an incident like this, the government may intervene and pass laws that would mandate the use of helmets off-road for children younger than age 18. Citing public safety as the reason, an enforceable change would be instituted. A change in the law actually moves the issue from reactive to proactive.

A proactive approach foresees risk and takes steps to minimize it. A great example of a proactive approach to safety is an insurance agency. Take homeowners insurance, for example. When a new home is built, many factors go into insurance rates. Besides the value of the home and where it's located, sometimes inspections are done. If an inspector finds safety violations, such as stairways without railings or an uneven sidewalk, the insurer may not insure the home or they may increase rates accordingly. Sometimes environmental studies are completed for underground fuel tanks, or verifications that there are no potentially violent breeds of dogs on the premises. Some life insurance policies require the insured to sign papers confirming that he or she will not partake in any potentially risky recreational activities such as sky diving or auto racing. Insurance companies gamble on risk every day, so a proactive stance is the only way to do business.

FIGURE 1-1
Fire and emergency services many times lag behind industry when it comes to a safety culture.

We would like to think that since we are in the risk management business and work in emergency situations daily, fire and emergency responders would be generally proactive (Figure 1-1). It's true that we wear steel-toed boots with our structural fighting gear to prevent smashed feet, and we wear hoods to protect our necks in fires. We stripe brightly colored **chevrons** on the rear of our apparatus and wear high-visibility vests or jackets on highways to make ourselves more visible. In fact, we tend to be quite proactive when necessary. **National Fire Protection Association (NFPA)** standards and DOT guidelines require us to follow certain rules and use specific safety equipment. Policies and equipment that aren't mandatory are sometimes ignored. In effect, we aren't much better than the kid who has to wear a helmet on a dirtbike because it's the law. Many times we electively change procedures as a result of an injury or death. Using firefighter fatality reports from the **National Institute for Occupational Safety and Health (NIOSH)** is a great way to forecast the next fatality, but it's still initially reactive in nature. Our primary goal in creating a safer work environment should be to forecast what *could* happen as opposed to what *did* happen. Steps should then be taken to actively prevent it from ever happening. Additionally, most changes should be industry-wide as opposed to local. It's not a very proactive "brotherhood" if firefighters 400 miles away don't get the message. But are completely proactive approaches to safety even possible?

Changes for Safety

Many high-risk industries use a near-total proactive approach to safety. Take, for instance, tunnel construction. Making a mistake in this type of work has the potential for catastrophic failure that results in death, but it is also capable of creating immediate financial ruin. For hundreds of years, a 30-mile-long tunnel burrowed under the English Channel between France and England was just a dream. The English Channel is a commonly traveled inlet of the Atlantic Ocean that has some of the fiercest storms of any body of water. Gale-force wind, high waves, and dense fog have caused numerous maritime disasters. The first proposal for a tunnel came in 1751, with many more after that. The early plans addressed engineering obstacles as well as safety concerns ranging from collapse and flooding to lack of oxygen. Even with obvious improvements in equipment and design, tunneling under part of the ocean was a serious undertaking in 1987. The potential for a leak was always present, which would result in near instantaneous flooding of pressurized water that would kill any workers in the path and destroy work already completed.

Unique to the project was the fact that the French and English governments didn't finance the construction. Investors, in the form of 220 banks from around the world, carried the note to the tune of $9 billion. Because of their investment and potential loss, the financial institutions insisted on perfection in construction and safety. Tunneling for each of the three tunnels began from both sides, using specially designed lasers to keep them on course. This required the two crews to meet in the middle with little room for error. The workers knew that if the two boring machines were 30 feet off in any direction, it would be disastrous because the boring machines were unable to back up and the ability to fix the mistake was questionable. In fact, the machines were actually designed to be gutted of all components when they met in the middle, and their shell was left permanently as part of the tunnel wall. The safety and construction plans were evaluated and reevaluated to ensure that no mistakes were made. This proactive approach from the start ensured that the boring machines would meet head-on more than 200 feet below the Atlantic Ocean, not once but three separate times without a leak.

The biggest difference between emergency scene operations and tunneling under the English Channel is time. When an emergency occurs, responders don't have years to lay out an effective incident action plan (IAP) with zero tolerance for injuries or mistakes. There will always be hasty decisions with limited information under stressful situations. Individually, any of these three components could spell disaster. Therefore, in order to accomplish a proactive safety initiative, we must continually evaluate standard procedures for emergencies while we eliminate nonemergency mistakes. If 100% of firefighter injuries and deaths were a direct result of **dynamic** emergency scene effects, it might be more difficult to execute a proactive stance. However, we know that most of our injuries and deaths occur in foreseeable situations on emergency scenes and preventable causes during nonemergency activities. By continuing to update SOPs, we create templates that include the risk benefit in all decisions. These proactive changes for the fire service as a whole can prevent many of the dynamic scene injuries that we haven't even seen before. Although proactive safety initiatives will never be 100% effective in eliminating LODDs, we can see a drastic reduction in their incidence.

Reasons for Change

Generally speaking, change in the fire service occurs for two different reasons. The first is because change is desired, such as the PPV on attack, for example. The desire to change can be critical, as a result of a problem, an injury, or a death. Or it can happen for no particular reason, such as a desire to create a new department patch or paint scheme. In either case, it's a result of the basic human desire to become better at what we do acting as the catalyst. Perhaps we go to a class or watch another department in action. Maybe it's a result of reading an article or a book that perks our interest, or studying a job-related injury or LODD report that makes us rethink our procedures. A desired change has a good chance of being implemented because it involves passion: If the member who is the instrument of change has access to powerful tools such as motivation and leadership, others want to jump on board.

The second reason change takes place in the fire service is because it is obligatory. Whether laws have been enacted that affect our job, or orders come down from the fire chief, we have no choice but to institute the change. We can choose to fight the change or do our best to implement it as painlessly as possible, but the fact is that it must take place. This type of change tends to encourage the most resistance, but the atmosphere is ultimately controlled by leadership. Accountability in situations such as this is covered in depth in Chapter 2, "Enhancing Accountability." An example of this type of change would be the federal law that was passed in 2008 regarding high-visibility clothing in the right of way of federally funded highways. The new law specifically identified firefighters and **emergency medical services (EMS)** workers in regards to their apparel. The law cites **American National Standards Institute (ANSI)** II as being the minimum visibility, which at the time neither normal firefighting nor EMS squad coats met. The only responders exempt from wearing high-visibility coats or vest are firefighters actively involved in firefighting or **hazardous materials (hazmat)** operations. Although it's not perfectly clear, a good rule of thumb is if you don't have an airpack on, you probably should be wearing a vest (Figure 1-2). This mandatory change was surely not embraced by all organizations, and some may still not have the required protective gear. The cultural change under way in the fire service is a combination of both reasons above. Not only is the fire service attempting to reduce LODDs with the 16 initiatives, but evolving NFPA standards and increasing legal liabilities are making more and more changes prudent.

LEADERSHIP DURING CHANGE

Periods of change are a time for management to shine, but sometimes leaders instead ignore the opportunity. Although there is usually resistance to change, some will accept it. In fact, the ability to embrace change is one aspect of an effective leader. One reason is because leaders are in a position to either make a change successful or to stomp it flat. If leaders do take on the challenge and support the change, it becomes a reality

FIGURE 1-2 Some safety changes are mandated by governmental agencies.

much faster. As change is successful, other members will find it exciting and motivating. A manager's history with change becomes an issue to employees that were part of the manager's past failures. Trust is an issue that follows leaders throughout their career. A lack of trust can lead to a fear of hidden agendas or questions about how committed the manager is to the change. If subordinates get the feel that the change is just the "flavor of the month," there likely won't be much support for it. On the other hand, if they get the impression that the change is here to stay, they're likely to make it work.

Managing Change

Initiative 1 calls for the incorporation of leadership, management, and supervision in adopting a safety culture. The **National Fire Academy (NFA)** curriculum for Managing in a Changing Environment[1] sees leadership during times of change as an opportunity to grow. They further describe the four main causes for change, each with unique challenges and opportunities for leaders. These causes include economic, social, political, and technological impacts.

Economic Impacts

At no time in the history of the fire service have economic impacts been so evident. The steady strides gained in equipment and manpower since the Great Depression recently ground to a screeching halt, with many fire departments and EMS agencies back where they were 20 years ago. Finances affect fire and emergency services because income is a direct result of how well business, industry, and families are doing. If income is lost or homes are foreclosed on, personal budgets take a hit. Because of the economic delay of taxes, there is usually a warning of impending financial reductions. At the time someone loses his or her job or home, most governments are still using tax money collected during the previous year. Leaders who see the crisis coming can reduce their financial obligations in anticipation. The economic delay also means that agencies will not recover until a year or two after everyone else recovers.

Social Impacts

Society has an impact on responding organizations based on the types of neighborhoods they serve. Longer driveways mean more hose, taller buildings mean taller ladders, and barricaded districts mean more forcible entry tools. Some departments need to truck in their water, whereas some need to wait for the police (Initiative 12). Other social impacts affect us inside our organization. Hiring practices, in response to federal laws, and cultural diversity affect the makeup of the personnel. There is a constant rise in the educational levels of new hires, with many already having degrees. This, along with differences in age, leads to generational differences between members, which has the potential to raise personnel issues.

Political Impacts

Decisions by governing bodies affect us in many ways. Local politicians drive what we do and what we do it with. New laws and standards influence how we do it. Some organizations have great relationships with their political leaders and others seem to have continuous problems. Unfortunately, the political climate can change any time elections take place.

Technological Impacts

With new technology coming out constantly, it presents continuous changes in emergency services. New technology is also unique in that it encourages new rules and standards for equipment. Equipment not available 20 years ago, such as thermal imagers, laptop computers, and **automated external defibrillators (AEDs),** are standard now, which in turn requires the expenditure of more money.

Resistance to Change

Human beings have a natural inclination to resist change, so one of the most crucial components of effective leadership is anticipating and reacting to that resistance. First of all, change takes energy. Isaac Newton's first law of motion states that an object in motion tends to stay in motion unless acted on by an outside force. It's pretty easy to get in the rut of "that's what we've always done" and act out this theory. Take, for instance, a lieutenant who is newly appointed to a busy truck company. In his first month, he spends much of his time learning his new equipment, response area, and crew. He notes that Antonio, his normal "irons" man, keeps the irons at his feet in the jumpseat. When asked about it, Antonio responds that it's a way of saving a couple seconds when he gets a "work" so he can force a door before the engine company needs it. The lieutenant asks what would happen if the apparatus was in a wreck, and if Antonio was worried about it becoming a missile. Antonio insists that it's been on the floor as long as he's been there and that he holds it down with his feet. He adds that lots of other guys do it, and that they learned how to do it in the academy. If the lieutenant sees it as a dangerous action and wants to make a change, he's bound to run into resistance with his new shift as well as the other shifts. Sometimes firefighters find it easier to overlook a "bad habit" rather than try to change it.

Resistance to change sometimes surfaces as a bad habit. For example, procrastination is something that many of us do, and most learned when we were young. It's common for children to put off chores that are undesirable, but learning how to buckle down and complete the task at hand is a useful skill to have as an adult. Unfortunately, some individuals never become efficient. Any bad habit, from biting fingernails to smoking, arises from some sort of stimulus and is predominantly subconscious. Some might chew their nails when they are stressed or smoke when in the company of others who are smoking. Procrastinators will develop excuses why a specific job needs to be delayed if it's difficult or they feel uncomfortable doing it. Many suggest identifying the trigger for a bad habit as the first step to change, followed by focusing on the benefits of not falling into the trap.

Bad habits are commonplace in all occupations, and emergency responders are no exception. Some don't create much of a safety issue, but others are "an accident waiting to happen." Consider if you moved to a rural area and joined the volunteer fire department. Although not very busy, you figure they seem like a pretty nice group of people, it's fulfilling a community service desire that's important to you, and every once in a while they get an exciting call. You notice that they are pretty well trained, but when there is an emergency call it turns into a race to drive the fire trucks. In fact, some who miss the driver's seat on the first rig out run to the second out and take that driver's seat. It's not uncommon for all six apparatus in the station to roll with only one or two people in each. Later-arriving members are forced to drive their personally owned vehicles directly to the call, which creates mayhem at the scene. Somehow over time, the members have been seasoned to think that the emergency is better handled by a rapid response by more vehicles and less manpower. The culture of the department appears to hold emergency vehicle drivers in a higher esteem, creating a greater desire to drive. Drivers tell stories back at the station of their "fastest" response times and driving abilities that defy physics. It's clear that the bad habits some members have with driving have become deeply rooted in the culture of the membership.

Similar to procrastination, bad driving habits will take commitment and time to change. After identifying and eliminating the stimulus of driving recklessly—in this case, esteem—alternative benefits with positive outcomes must be introduced. Solutions might include driver's training classes that emphasize skill levels in maneuverability rather than speed. It could even include a competition for the best driver (not the fastest). The class could cover emergency vehicle accidents and their **investigations,** causes, and prevention. Drivers could be required to take the training and pass a test to be "certified" to drive. Likewise, there could be classes on pump operations that would require drivers to be proficient at basic fire hydraulics in order to operate the vehicle. Other training classes could include fire attack jobs based on seat positions. By making the **officer in charge (OIC)** and jumpseat positions more interesting, some might prefer them over driving. A push around the firehouse to promote stories of the firefighters that drive the safest or the firefighters "making a good stop" can also be beneficial. Finally, a stance by the chief that unsafe behavior will not be tolerated should take care of anyone else who doesn't learn the importance of this.

Historically, firefighters are no stranger to change. It's just that some departments are a little better at it than others. Chief Brunacini claimed that his Phoenix firefighters had been eating change for breakfast for a long time.[2] With all of their theories on the fireground command system, they went through as much experimentation and change as any department anywhere. In fact, open up any book on the history of firefighting and take a good look at a fire scene. What do you think firefighters thought of replacing horses with motorized apparatus when that change happened? What about the introduction of those "leather lungs"—were they happy to start using them instead of breathing through a wet beard? Speaking of beards, you can just imagine how happy they were when they were told they had to shave them off. The fact is, change has always been a part of the fire service and usually has not been embraced. However, history also proves that when change has been better for operations, service, or safety, it almost always prevails.

So why do we resist change? Type it into any search engine on the Internet and you'll find plenty of explanations. Besides the fact that we are wired to remain skeptical and resist change, we tend to fight change for several other reasons. For instance, let's say that your battalion chief has just issued an order that all first-arriving officers on his shift will conduct a 360-degree view of a fire building to see all four sides prior to commencing an attack (Figure 1-3). His reasoning is that many LODDs were a result of not getting a good look at the building or fire progression. You may resist the change because of the following:

- **THERE'S NO NEED FOR THE CHANGE:** *"The 360 is not vital to initial operations and can be done by a later-arriving officer."*
- **LOSS OF CONTROL:** *"He doesn't trust me to make good decisions so he keeps me busy until he gets there to give orders."*
- **CLOSED MIND:** *"It's a waste of time. Should I just tell the victim in the window to wait, that my boss wants me to lap the house first?"*
- **NOT WANTING TO LEARN:** *"I've been fighting fires since the BC was in KG, and now he wants to 'teach' me how to fight a fire."*

FIGURE 1-3 Many fire departments now require the completion of a 360-degree view of a structure during size-up.

- **Connection with the old way and people who did it that way:** *"Lt. Edmond never did a 360 and he never had an issue with it."*
- **No role models for the new way:** *"Why aren't the other shifts doing it? The same reason nobody in the country is doing it!"*
- **It's too overwhelming to change:** *"It's just one more thing I have to do during the busiest period in a structure fire's timeline."*
- **Bad experiences:** *"It's just the flavor of the month and won't last. Remember last year when he had us take search rope with us every time we wore an airpack?"*

Firefighters in particular are very passionate about their profession, and therefore aren't easily convinced of something new. Part of this is because they respect those who have come before, and feel a strong tie to tradition. For instance, European firefighters wear helmets with their turnout gear that are reportedly lighter, recess the shield, and offer more protection than some traditional helmets. On safety alone, there is a good reason to consider a change. But most firefighters take one look at them and scream. The traditional American fire helmet has a unique shape that was designed a long time ago to keep hot embers from falling into the collar of a canvas or rubber coat. With the protective ensemble used today, the rear brim is more traditional than practical. A national change to the European-style helmet would be one of the most difficult changes in the history of the fire service. As long as traditional helmets can meet the safety standards required, it's a change we probably won't ever make.

Other reasons to resist change could be ego or personality driven. An officer might not want to try something new for fear of failing, or because he or she feels that the risk of change outweighs the risk of doing nothing. Sometimes there is a lack of trust between a supervisor and an employee being asked to change. This can lead to a fear of hidden agendas. A supervisor can reduce the risk of resistance due to personalities by identifying potential issues ahead of time, and communicating the change and its intentions. The employee with concerns should be able to voice an opinion, but must ultimately make a conscious effort to adopt the change. Although resistance to change must be anticipated, progressive personnel and organizations must always be on the lookout for opportunities.

Opportunities for Change

Retired Chief Dennis Compton of the Mesa, Arizona, Fire Department described changes in the fire service as "waves."[3] He explained that every 10 years or so, the fire service is exposed to a new wave. When the wave comes, a department can either get up on the board and ride the wave, or let it pass. Whether a department chooses to ride it or not has to do with the ability to see opportunities and a desire to improve. If an organization misses the wave, it's not uncommon to see them work twice as hard later to catch up with everyone else. For instance, in the 1970s, the wave was EMS. Some fire departments saw an opportunity to improve their service to the community, whereas creating jobs and assigning a better value to their service. EMS was a good fit for many fire departments, whether they simply provided first responder assistance or ran **advanced life support (ALS)** transports. EMS billing brought in income without additional taxes, which made the department more valuable. Most fire departments that fought off EMS like a terminal disease are now at least partially involved in medical responses. Following are other perceived waves in which the fire service has been involved.

- 1980s—Hazardous materials response
- 1990s—Technical rescue, public education
- 2000s—Terrorism, safety initiatives
- 2010s—Budget cuts, safety culture, ???

Although we may not know what the next wave will be, it's up to progressive firefighters to identify and embrace it. The one thing for sure is that changes will continue to come, and those who can react and accommodate the change will be the most successful. Many believe the future includes becoming more involved in prevention activities, such as home inspections. When residential sprinklers become more commonplace (Initiative 15), fire departments could become more involved in new construction inspections. With terrorism and natural disasters an ever-present danger, we could see an increase in regional planning and disaster training. It could be that, in the future, we expand the amount of medical services provided to a patient such as sutures or x-rays, or assist in mass immunizations. Many emergency management, police, fire, and EMS organizations were heavily involved in the H1N1 virus immunizations in 2009. The experience in medical response and handling of large volumes of people greatly benefited health departments. Businesses have been involved in disaster mitigation and business continuity plans. Whether they sustain a natural disaster such as fire or flood, a human disaster such as a disease, a bomb threat or hostage situation, or even a technological disaster such as a power outage or widespread computer virus, many want to have a plan. Investors and customers want to be assured of sustainability in the event that the unthinkable happens. Fire and emergency services are equipped to provide the leadership necessary if there is a need in the community and a desire in the organization. Sometimes we don't have a choice but to be involved in the change. Some industries have allowed changes in society to alter their business model. For example, the Internet has changed the way many consumers purchase goods. When the Internet began to take hold in the mid-1990s, many retail businesses refused to ride the wave. They saw Internet sales only as competition to their bottom line, and continued to rely on phone books and local advertising to bring in business. Others saw it as a way to make additional sales and to reach more customers. They built websites to advertise their products, providing prices, specials, and even maps to the store. Email addresses were made available for customers to ask questions anytime, virtually eliminating store hours. Some even offered secure checkouts so purchases could be made safely online and shipped, or held at a local store for pickup. Some small package shippers saw the change, and made residential shipping easy for stores, even offering free pickup. Others created shipping calculators as a simple way to get an immediate quote on shipping, and even allowed customers to pay online and print the bar code shipping labels directly from their printer. Like the fire service, many individuals that fought the wave are now struggling to catch up. Ironically, several of the stores that held on tight to the phone books aren't even in them anymore.

Once a new opportunity presents itself, the goal of any change process must be to succeed. We've already discussed how and why changes come about, and the reasons for resistance to change. The inability to identify key players and sources of resistance is the most common cause of the failure of a change. Therefore, an important strategy is to create a team of individuals from different groups who have a common interest in the change. It's important to pick members who are open-minded and who have the ability to lead. Don't mistake this for a group of "yes men." It's important to bring different opinions to the table, but they must be able to leave with a plan. This group must be able to take the plan back to the members and communicate the purpose and advantages of the change along with a timeline for implementation. They must collect any concerns and complaints that arise, and bring them back to the group for discussion. The plan must be continually evaluated and adjusted as needed throughout the implementation period.

In an effort to initiate a change for safety in the fire service, the **Everyone Goes Home (EGH)** campaign took a similar team approach by holding life safety summits in 2004 and 2007. Invitees included those from the fire service and industry, some identified through their involvement in the **National Fallen Firefighters Foundation (NFFF)** and the NFA. After identifying concerns about safety and creating the initiatives, they developed a training program to build a workforce of advocates to spread the message around the country. Advocates visit fire stations and present the 16 life safety initiatives to firefighters and officers on a local basis. Fire departments are then supported in their quest to implement the safety improvements.

SAFETY CULTURE

The goal of adopting a safety culture in fire and emergency services is a daunting yet essential task. In order to define a safety culture, we first have to examine the causes of injuries and death. Emergency responder injuries can be caused by mechanical failure (a rope breaking), human failure (overloading the rope), or an organizational failure (not inspecting or replacing the rope). Organizational failures are always a result of not having a safety culture. There are many informal definitions of safety culture, but it is probably best described in a technical report prepared in 2002 for the **Federal Aviation Administration (FAA).**[4] Researchers identified the presence of a safety culture when the following were in place.

- *Organizational commitment.* Management must view safety as a core value or guiding principle. It must actively support and reevaluate safety through equipment, policies, training, and financial support. (Initiative 1)
- *Management involvement.* Management must be involved in safety seminars, training, and planning. This includes presence as well as contributions to safety efforts. (Initiative 1)
- *Employee empowerment.* Employees are "the last line of defense" for errors that can cause injuries. As a result, they must be involved in safety decisions and empowered to stop unsafe acts. (Initiative 4)
- *Reward systems.* Reward systems must be in place for safe practices, while punishment is reserved for unsafe acts as opposed to mistakes. (Initiative 1)
- *Reporting systems.* Employees must be able to report safe and unsafe practices with no fear of discipline. (Initiative 4) This is a vital component of a safety culture when attempting to identify risks and prevent injuries. (Initiative 3) Recommendations also include near-miss reporting. (Initiative 9)

This recipe for creating a safety culture is as applicable to fire and emergency services now as it was for the FAA when it was written. In effect, maintaining a safety culture is evidenced by management's commitment to embrace safety procedures and employees' commitment to safe practices even when nobody is watching. A fire department that truly buys into a safety culture includes it in their mission or vision statement, and makes it clear through policies, training, and actions that firefighter safety is the first priority. Rewards are reserved for heroic actions when making a rescue attempt as safely as possible under the circumstances. In other words, risk management is applied to all emergencies, ensuring that if a firefighter dies in the line of duty, it was part of a calculated risk in an effort to save a life (Initiative 3).

Fire and emergency services are not alone in this endeavor. Many other dangerous occupations, such as steel workers in the construction industry, are actively pursuing a

FIGURE 1-4
Integrating prevention, control measures, and actions to adopt a safety culture takes a significant commitment by an organization.

safety culture. They believe that a safety culture doesn't specifically eliminate a hazard, such as an iron worker falling from a building or firefighters falling through a floor. Instead, the National Construction Agenda for Occupational Safety and Health establishes broad beliefs that "prevention, control measures, and action" are three specific ways to use a safety culture to counter hazards.[5] If we apply their belief, firefighters have a lesser chance of falling through the floor if *prevention* (training, preplans, and residential sprinklers), *control measures* (policies and procedures), and *action* (size-up, 360s, and situational awareness) are standard procedures at all structural fires (Figure 1-4). To create a safety culture, we need to utilize all the tools and experience we have gained over the years and implement them into every function we participate in.

The Evolution to Safety

Did you ever wonder why discussions regarding firefighting safety are often compared to the airline industry? Besides the fact that they are a high-risk profession like fire and emergency services, they also have the potential for large losses of life based on human decisions. Many times the information they have is incomplete or inaccurate. However, the biggest reason could be that they identified a lack of a safety culture and actually evolved into one. They have a solid track record with change, and in the past 30 years have rewritten the manual based on safety procedures. There are more lessons for emergency responders to learn from the airlines than any other industry, primarily due to their adoption of **crew resource management (CRM).** Chapter 4 explores the concept of CRM, a program that has had an enormous impact on reducing incidents.

On a recent flight to complete a final inspection on a new piece of fire apparatus, a couple of firefighters and their salesman left Chicago O'Hare on Flight 6019 heading west. As the airplane gained altitude, it suddenly leveled off and took a hard 180-degree right turn. One firefighter turned to the salesman and said "either our pilot just figured out we were heading east or we've got troubles." It didn't take long for the captain to get on the intercom and apologize that there was a little problem that didn't involve passenger safety but warranted a return to the airport and a change of planes. When the plane landed, there was no fire, no crash, not even a response by the airport fire department (much to the dismay of the salesman who sold some of the **aircraft rescue and firefighting (ARFF)** equipment). What was it that was enough of an emergency to make the pilot turn around, but not enough to create an emergency response or standby? The fact is, returning to the airport was expensive. Besides extra fuel, scrambling to get another plane ready, getting an open terminal, and changing flight plans at one of the busiest airports in the country, they made a lot of customers unhappy. If there were connecting flights, they had to accommodate those people, possibly covering meals and hotel costs. One thing for sure is they made no money that night.

Whatever the cause, probably a warning light or gauge reading outside normal limits, it was treated as a true emergency that required immediate attention. Twenty years ago, that airplane probably would have continued west and probably would have landed at its destination without any problem. Compare that with an engine company that is responding first due to a reported garage fire. The vehicle pulls out of the station with the low air alarm buzzing, but it slowly builds enough pressure to release the brakes and turn the alarm off. Now, only a half mile out and a header visible in the sky, the air alarm goes off again telling them there is a probable leak. The gauge reads 50 psi. Do they call dispatch and cancel the response? A cultural change for safety in the airline industry shifted its primary goal of delivering passengers and turning a profit to delivering passengers safely 100% of the time and turning a profit. Because their safety culture was initiated, airline crashes have continued to drop, with only 28 worldwide in 2009.[6] Compare that statistic with the 392 fire apparatus crashes requiring a police report in 2007—just in the state of New York.[7]

Tips for Creating Change

In order to successfully institute a change, there are many components that help make it work. It is possible to create a cultural change for health and safety in every emergency services department. Just as each airline instituted changes in their thinking, we can as well. Some tips we can borrow on a department level are noted below.

Take It for What It Is

If the alarm says there is a problem, believe it. Safety devices are on equipment for a reason and should not be circumvented or disabled. If the mechanic can't fix it, find someone who can. If a specific safety device is inoperable and a suitable replacement is available, use it. For instance, a vehicle found to have a broken backup alarm can be used safely by having a backer every time reverse is used until it is fixed. If a suitable replacement is not available, park it. An **antilock braking system (ABS)** warning light that suddenly comes on is pretty hard to diagnose responding to an alarm while passing cars left of center. It has a best-case potential of being a faulty sensor and a worst-case potential of resulting in a fiery crash. A safety culture says the gamble isn't worth it; let second due handle the alarm and you call for a tow.

Make It a Line-Level Effort

Besides the fact that those toward the bottom of the chain of command are the ones who actually change their behavior, many times they know how to make the transition smoothly and successfully. They can see the benefits of making it happen, and are most likely to be unhappy with the current situation. They're also the ones with the highest rate of injuries and death. Although safety cultures require management involvement, a conscious decision to implement safety can be achieved at any level, and many times without anyone else even knowing. In fact, the fire chief is not even involved in most decisions that affect firefighter safety on a daily basis (Figure 1-5).

Communicate the Issues

Communicating safety concerns and near misses is an excellent way to start the process. Rather than allow the conversation to turn toward how management allows unsafe practices or won't dedicate the resources to safety, honestly evaluate the things that can be changed. Once change begins to take place, ensure that everyone knows the intent of

FIGURE 1-5
It doesn't take an organizational commitment to begin the adoption process of a safety culture.

the plan. The worst thing an organization can do is to fail to communicate change among all members. Change makes some people nervous, causing them to jump to conclusions. Rumors can get out of hand and can prevent participants from making an informed decision on their own. As a change is taking place, progress has to be communicated along with potential problems or oversights.

Make It Easy

The change has the best chance of success if it requires minimal effort. If it is a big change, investigate any other options that may improve its chances of success. For example, one shift decides that they are going to make a conscious decision to eat healthy for the next month, following strict health guidelines. Discussions prove that there are many different opinions by the members of exactly how strict it should be. A simple survey may show that if the strict plan is enacted, there will only be 65% involvement. By loosening the plan slightly, it may jump to 90% participation. Mathematically, the fat intake of the shift is reduced by simply allowing a little fat in the diet. If the change is going to require a significant amount of effort, participants must be warned of what is involved. It should be manageable and made as painless as possible. Hurdles should be identified and targeted ahead of time so they aren't a surprise.

Break It Down

It's been said there's only one way to eat an elephant: one bite at a time. Try to break down the change into easy steps. A mandatory change to add cardiovascular training to the daily routine would not be well received by members who are out of shape and have a sedentary lifestyle outside of work. Some of the baby steps could involve more walking on shift during preplans or low impact sports between emergency calls. If the shift occasionally gets together on days off, they could golf nine holes one day without a cart. Whatever the goal, any fitness program should be presented in phases.

Bosses on Board

The last way to improve the chances of a successful change is for supervisors to jump on board. If a department decides to start a wellness and fitness program, it boosts both morale and involvement if everyone is held to the same standards. Just like a chief wearing full turnout gear on a fire scene in the summer because everyone has to, a chief on a treadmill is a wonderful sight. It shows management's commitment to safety, with zero tolerance for not following policy. More importantly, it shows leadership during change.

Cultural Changes for Safety

You have just been assigned to an engine company that has a fairly long response time to a first-due industrial area. In fact, during rush hour on school days, it is common for at least one other company to beat you to your area. It's been subtly apparent for several years that companies either race or sandbag to the area based on dispatch information and even joke about it. As an engine operator, it has been made clear by your lieutenant that you are expected to be first in. You are surprised to find that each member does his or her part to make it happen, with a special "response procedure" specifically for that area. On dispatch, chauffeurs report directly to the engine without checking a map or donning gear. Firefighters scramble to get into whatever gear they can before the engine starts moving. The lieutenant gears up en route, while reading a map propped on the windshield. Although seatbelts are normally worn by all members, they sometimes neglect them for emergency responses to this area. It's not uncommon for the engine to arrive with no one but you buckled.

One afternoon you are dispatched to a forklift on fire in a plant in this specific area. The crew assembles and responds quickly as they normally do. You pause on the pad to close the bay door and the officer yells "Go, go, go!" As you approach the school zone, he appears to be pushing you through traffic with little regard for safety. You do your best to weave at a brisk pace while watching for pedestrians and not causing an accident. At one point, you are forced to lock up the brakes when a car pulls out of a gas station in front of you. The lieutenant curses and gestures to the driver. Despite the harrowing drive, you make it to the building just ahead of the second-in unit and pull to the rear, where black smoke is emitting from an open overhead door. A plant manager approaches your window and tells you everyone is out of the building, but a propane-powered forklift is on fire inside.

Your lieutenant gives an on-scene report and assigns duties to the firefighters. You engage your pump and pull a section of **large-diameter hose (LDH)** to a hydrant nearby. You feel a burr on the aluminum coupling slice your finger and look down to see blood running down your hand. You then remember that you don't have any gear on, so you drop the line to don your gear before the captain arrives. By the time you gear up, you hear someone yelling and see that the crew is ready for water. It's then that you realize the portable radio you usually grab when exiting the cab is still in its holder. The line is charged and you climb back in the cab to get the radio. Now you return to the hydrant to make your connection.

The crew inside finds the forklift on fire and extinguishes it before the propane tank breaches. As they exit the building, one of the firefighters approaches your engine and opens a pump panel discharge on his left hand. When you get to him, you notice a slight second-degree burn to the back of his hand. He tells you he apparently lost his glove at some point and received the minor burns when steam blasted back off the machine. You try to convince him to see EMS but he refuses. When you get back to the station, you find the missing glove on the bay floor just outside the firefighter's jumpseat door.

What kind of culture exists in an organization such as this? It resulted in no LODD, but it's obvious that there are plenty of brush fires burning in the department. It's only a matter of time before someone is seriously injured or killed. The first step to creating a safety culture is to identify our current culture. By defining the existing culture, we can identify the attributes that we want to keep. It's pretty clear that our culture is one that firefighters are very passionate about. It's a perception we have of ourselves, and one of how the public sees us. As a result, most would hate to give up any of our assets. Some of those assets include the "brotherhood," the apparatus, the uniform, the fire station, the job, and the "hero."

The "Brotherhood" (Both Men and Women)
Whether it's the teamwork at a structure fire, loading supply hose back onto the pumper after the fire, washing the ladder truck at the station, or helping a member move into a new home, the picture is solidarity. Lawyers and accountants might not get together away from the fire station, but we do.

The Apparatus
It doesn't matter if it's an engine, an ambulance, or an aerial ladder. They're always clean and usually red. They have lots of chrome and more lights than a Christmas tree. They are loud and fast and know every street in the town. They belong at the front of a parade. Children stare as we go by, and so do the adults. Most kids want to climb inside (and so do adults).

The Uniform
From Class As at the funeral of a retired captain to duty pants and a navy blue t-shirt when we're washing windows at the fire station, our uniform turns heads. Sometimes it's turnout gear or just a helmet and gloves rappelling down a rope. When we turn down the aisle of a grocery store pushing a shopping cart or when we pull up at a block party, we stand out from the shorts and the flip flops. There's no secret who we are.

The Fire Station
It's a home like no other. The American flag is always flying, but sometimes only at half mast. The fire trucks are always backed in and they get washed every day. If the weather is nice, it happens out front in the driveway. It's got a huge garage, a big kitchen, and lots of recliners. Paper towels serve as table linen and concrete blocks are the walls. Kids are welcome to stop by and so are dogs.

The Job
People know they can call us if an airplane crashes or if there is a snake on their porch. If they don't know who to call, they call us. Everybody knows our phone number. We won't make fun of them or give them a bill for calling us out. We'll get them the help they need and make things better. They need us, but we need them. They're our customers, but usually it's more than that.

The "Hero"
We'll always be the hero. Whether we personally earn the title or not, we are what we represent. We don't get scared. Some people are aware we saved their life in the back of

the ambulance or in the back of their apartment. Some will never know it, or will blame us for things we didn't do. Still, people look up to us because we go into buildings others run from. Some of us will pay the ultimate price.

The list goes on and on because we have a lot to be proud of—and a lot to hold on to. However, there is one thing that doesn't have to be part of our culture. We must reduce the number of our family members that pay that ultimate price needlessly. Although we know we'll never eliminate LODDs, we know a cultural change for safety will make a huge difference in the number of funerals we go to. We put LODDs on a pedestal, which is exactly where they should be. But if we reduce the total number of deaths per year by eliminating the needless ones, it actually increases the value of the pedestals we have. The best way to honor those who gave their lives is to make sure it never happens again. This leads us to a question: Can effective firefighting tactics and improved safety go hand in hand? Can a high-risk organization adopt a safety culture? Once again, some investigation outside emergency services might shed some light on the debate.

Components of a Safety Culture

When the U.S. navy began designing nuclear-powered vessels, such as aircraft carriers, in the Nimitz class, Admiral H.G. Rickover was forced to adopt a safety culture.[8] The world had little experience with nuclear power, and problems in the commercial reactor industry heightened public concern. In fact, trust with nuclear power was so low that Navy leaders insisted on perfection. A single reactor problem on any of more than 100 nuclear-powered ships would shut down the entire fleet, leaving questions about national security. Although it was a revolutionary theory in management, the adoption of a safety culture by the navy's Nuclear Propulsion Program was effective in designing a safe nuclear-powered fleet that is still in operation today.[9] Leaders instituted a safety culture based on three specific changes: no-fault management, supervisor responsibilities, and employee responsibilities.

No-Fault Management

The first step in creating a safety culture is to have **no-fault management (NFM)** for most issues. NFM is based on the principle that management is responsible for both the successes and the failures of the organization. Therefore, any failure of a worker or a system component belongs to management. If a failure of any kind is hidden from management in an effort to avoid discipline, it will cost the organization in the form of money, time, and maybe even lives (Figure 1-6). Problems must be identified early without fear of repercussion. Rather than identifying and punishing the person responsible, effort must instead be focused on fixing the failure and ensuring that it doesn't happen again. The actual person responsible may already know where the root of the problem is and how to fix it. That person will be much more willing to help if a supervisor isn't micromanaging. In fact, the main fault in NFM is a failure to communicate potential issues.

FIGURE 1-6
No-fault management seeks to expose near-misses and potential problems without fear of discipline in many instances.

NFM can be especially effective in emergency service organizations in which split second decisions are necessary. We have a unique group of personnel, and the very best employees society has to offer. Take Dave, for example, a paramedic working for a large city EMS system. He was a high school graduate who decided to enter **emergency medical technician (EMT)** school. His class included anatomy, physiology, and basic medical procedures, and lasted approximately six months. Clinical portions were completed in the hospital as well as during ride time in ambulances. Dave had to pass the class final in order to sit for the **National Registry** exam. The National Registry test is made up of two separate components: the computer-based knowledge exam and the practical applications test. When Dave passed those, he became a certified EMT and became eligible to enroll in paramedic school.

Dave's choice for paramedic school had an entrance exam and two prerequisites to be accepted, which he completed. When paramedic school began, he found himself in class 12 hours a week and averaged an additional 12 hours of clinical time. He had to maintain a passing score through advanced anatomy, acid-base balance, pharmacology, and cardiology. He completed certified courses in both pediatric and adult advanced life support. After approximately a year of school, he passed the class final and once again sat for the two-part National Registry exam. He passed both components and became a paramedic. Dave applied for a job at several EMS agencies. When he heard that a large city in the southern portion of the state was taking applications and giving a civil service exam, he made the trip. He scored well on the exam, and found out his ranking was fifth out of the 64 paramedics who took the exam. He did well in his interviews and his background check showed no criminal history. They checked his driver's license record, and found that clean as well. Dave was hired and assigned to the midnight shift on Medic 36.

If we look at all the steps Dave took from high school to being hired, it's obvious that fellow candidates dropped off at each point along the way. He passed hundreds of tests before he could even get an application for the EMS agency. Civil service testing is designed to identify the very best candidate of all who are qualified and rank the rest behind number one. Even at that point, many high-ranking and qualified candidates have had a "skeleton" in their closet exposed when it came to background or driver's license checks. Compare Dave with someone who dropped out of high school and filled out an application at a convenience store. After one interview, he gets the job because he has a pulse and is available to close on the weekends. It's obvious that Dave is the cream of the crop when it comes to the pool of possible employees. We would expect Dave to not make any mistakes or be involved in any mishaps during his career. He is, however, human.

Dave's been on the job now for 18 months when he and his partner Kelly are dispatched to a nursing home at 5:00 a.m. for a patient with shortness of breath. They assess a dementia patient and begin treatment in the patient's room. They lift the patient onto their cot and head outside. As they roll the cot down the sidewalk to the medic unit, it hits a bump and the cot suddenly drops 3 feet to the collapsed position. The patient screams from the jolt, but appears to be uninjured. Kelly barely gets her fingers out of the way from being crushed, and Dave tweaks his back slightly from trying to slow the fall. They reassess their patient and find no obvious injuries. They load the patient, who is unaware of what just happened, into the ambulance. At the hospital, they examine the cot and find no problems with it. They decide they must not have had it completely in the locked position but agree to keep an eye on it in the future. They debate whether or not to report the incident to their supervisor.

Policies and procedures would dictate that a report must be filed with statements by both crew members. Human nature would theorize that no harm was caused and that there is no need to bring undue attention to a meaningless event. History might remind the paramedics that the last crew to drop a patient was written up, with a permanent copy in each of the employees' files. Most would agree that even though Kelly and Dave are both phenomenal employees, they may very well keep the incident quiet. Under NFM, policies and procedures would explain the steps to take when an incident such as this occurs. This alleviates the fear of being written up and is replaced by the desire to investigate what went wrong, and to prevent it from ever happening again. The gurney would immediately be taken out of service until it could be inspected. If it was determined that it was a mechanical failure, it would be repaired or replaced. If no problem was found with the cot, the training deficiency would be identified and used to ensure that all employees would know how to avoid a similar occurrence.

NFM does not necessarily mean no discipline. It would not pertain to incidences involving the dereliction of duty or a violation of safety policies. A pattern of unsafe acts would still bring discipline, as would events completed under the influence of drugs or alcohol.

Supervisor Responsibilities

As a result of NFM, supervisors have specific responsibilities that encourage a safety mentality on the part of the workers. NFM necessitates a change in management style, sometimes referred to as the "upside-down pyramid." Managers take responsibility for assisting the workers in completing their duties, while ensuring that communication errors do not occur. Additionally, two specific duties include eliminating **problem filtering** and removing barriers.

Eliminating Problem Filtering. One of the main roles of a supervisor is to ensure that problems that do occur are not filtered. Rather than trying to place blame, NFM concentrates on what went wrong and how to prevent it from happening again. The military found that problem filtering had a negative effect on investigations. For instance, when investigators look into the cause of an airplane crash, they usually identify several contributing factors. A **contributing factor** is something that did not necessarily cause the plane to drop from the sky by itself, but definitely could have been "the straw that broke the camel's back." As contributing factors are discovered, they are collected. They continue digging for what they really want—the **root cause.** NIOSH reports for line-of-duty deaths and fire investigations operate in much the same fashion. In no-fault management, a fact-finding operation is exactly the same. Problem filtering occurs when contributing factors are covered up in an effort to hide minor mistakes or procedures. Thus, eliminating problem filtering is a way to ensure that all the contributing factors are identified.

Why is collecting all contributing factors so vital to a safety culture? Simply put, any combination of contributing factors under the right conditions will result in another incident. Obviously, the next incident could be more severe. It is possible that there was no root cause at all, just contributing factors. Take Dave and Kelly's incident at the nursing home, for example. A fact-finding investigation under no-fault management might turn up contributing environmental factors, such as darkness, uneven sidewalk joints, or extraordinary travel distance to the ambulance. It could expose operational factors such as discovering a seatbelt was wedged in the linkage, or that the lock was worn and prone to slipping. Maybe the investigation finds that Kelly's hand position had the potential for injury and that both of them were at the end of their shift and tired. Problem filtering can occur because someone thinks a fact is unimportant or because they don't want to have to explain why they perform their duties in a certain way. Either way, problem filtering must be avoided. It should also be obvious that without NFM, Kelly and Dave would never have supplied the information needed to expose these contributing factors and possibly prevent the next occurrence.

Removing Barriers. A manager's job should be to provide employees what they need to complete their job and eliminate barriers as they are discovered. If a supervisor spends the majority of his or her time looking over the shoulders of workers (micromanaging), it's both counterproductive and gives the wrong message about NFM. During the construction of nuclear vessels, managers identified their main job as clearing obstacles for those doing the work. Under this theory, a lieutenant in charge of fire department fleet maintenance might spend more time ensuring that parts are available as they are needed and that rigs are rotated through preventative maintenance at regular intervals. Sometimes his duty is more technical, like locating an alternator that holds up to higher loads. Regardless, he would spend less time watching mechanics work and more time streamlining the process of work. A manager has to figure out how to clear the path. A mechanic ensures vehicles are safe only when he is working on them. If an emergency alarm delays getting Ladder 1 to maintenance or there is no reserve apparatus for Ladder 1's crew to use, efficiency suffers. In a safety culture, efficiency = safety.

One serious safety concern in the past several years has been a reduction in staffing. Shortly after NFPA established 1710 and 1720 standards, the economy took a downturn. The vision of staffing improvements for career departments from the new standards soon faded away. Many saw no increase in personnel, but instead saw a further drop. Volunteer agencies also found a reduction due to members having to get another job, or even to move to find work. Not having sufficient personnel on hand is a twofold problem, having both operational and safety effects. If a successful safety culture is dependent on managers removing safety obstacles, sufficient staffing is at the top of the list. A manager working on this safety issue only has two ways to make it safer. The first way is to increase personnel, which may not even be an option for some departments. In some cases, employees might be hired with variations in qualifications or additional responsibilities. Automatic aid agreements or joint staffing on jurisdictional borders could be an option. If increasing personnel is not an option, the only other way is to reduce the risk to the limited staffing you have left. This doesn't mean that all interior firefighting will cease to exist, but, rather, that every job must have a risk-benefit factor assigned to it. The fire service has always been able to adapt and overcome, and working with limited personnel forces us to do a better job. You can't do more with less, but you can be more efficient. Once again, efficiency improves safety. At a fire scene, it's always a little more reassuring when an **incident commander (IC)** has a rapid intervention crew standing by outside and four apparatus not committed. Without the extra personnel available, the IC is forced to be a little safer. Even gamblers bet a little smarter when their chips are low.

Fortunately, there's more to clearing safety obstacles than large problems such as sufficient personnel. Everyday decisions by company officers can be made to clear safety obstacles. If firefighters won't wear a seatbelt because it interferes with the airpack strap, some chiefs have been known to pull out the packs. Officers and firefighters can train daily on putting their **self-contained breathing apparatus (SCBA)** on under a seatbelt until it becomes second nature. If they won't learn to do it, they can put on the SCBA before the vehicle moves or after it stops at the scene. It could be that the way the airpack is mounted interferes with the seatbelt and that simply moving the bracket fixes the problem, or that the straps and belts are too similar to differentiate in the dark. Maybe new seatbelt or airpack designs would solve the problem. Whatever the issue, management must provide the resources necessary to eliminate the distractions in order to allow a safety culture to exist. By providing solutions to fix the seatbelt issue, management is clearing the path.

Employee Responsibilities

The final component of a safety culture is that employees must learn every component of a job. On the whole, fire and emergency services are pretty good at this one. The natural desire to learn and become better is an inherent trait of most first responders, especially young ones. As some progress through their career, they become disenchanted, discouraged, or downright disgruntled. Although there are no sure-fire ways to bring back the members who have given up on their department, communication can sometimes unveil where the downturn occurred so it doesn't happen to someone else. Many times it's simply the fact that the job has become mundane. An opportunity sometimes overlooked is **mentoring.** By matching up someone with experience and knowledge to someone who has a strong desire to learn, they can actually help each other. For example, in recruit school, the basic operation of forcible entry tools is learned, but what's key is when those emergency responders use those tools to solve problems on the scene. Good mentoring techniques would allow the employee providing the mentoring to put the rookie on the spot by throwing "what-ifs" at him or her, things like using the tool for breaching walls, lifting manhole covers, opening the trunk of a car, breaking a window on the floor below you, or even anchoring an escape line. Mentoring utilizes every person on the department to improve training. It gives a rookie a goal to shoot for and someone at the end of their career a reason to get enthusiastic about coming to work.

Safety as a Certification Level

Everyone in the fire service has his or her own strong points when it comes to interest, knowledge, and abilities. Take ropes, for example. All firefighters and EMTs must be able to tie some basic knots. From tying off a Stokes basket for pulling a victim up a small hill to hoisting a hydraulic cutter onto a bridge overpass, we never know what the next call might bring. We expect everyone to be able to use the tools provided in a safe and appropriate manner, including ropes. Some departments have technical rescue or rope rescue teams that specialize in advanced practices of haul systems or high-angle lowering systems. Hazardous materials teams may have to tie a knot in a Level A hazmat suit in order to hold open a trailer door or secure a pallet of leaking materials. Dive team members need to be able to tie knots under water or in ice rescue suits to assist in the removal of victims or evidence. Emergency responders are generally trained to different levels for a given subject based on their job description. A successful safety culture should follow this same tiered format when it comes to the subject of safety (Figure 1-7). Although it's covered in detail in Initiative 5, the basics include awareness-, operations-, and technician-level aspects.

FIGURE 1-7 Integrating a safety culture means applying safety at all levels.

Awareness Level

The awareness level is an introductory level that identifies basic principles and risks. Awareness classes usually target new employees and other agencies involved with first responder activities. In a safety culture, the awareness level would introduce the 16 life safety initiatives and a brief description of how each is being implemented.

Operations Level

The operations level is the standard by which all firefighters and EMTs should be trained. This actually gets into the theories behind the practices and introduces the hands-on portion of training. A safety culture would identify certain theories of risk management in the operations level and begin incident safety officer training. Not only would operations-level members identify safety issues, but they would also begin to fix them.

Technician Level

Technicians have been trained to the highest level of safety, and have become proficient in setting up safety systems. They become certified incident safety officers and ensure that all other members are trained to the operations level as a minimum. Their focus would be forecasting potential issues and building safety into each skill performed by firefighters.

CULTURAL COMPLIANCE

Cultural compliance is not just having rules and regulations, or policies and procedures. It's putting your seatbelt on not because it is a state law, a department policy, or even a cause of firefighter death. Cultural compliance is putting it on because you can't imagine riding in a vehicle without it on. The U.S. Navy instituted these rules with their Nimitz program described earlier in this chapter. Charles R. Jones, a nuclear safety consultant who retired from the U.S. Navy and was actively involved in creating the safety culture, stated that the design of the nuclear reactors was given to the engineers to construct because "it was possible, not because we knew how to do it."[10]

It's a long stretch to say that instituting a safety culture in fire and emergency services is impossible. We know it's possible; we just have to figure out how to do it. It's no longer an option to maintain the culture we have.

SUMMARY

Defining a cultural change for safety is a vital component of reducing injuries and line-of-duty deaths in the fire service. In the introduction, you can see the 16 initiatives depicted as branches on a tree. Although trees have been used for countless illustrations in the past, you might notice in this example that a cultural change for safety is not even on the tree, but instead makes up the roots. It's not always easy to see or identify a safety culture. It's not something tangible, but without it, nothing else will survive. For instance, if you ever saw a full and healthy tree dangling perilously over a stream or river with its roots exposed, you may have wondered how long it will continue to hang on. If you didn't look at the base of the tree, it might look as if it would be there for another hundred years. The trouble is that we all know the waters will eventually rise and the wind will blow. The tree might try to resist, but without deeply anchored roots, it will simply topple into the water.

Similarly, emergency service organizations must adopt a safety culture as a foundation for their operations. It doesn't necessarily require the abandonment of existing traditions or culture, just a primary emphasis on safety. We still need to do our job—we just need to do it smarter. We've been writing our own job description for more than 200 years and can start writing safety into it anytime we are ready. Other industries were ready a long time ago and made the changes they needed to reduce their line-of-duty deaths. The military still fights wars, pilots still fly, and miners still mine. They just moved safety a little higher up the priority list by adopting a safety culture. In many cases, this will require a change. History shows that leaders who are able to accept and manage change can use it as an opportunity to improve safety and allow the organization to grow. Our job is a dangerous one, and without a deeply rooted safety culture with several layers of protection, our greatest assets are at risk.

KEY TERMS

ABS - Antilock braking system.

AED - Automated external defibrillator.

ALS - Advanced life support.

ANSI - American National Standards Institute.

ARFF - Aircraft rescue and firefighting.

chevrons - A method of diagonal striping consisting of opposing colors utilized on the back of vehicles intended to provide more visibility.

contributing factor - A tertiary component of an event that may have encouraged an event to occur or worsened the outcome.

crew resource management (CRM) - A procedure adopted from the aviation and maritime industries that recognizes the value of all personnel.

culture - The values, customs, and traditions of a group of people or an organization.

data - Information, usually in the form of numbers or statistics.

DOT - Department of Transportation.

dynamic - Changing or moving.

EMS - Emergency medical services.

EMT - Emergency medical technician.

Everyone Goes Home (EGH) - A prevention program created by the National Fallen Firefighters Foundation in an effort to reduce future line-of-duty deaths. One of the major accomplishments was the creation of the 16 initiatives, the basis of this text.

Federal Aviation Administration (FAA) - A division of the United States Department of Transportation responsible for civilian aviation oversight and safety.

hazmat - Hazardous materials.

IC - Incident commander.

investigation - A review of an event in which fact finding provides insight as to the root cause and contributing factors with the intent of preventing future events.

LDH - Large-diameter hose.

line-of-duty death (LODD) - Fatalities that are directly attributed to the duties of a firefighter.

mentoring - The process of an experienced person counseling someone else who is new to an organization or career.

National Fallen Firefighters Foundation (NFFF) - A nonprofit organization created to honor and assist families of firefighters who die in the line of duty, and create programs to prevent future events.

National Registry of Emergency Medical Technicians - An organization that establishes standards for the training and certification of EMS providers.

NFA - National Fire Academy.

NFM - No-fault management.

NFPA - National Fire Protection Association.

NIOSH - National Institute for Occupational Safety and Health.

OIC - Officer in charge.

positive-pressure ventilation (PPV) - A technique of forcing pressurized air into a structure or enclosed space in an effort to clear the area of smoke or gasses. It can also be used in conjunction with a fire attack in certain situations.

proactive - A method of making changes to avoid an event before one can occur.

problem filtering - During an investigation, the process of eliminating or not identifying contributing factors. It can occur inadvertently or in an effort to prevent fault.

reactive - A method of making changes to avoid future events after one has occurred.

risk management - Identification and analysis of exposure to hazards, selection of appropriate risk management techniques to handle exposures, implementation of chosen techniques, and monitoring of results, with respect to the health and safety of members.

root cause - The primary cause of an event. Without it, the event would likely have not occurred.

safety culture - A philosophy that prioritizes safety as a paramount value and relies on it to guide many of an organization's decisions.

SCBA - Self-contained breathing apparatus.

SOP - Standard operating procedure.

static - At rest, or unchanging.

REVIEW QUESTIONS

1. What is the definition of culture?
2. Why is there a need for a cultural change in the fire service relating to safety?
3. What are some of the components of a safety culture within fire and emergency services?
4. Why does change occur in an organization? Give an example.
5. What are some of the reasons why people resist change? Give an example from fire and emergency services.

FIREFIGHTING WEBSITE RESOURCES

http://everyonegoeshome.com
http://cdc.gov/niosh/fire
http://www.iso.org/iso/home.html
http://www.isomitigation.com/ppc/0000/ppc0001.html
http://technidigm.org/Technuke/nsctests/training/nscdisc.htm

NOTES

1. Managing in a Changing Environment. 1995, September. FEMA, USFA, NFA MCE-IG.
2. Brunacini, A.V. [date]. Essentials of Fire Department Customer Service.
3. Chief Dennis Compton, FDIC speech, 2002.
4. Wiegmann, D.A., H. Zhang, T. von Thaden, G. Sharma, and A. Mitchell. 2002, June. A Synthesis of Safety Culture and Safety Climate Research, Technical Report ARL-02-3/FAA-02-2.
5. National Occupational Research Agenda (NORA). 2008, October 27 revision. National Construction Agenda for Occupational Safety and Health Research and Practice in the U.S. Construction Sector.
6. http://www.flightglobal.com/articles/2010/01/11/336920/global-airline-accident-review-of-2009.html
7. New York Department of Motor Vehicles, 2007.
8. Written correspondence with Charles Jones, Germantown, MD.
9. http://technidigm.org/Technuke/assess/Jones%20ANS%20NSC%20Presentation_files/frame.htm
10. Ibid.

CHAPTER 2

Enhancing Accountability

LEARNING OBJECTIVES

- Define personal and organizational accountability and list their advantages.
- Discuss ways that accountability can affect health and safety within fire and emergency services.
- Explain the process of using NFPA 1500 to improve the accountability related to the health and safety of an organization.
- Discuss implementing the combination of accountability and no-fault management together.
- Discuss the need to create health and safety parameters for organizational accountability.

INITIATIVE 2. Enhance the personal and organizational accountability for health and safety throughout the fire service.

Did you ever wonder what makes one fire department better than another? How can two fire departments that have similar equipment, finances, and response areas consistently come up with different results? Take two departments with a common border that even have personnel who are members on each department and yet they still operate differently. Why is it that some organizations just click and some don't, and why is it that some members might leave one organization for another? Words like *leadership* and *accountability* are often used to describe the difference. It's easy for members of a department with problems to play the "blame game." The fact is that a lack of accountability is one of the biggest complaints of employees in any profession, and firefighting is no exception. With so many employees in so many industries throwing out the "no accountability" flag, they must be onto something. So what exactly is accountability, and why is it so hard to find?

ACCOUNTABILITY

Many times the words *accountability* and *responsibility* are used interchangeably. They are similar in meaning, but are actually quite different from each other. Let's say a fire chief has been given the order to reduce his operating budget by 20% for the remainder of the year. With this directive, he is now **responsible** to cut costs, which then makes him **accountable** to everyone involved. In other words, responsibility is assigned to you by a superior, which then makes you accountable to your superior, your subordinates, the general public, and even yourself. The fire chief has several ways he can accomplish the cuts and must make tough decisions based on his vision, opinions, and values. He may decide to cut staffing through attrition or layoffs. He may decide to close a fire station, combine companies, or delay the replacement of apparatus. Possibly the cuts will be made to maintenance, training, or fire prevention. One thing is sure: If he fails to cut the budget as ordered, the mayor will hold him accountable to the fact that he didn't do his job. If the decision he makes is unpopular with the firefighters, the chief is accountable to them. If the fire department fails to respond to an emergency in a timely matter due to his decision to close a company or station, he is accountable to the public. Finally, he is most accountable to himself. Any decision he makes will have immediate effects on lives, and may even have unforeseen future results.

In this case, it would be very easy for the fire chief to claim that the city administration is actually responsible for the cuts. Although this would be an accurate statement because the mayor is responsible for balancing the budget, the mayor did not actually cancel the order on the new truck and decide to run Engine 33 into the ground. This is where the fire chief either exemplifies accountability or chooses the easy way out and plays the blame game. If he is truly accountable, he explains to his members that his decision was the best option at the time. He shows leadership by creating a team approach to making E-33 the best it can be. He asks for the firefighters' suggestions and assistance in making the existing vehicle both safe and functional for another year or two. He also offers to do whatever he can to provide the resources they need to keep E-33 in a ready condition.

If the chief leads by example and stays accountable, his decision creates a string of accountability. By explaining to the personnel that they would not be replacing the engine, the *responsibility* to make old E-33 ready to serve the citizens in a safe and efficient manner for a couple more years is issued to the officers, firefighters, and maintenance crew. They are now *accountable* to make sure that it happens (Figure 2-1). Imagine the feelings of a firefighter who was on the truck replacement committee and helped write the specifications for new E-33. His natural reaction would be that all the work he did was wasted, and could become very bitter as a result. He could blame the fire chief for "not standing up to the mayor" and "making us drive an unsafe piece of junk." He may even be tempted to sabotage a component of the old truck just to prove his point. If he chooses this road, he essentially breaks the string of accountability at the shift level and destroys accountability for everyone. Interestingly enough, a person like this that throws

FIGURE 2-1
Older equipment doesn't necessarily mean unsafe equipment, but the maintenance may require more accountability.

out the claim that the supervisor lacks accountability is usually the real person who is unaccountable. Intentionally sabotaging not only shows no ability to be accountable, but also demonstrates a destructive attitude and a potentially dangerous character flaw.

The accountability string goes up the chain of command as well. If the mayor believes in accountability, she defends the fire chief for his decision to delay apparatus replacement; she doesn't blame the taxpayers for a decrease in revenue, or even the economy. Accountability was a tribute that some politicians, such as President Harry S. Truman, personified. He even had a plate made for his desk that said "The buck stops here."[1] It derived from the slang phrase of "passing the buck," which meant to blame someone else. Rather than passing blame, the quality of accountability instead looks at a problem and asks what "I" can do about the situation. Effective leaders look at a hurdle or a failure by evaluating all the causes of the situation or "breakdown," starting with their own contributions. As discussed in Chapter 1, "Defining a Cultural Change," the chain of command dictates that every boss is responsible for every action initiated under his or her position. Therefore, blaming a subordinate has no benefit.

BLAME

Besides providing a great excuse, blaming someone or something has some nice side effects. Blame deflects faults or weaknesses, which in turn boosts egos. It makes some people feel better about making a mistake or having a poor performance because they can convince themselves that they were not at fault. Rather, they can conclude that they were simply a victim of circumstance. For instance, one evening at work, you walk into the ambulance bay after dinner and see that your ambulance is parked outside. You know you left it inside earlier, but it appears your partner pulled it outside to clean the bay floor. It's getting late, so you decide to put it away. As you back into the bay, you feel resistance and hear a loud crash. When you get out and look up, you see the crushed bottom panel of the bay door wrapped around the light bar. Apparently the overhead door was not all the way up when you backed into it.

What's your first reaction? It is easy to blame your partner. After all, he put the ambulance outside and didn't put it back. He left the door halfway down, just asking for someone to hit it. There may even be a policy he broke about doors being partially closed. He does it all the time, he never finishes anything. And where is he now while you're working? He's probably on the phone or asleep in a chair. Now that you think about it, the only thing that points to you at all is the fact that you did it. Maybe it's anger, maybe it's embarrassment, or maybe it's a fear of discipline. Whatever it is, it probably will lead to blame. By blaming someone else for the failure or loss, you essentially make yourself the victim. Some people enjoy being a victim, and rely on it for esteem issues. Some just use it to keep from becoming the villain. If you're able to convince others that you are the victim, they tend to look for and turn on the villain. Victims always receive more sympathy than villains.

One interesting aspect of this **victim syndrome** is specific exclusively to emergency services. It's a well known fact that when call volume drops, so does morale. More than one chief has claimed that nothing cheers up firefighters like a fire. It's not that firefighters enjoy seeing someone lose their property; it's just that periods of inactivity tend to be breeding grounds for personnel issues. The biggest reason is probably because after they joined, they were trained and equipped to handle serious emergencies. Not much of recruit school is dedicated to odor investigations and fire alarms. Then when all they seem to get is "smells and bells" and there is a prolonged lack of significant calls, they feel underutilized and become disappointed. Having all the emergency equipment and gear may lead to the feeling of "all dressed up and no place to go." They may see nonemergency functions such as fire safety inspections or hydrant flow testing as "busywork" during these times.

This is even more prominent if the fire chief looks to expand prevention services or improve the department's public image with new programs. Maybe the department begins offering residential fire inspections or door-to-door smoke detector tests. It could be reading a fire safety book to children at the library on Tuesday nights or performing blood pressure checks at the mall on Saturday mornings. Emergency service personnel don't like being bored. For generations, groups of firefighters sat around the table drinking coffee and planning their next coordinated attack on the enemy. This presents an issue if they can't find an enemy. They joined for the excitement, and if they don't get it, they create it. Some departments plan for downtime by designing stations and assigning personnel based on a special skill or interest. A fire station may house **self-contained breathing apparatus (SCBA)** repair with all the necessary testing equipment and tools, while technicians on each shift are trained to make repairs. Other stations might have a wood shop, welding shop, or sewing shop. Sometimes pump testing pits or lakes are adjacent to stations to facilitate pump testing and others have adjoining public education facilities, such as a safety village (Figure 2-2). Besides the fact that members keep busy doing something they enjoy or even learn a new skill, a properly designed specialty shop can save the department money. Fire departments are commonly looking at new ideas to share resources and save money, and a specialty shop is a great way to do that on a regional basis. Even if the department doesn't offer specialty shops, the personalities of some people tend to focus their unused energy on other suitable projects. They might join a special team and devote slow times to that. A member who is active on the dive rescue team could spend Saturday afternoons on shift organizing the dive trailer or inspecting dry suits. A person interested in the history of the fire department might spend downtime clipping old articles or researching the history of Station 2. Sometimes being involved simply means reading online articles of firefighting tactics or **National Institute for Occupational Safety and Health (NIOSH) line-of-duty death (LODD)** reports.

Unfortunately, sometimes firefighters just feed off each other negatively in slow times. It could begin with nitpicking and blaming between them, and potentially evolve into an all-out shift war. They might fight changes from administration and file

FIGURE 2-2
Specialty shops such as pump testing facilities not only can save a department money, but also can keep firefighters trained, active, and interested.

complaints or grievances over situations with which others would not have a problem. They might blame their problems on the attitudes or training level of the new firefighters joining and relish "the good old days." They could think the old apparatus was garbage and the new is built poorly by "low bid." Whatever the situation, they paint the picture of being constantly victimized by their coworkers, their shift captain, and "greenhorn" chiefs. Everything they discuss points the finger at someone or something else. It's not fair, and none of this would have ever happened back when they joined. Their audience is anyone who will listen, and the stories of how bad it is continue to exaggerate each time they are told. You might actually think that things have gotten pretty bad in this department.

Compare this crew with the discussion in Chapter 1 about Dave and Kelly working on Medic 36. Remember that they beat out hundreds of students and potential hires for their jobs and are essentially the cream of the crop. They passed tests and screenings most professions don't even use. What could make the two groups so different? The answer is a lack of accountability. They aren't necessarily bad employees who need to be fired, but they have decided to embrace blame. In this case, a lack of activity at work and some potentially bad attitudes have been fertilized, cultivated, and harvested into one bad situation. It may not be too late to save them. In fact, at the next major incident, they may actually shine and impress everyone. One theory at work here has to do with the human psyche and the victim syndrome. If the personality of an emergency services worker is caught in a victim syndrome cycle, then a conflict occurs when there is a real victim. During an emergency, responders automatically identify and help the victim. Because the victim role has already been filled, the emergency responder takes the role of hero. Victims receive attention, but heroes get more. That's why there is very little arguing and blame on emergency scenes. In fact, most emergency responders agree that personnel interact much better on scenes than back in quarters. You could say it's hard to be the victim when you're helping one. The only way to prevent blame from taking over is to make a genuine commitment to being personally accountable.

PERSONAL ACCOUNTABILITY

The term **personal accountability** refers to the ability of an individual to take responsibility for the present situation, and to react to the circumstances without attributing faults to other factors or people. In other words, you have the ability to account for all of your results. This definition points out several key components. The word *ability* infers that someone learned the art of personal accountability, much like learning other skills. The ability of an engine operator to estimate friction loss at 4 a.m. while supplying three handlines and a standpipe is not a God-given gift. An honest effort to become proficient at hydraulics is the only reason a pump operator can pull it off. For most, accountability makes hydraulics look like a piece of cake. Some could be simply born with it, but most would agree that being accountable is not the way most humans are wired. From childhood to adulthood, the typical response when something goes wrong is to pull out an excuse. Sometimes the defense mechanism blurts it out so fast that we don't even realize how ridiculous of an excuse it is until we hear ourselves say it. It takes a conscious effort to learn to put on the brakes and *not* to blame.

The technique is simple in theory, but it's difficult to master. Let's say you are a newly assigned firefighter on a busy engine company. Your first three months have gone well, and your lieutenant is happy with your progress. One day you get dispatched second-due to a house fire in a two-and-a-half-story wood frame building from the 1930s. You pack up en route while Gene in the other jumpseat tells you to grab the line if needed and he'll take the tools. Your engine is assigned to pull a backup line through the kitchen on the **Delta side.** After entry, it's clear that first due put a pretty good hit on the fire in an adjacent room and you hear the incident commander assign your lieutenant to change to overhaul operations. Your boss gives you vague instructions to retrieve more pike poles. He leaves the kitchen while Gene begins to pull ceiling. You start to lay the line down on the floor, but second-guess leaving it on the steps and in the doorway where someone could trip over it. You ask Gene what he thinks and he tells you to pull it back out to the driveway.

As you return to the house with more tools, you are greeted by crews yelling and scrambling for a line. Gene apparently had opened up a hot spot that erupted into fire when it got oxygen. The fire is now visible in the kitchen windows and is working its way down the hall. You pick up the abandoned hoseline and feed it to firefighters at the door who push the flames back. They extinguish the fire and account for all personnel. Your lieutenant exits the building visibly upset and heads right for you. He tears off his mask and gloves and begins to berate you in front of everyone, including the neighbors. With a generous use of choice adjectives and adverbs, he basically asks why the line was removed from the building. This is it. This is the point where the hair stands up on the back of your neck and blood rushes to your brain. You flash back to seeing your father holding a baseball by a broken window. You scan the scene for excuses and prepare your list: "It was a trip hazard, there was a line inside with the other crew, you didn't tell me to leave the line there, I thought you wanted me to take it out when you told me to chase tools," and the big one—"*Gene* told me to." The ability to take responsibility is the first part of the definition of personal accountability. The act of not answering with an excuse is an ability that can be learned, and is a vital step in becoming accountable. If you had learned this technique, your first reaction would be to hold your tongue, and investigate all the influences you had on the situation and its progression.

The second part of the definition worth noting is in regard to "taking responsibility for the situation." As you were being yelled at by your lieutenant in the driveway, it's hard not to defend yourself and keep your cool. Take a deep breath and start to prepare a mental list of how *you* could have been responsible for what just happened. After reviewing the facts:

1. *You* pulled the hoseline.
2. *You* were assigned another task, but were not told to remove the line from the house.
3. When they needed it and it was gone, it was because *you* put it in the driveway.

This is exactly how everyone else sees the facts, which is very similar to the old technique of trying to "put yourself in someone else's shoes." By looking at this list, you agree that it was at least partially your fault, but that there were contributing factors. Your lieutenant and partner should take some of the credit for the mistake. To prove it, you put together a list of contributing factors that include the following:

1. Not getting clear orders about what to do
2. Not anticipating the potential of fire spread during overhaul
3. Listening to Gene instead of your lieutenant

Sure, you could blame your supervisor for muddy orders, but it is up to you to ask for clarification. He knew what he meant when he said it. Similarly, nobody said Gene was the right person to ask. You chose to do that. Actually, if you look at each contributing factor, they too are your fault. Blaming anyone but yourself for what happened is wrong. Therefore, the first words out of your mouth would consist of an apology, followed by an explanation of misunderstanding the orders, capped off by an honest promise to learn from the event to ensure that it never happens again. Not blaming someone or something else is not necessarily "jumping on the hand grenade" and taking blame for something you didn't do.

A **postincident critique** at the fire station should be used to explain all the reasons for your actions without blaming anyone. If Gene and your lieutenant are both accountable, they will offer their responsibility as well. The training that could come out of a situation like this is that your lieutenant will do a better job of communicating orders and you will ensure clarification. It could evolve into discussions of manned hoselines being present during overhaul and ways to reduce trip hazards at fire scenes. If you really commit to getting better at personal accountability, it's amazing (and entertaining) how many times you notice others blaming their circumstances on other people or even inanimate objects. We all end up in situations we would rather not be in. The situation can be uncomfortable, embarrassing, or even downright horrifying. Regardless of the circumstances, personal accountability can work.

Examples of personal accountability, both good and bad, are found every day in professional sports. In no other occupation is a person's performance at work under such a public microscope than in sports. Players are constantly compared to their peers, and even ranked by their strong points and weak points. Baseball even keeps track of mistakes in the form of errors committed by a fielder. Imagine for a minute, the levels of this accountability being subjected to firefighters. Take the greatest published fire officers of all time: Bowman, Norman, Carter, Mittendorf, Brunacini, and Dodson. Picture them lined up in their Class A uniforms sitting behind a wide desk with bright lights on them and cameras rolling. They replay your performance at a structure fire last night in slow motion and question why your chief is even "playing" you. Everyone sees the replay on the news of you lapping the aerial twice, opening every compartment door looking for a salvage cover. "Nope, still not there" they joke. They talk about how you lost your temper and threw your helmet in the locker room after the fire. They point out that you're usually the last one to drill and the first to sneak off, and it's showing in your performance every week. Imagine your mayor reading the reports in the paper about your waning performance: 2 for 12 on successful intubations, or three backing accidents in the past 14 months! Maybe the citizens even call in and say you're a washed-up has-been. "Trade him before we have to pay him his clothing allowance or he gets promoted," they write.

It may be a little off the wall to even imagine it, but luckily we aren't subjected to this level of public scrutiny. It is clear that athletes are held accountable to the public for their achievements and actions both on and off the field. Whether the player is truly accountable or not is another subject. In his autobiography, *Get in the Game: 8 Elements of Perseverance that Make the Difference,*[2] Hall of Famer Cal Ripken Jr. reinforces the importance of personal accountability as he tells a story from his career. He was thrown out of a game in the first inning for arguing with an umpire. He later found out that a fan had brought his young son all the way to Baltimore to see his favorite baseball player play, but never saw him take one swing. The boy was heartbroken, and the press ran with the story. Cal felt horrible. He began evaluating what was important to him and where he needed to improve. He settled in on the subject of personal accountability. A short time later he was playing a game in Toronto. As he took the field in pregame warm-ups, he noticed the sod where he played shortstop was raised up to a hump. Apparently the irrigation or drainage system had leaked the night before and froze under the turf. His first thought was to complain to the umpires and groundskeepers. He could have chosen to ask them to delay the game to fix the field or he could have used it as an excuse if he had a bad game. Instead, he chose to suck it up and play a couple feet in from his normal spot, making the most of it. He decided he wasn't going to allow outside circumstances to affect his performance. He played that game without a complaint, and went on to continue playing every game for the record of most consecutive games played in the history of professional baseball.

If only everyone involved in sports was so accountable. When confronted about their performance, many players blame their teammates or the coaching staff. Some players complain that they are carrying the team or not getting the play time they deserve. The referees or umpires are always wrong. When a player is called for a penalty, most deny it ever happened or complain that it was actually someone else's fault. Even with proof on instant replay, some continue to argue. Not many players who test positive for performance-enhancing drugs come clean with their fans. It's usually blamed on vitamins, herbs, allergies, or some sort of interaction from something a trainer gave them. Some teams blame a losing season on injuries or on an unfair schedule. Some owners blame it on an old stadium and move the team to another city. Coaches blame it on poor farm teams, draft picks, or even opposing fans. Sometimes it's a salary cap, or outdated league rules. Even when they seem to take a little accountability, they use obviously vague phrases that show no accountability like "we failed to execute" or "we weren't mentally prepared."

IMPROVING PERSONAL ACCOUNTABILITY

Improving personal accountability involves making the decision to be accountable, becoming an active member of your department, speaking up, taking responsibility for the outcome of calls, and taking responsibility for safety.

Make the Decision to Be Accountable

The first thing to remember about personal accountability is it's a personal decision—you cannot force it on others or take it on with half interest. Changing may not even be noticed by others right away. It could be considered an internal struggle, trying to break free of the addiction of blame. It takes determination to make it work, just like a choice to eat healthier or exercise more. The most important point is that there is nobody stopping you from making the change, and anyone can do it. Some think you have to be an optimist to achieve accountability. Although the argument could be made that it might make learning accountability easier, it's certainly not necessary. An optimist looks at his or her garage that caught on fire and vows to build a bigger and better one. An accountable person looks at how he or she may have contributed to the cause or spread of the fire and vows to never do it again.

A great example of a decision to become personally accountable for health and safety is a commitment to being healthy (Initiative 6). Personal accountability can strengthen the desire to make conscious decisions to improve your own health.

Although it's not possible to change someone else's lifestyle, you can make alterations to your own and, for example, can encourage others to eat healthier and to exercise. A person who doesn't know how to operate a treadmill has little effect on others.

Become an Active Member of Your Department

One obvious way to improve your accountability is to practice by being an active member of your department. By being an active participant in change, you can make the department better and work on accountability at the same time. Someone who "does eight and hits the gate" or "does twenty-four and hits the door" with only the bare minimum while on duty is likely to circumvent opportunities to be accountable. Volunteers can choose to be more active, or can just meet the minimum requirements. You could say that being involved puts you in a position to practice accountability (Figure 2-3). Whether it's a promotion or involvement with a committee, everyone tends to keep an eye on the one who is attempting to make a change. It could be that they support the change or that they are truly worried about how the change might affect them. Maybe they just like to root for the underdog. For whatever the reason, you have their attention. Adopting personal accountability has some great motivational affects on others that witness it. As discussed in Chapter 1, some embrace change, especially if there is success.

Suppose you are a lieutenant assigned to the training division and you want to work on implementing Initiative 1 in your department. With a broad goal of putting firefighter safety first in a true **safety culture,** you decide that the first step is to involve all of management in your safety training programs. By having chief officers actually assist with

FIGURE 2-3
Staying involved in your organization can help improve accountability.

training, you can show the department's commitment to changing the culture and increase member buy-in. Your first meeting with the chief goes well, and she seems somewhat interested in teaching the drill. She asks you to put together an outline for the presentation on what you envision. You leave the meeting excited, and even tell others about the plan. Most of them tell you you're wasting your time, and that nothing will come of it. Undeterred, you check out the **Everyone Goes Home (EGH)** website and download a great safety culture lesson plan. It's full of examples from new equipment to sample **standard operating procedures (SOPs)**. It's even got video scenarios for group discussions that should fit your department nicely. You transform it into a slide presentation, complete with pictures of your apparatus and personnel, and lay out the department's safety plan. You make a copy for the chief and set up a time to go over it with her. It doesn't take long for you to notice that she is less than impressed with your lesson.

It turns out that some of your plans are a little more drastic than what she envisioned. She feels safety is a subject that has more to do with firefighters' attitudes and actions, and less about increased funding. She believes that if existing SOPs were followed, a safety culture would already exist. She wants to keep firefighters safe at fire scenes, but she doesn't believe a change in policy is necessary. For instance, your suggestions about **risk management** and rapid intervention teams would involve commitments she doesn't agree with. You realize that some of her points are valid, but that there doesn't appear to be any improvements she would be willing to make in administration or operations. She tells you what she wants removed from the program and sends you back to the drawing board. As you walk back to your office dejected, you try to take in all that she said and try to figure out what can be salvaged with the drill. Even more difficult is how you could possibly remain accountable in this situation.

It appears as though your marching orders are clear. She outlined what the lesson plan would include and what it wouldn't. You have been given the responsibility and are therefore now accountable for designing the lesson plan she wants. It would be easy to throw the project down and say forget it. It might be tempting to use sabotage and set the presentation up in a way to embarrass the chief. You might consider talking to the firefighters ahead of time and giving them ammunition for arguments when she tries to present it. It's easy to blame the chief for not wanting to change her old-fashioned ways or not really caring about the safety of her crews. But accountability doesn't allow you to follow the easy paths. The fact is, the drill could still be a success, and could very well be the first step in defining a safety culture in your department that you desire. Don't throw out your original presentation. Look at it as the second or third step in the progression that you can use in the future. Personal accountability doesn't blame the chief's opinion—it considers the impact *you* had on the failure of her approving the proposal. In other words, maybe it was a little too drastic. Maybe your idea of what the drill would be was never what she saw. It could be that your presentation would have actually pushed firefighters away, and they simply weren't ready for that big of a change yet. It's time to pick up the pieces, start over, and create a presentation that will meet the expectations of the chief, as well as be acceptable to you.

Speak Up

Another way to improve personal accountability is to speak up if you see something that doesn't make sense or is just plain wrong. Allowing a questionable activity to continue just because it's always been that way does not make it acceptable (Figure 2-4). Ambulances regularly exchange linens at hospitals, including towels, sheets, and blankets. Everyone agrees that the intention of the service is to exchange dirty linens for clean after they are used during an EMS call. Most **emergency medical technicians (EMTs)** and paramedics feel that using hospital towels to decontaminate the inside of the ambulance after a call is also an appropriate use of linens. Some feel that keeping towels at the station to wash the ambulance exterior is acceptable, and others think sleeping on hospital sheets while on duty is permissible. A select few think it's okay to take hospital linens home with them for personal use. So what exactly is acceptable use of hospital linens? Morals and opinions guide us in everyday decisions, but personal accountability is what holds us to what we do or don't do. If personal accountability was easy, we'd all be doing it. That is precisely why we have to practice accountability.

Take Responsibility for the Outcome of Calls

Another way we can improve our personal accountability is to actually take responsibility for the outcome of the emergency calls we respond to. We've always taken the approach that we didn't create the emergency, and that buildings will continue to burn down and people will die despite our best efforts. These are true statements and we

FIGURE 2-4
Speaking up when something isn't right should be encouraged in every fire and emergency service organization. "Matt Henk," courtesy of Rob Gandee.

have to realize that we're not here to change the world. However, it's difficult to make an argument for personal accountability and then dismiss the most important component of our job as "out of our control." It's true that deciding when personal accountability can be applied is directly tied to the amount of control we have over a specific situation. Fire and emergency services have three specific components of an emergency that we deal with that dictate how much control we have over them. They are uncontrollable, semicontrollable, and controllable components.

Uncontrollable Components

Emergency calls are initially uncontrollable. If a man has a diabetic emergency while driving and ends up parking his Buick on a crowd at a street fair, there's not much we can do about it. That's why fire and emergency responders will never "prevent" themselves out of a job.

Semicontrollable Components

Immediate response to the pedestrians struck is partially controllable. In other words, we may have influenced getting enhanced 911 in the town. We could have taught first aid classes and pushed to have **automated external defibrillators (AEDs)** at public events. We may have met with the festival committee and ensured there were good maps for us and that fire lanes were kept open. We've worked in the past with the police who arrive on scene before us, and they give us the information we need to make decisions.

Controllable Components

Based on dispatch and police reports, we call for extra equipment. We pick our route to the call, and decide on the speed and style of our driving. Training has taught us how to rapidly triage and handle a mass-casualty incident. In no time, we have all the manpower, ambulances, and helicopters we need to give the best possible medical care.

Take Responsibility for Safety

The safety aspect of situations follows the same model as the outcome of calls. We don't have much influence over safety when the emergency occurs. Our influence on safety increases when we are responding, and is entirely our responsibility when we get on scene. That's because if we examine emergency calls, we discover that they are not a single event waiting for us to arrive and take command of. They are actually a progression of foreseeable and unforeseeable independent events. At any point during this dynamic process we can make a conscious decision to act or not act in a specific manner that either changes or doesn't change the outcome. Even on significant incidents where numerous emergency responders converge, a decision for personal accountability has an influence on the potential outcomes of the incident.

Personal Accountability to Safety

The first component we must look at is our accountability to safety: for us, our crew, our victim, and the general public. Taking responsibility for our own safety is the main component of a safety culture. If we remain uninjured on the scene, we can continue to be a resource rather than another liability of the emergency. Our seatbelt, our hood, and our crew integrity are our concerns. Without personal accountability to safety, the incident is destined to have a bad outcome. Everyone is responsible for safety.

Personal Accountability to Training

Nothing affects the outcome of an emergency call more than how prepared we are for the emergency. This can be evident in both common and rare emergencies. Passenger train derailments are not necessarily a common occurrence for emergency responders. Even if we have rail service through our community, chances are that most responders won't ever work on a derailment. Knowing the location of fuel shutoffs as well as the existence of high-voltage power and extrication techniques are things we should know. Practice is the only way to perform effectively for the rare occurrence of high-risk incidents. Just as important are the everyday events we respond to. Lack of training on the common occurrences leads to bad habits, the use of antiquated techniques, complacency, and, often, injuries. The old adage "If you don't use it, you lose it" applies to emergency responders. Training reinforces our actions and knowledge of how to handle a given situation, and how to improvise when we've never seen it before. Without training, we are forced to improvise every day. A well-trained responder has a distinct effect on the outcome of a call.

Personal Accountability to Equipment

Improving equipment is one of the most obvious ways to improve the outcome of emergency events. The availability of gas meters, thermal imagers, and 12-lead EKG monitors have allowed responders to improve the quality of the job performed. This could be by doing our job quicker, more safely, or by saving more lives and property. Being aware of what new equipment is on the horizon and persuading those capable of providing the equipment to do so is everyone's job. Some choose to simply complain about outdated or lacking equipment rather than solving the problem. There's no argument that AEDs on all emergency response vehicles save lives; it's just a matter of figuring out how to pay for them.

Personal Accountability to Services

If we never improve our level of service, we remain stagnant. We must be cognizant of what needs evolve in our jurisdiction and how the rest of the industry is reacting to their needs. We must reevaluate what we do and how we do it, and make appropriate alterations in our techniques. Much like lobbying for better equipment, we must also lobby for improved services.

Personal Accountability to the Organization

By increasing our personal accountability, we motivate others to come up to our level. This is already evidenced by groups of emergency responders who seem to be more proficient and act as a team more than others.

ORGANIZATIONAL RESPONSIBILITY

The next logical progression in improvement from personal responsibility is organizational responsibility. So far we've been covering various aspects of a safety culture, have explored the advantages of **no-fault management (NFM),** and have explored the need for personal accountability. We discussed blame, and how it's easy to use it as an excuse when things go sour. Now let's weld these theories together and see what they look like in a work environment with organizational responsibility. Remember when you were putting the ambulance away after your partner left it outside and you smashed the bay door? The following is what that situation would be like with the safety culture model we have been developing.

When you get out of the ambulance and realize you broke the door, your first thought is not fear that you are in trouble, but rather fear that you may not be available to respond to an emergency call. You feel disgusted it happened, but because of no-fault management, you do not blame your partner. Instead, you remain accountable. When your coworkers come out to see what the noise was and ask what happened, you simply respond, "I backed into the door." Your coworkers ensure that the ambulance is safe to keep in service, and the door is bandaged up. Statements by you and your partner concentrate on the contributing factors to the failure, as opposed to who is at fault. No-fault management forces the involved parties to identify system failures. System failures in this example are communication, policies, and procedures. By being involved in the investigation, you and your partner decide the first step in preventing future incidents is better communication. Besides discussing things such as moving the vehicle, you recall times in which equipment was forgotten at a residence or the hospital, and agree that communication can be improved in several areas. You even talk about putting these ideas into a communications drill for the entire district. You ensure there is a policy about doors that are all the way open or all the way closed to prevent future accidents, and look at an electronic system that opens the door all the way if it is left partially closed for a certain amount of time. You even talk about suggesting a walk-around policy to administration and you both agree to back each other in it when possible.

Some might consider these theories of discipline as ridiculous, and the equivalent of a "time-out for adults." The fact remains that you are a highly trained, valuable professional that made a stupid move when you backed through a door that wasn't fully open. It was an expensive mistake with the potential for injury, but really no more of a mistake than forgetting an axe at the scene of a structure fire or dropping a gas detector into a manhole. You're human, and humans always will make mistakes. You didn't intend to do it, and wish it never would have happened. In fact, the chance that you ever will do it again is slim, whether or not you receive any form of discipline.

Following the traditional method of discipline, you would be written up and the letter would be put in your file. It's actually just a slap on the wrist compared to how embarrassed you feel already. In fact, the only organizational benefit to writing you up for an incident like this is a paper trail to keep track of those who have a pattern of poor decision making. If that's a concern for the organization, they can always file the investigation report rather than a letter of discipline. The paper trail is not very accurate anyway. Very rarely do near-misses make it to a personnel file because most of the time they're not reported. You may have been the one who hit the door this time, but other people have just been lucky.

On the other hand, the benefits of no-fault management to instituting a safety culture are very effective. Rather than hiding from the facts, the members are directly involved in the solution, and thus prevent future incidents. Critiques should occur after department incidents, just as they do for emergency incidents. There should never be blame, only findings of what happened and how we prevent it from happening again. Nobody got hurt this time, but the next accident that was prevented by exposing the contributing factors may very well save a life. Because the members were less involved in pinning blame on each other, morale was improved. Teamwork was encouraged throughout the fact-finding investigation and solution-developing process. Finally, identifying and eliminating contributing factors affects an entirely different realm of accidents that have nothing to do with bay doors. The discussion you and your partner had about communication and avoiding leaving equipment at the hospital or patient's home may have resulted in a district-wide communications and accountability drill. That training may spark the interest of members who would apply the skills to other aspects of their job. Once the predominant trait of the members is personal accountability, the scales tip and the department can't resist becoming accountable. Personal accountability takes time, but organizational accountability is suddenly an overnight success.

By mixing personal accountability and successful management techniques, an organization can grow exponentially. **Organizational accountability** involves everyone working together to achieve the vision and mission of the organization. It has less to do with individual job responsibilities and more to do with the overall picture. Breaks in communication, however, can lead to failure. For example, let's say you are a firefighter on an engine company that responds to a fire in a small bungalow and arrives second-in. The family is safe outside, and the first-in engine crew is upstairs fighting a room and contents fire. Your lieutenant receives the order to stand by on the front porch to relieve the upstairs crew if they run out of air or need anything. You look in the front window and see a neatly kept house with surprisingly nice furniture in the living room. There are numerous pictures on the walls, as well as statues and various pieces of art amidst the white and gold-trimmed furniture. There are no signs of fire extension, smoke, or water in the area, but you can hear the handline operating upstairs and figure that it's only a matter of time before dirty water works its way down. Your lieutenant points to the now sagging ceiling and asks you how long you think it'll be before it breaks through. You ask if he wants some salvage covers so you can make a water chute out the window or collect some of the belongings and move them to the garage. He denies your request, explaining that you were assigned to stand by for relief of the crew upstairs, not to perform salvage operations. Freelancing is not permitted on the fireground and will not be tolerated.

Our fire service culture has done a great job over the years of creating rules in an effort to make a safer scene and more efficient operations. One of the biggest obstacles to creating organizational accountability or establishing a safety culture is that some of these improvements seem to fly in the face of established rules. There is no doubt that freelancing is an unsafe practice that cannot be tolerated on a fire scene. However, the order to stand by was given by someone who may not have all the information. If this situation had been part of a "skull session" in a training room, the assistant chief who is incident commander, your lieutenant who is assigned to stand by, the lieutenant who is upstairs fighting the fire, and even the viewers watching at home would say grab the pictures or a salvage cover. The department mission statement probably alludes to something about protecting lives and property (if a life isn't an issue, save the property). With personal accountability, the crew standing by sees the events unfolding and forecasts that a bunch of valuables are going to be trashed in a matter of minutes unless they act now. If the lieutenant was personally accountable, he would forward the newly discovered information to the incident commander who could then make an informed decision. The incident commander may see from his vantage point that the fire is spreading and that he may need the backup crew to ignore salvage and pull another line. At least he will have all the information he needs. Again, the biggest mistake that can be made in no-fault management is failure to communicate a potential problem.

It's interesting that the developers of the 16 life safety initiatives put culture changes and accountability right up front when it comes to saving firefighters' lives. So many times we blame the trusses or the lack of the seatbelt as the cause of our problems. What we forget—or simply ignore in some cases—is that by the time the trusses are collapsing or the seatbelt is needed, we've already had several failures in the system (Figure 2-5). Author Dave Dodson has said that structural firefighting gear is the last line of defense for a firefighter. This means that we can't blame a burn injury on failure of the gear. Several failures have already allowed a firefighter in the gear to get to the point of it being tested. In other words, the events that lead to an injury or death are usually a failure in both accountability and lack of a safety culture. Before we can even think about risk management or situational awareness at emergency incidents, our organizational accountability for health and safety must be shored up.

HEALTH AND SAFETY ACCOUNTABILITY

For an organization to implement accountability in health and safety, it should already be proficient in the practices of personal and organizational accountability. It's an ongoing process that eventually works its way into every aspect of the operations of an emergency services organization.

FIGURE 2-5 Many times we view undesirable events as the failure of a component, when the root cause can be traced to accountability.

NFPA 1500

National Fire Protection Association (NFPA) 1500,³ the standard for fire department health and safety, is a great reference and standard to where we should be. Although many departments cannot financially meet all the components of the standard, several sections can be applied to everyone. Many departments use NFPA 1500 as a guide to making continuous improvements.

The standard comes with a checklist so that a fire department can establish where it stands with regard to compliance. Many significant steps can be taken toward compliance without spending additional money. There are also several components that have become commonplace. One standard referenced is NFPA 1403, the live fire training standard. There literally is no excuse for not complying completely with the standard. A fire chief may not "choose" a budget reduction that doesn't include annual physicals for each member but they certainly can "choose" to not use live victims and gasoline in an abandoned house used for live fire training. Once the checklist has been completed, a list of noncompliant items out of NFPA 1500 can be assembled in one of three different categories: those that require no effort, those that require some effort, and those that require long-term planning.

No Effort

These changes will have no impact on the budget of the department, and will not have a significant effect on operations. They can be completed quickly and with little effort.

An example is the section (6.2) regarding apparatus drivers and operators. By simply combining some outdated policies, adding a couple new ones on apparatus operation, and establishing a better training program, the department becomes compliant.

Some Effort

Chances are that you'll come across a section that will take a little more work. Breathing air compliance might be a good example. You may already have your airpacks bench tested annually and ensure that bottles are hydrostatically tested when due. However, your department just can't swing buying a fit test machine (7.12). Your efforts might work toward saving money to buy the needed equipment, hiring an outside company to complete the testing for you, or purchasing the equipment on a regional basis.

Long-Term Planning

Other elements of the checklist are far more difficult. For instance, your dispatch center might be shared with the village police department and be quite antiquated. It may even be controlled by the police department. Section 8.1.9 cross-references dispatch and communications systems to be compliant with NFPA 1561 and NFPA 1221. For the low introductory rate of several million dollars, you could bring your dispatch center up to the standards and comply with both. It's obvious that you may never see compliance with this section. Instead, it is important to relay the deficiencies to the police chief and village council while concentrating on smaller components of 1561 and 1221.

Each section in NFPA 1500 has the capability of saving lives. Each should be used as a safety culture reference that assists fire departments in achieving their goals. Section 4.5 is one of the most important when it comes to health and safety accountability. By creating an occupational safety and health committee that meets regularly, you preload the organization for accountability. Safety committees are an open forum to bring in concerns as well as to work toward compliance with NFPA 1500. If an issue is brought up in this environment, accountability prevents blame from taking over and forces the members to address the concern from each level. For instance, let's assume you are a member of your department's safety committee, along with members representing various ranks. Someone brings up an issue of loose contaminated IV needles in EMS jump kits on the engines. Apparently the EMS bags contain a sharps container for collecting used needles from EMS scenes in one pouch of the bag. Dirty needles were found outside the container but still in the pouch, either from carelessness or the fact that the pouch is much bigger than the sharps container. Accountability forces each rank to take responsibility for the loose needles. Paramedics must ensure proper disposal while firefighters carefully check the bag each day. Line officers work on better plans for needle disposal by providing separate sharps containers or ones attached outside the bag. Training increases the crew's knowledge of the flaw in the system while staff officers put out safety bulletins. The safety committee lays out a plan with a deadline to fix the problem and promptly reports the condition as a near-miss (Initiative 9).

Rating Your Department

Organizational accountability can also be evaluated through the use of rating systems. They can be effective in monitoring the progress of an organization in achieving goals. Additionally, organizations that receive a good rating tend to take pride in it. One of the problems with the evaluation of the fire service, and safety in general, is that our main rating system achieves the goals of the insurance industry rather than our own.

Insurance Services Office

The most identifiable measure used by the fire service is the **Insurance Services Office (ISO) Fire Suppression Rating System (FSRS) Public Protection Classification (PPC).** This number is assigned to fire departments by a fire insurance industry organization that rates their fire suppression capabilities based on several components, including manpower, equipment, and water supply. Theoretically, a department that has a PPC of 5 would protect the insurer's investment more effectively than a department with a 6 rating. Therefore, many departments devote significant resources to drop their PPC and thus the insurance rates for their community. This third-party rating has a significant impact on both public and political awareness of what the department has and what it needs. Sometimes ISO ratings convince governmental boards of the need for increased funding. For instance, their suggestions may lead to an increase in staffing or construction of a second station, something the fire chief may have been fighting to get for years. Other times they put up a strong argument for a piece of apparatus such as an aerial ladder.

Unfortunately, ISO grading schedules have very little to do with a department's accountability to health and safety. The scoring system[4] is not always indicative of specialized equipment that some departments need for their local hazards, such as radiological monitoring equipment. They don't consider the strength of command and control at emergency scenes (safety officer or SOPs), national standards like the **National Incident Management System (NIMS),** or specific safety procedures (personnel accountability reports). Although they do take into account training hours and time of day, subjects get less attention. Some of the hottest training topics in firefighting these days are related to firefighter safety and survival, rapid intervention, air management, situational awareness, and risk management. Teaching subjects like these means nothing to the rating system compared to a having a drill tower behind the station that may not even be used. An ISO rating for a community is a fire insurance industry standard, not a fire department standard. The PPC should be a tool to measure cost effectiveness related to an investment in fire protection (e.g., "An annual investment of an additional 15% of the fire department's budget would reduce the average commercial property insurance rate by 12%."). Although some fire departments are proud of their PPC rating, other industries might have more to be proud of from their rating systems.

International Organization for Standardization

The familiar **International Organization for Standardization (ISO)** creates standards for industry that can be used worldwide. This ISO[5] is based in Geneva, Switzerland, and serves 162 countries. Much like NFPA, once a need is identified, a technical committee takes suggestions for the standard and develops drafts. Drafts undergo revisions until they eventually become standards. The advantage for a business is that when a component supplier for a manufacturer is certified in a certain standard, you can trust that their systems are effective. You probably have seen an ISO 9001 flag flying over a business in an industrial area. The flag simply makes a statement of accomplishment; in this case, it's one for quality management (Figure 2-6). ISO has hundreds of standards, and it's apparent by their standards under development that the hottest topics for them are in the medical fields. Although ISO creates the standards, they do not "certify" compliance. Companies that wish to claim they are certified can hire an outside third party to certify them, or simply use the guidelines and complete an in-house audit. The company claiming to be ISO registered or compliant maintains records in case someone disputes their claim or an issue surfaces. This desire for organizational accountability is obvious by every company that flies the ISO flag. Is it possible the same theory could have a place in the fire service?

FLSI 16 Safety Rating

A campaign is under way to create a more effective rating for the fire service. Similar to the ISO 9001 program, an **FLSI 16** registry would allow departments to display their organizational accountability based on meeting specific safety criteria, including

FIGURE 2-6
An ISO flag is a common way for industry to show the world they are accountable.

adoption of the 16 life safety initiatives. A registered department would also have to complete specific components of NFPA 1500. The advantages of having such a system are twofold.

Besides the dedication to a safety environment that is both measurable and achievable, the rating system could be a source of tradition and pride in the fire service. Until now, the best way for a department to judge itself was with the ISO PPC. Many times the number was painted on the side of their apparatus or printed on their department letterhead. As discussed earlier, the ISO PPC is not necessarily the best way to rate ourselves. Just like the machine shop in the industrial park flying the "ISO 9001 Registered" flag, fire and emergency services would be able to fly the "FLSI 16" flag.

Successful Implementation

In order for organizational accountability to health and safety to successfully take root in fire departments, it must be established continually throughout every aspect of the fire service. These include recruit schools and fire academies, firefighter orientation, daily operations, promotional exams, training, and effective emergency scenes.

Recruit Schools and Fire Academies

Textbooks must be written from a safety culture's viewpoint. Rather than devoting a chapter or a sidebar to safety, it must be mixed into the lessons. For instance, eye and hearing protection is actually one of the steps to starting a chainsaw, rather than a so-called safety tip. Rookies need to know how to address safety concerns with a supervisor without appearing insubordinate (Initiative 4). Practical applications must design safety into the practical skills. Instructors should be trained to the technician level in safety systems and ensure that evolutions are done under strict compliance to standards. Even testing situations should comprise a fair amount of safety questions, including scenarios in which they must make task-level decisions based on risk analysis.

Firefighter Orientation

Once a firefighter completes basic training, safety should be continued through orientation and probationary periods. The candidate's commitment to safety should be included in written evaluations. Although most evaluations have only one or two lines about the person's attitude toward safety, it should be an entire section that covers the expectations for specific aspects of personal accountability and learning the safety culture. Physical training and healthy lifestyles must also be part of the firefighter's orientation.

Daily Operations

Whether a firefighter is a volunteer who enters the fire station once a week or an employee who walks in every day, the safety culture must be applied religiously. Using personal accountability as a stepping stone, constant effort must be put forth until safety is a normal part of daily operations. Safety practices range from using ladders to change light bulbs in the station to using power tools to fix a broom bracket on the wall. Safety in the fire station leads to safety on the fireground. Time set aside daily for physical activity must be encouraged. Whether it is time in the weight room or time spent in group activities, maintaining a healthy, active lifestyle must be lived every day.

Promotional Exams

If an organization is truly operating under a safety culture, they will not promote anyone that doesn't have personal accountability to safety. Promotional exams must include reading material and testing situations involving safety information as well as scenarios to ensure that the leaders of the organization understand the importance.

Training

Drills are no exception and must follow the same safety culture aspects. Far too many firefighters are injured or killed every year in training accidents. This may be because we tend to be less cautious in nonemergency situations where there is less apparent danger. More importantly, we tend to work like we train and we need to work safely. Trainings should mimic firefighting skills as much as possible. For instance, it does no good to simulate roof ventilation without having full gear and SCBA in place. The body has to be seasoned to increased temperatures in firefighting gear on a regular basis to be ready for fire scenes.

Effective Emergency Scenes

Emergency scenes are where the rubber meets the road, and the results of a safety culture pay off. After a high-stress emergency call, responders sometimes claim that they really didn't think about what to do or how to do it. They say they didn't really think about the danger either—they just did what needed to be done, what they were trained to do. If we've spent our nonemergency time practicing a culture of safety, when the emergency does come, then safety is already second nature. Take running a chainsaw, for instance. If safety glasses are an afterthought at the station, you won't remember them at an emergency when stress levels are up. The best way to perform emergency scene duties safely is to not know how to perform them unsafely.

Learning the aspects of personal accountability is a skill needed to develop a safety culture, as well as to grow as a person. Looking back on your best mentors, teachers, or leaders, you will see aspects of personal accountability. Likewise, if you ever worked for a great company, department, or division, you probably saw what organizational accountability looks like. Because these people and organizations are led by human decisions, even the best will trip up occasionally. Virtually every time this happens, lack of accountability is at the root. If it happens, simply recognize it for what it is, pick yourself up, and take responsibility. These attributes are vital if fire and emergency services are going to be truly accountable for health and safety.

SUMMARY

Enhancing accountability is an effective way to improve safety, and starts with *you* in the form of personal accountability. Decisions you make everyday not only can make your job more productive and safer, but can even raise your job satisfaction. Some people might be unaware of the importance of accountability, and others may instead choose to attribute blame to others or the organization. Many risks to a firefighter are affected by personal decisions made at emergency scenes and on the training ground. We therefore are charged with the choice to be accountable for our own safety and the safety of our peers, or simply blame someone else for risks, mishaps, injuries, and death.

In Chapter 1, we identified the need to establish a safety culture. One of the first steps to making that change is to accept the responsibility associated with accountability for health and safety. A decision to remain accountable has the ability to spread to others throughout the organization. For instance, a firefighter that stays accountable by eating right and exercising tends to encourage others to do the same. As he or she begins to pursue his or her own accountability for health and safety, the organization begins to tip the scale in favor of organizational accountability. Other industries have utilized components of no-fault management as a way to increase personal accountability for safety and should be looked at as a possible solution.

Organizational accountability can best be accomplished through a conscious decision by the leadership of the organization. Besides instituting policies and procedures to ensure that safe practices are followed by the members, personal accountability can be demonstrated by a fire chief who adheres to a strict personal accountability for safety. The fire service as a whole needs to have a registry of organizations that put an emphasis on safety and are therefore accountable. An FLSI 16 registry would provide information and checklists to provide departments with criteria, and could be used to qualify for certain financial incentives such as grants. A combination of organizational and personal accountability for safety has the best opportunity for reducing injuries and deaths.

KEY TERMS

accountable - Similar to the term *responsible*, it describes a person who is held liable for completing a specific duty.

AED - Automated external defibrillator.
Delta side - The right-hand side of a structure.
EMT - Emergency medical technician.

Everyone Goes Home (EGH) - A prevention program created by the National Fallen Firefighters Foundation in an effort to reduce future line-of-duty deaths. One of the major accomplishments was the creation of the 16 initiatives, the basis of this text.

Fire Suppression Rating System (FSRS) - A system of grading used by the fire insurance industry to establish rates.

FLSI 16 (Fire and Life Safety Initiatives 16) - Proposed rating for fire departments and EMS agencies based on their dedication to safety; could be used for grant eligibility.

Insurance Services Office (ISO) - A fire industry organization.

International Organization for Standardization (ISO) - A global organization that provides registries of companies that comply with certain standards.

line-of-duty death (LODD) - Fatality that is directly attributed to the duties of a firefighter.

NFM - No-fault management.

NFPA - National Fire Protection Association.

NIMS - National Incident Management System.

NIOSH - National Institute for Occupational Safety and Health.

organizational accountability - A term used to describe the ability of an association to be held to certain standards.

personal accountability - A term used to describe the ability of a person to be held to certain standards.

postincident critique - An evaluation of an incident after it occurs, specifically examining successes and areas for improvement.

Public Protection Classification (PPC) - A number assigned by the insurance industry to fire departments and districts based on several components including equipment, staffing, and water supply.

responsible - Similar to the term **accountable,** it describes a person who has been given the authority to carry out a specific duty.

risk management - Identification and analysis of exposure to hazards, selection of appropriate risk management techniques to handle exposures, implementation of chosen techniques, and monitoring of results, with respect to the health and safety of members.

safety culture - A philosophy that prioritizes safety as a paramount value and relies on it to guide many of an organization's decisions.

SCBA - Self-contained breathing apparatus.

SOP - Standard operating procedure.

victim syndrome - The desire for an individual to be portrayed as a victim, usually for esteem benefits.

REVIEW QUESTIONS

1. What are the definitions of personal and organizational accountability? List an advantage of each.
2. What are some of the ways that accountability can affect health and safety within fire and emergency services?
3. How does the process of using NFPA 1500 improve the accountability related to the health and safety of an organization?
4. How you could implement the combination of accountability and no-fault management together?

5. Why is there a need to create health and safety parameters for organizational accountability?

FIREFIGHTING WEBSITE RESOURCES

http://everyonegoeshome.com
http://cdc.gov/niosh/fire
http://www.iso.org

NOTES

1. http://www.trumanlibrary.org/buckstop.htm
2. Ripken Jr., C. and D.T. Phillips. 2007. *Get in the game: 8 elements of perseverance that make the difference.* New York: Gotham Books.
3. NFPA 1500 Standard on Fire Department Occupational Safety and Health Program, 2002 ed.
4. ISO Mitigation Equipment List 2010. http://www.isomitigation.com/ppc/3000/ppc3003.html
5. International Organization for Standardization. http://www.iso.org

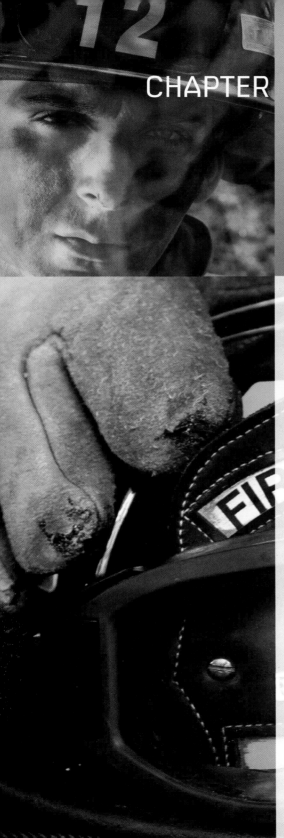

CHAPTER 3

Applying Risk Management Techniques

LEARNING OBJECTIVES

- Define risk management and why it is an important component of incident management.
- Describe a risk-benefit analysis and its application to emergency services.
- Apply the United States Coast Guard's risk management model to emergency services.
- Identify the features of recognition-primed decision making (RPD).
- List the five safety features of the incident management model.
- Explain the importance of risk management in the incident action plan (IAP).
- Describe OSHA's "two-in, two-out" rule and its effect on rapid intervention.
- Identify risk management practices at the strategic, tactical, and task levels.
- Explain the components of effective communication from the aviation industry.

INITIATIVE 3. Focus greater attention on the integration of risk management with incident management at all levels, including strategic, tactical, and planning responsibilities.

An emergency scene is usually the first location that comes to mind when we think of a line-of-duty death. The dynamic changes that occur during an emergency incident put a constant stress on emergency responders. Additionally, the chance of injury or death is increased due to a genuine or perceived urgency in managing the incident. This feeling of compressed time creates additional stresses and forces quick decisions. Many times, the decisions are based on unknown factors and unknown dangers. The weakest link in the system is the human decision. In a recent CBS documentary,[1] author Joseph T. Hallinan claimed that humans make hundreds of mistakes every day. The brain is too "clogged up" with attempting to decipher all the messages it's receiving, especially when multitasking and considering options. Unfortunately, an emergency scene is a breeding ground for such mistakes. One effective way to maintain a safety culture on emergency scenes is to treat each job with a risk versus benefit analysis.

RISK-BENEFIT ANALYSIS

Let's say you're heading home from work one day on the interstate. Suddenly, something catches your eye between the lanes ahead. As you pass it up, you see it's a 48" pipe wrench that apparently fell off a work truck right in the middle of the freeway. It doesn't look like one of those cheap ones either—it's probably worth almost a hundred dollars. Chances are it's perfectly fine because a tool of that caliber could have survived that fall without damage. You're no plumber but it sure would come in handy to have (and it would look pretty good on your pegboard, too). There's nobody right behind you so you pull to the shoulder and back up, adjacent to where the wrench is to get a better look. It appears to be in great condition. It's only 12 feet away, but there's a lot of traffic. As you contemplate grabbing the wrench, your brain does a little exercise called risk-benefit analysis.

Risk-benefit analysis is simply the weighing of the facts by determining the advantages and disadvantages of a certain activity. The advantage of going for the pipe wrench is pretty clear. If you didn't want it, you never would have stopped. The disadvantage is you may get hit by a truck. So basically your best-case scenario allows you to go home with a wrench, and your worst-case scenario gets you killed. Risk-benefit analysis would be pretty easy if it was that cut and dry because the wrench is clearly not worth trading your life. You might reason that you wouldn't be trading your life, that you would simply wait for a break in the traffic. If the gap was sufficient, you would have time to run to the centerline, grab the wrench, and get back. This would reduce the risk to zero, and your analysis probably stops right there. However, if you perform a complete risk analysis, you take into account all the potential risks. In this case, what if you misjudge the speed of a vehicle, or trip on the way to the wrench? What if a car in the high-speed lane suddenly changes to your lane after you've committed? Someone could be looking at your vehicle on the side of the road and never expect you to cross into traffic. It's clear that your risk is not zero. It's really dependent on the drivers headed your way, which you know absolutely nothing about. There could be a drunk driver approaching, a driver who is falling asleep, or even someone running from the police.

In order for risk-benefit analysis to be effective, all the possibilities must be considered. Unfortunately, creating a list of all the possibilities is time consuming in the initial stages of an incident, and sometimes the list lacks all the information necessary. Preincident plans allow responders to create "boilerplate" lists when time is not an issue. Many times we instead rely on our experience and training to complete risk-benefit analysis. Take, for instance, that you are the captain assigned to a heavy-rescue company stationed downtown when you receive a call for a motor vehicle crash on an interstate overpass that involves a tractor trailer. While en route, dispatch updates the responding companies that the truck is hanging from the "high bridge." You know the "high bridge" refers to the uppermost bridge of an area where two bridges cross the interstate below. The first-due engine company reports that they are stuck in traffic approaching the high bridge and that a black smoke column is in the area of the bridges. They call for another alarm, with the best approach from the opposite direction. You decide the best route will take you under the crash on the second bridge, where you can pass underneath and backtrack to the high bridge. As you arrive "under" the crash, you are the first fire apparatus close to the scene. You report that there is a semitruck on the high bridge with the tractor hanging off the edge 25 feet above the bridge you are on. There is also a fire above but you can't tell what's involved or how many people are trapped or injured. You can talk to the driver from your position and it is clear he is unhurt but ready to jump. He tells you that he is carrying an enclosed trailer with pallets of 5-gallon pails containing hydraulic oil. He's already removed his seatbelt and is half out of the window; he is scared his truck is going to blow up and fall. It's time for a risk-benefit analysis.

To perform the analysis competently, you must be proficient at the process and examine all possible outcomes. Your gut reaction is to pull your 12-foot-high rescue truck under the victim. That cuts the distance of his fall in half, meaning he probably will survive with little injuries. One of your crew members guesses what you're thinking and suggests putting the folding A-frame ladder on the roof to reach him. It wouldn't be very secure, but another firefighter suggests putting up the 15-foot light tower and tying off the ladder to stabilize it. Some of the crew has reservations. Being under the victim means being under the truck, which is not the safest place to park. With no report of what's burning or what other dangers are above, you don't feel very comfortable with it either. Best-case scenario, the driver climbs down with assistance. Worst-case scenario, the tractor trailer breaks loose and crushes your rescue rig and everyone close to it. Fire and emergency services has relied on a simple risk-benefit model preached by Chief Alan Brunacini for many years to guide your decision:

- *We will, in a structured plan, risk a lot to save a lot (human lives).*
- *In a structured plan, we will risk a little to save a little (property and pets).*
- *We will risk nothing to save nothing (or something that is already lost).*

In this case, the truck driver's life is worth a lot, and therefore worth risking a lot. You just need to formulate the plan, and make it as safe as possible. You decide that the worst thing that could happen is for the truck to fall. Other issues are related to debris from above, as well as the truck driver or firefighters falling during the rescue. Although you can't take any specific steps to secure the truck, you can limit how much time you spend under it and secure firefighters and the victim to keep them from falling. You reassure and calm the truck driver while they raise the light tower and secure the ladder to it, still outside of the range of the falling truck. Your team ties off to the light tower with safety lines and have a third line ready with a hasty hitch for the truck driver. When everyone is ready, you guide the rescue rig forward under the perched semitruck to a position where the ladder is adjacent to the truck window. A firefighter maintains contact with the truck driver and hands him the safety line. The driver climbs from the window and down the ladder. You guide the rescue truck away to a protected area where everyone climbs down without injury.

Fire and emergency responders have been completing risk-benefit analyses in their heads and on the fly for a long time (Figure 3-1). Unfortunately, the analysis is only as good as the person completing the assessment. Experience goes a long way to success, but it only comes with the expense of making mistakes. Mistakes while learning are okay when they don't cause damage, but the risk-benefit mistakes we're talking about here can kill someone. As stated earlier, everyday emergency scenes aren't necessarily the ideal location to pull out a pad of paper and a pen to list all the hazards. The time-sensitive example above uses dialogue between the crew members that brings to light concerns as well as the fastest, safest way to accomplish the goal of rescuing the trapped driver. However, that pad of paper does come in handy for large incidents, pre-planning, and training. Some emergency responders might believe that they are proficient at (mental) risk-benefit analysis, and that the method is sufficient. They may not be entirely correct. A similar situation occurred in the 1980s when the fireground command system was being introduced. Many felt that it wasn't needed, because everyone knew who was in charge at a fire and understood the chain of command. Calling someone "command" was a crazy idea. These responders had a valid point when they said the existing system in place was sufficient for most events. What they missed was that

CHAPTER 3
APPLYING RISK MANAGEMENT TECHNIQUES

FIGURE 3-1
It's not always easy, but dangerous scenes with no life hazard should consist of fire line tape and exterior hoselines.

when an incident continued to grow until the resources outnumbered the supervisors, the chances of a successful outcome diminished and an expandable command system was needed. This was especially evident in campaigns that involved numerous agencies over several days. Similarly, a simple risk-benefit analysis is usually sufficient for most incidents. However, during highly volatile incidents, a thorough system must be initiated to ensure that all potential factors are accounted for. Other industries refer to this process as **risk management.**

RISK MANAGEMENT: OVERVIEW

It's important to understand that risk does not necessarily mean "bad." The **United States Coast Guard (USCG)** believes that a substantial part of their **Operational Risk Management (ORM)**[2] program is using risk to advance the mission. When the coast guard deploys surface vessels or aircraft for emergency **mayday** responses, they don't usually have weather on their side. In fact, many air and water search and rescues occur when other industries would not fly or sail. They can't avoid risk, but instead are forced to work with and around it. Much like the USCG, fire and emergency services operate under risk all the time. The trick is identifying as many risks as possible so you can address and minimize each one. One of the methods the coast guard uses for identifying the risks is referred to as the **SPE (severity, probability, exposure) model.** The SPE model uses three components to calculate risk as a number: severity, rated between 1 and 5; probability, rated between 1 and 5; and exposure, rated between 1 and 4. By multiplying our three numbers together, we get a number between 0 and 100 that we could consider a percentage of risk. Anything more than 80 should be abandoned, whereas anything between 50 and 80 should be reassessed.

In a postincident critique or after-action report, we can calculate the risk involved with our semitruck hanging off the overpass by rating the three factors of risk individually:

- **Severity (1–5):** Severity is a rating of how bad it would be if our worst-case scenario actually occurred. If the truck fell at the worst possible time, it could crush the rescue truck and kill six people. We might rate this severity a 5.
- **Probability (1–5):** The probability that it will fall during our rescue is not high. If it was going to fall, it probably would have already; but the longer the fire burns above, the better the chance of the truck or other debris falling. If we had resources on the bridge above, they might have been able to give us a more accurate rating about how well the truck was secured. They might give it a 1 with very little chance of falling or a 5, meaning it's barely hanging on. With the limited information we have, we might give probability a 3 as long as we complete the rescue immediately or a 4 if we wait for an aerial ladder.
- **Exposure (1–4):** By factoring in exposure, we look at how long or how many times we expose our rescuers to the risk. We may decide that if the personnel and equipment are ready before pulling under the risk, and the rescue is fairly quick, we can keep exposure to a 3.

For our example, with as many safeguards in place as possible, we are estimating a risk of 45. If we were to wait for an aerial ladder to make the rescue, risk would rise to 60.

An obvious component of this risk management scoring system is to take into account the equipment and personnel available for the task. If your rescue truck only consisted of you and a driver or if there was no light tower to secure to, the score would probably be closer to 75. Another aspect is environmental concerns. A heavy wind or explosions above could push the number to 90. In that case, we have to look at another plan. The plan might instead be hasty placement of the rescue truck and allowing the driver to jump to the roof. Possibly we throw the driver one end of a rope with a harness and pulley and explain how to hook the carabineer on the steering wheel so that we can lower him down with a belay. These plans would be quicker and would significantly lower the risk. Once we have a number to assign to the risk, we must look at what it is we are trying to save or gain. In this case, the life of the truck driver is well worth the 45% risk. If the risk was just to identify the cargo, it wouldn't be worth it. The next step is to make a decision.

DECISION MAKING

All risk-benefit models are dependent on decision making. Emergency responders learn how to make decisions quickly, based on limited and sometimes inaccurate information. At times, the decision they make is based on actual training that matches the situation at hand. Other times, decisions must be more problem solving in nature, based on components of training or theories that the responder has developed. The captain in the previous example had never considered tying off a step ladder to a light tower before, and the crew had never heard of it being done. The light tower is not rated for the potential lateral load it could have experienced, and it's never safe to move a piece of apparatus with a step ladder and personnel (much less a victim) on the roof. Some might chastise the crew for their decisions, whereas others applaud them for their ingenuity in a clutch. The fact is they made a decision to risk a lot to save a lot with the limited information they had available at the time. They performed the rescue at great risk to their own lives but used as many safety precautions as possible, as quickly as possible.

How well do we actually make decisions under stress? The general impression is that stress increases mistakes in decision making, but several studies show that this is not always the case. One such study subjected participants to evaluation in a forest firefighting simulation.[3] The simulation lasted five hours, and 20 volunteers were allowed to complete it with no interference. The remaining 20 were subjected to a disrupting environment the entire time. Results of the study showed that those without outside stress used deeper levels of analysis, whereas the stressed subjects made quicker broad decisions. As a result, the first group completed each component of the exercise thoroughly, whereas the second group ignored minor components and put more emphasis on prioritizing strategies. Interestingly, there were no significant differences in the results of their decisions or how much fire consumed the forest, just in how they made their decisions. In fact, it's common knowledge that many people, such as athletes, tend to make better decisions while under stress. Individuals use two distinct decision-making processes when faced with a dilemma in a stressful emergency situation: training based and recognition primed.

Training-Based Decision Making

Training-based decision making utilizes skills learned as a direct result of training, especially when an incident unfolds "by the book." When Captain Chesley Sullenberger ditched an Airbus A320 into the Hudson River in 2009, he attributed it to skills he had learned. He stated in an interview[4] that he had been making "small deposits" in an "experience account" for 42 years as a pilot through training and education. When he needed it, he made a withdrawal from the account and pulled out a perfect water landing. Although many pilots are taught how to ditch an aircraft, it very rarely occurs. Fire and emergency service workers perform duties daily that rely on training and education. Many times after an extraordinarily stressful incident, responders are surprised that they were able to keep their cool and act in an efficient manner. Many times they attribute it to the training that they have received. In fact, a good training program is designed to teach proper methods in a repeated fashion to accomplish this very goal. Similarly, bad habits that are permitted during drills tend to surface later at incidents (Figure 3-2).

This method of decision making is the way inexperienced emergency responders learn new skills. It's also the best choice for workers in risky professions who practice

FIGURE 3-2
Bad habits must be forbidden during training sessions or they are likely to reappear on-scene.

for emergencies but hope they are never involved in one. Skills learned during emergency drills are effective during periods of stress, and have been observed in mine workers involved in underground fires when they were forced to make decisions to escape a fire.[5] Most followed predetermined plans that matched their training. As to be expected, panic set in easier for miners with less experience in donning a **self-contained breathing apparatus (SCBA)** and escaping, but those with more experience and training performed effectively. All workers carried out the emergency action plans better when their supervisor was present to give orders. This type of decision making is the backbone of the "fire drill" that we have used successfully in elementary schools for the past 50 years. Teachers and students don't necessarily expect a fire to occur, but are ready if one does.

This type of training has an added benefit. Practicing *predetermined* plans helps in *undetermined* plans. For example, in 1988, P. X. Rinn was commanding the USS *Samuel B. Roberts* in the Persian Gulf when it found itself in the middle of a minefield. The commander stopped the boat, but he recalled not being trained in this specific situation. Procedure was either to steer away from or blow up the mines. Unfortunately, he wasn't sure which way to turn because he couldn't see all of them under the water and they were too small for radar. Detonating the ones he could see could cause others to explode. He chose to back up slowly, but still hit one. The blast rocked the ship and ignited fires as it began to take on water. Navy training at the time called for extinguishing fires before dealing with the incoming water. Commander Rinn watched the firefighting efforts, and decided that at the rate she was taking on water, the fires would be extinguished by dropping below sea level long before the crew's hoselines had any effect. He therefore ordered his crew to drop their firefighting handlines and concentrate on keeping the ship afloat. This involved plugging cracks in the bulkheads with clothing, pillows, and anything they could, to slow the influx of water. They committed to getting generators online and pumps running to displace the water they couldn't stop. When the ship was finally stabilized, his crew then extinguished the fires. In the book describing the events, *No Higher Honor*,[6] Rinn acknowledged that his decision was a combination of his experience and training with the limited information available at the time.

Although Captain Sullenberger followed his training to the letter and Commander Rinn abandoned everything he had been taught when he discovered it was wrong, they both made extraordinary decisions based on their training. Both leaders are now textbook examples in their respective fields of training, and changes in equipment and policy have likely been enacted based on their decisions. It's important to remember that all policies, procedures, and training we use were developed by humans. Early in this chapter, we established that humans are not flawless, but continue to make mistakes. Therefore, it's not plausible to argue that the policies, procedures, and trainings we follow are always the best way to accomplish our goals. Instead, we need to constantly reevaluate how we do things and what changes need to be made, especially in our training (Initiative 5). One of the most ridiculous and prehistoric remarks we can make is, "But we've always done it that way." However, it's important to carefully examine predetermined plans and their intent before dismissing them as being simply outdated.

Recognition-Primed Decision Making

The second method of emergency decision making is a form of recall. Have you ever been riding in a car when the driver had to slam on the brakes? Sometimes when that happens, the driver reacts by reaching in front of the passenger as if to hold him back. This "human seatbelt" reaction has little chance of really helping in case of an accident, but is a good example of the way some people make split-second decisions. A similar theory that has emerged recently has been identified by Gary Klein[7] as **recognition-primed decision making (RPD)**. Although most people agree that experience assists in decision making, this research takes the idea one step further. Studies show that as emergency responders gain more experience, they save mental snapshots of what works and what doesn't work in a given situation. Experience also records how specific incidents progress. This snapshot is held in the memory as a template that can be used for future emergencies. As opposed to weighing the options, this technique instead assesses the present situation for a workable solution. In the case of our well-intentioned driver, another car suddenly pulled out in front of her. As her foot was moving to the brake pedal, she began mentally preparing for the sudden unanticipated stop and checked her memory banks for a response template that would fit. She grabbed a template that had prevented her son from being thrown forward in the seat several years ago. Because it worked well, she saved it as a valid option. Five years later when she gave you a ride to get your car, she reacted by reaching out to hold you back. She clearly didn't weigh the options, but instead reacted in a way that was a workable solution that had been effective in the past.

One of the advantages to RPD is that quick decisions can be made. Emergency scenes are not conducive to long, drawn out comparisons of options. As discussed earlier, many decisions are made with little information about questionable accuracy under stressful situations. A related benefit to RPD is that templates are pulled to find a workable solution. In other words, a solution is taken out and held up to the incident presenting itself. If it looks like it fits, it's used immediately. If it doesn't match, it's thrown aside for another. It's like looking in the kitchen cupboard for the lid to a pot. Most people don't get a tape measure to see what size they need then compare all the lids. Rather they grab what looks close and give it a shot. If it's the wrong choice and it doesn't fit, not much time was wasted. That does bring us to one significant disadvantage of RPD: It's prone to failures. The success rate has a direct correlation to how much experience you have with the specific situation. You can probably find the lid to the skillet quicker than a 6-year-old can. That could be because you recognize the lid from the last time you used it, or because you can better estimate the size. Either way, experience is what makes RPD work.

Templates are also stored when things don't go well. In his book *Blink: The Power of Thinking Without Thinking*, author Malcolm Gladwell introduces a similar theory to RPD.[8] Citing several different examples, he explains that many times the unconscious mind solves a problem long before the person is even aware of it. Many experienced firefighters can recall examples of similar situations. For example, let's say it's 3:30 a.m. and you are dispatched to a possible building fire at a fairly large one-story convenience store. You are part of a three-person crew riding jump seat on the quint behind your lieutenant. Jim is the smart and steady senior man and Lt. Evans has been a wealth of knowledge with no fear. As you arrive on scene, you notice a slow grey smoke emitting from the soffits above and cracks around the front door. Your memory jumps back to a similar smoke condition you had in an apartment fire last year. It turned out to be an electrical fire behind the stove that was easily extinguished. You decide it probably is the same type of fire. Lt. Evans tells you to stretch a line while he checks the back of the building. You pull the 1-3/4" preconnect and advance it to the front door. You grab the "through the lock" forcible entry tools assuming you'll force entry with as little damage as possible just as your lieutenant makes it back. Lt. Evans looks nervous and calls for additional alarms, then tells you to drop the line and pull a 2-1/2" but to stay back. Although you're surprised at the apparent chink in his armor, you are more confused by his choice of hoseline. You pull the larger line without argument while he assists Jim in cutting the utilities. As the second-due quint rolls up, Lt. Evans assigns them to ventilate the display windows on the **"D" (Delta) side** while the two of you apparently just sit and wait. As they take out the windows, smoke billows from the opening and turns instantly black. Lt. Evans yells "NOW" and breaks out the glass on the door as you watch the entire store flashover before your eyes. You open the smooth bore wide open through the door and begin what turns into an all-night job.

For the next several hours you think back to what happened. Your template wasn't even close to what reality was, whereas your lieutenant's was spot on. You ask him what he saw on the **"C" (Charlie) side** that gave it away. His response was not much more than what you saw in the front. He says he knew the store was unoccupied and that it wasn't worth entry until backup arrived. He explains that visibility through the windows was nonexistent and that smoke was emitting from small cracks and leaks around doors. His template was that of a deep-seated fire that hadn't completely run out of air yet or "blacked up," but wasn't very far off from a backdraft. He remembered fires in large buildings like these many times appear small just because of the store size and layout. His decision was to vent away from the attack position and clear of the exposures, then hit it if it took off. If not, he could advance the smaller line to find the seat of the fire. Additional alarms were pulled to ensure sufficient personnel for effective incident management.

INCIDENT MANAGEMENT

The **National Incident Management System (NIMS)** was established in 2003 as a result of Homeland Security Presidential Directive 5.[9] NIMS was designed to be an expandable approach to incident management, allowing federal, state, and local authorities to work together effectively and efficiently. It also was intended to assist in planning for and responding to domestic incidents. Composed primarily of material from the National Fire Service Incident Management System Consortium, it combined the simplicity of the Fireground Command System and the expandability of the **Incident Command System (ICS)** from **Firefighting Resources of Southern California Organized for Potential Emergencies (FIRESCOPE),** bringing them together under one all-risk system. In 2000, the group released the second edition of *Model Procedures Guide for Structural Firefighting*.[10] The guide provides specific processes to assist an organization in developing an incident from a single-unit response to a complex campaign. There are specific implications for the four sections of the general staff: operations, logistics, planning, and finance; as well as the three levels of function—strategic, tactical, and task. Additionally, it outlines five specific areas of safety in regard to a command structure: safety officer, scene accountability, emergency traffic, rapid intervention, and responder rehabilitation.

Safety Officer

One of the most important positions to fill in the command system at a growing or significant incident is the position of the **incident safety officer (ISO).** The ISO is a member of the command staff who reports directly to the incident commander and assists in observations and evaluations of scene safety. The ISO ensures that the actions of the crews are matching the strategic decisions of the **incident action plan (IAP),** and that feedback and recommendations are communicated to command. In his book *Fire Department Incident Safety Officer*, Dave Dodson suggests that ISOs must be able to "read-risk."[11] This involves collecting information about the scene, analyzing the hazards and opportunities, and judging the risk involved.

When the safety officer position was new to the command system, it wasn't uncommon for any available officer (or even a firefighter) to arbitrarily be assigned to the position, regardless of training level. This is no longer the case. The **National Fire Protection Association (NFPA)** established NFPA 1521 to define the duties and requirements for an ISO. It also defines and advocates the use of **assistant safety officers (ASOs)** as needed on larger incidents. A professional qualification[12] standard has been developed for certification of ISOs to ensure that a certain level of knowledge is reached.

Scene Accountability

Maintaining incident scene **accountability** means utilizing a system to track the identification, location, and function of all crews on a fireground at a given time. This can be accomplished in different ways, and no system is dramatically better than any other (Figure 3-3). The key to effective accountability is to ensure that every member of the

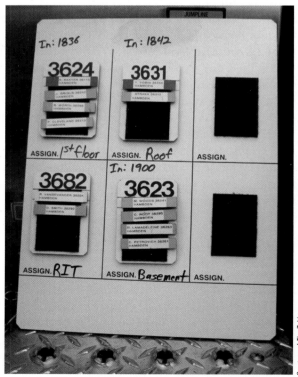

FIGURE 3-3
Regardless of the type of accountability system used, it must track much more information than just "who is there."

department complies with the policies, and that companies and commanders continually practice with the system. This is true whether the system is simply dog tags in a coffee can or a computerized barcode scanning-and-tracking system. No system is effective if its rules aren't followed. The system must also be able to identify when a member is unaccounted for, usually through a **personnel accountability report (PAR)**. PARs can be called for at specific time intervals, fire control benchmarks, or when there is a change in strategies. Four of the most common current accountability systems are the tag system, the passport system, barcode scanning software, and span of control.

Tag System

The tag system consists of a tag on a key ring issued to all firefighters with their personal protective equipment and labeled with their name, rank, department, and ID number. Additionally, each responding vehicle has an apparatus ring and clip with the vehicle's number on it, and an accountability board of some sort. As members arrive on scene, the tag is removed from the individual's gear and is given to the engineer, who places all the tags on the vehicle ring. The intent of giving up your tag is basically your ticket to the hot zone, and must be promptly retrieved on exiting the building. In some systems, the first-in engineer is assigned as the accountability officer until a company officer or permanent accountability officer is assigned. At that point, accountability becomes a sector (group) that uses the accountability board to keep the rings organized. The board can be written on, identifying the function and location that crews were assigned to, as well as a time they went on air. An accountability officer is usually staged outside the hot zone at or near the point of entry. Multiple entry points require multiple accountability officers, and high-rise scenes usually assign accountability as lobby control.

Passport System

The passport system is similar in operation to the tag system, but instead uses a Velcro-type fastening system. All members are assigned at least two passport tags with their personal information on it, which are usually stored under the brim of their helmet. At the beginning of shift or when firefighters respond from home, they place one of the passports on the riding board in the vehicle. The accountability officer collects the boards, and members use a second passport to gain access into the hot zone.

Barcode Scanning Software

An accountability system is simply a method of tracking who is in the hot zone, where they are, what they're doing, and when they went there. The tag or passport system can be upgraded to a simple database, utilizing scanners or wireless chips attached to airpacks. Timing devices remind the officer when to call for PARs. Technology under development (Initiative 8) could help track locations through technology such as global positioning satellite (GPS), as well as the individual SCBA bottle level, and even firefighters' vital signs.

Span of Control

The combination of sound incident management and strict span of control rules can maintain accountability. **Span of control** is the number of personnel a supervisor can adequately manage or track, which is generally considered three to seven individuals. Some sources limit the number to five individuals. For example, the incident commander orders your engine to complete a search of the second floor. The incident commander writes down E-4 on the tactical worksheet for search and rescue on the second floor, and documents the time. If your crew doesn't stray without notifying command, accountability can be maintained. As the system expands, so does the span of control. As a result, your group leader now keeps your crew's accountability. The system works great on paper, but has the potential to be less effective when there are mixed crews or when members respond from home. It important to note that regardless of the accountability system you are using, span of control is an important rule that must be followed. Even if a specific accountability system is utilized, it doesn't relieve the supervisor of the responsibility of knowing where the crews are at and what they are doing.

Emergency Traffic

Incident communication systems must plan for priority communications when a firefighter or operational safety emergency occurs. The term **emergency traffic** should be reserved for situations that demand complete control of the communications system. The term should be well known by all members and dispatchers through policies and practice. Some departments have specific actions that occur when the announcement is made. It could be that command or dispatch calls to all units and announces that E-4 has emergency traffic before yielding the channel to E-4. Sometimes dispatch uses a special tone to declare that emergency traffic will follow. Whatever the system, it must be practiced. Examples of emergency traffic include the following:

- Trapped or lost firefighters' change of strategy from offensive to defensive
- Critical incident information, such as collapse potential or other evolving hazards

One important note related to emergency traffic is the use of the term *mayday*. Historically used in the aviation and maritime industries, the term has been used by many firefighters to declare a firefighter emergency. Unfortunately, NFPA 1500 specifically states "the term *mayday* should not be used for fireground communications."[13] FIRESCOPE had previously defined the term *mayday* as "[a]n international distress signal that will not be used for fire ground communications" in their 2004 Field Operations Guide (FOG) but removed the word totally from their 2007 edition.[14] Some sources suggest that fire and emergency services are slowly accepting the word *mayday*, and will eventually approve it due to its frequent use and immediately recognizable definition.

Rapid Intervention

In 1998, the **Occupational Safety and Health Administration (OSHA)** developed the respiratory protection standard known to the fire service as the "two-in, two-out" rule.[15] It starts off as a checklist for any occupation in which employees must work in an atmosphere that is **immediately dangerous to life and health (IDLH)**, such as firefighting. It then adds some components specific to the fire service. The standard as written states that for all IDLH atmospheres, the employer shall ensure that:

- *One employee or, when needed, more than one employee is located outside the IDLH atmosphere;*
- *Visual, voice, or signal-line communication is maintained between the employee(s) in the IDLH atmosphere and the employee(s) located outside the IDLH atmosphere;*
- *The employee(s) located outside the IDLH atmosphere are trained and equipped to provide effective emergency rescue;*
- *The employer or designee is notified before the employee(s) located outside the IDLH atmosphere enter the IDLH atmosphere to provide emergency rescue;*
- *The employer or designee authorized to do so by the employer, once notified, provides necessary assistance appropriate to the situation;*
- *Employee(s) located outside the IDLH atmospheres are equipped with*
  - *demand or other positive-pressure SCBAs, or a pressure demand or other positive-pressure supplied-air respirator with auxiliary SCBA; and either*
    - *Appropriate retrieval equipment for removing the employee(s) who enter(s) these hazardous atmospheres where retrieval equipment would contribute to the rescue of the employee(s) and would not increase the overall risk resulting from entry; or*
    - *Equivalent means for rescue where retrieval equipment is not required*

In addition to the requirements set forth under paragraph (g)(3), in interior structural fires, the employer shall ensure that

- *At least two employees enter the IDLH atmosphere and remain in visual or voice contact with one another at all times;*
- *At least two employees are located outside the IDLH atmosphere; and*
- *All employees engaged in interior structural firefighting use SCBAs.*

One of the two individuals located outside the IDLH atmosphere may be assigned to an additional role, such as incident commander in charge of the emergency or safety officer, so long as this individual is able to perform assistance or rescue activities without jeopardizing the safety or health of any firefighter working at the incident.

Nothing in this section is meant to preclude firefighters from performing emergency rescue activities before an entire team has assembled.

The standard never says that "outside the IDLH" means "outside the building." It does, however, say that the crew outside must stay in "visual, voice, or signal-line communication," but is unclear whether radio communications would even meet this requirement. Although this specific element is difficult to achieve, other components are easy to comply with and are even common sense (entering as a team with airpacks on). As a result, the fire service has adopted its best version of compliance, and uses the term *rapid intervention* to describe a group of at least two firefighters ready and uncommitted to other duties to rescue a firefighter who declares an emergency. Some departments refer to them as the **rapid intervention crew (RIC), rapid intervention group (RIG), rapid intervention team (RIT),** or the **firefighter assist and search team (FAST).** Because *rapid intervention* has become the most common reference and NIMS uses the word *group* to identify personnel completing a common function, the abbreviation RIG (rapid intervention group) would be the most technically correct designation when planning for possible expansion. RIC/RIT/FAST are adequate terms for the initial setup, but would quickly exceed the span of control rules if the team was deployed. As author of the "Ten Commandments" series of articles, Mark Emery suggests assigning a rapid intervention group supervisor (RIGS) as head of the first assigned RIC/RIT/FAST.[16] In the event of a firefighter emergency, a group supervisor can easily be responsible for more teams to accommodate the rescue. This is an effective way to expand rapid intervention quickly while maintaining strict span of control rules. Regardless of what a department refers to them as, they are an important function at structural fires where firefighters are in an offensive strategic mode. Although references have opposing opinions on where rapid intervention should stage and what they are responsible for, there are several common theories. Once again, department policies must describe the functions and training must be continuous. Specific questions of theory include: Should RIG be a special assignment? Where is the best location for RIG? What equipment should RIG have? What is RIG allowed (or not allowed) to do?

Should RIG Be a Special Assignment?

As rapid intervention theories developed, two thoughts emerged regarding the assignment of a RIG. The first idea was that rapid intervention should be a specially trained group of firefighters that are specifically assigned as a RIG. They would remain uncommitted for any other fireground assignments and crews would be rotated *around* them for firefighting tasks. The second theory is that crews should be rotated *through* RIG. As a fresh company arrives on the scene, they would take the place of the existing RIG, who would then be assigned firefighting tasks.

Where Is the Best Location for RIG?

Because rapid intervention is responsible for the potential rescue of crews inside, it follows that the group should be familiar with the structure as well as the location and function of each crew. Obviously, the best geographic location for this to occur outside the IDLH environment is adjacent to the individual who is keeping track of accountability, either the incident commander or the accountability officer. If the scene dictates multiple accountability officers, it generally requires multiple RIGs. The best way to learn the layout of the building is to ask crews that are coming out of the building, even drawing a sketch if a preplan is not available. If an accountability officer is using tags or passports, it's clearly the best location for rapid intervention. This has the added benefit of briefing crews as they tag in and enter the hot zone, and even give them a "once-over" on their **personal protective equipment (PPE).** Because RIG is responsible for saving the troops that are entering, it only makes sense that they brief and check (approve) them before they enter. Some departments may use incident safety officers or assistant safety officers to check personnel prior to entry (task level), whereas other departments believe that an ISO or ASO should concentrate on broader safety issues (strategic and tactical levels).

What Equipment Should RIG Have?

Most sources agree that a complement of hand and power tools are necessary, including a rescue air pack or spare cylinders, a portable radio, rope, and handlights. NFPA 1407, Standard for Fire Service Rapid Intervention Crews, provides guidance as to what specific tools should be provided. Many departments keep the tools in a special bag with a tarp to lay out the tools. Progressive departments have a checklist included that specifically lists their equipment and duties.

What Is RIG Allowed (or Not Allowed) to Do?

Incident management is clear that the RIG is not permitted to actively participate in firefighting activities, and must be readily available to perform a firefighter rescue if needed. It does not say that RIG members have to stand on the tarp with the tools. In other words, consider RIG as another layer in the firefighter's PPE that we'd prefer not to test. If RIG prevents a firefighter from being trapped, they will never have to rescue the firefighter. Therefore, many believe that a RIG should maintain a constant state of readiness, but should also perform the following duties:

- Monitor the fire frequency.
- Check in regularly with accountability to ensure PARs are completed on a consistent basis.
- Ensure ladders are placed at windows.
- Clear debris from doorways and exits.
- Keep bystanders (including curious firefighters) away from the hot zone.
- Brief and debrief firefighters entering and exiting.
- Ensure that firefighters are rehabbed before returning.
- Assist the safety officer in looking for hazards.

It's pretty clear that when OSHA established the "two-in, two-out" rule, the intent was just as much tracking firefighters as providing for firefighter rescue. To say that accountability and RIG are two distinct and separate sectors is not an effective use of personnel. If RIG performs additional functions, they become proactive rather than reactive, which always has better outcomes. It could almost be said that RIG should join the safety officers on the enforcement side of the fireground until they are needed for a rescue. In fact, if they are doing their job adequately, they reduce the chance they'll even be needed. Whatever they do on the scene to improve safety is important, but they must always be ready for immediate deployment.

Responder Rehabilitation

Establishing **rehabilitation (rehab)** is an important part of incident management. NFPA 1584, Standard on the Rehabilitation Process for Members During Emergency Operations and Training Exercises, was developed to detail the process of rehab. Personnel are clearly the most important logistic at a fire scene, and need to be rotated through rehab to ensure that they are effective in the IAP. Health and wellness is vital for firefighter safety (Initiative 6) and dehydration is one of the most important factors that need to be addressed on the fire scene. Athletes prepare for a game or a marathon by prehydrating, something emergency responders don't always have the luxury to do. Therefore, the increased ambient temperatures, a lack of ability to lose heat through convection, and physical exertion lead quickly to dehydration. With higher pressures of SCBAs, firefighters are able to work for longer periods of time, almost guaranteeing that when they do change bottles they are ready for a break. As a result, firefighters must periodically take a break from firefighting and "open up" their gear to cool down their core body temperature while replenishing with liquids. In "Rehabilitation and Medical Monitoring," an implementation guide for NFPA 1584, the author makes note of the fact that gear can contain toxic materials that could potentially off-gas. As a result, the guide suggests removing firefighting gear completely during rehab.[17]

Additionally, firefighters should be checked by medical personnel. Departments need policies that dictate when rehab is necessary and how often firefighters should be medically monitored. Checking vital signs is an easy process, and most departments have both the equipment and personnel to perform such checks. A firefighter found outside the parameters of normal vital signs should be pulled from going back in, and given more time to rehab. If the firefighter's vitals fail to return to normal levels, then this firefighter should be transported for a full medical evaluation. Too many **line-of-duty deaths (LODDs)** occur hours after the incident has ended. It would be a fair assumption to say that many of those deaths may have been prevented with proper rehab and evaluation before the firefighter at risk ever left the scene.

RISK MANAGEMENT AT THE STRATEGIC LEVEL

The first decision the incident commander must make is a strategic one. Whether it's a structural fire, a wildfire, or a snowmobiler through the ice, a decision to be offensive, defensive, or marginal is in order. The evaluation must consider life safety the first priority, and base the decision on it. Once the operation has commenced, the incident must be constantly reevaluated as it progresses to ensure that the proper strategy is reaching the goals and benchmarks of the IAP. Risk management has to be part of every valuation.

Urban Strategic Risk Management

For instance, you are the driver for a battalion chief who is dispatched to a reported fire early one morning in a commercial/residential mixed-use complex (Figure 3-4). The first companies on the scene report a working fire involving the basement of a closed restaurant with occupied upstairs apartments. On your arrival, you see that the building is a three-story building of four units that are connected, and fire is venting from the basement windows of the restaurant. Handlines have been stretched from the first engine to the fire building, and a tower ladder is being set up. The battalion orders you to drive to the rear alley, where heavy black smoke blocks your view of the building. You return to a parking lot across the street from the fire building and set up a command post.

Interior crews report difficulty accessing the basement, and command calls for a second alarm. As per policy, you exit the vehicle and attempt to find the location and function of all units. You return to the command post and report that Engine 1 (E-1) is

FIGURE 3-4
Busy companies must remain vigilant in an effort to avoid becoming complacent.

attempting entry to the basement while Tower Ladder 1 (TL-1) crew is working on a search and rescue of the restaurant and the TL-1 operator is setting up to the third-floor windows. E-2 (supplying E-1) and TL-1 have both obtained water supplies, and E-2 has begun to search the second-floor apartments. Command orders incoming Squad 1 to search the third floor, E-3 to the rear, and TL-2 to FAST truck (RIG). Your battalion starts telling you that it looks like things are getting worse when E-1 reports they are unable to advance down the stairway and are backing out. Your battalion chief has a serious decision to make. E-1 was probably thinking offensive attack when they parked the engine. The battalion was leaning more toward marginal when he assigned his incoming units to search above. With no progress on a well-involved basement fire, the battalion chief has to be considering defensive. The crew most in danger right now is TL-1, searching a restaurant that is closed. A simple SPE of a first-floor collapse shows:

- **SEVERITY (1–5):** The crew of TL-1 falling into the basement, which would be catastrophic—5
- **PROBABILITY (1–5):** The probability that the fire is close to eating the first floor—4
- **EXPOSURE (1–4):** The amount of exposure TL-1 has to falling into the fire—4

Risk = 80, for no apparent life hazard

The battalion chief has years of experience, and he doesn't necessarily have to "do the math" to realize that TL-1 has to be pulled immediately in this situation. The real question is how long he gives the crews above the first floor to finish the apartments. A fire early in the morning in an occupied apartment building dictates a search. If TL-1 exits the restaurant safely, it's time to reassess the next crew in danger (E-2 on the second floor). The strategy is still marginal, with a goal of protecting the next crew in danger and possible victims. Tactics for this strategy include holding the fire in check defensively for as long as possible, to complete all searches. Just because everyone is out of the fire building doesn't mean risk management is over at the strategic level. If the strategy goes defensive, other hazards, such as building collapse, must be considered.

Rural Strategic Risk Management

Emergency scene risk management at the strategic level is assessed by the incident commander, whether he or she is pulling an attack line into the living room or is seated in a warm vehicle three blocks down the street. Once again, the biggest decision to make is how much is going to be put on the line, and for what gain. The 2008 edition of NFPA 1561, Standard on Emergency Services Incident Management System,[18] emphasizes the value of strategic risk management, and stresses the safety of responders through decisions made by the incident commander. It doesn't matter the size of the department or the size of the incident, strategic decisions must take into account personnel, equipment, and water supply (Figure 3-5). Take, for example, a fire department that is dispatched to a house on fire in the middle of a summer afternoon. You are a

FIGURE 3-5
Rural fire and emergency services have more components that must fall into place including personnel, specialized equipment, and water supply.

firefighter in the officer's seat of a pumper responding first due. Besides Danielle, who is driving, your only other help is automatic aid coming from three other stations that are at least 10 minutes behind you. There are no hydrants in the area. As you turn onto the street, you see a column of smoke at the cul-de-sac.

The structure involved is a 3,500-square-foot colonial with a high-pitched roof, set back about 600 feet off the road. Fire is showing from the second-floor window on the **"A" (Alpha) side** at the Delta corner, with dark brown smoke pushing from the ridge vent over the fire. You give your on-scene report, including a first alarm for a water shuttle, and go to work. You order the engineer to stop at the end of the driveway so you can wrap the mailbox and lay in. An off-duty firefighter from an urban department who lives across the street tells you there is nobody home but "you better get in there and keep it from getting into the attic or you'll lose the whole thing." He's got a good point. Your knowledge of building construction and the houses being built in your response area tell you that attics like these are "basketball courts" with plenty of unprotected fuel and oxygen. Additionally, the steep slope of the roof allows the fire to climb the inside of the roof like it's on a ladder. In minutes you're going to lose this home from the top down.

Your strategic decision is a little cloudy at this point. Your choices are one of two options:

1. *Pull a 1-3/4" preconnect and make an offensive interior attack through the front door.* You could knock down the fire on the second floor, then hit the attic before it takes off. Room and contents fires are a piece of cake in these types of houses and it's already vented to the attic so visibility will be fine. The off-duty fireman could feed you line at the door, and Danielle has 1000 gallons of water which

should be adequate for your attack. Automatic aid departments are on their way to assist. You can save the house single-handedly and impress the big-city firefighter. The fire chief will surely promote you to lieutenant. In fact, this is exactly what you joined the fire department to do.

2. *Pull a 2-1/2" line. By going defensive with a blitz attack, you could knock down the fire in the bedroom from the outside.* If the bedroom door is closed and hasn't burned through, you might get lucky and force the vent into the attic, taking steam with it to slow the spread of fire in the attic. By then, the reinforcements should be here to pull a handline and help you extinguish the fire while you get that attic opened up. However, if that door is open or breached, there's a good chance you could spread the fire across the second floor, and push the fire down into the rest of the home, causing extensive smoke and heat damage.

It's no secret that the fire service has evolved significantly over the past 50 years with regard to tactics and equipment. We learned that a fire in a building is basically a fire in a box, and the most effective way to extinguish it is to climb into the box with it. By opening the lid on the box, we can let out the heat and smoke, permitting a quicker resolution to the problem. We've been trained to "get in there" without making any type of risk analysis decision and fight the fire from the unburned side. A complete ensemble of protective gear and an airpack allow you to enter the box with a hoseline and come out victorious, dirty, and wet 20 minutes later. Our training and culture have taught us that simply spraying water into the windows from the outside is a primitive tactic that spreads the fire. It's true; the best way to stop the fire is to get in there and stop it. Although you are trained to, are capable of, and are brave enough to go in by yourself, you simply can't do it. A risk analysis proves to you all the reasons to go defensive:

- *Nobody is home.* We will risk a lot to save a lot—in a structured plan. We will risk a little to save property and pets. Going in by yourself is risking a lot to save a little.
- *Rapid intervention.* If you got into trouble, there is nobody to come help. Nobody else on scene even has on an airpack. If they saw you in trouble, they may run in unprotected and make the situation even worse. It's hard to believe that the off-duty firefighter will wait at the door and feed you line (you'll hit the bedroom door and guess who's behind you).
- **Situational awareness.** We should be aware that there's a room and contents fire on the second floor with obvious extension into the attic. Fire loading in a bedroom is high with synthetics and doesn't take long to reach flashover. Lightweight trusses are propped up above the fire and the building materials in the attic are involved as evidenced by the brown smoke. Fire overhead in truss systems is a risky situation, even with sufficient manpower.

The first option is not an option at all. Your only choice is to hit it from the outside and hope for the best. As you launch water into the upstairs window, limit your stream to the heaviest fire and think about your plan of attack before help arrives. Continue to monitor the spread of fire and smoke from your position outside, and maybe even send the other firefighter to do a 360 and report back on the fire spread. The off-duty firefighter can pull the 1-3/4" line and assemble forcible entry tools and pike poles at the front door. He can even set ladders to the second-floor windows. Once the cavalry arrives, assess the structure and make entry if possible. Attacking it from outside has nothing to do with bravery. In fact, it actually takes more courage to *not* go in.

RISK MANAGEMENT AT THE TACTICAL LEVEL

Tactical risk management involves choosing operational methods to meet the strategic goals that have been established. Some might think that once the incident commander declares a strategy, crews are forced into tactical decisions that do not require risk management. Nothing could be further from the truth. Just because you are ordered to deploy a rowboat to rescue a victim from the roof of her car in a rain-swollen creek doesn't mean you can't make tactical decisions based on risk. Those decisions could involve using a dynamic ferry to help control the boat from upstream. It's possible your assessment could reveal that any ropes at all are too risky or time consuming. Should you approach the car from upstream, stern first with ropes and no motor, or is it safer to approach by drifting in from the side using the outboard motor? Choices may also take into account anchor points or whether the car is still moving. Decisions about how many rescuers to send in the boat or if any will be in the water are yours to make. The incident commander wants a successful rescue with the least risk, so suggestions are usually welcome. Requests could be made for safer equipment, such as a better-suited boat, or possibly lowering a rescuer from a highline or a helicopter. Every idea must be analyzed at the tactical level, and if it doesn't lower risk, it should be abandoned. Reducing risk is important, whether it affects two rescuers or several hundred.

Let's say you are a captain assigned as a division boss to direct two task forces and a strike team at a wildland fire operation. After receiving your briefing, you set out to protect Tobyville, a small town in the path of a 10,000-acre fire. The town sits at the top of a ridge south of the main body of fire, and is surrounded on three sides by thick brush and pines. The only open side is a meadow to the north, between the town and the head of the fire. Two roads reach Tobyville, one from the east and one from the southwest. The eastern approach provides the best tactical position, with a lake just southeast of the main road at the bottom of the drainage. Besides having a great water supply, it's the most likely path the fire will travel. You meet with your task force and strike team leaders and agree that ground crews should begin clearing the eastern flank down slope, while others cut control lines and start backfires at the top of the ridge to provide a firebreak. Crews are confident that if they can clear enough of the slope, the fire will be forced to jump the drainage heading due east with the wind and completely miss the town.

While work continues, you drive to the western end of town and look for spot fires and potential safe zones. You make note that the meadow to the north will make a good, safe zone, if needed. You drive your SUV north, deep into the meadow to get a good look at the

FIGURE 3-6
Once a strategic decision is made to put personnel in dangerous positions, tactical choices must be based on risk versus benefit.

town from the fire side and are a little worried by what you see. The dozers working at the bottom of the drainage by the bridge are making progress clearing the slope, but the pine and oak growth further down slope from them is thicker than first thought. Although the break would normally be sufficient, the afternoon wind that's starting to pick up could outrun the crews if the fire doesn't cross the drainage. It appears that it still has plenty of fuel and could jump the line, trapping the crews on the ridge. You start to rethink your tactics and consider pulling crews out and redeploying to the southwest side. You return to the ridge in town at the highest point to get a look at the fire. The head is now 0.5 mile wide and 4 to 5 miles away on a path for the meadow. You estimate that with the wind picking up, you could see fire in as little as 20 minutes. Your tactical decision has placed unknowing crews in a position of potential danger. You predict the fire will do one of three things:

1. The fire will stop if crews continue to clear and the break is sufficient in size.
2. The fire will jump the drainage to the east regardless of any more work and miss the town.
3. The fire will build steam at the bottom, overrun the crews, and head for town.

Your risk assessment shows that there will be little effect on the outcome if the crews continue to work. The swirling wind at the bottom of the slope with plenty of fuel is a recipe for a conflagration that will take whatever it wants. A tactical decision to withdraw the crews immediately is in order. You can only afford to risk a little to save this town, and the only place to do that is in the meadow. If they move now, the crews have enough time to evacuate to the safe zone and prepare to deploy their shelters if needed (Figure 3-6).

RISK MANAGEMENT AT THE TASK LEVEL

Risk management at the task level is often overlooked as a contributing cause of the overall success of tactics and strategies, but is actually a significant component. If you consider that most injuries occur at the task level, and the worker injured is attempting to complete tasks to accomplish strategic and tactical goals, it only follows that those injuries will directly affect the outcome of the incident.

Safe Search

For example, you and another firefighter are completing a primary search on the second floor of a single-family home late one night. There is a working basement fire that another crew is presently attacking, and you have no information about any occupants. Specific decisions you make on the search team affect whether or not a trapped victim survives. Some important components include accountability, building aspects, personal protective equipment, methodology, and emergency procedures.

Accountability

Were you assigned to search or did you self-deploy? Did you report to your supervisor or accountability officer your exact location? Is there a rapid intervention group available yet or another crew searching? Are you supposed to take a handline with you?

Building Aspects

Where are your exits? Are ladders deployed and, if so, to which windows? Are there bars on the windows? Is there only one stairway? Is there a porch roof outside any windows? Is there a cliff outside any windows? Is the power still on? Is there any chance the fire is above you? Where are you? Are you maintaining an orientation to your location? How much clutter is in the house? Can you get trapped? Who are you searching for? Do kids live in the house? What about a baby?

Personal Protective Equipment

Did you remember to don all of your protective equipment? Was the air bottle full when you got out of the jump seat? How much time do you have left before you need to make your way back out? Did you and your partner both bring in portable radios?

Methodology

How are you searching? Did you remember the thermal imager? Are you following a specific method or pattern that allows a quick and complete search? Are you concentrating on areas you are most likely to find a victim or are you simply moving from one end of the hall to the other? What are you going to do if you find a victim? Do you have equipment to remove a victim? Are you and your partner proficient at different techniques? Is your partner panicking and becoming a possible liability?

Emergency Procedures

Do you know what to do if your partner goes down? What are your department's emergency procedures for a trapped firefighter? Do you have tools with you to breach a wall or a rope for a bailout?

Safe Tools

Firefighters have a vast array of tools to choose from. Although many tools will accomplish a certain task, some are better than others, and some are safer. Say, for example, you find an article about the best vehicle extrication tool ever invented. Being both efficient and powerful, it has the ability to slice quickly through metal, plastic, rubber, laminated glass, and even seatbelts. Steering columns, Nader pins, and hinges are no problem. Reinforced A posts can be cut in seconds. Busses and trucks are no match for it. It only takes one person to operate it, and sets up faster than any tool with hoses or wires. There are no pressurized cylinders or regulators to worry about. It takes less than two minutes to take off the roof, and you don't even have to remove the windshield first. In fact, it can split a car right down the middle from license plate to license plate without refueling, changing blades, or even becoming dull. The best thing is a brand new one costs less than $1000. It became available to fire departments around 1960 and is still carried on most apparatus. Rotary saws or K-saws are extremely effective in automotive extrication.

So why are these saws a terrible choice? First off, they are incredibly loud on the scene, and almost deafening for patients in the car. They send a spray of hot sparks that can ignite fires on spilled hydrocarbons, vehicle interiors, or even patients' laps. Rotary saws rip through a glass windshield quickly, but launch glass chips everywhere. The saws are known to kick when they bind, so cutting a B post by a patient's head is generally a bad idea. This line of tools is a perfect example of being versatile but not practical. Although the saws are effective, they are not remotely safe at a crash scene. It's clear that safety must be the primary concern when picking a tool for a specific job.

Situational Awareness

Although situational awareness is considered a new concept to the fire service, the theory is thought to have been introduced in Germany in 1832. The military manual *On*

War, by General Carl von Clausewitz,[19] describes the shortage of information and increase in stress which leads to actions being planned in a fog. This theory of the "fog of war" has surfaced periodically in the last 200 years in the military and has evolved into an awareness of surroundings. Situational awareness is not only knowing what is going on around you, but also being able to predict how the environment will change with time. For firefighters at the task level, situational awareness involves listening to what both the building and the fire are saying. Knowledge of building construction and reading the smoke are skills that promote situational awareness.

Mismatching Expectations with Realizations

One effective way to maintain situational awareness is to compare expectations with realizations and look for a mismatch. Let's say you are a lieutenant who is first due for a fire in a garden-style apartment building. Arriving on scene, you find a two-story multiple-residence building with black smoke issuing through an open sliding glass door on the second-floor balcony. You and your crew pull an attack line and enter the main stairway to the apartment door, where you force entry through a deadbolt and chain lock. Visibility is zero, and your crew knocks down the flames coming from the kitchen. It appears that nobody is home, and you suspect that food was left on the stove inadvertently and had spread to the cupboards and furnishings. Your expectations are that the fire is mostly extinguished, but in need of overhaul and ventilation. However, the flames immediately jump back with the same intensity. You have just discovered a mismatch.

Apparently there is more than what meets the eye, and it's not just food on the stove. Your recognition-primed decision making looks for another template, and you immediately reason that you have a natural gas–fed fire. You call an incoming unit to get the gas shut down as you make your way closer to the kitchen. As the smoke starts to clear, the light of the gas-fed flames exposes nothing on top of the stove, but all the burners are on and the oven door is wide open. Again, you note a mismatch. Your brain recalculates the new information and you start thinking it's a set fire. You now expect that the occupant may have turned on the gas and left the building. As the nozzleman holds the fire in check, you and the rest of your crew begin a search. In the back bedroom, you find an unconscious female victim behind the bed and remove her to the front yard. You know that the fire was still early in its progression, she was found in a remote location, and most of the smoke was venting out the patio door, so you expect that there's a good chance she can be revived. As **emergency medical services (EMS)** begins to assess her, they find multiple gunshot wounds to her head and chest.

Again, your expectations do not match the realizations of what is really happening around you. Suddenly the open sliding glass door and the dead-bolted, chained entry door make more sense. Your newest theory is that the fire was probably to cover up a crime, and the perpetrator left through the sliding glass door. It also hits you that the rest of your crew is still in the apartment, and who knows what other dangers might be up there. Maintaining situational awareness allows you to process new clues as they are discovered and develop new theories. Each theory gets you a little closer to what is actually happening around you. The real problem evolves when you ignore the clues at hand. Continuing to operate in a dynamic situation with static logic cannot be effective. When the building is trying to tell you that it can't hold on much longer or the fire is laughing at your futile efforts, you need to stop and rethink the present situation.

Air Management

In the past, when a low-pressure alarm sounded on a SCBA, it meant that it was time to exit the building. Unfortunately, we now know that if your bell alarms unexpectedly, you may not have enough to even make it back out of the building. **Air management** is the process of keeping track of your supply of fresh air, and is actually a component of situational awareness. Air management puts your bottle level closer to the front of your mind (both literally and figuratively). A **heads-up display (HUD)** takes the bottle level and passes it to a series of lights in the SCBA mask. Most regulator gauges glow in the dark and are at the end of a short whip of hose, allowing the operator to view the pressure reading even in darkness. Some trainers teach students to "charge" the glowing gauge with a flashlight before entering to make it even more visible. Current training encourages crews to check their air level often, and report back to their supervisor when any member reaches the halfway point of air supply. This not only lets the superior officer know that a relief crew will be needed soon, but tells the crew that they are halfway done with their assignment. In other words, if the first half of their air was used getting to their present location, they will need at least that much to backtrack out and they need to start moving now.

We need to view our fire building as having a point of no return. This arbitrary location is the spot where you cannot make it back out again. The fact is, if a firefighter runs out of air, there is a good chance of being injured or killed. Many present-day firefighters learned how to stick the low-pressure hose of their SCBA into their coat if they ran out of air. It's not something we teach anymore because it's not feasible with mask-mounted regulators and it just plain doesn't work. Your turnout coat is not an **N-95 filter mask,** and Nomex is not an **oxidizer.** Running out of air is simply not an option. Some training systems train students to remain "air aware"—to be able to estimate how much air they have at any given time. The low-pressure warning is not intended to be your exit air, but it's an emergency reserve if you run into problems. Anytime you use your last line of defense for normal operations, you are destined to fail. Divers who use **self-contained underwater breathing apparatus (SCUBA)** don't use their emergency reserve air to ascend to the surface; and skydivers don't use their reserve chute to slow their descent. Firefighters are no different, and must learn to remain cognizant of air levels at all times.

Radio Communications

For fire and emergency services, communications is probably our weakest link. Besides the fact that most of our portable radios are not designed for wet environments or transmitting from the bowels of a building, we are generally poor communicators

FIGURE 3-7
Many fire and emergency services need to improve on radio communications.

when it comes to portable radios. If you take offense to that statement, listen to a tape of your last fairly significant incident and pull out a pen and paper. Keep score on the paper how well the communication went (Figure 3-7). Communication can be described as a message being sent between a sender and a receiver. It also means that the message must be received as it was intended. When you hear command call interior, draw a dash (-) on the paper. When interior answers, make the dash a plus sign (+) and circle it. That was a complete transmission. You'll be surprised to find how many times you have lonesome dashes on the paper because people don't answer, they misunderstand, or they ask for an address or order to be repeated.

Ineffective Communications. Emergency scenes are hectic, and there are several reasons why our radio transmissions are so often missed. If we can identify them, they will be easier to eliminate. These include excitement, background noises, terminology, and multitasking.

- *Excitement.* When we describe an ideal dispatcher with excellent radio communication skills, one of the first words we use is "calm." A panicked voice on the radio instills panic. Being dispatched to a child struck by a vehicle, who is unresponsive and possibly not breathing, is excitement enough for responding units. A good emergency communicator will keep a calm and methodical voice throughout the incident. As the excitement level in a voice rises, so does the tone and volume. The higher and louder a voice gets, the harder it is to understand. Yelling into a radio does nothing for the volume, it just makes it harder to understand. Sometimes you simply need to take a deep breath before speaking.

- *Background noises.* From sirens to saws, emergency scenes have plenty of noise. Unfortunately, radios don't differentiate between background noise and voices, so they transmit them all the same. If you are trying to transmit in a noisy environment, figure out what you're going to say before you do, and keep it short. Sometimes shielding the microphone from the source of the noise is helpful. Speaker mics are better suited for shielding. At times, other portable radios are the cause of background noise in the form of feedback or echos. Some officers believe that every member should not have a radio for this very reason. Unfortunately, this opinion is ridiculous. It's like taking the true statement that "opposing hoselines are dangerous" and extrapolating it into the false statement "only one hoseline in the house is justified." Every firefighter on scene *should* have a radio, and everyone in the IDLH *must* have one. Radios should be kept at the proper volume: loud enough to hear over your environment but not so loud as to interfere with others on the team. **Personal alert safety system (PASS)** alarms must be treated as a true emergency, and never heard on the fireground unless someone is really trapped.

- *Terminology.* Fire and other emergency services students learn common terminology in their education. Textbooks and national exams for firefighting and emergency medical services use words and phrases exposed to emergency responders worldwide. The real problem occurs during on-the-job training. Some terminology is necessary simply based on the needs of the community. Some responders must know how many **roentgens** they can take in an hour; others must know how to get through an **eddy**. Sometimes departments have slang terms for words. For instance, the word *alarm* usually has to do with an assignment of apparatus for a specific emergency. So what does a *third alarm* get you? It depends entirely on where you are located. Ten-codes were once a lot more popular than today, but some departments still use them. They were supposed to make communicating easier, but made it worse. You could be 10-17 to the 10-20, but it's far easier to say you are en route to the scene. Misunderstanding the codes was a common problem, especially when different responders used different codes. Some hear 10-75 and think of a working fire; others hear 10-75 and think of LP gas. You could respond code 3 to a code 4, or simply respond with lights and sirens to a car crash. For many departments, plain radio communications have become the standard to eliminate miscommunication and potential incidents.

    The National Incident Management System (NIMS) has done a good job of trying to standardize terminology, to the point of coming out against using 10-codes in favor of common terminology; however, regional terminology is a hard habit to break. Call for a tanker in McMinnville, Oregon, and a 747 with more than 20,000 gallons of fire retardant will clear the trees. Call for one in Canaan Valley, West Virginia, and it will crawl up the mountain on an all-wheel-drive 2500-gallon Freightliner. And a tanker is the last thing the Port of Houston

Fire Department wants to see at one of their fires. Initiative 11 works to get standard operations for fire departments. Standard terminology is key.

- *Multitasking.* Probably the biggest reason our radio communications are so bad is because, to a firefighter, the radio is simply a tool, like a Halligan bar. Although it's one of the most versatile tools on the fireground, you don't need it the whole time at a fire. Say, for example, you are helping advance a line into a smoke-filled grocery store, but you have the Halligan in your hand. It's not needed now, so you lean it against the wall while you feed line to the crew inside. They yell that they need more line, but you've got a pretty good loop inside. It appears as though it is stuck on something, so you grab the tool and follow the line in. As you crawl to the first aisle, you note the hose has become wedged under the toe kick of a freezer and can't be freed by hand. You make short work of it by using the tool to pry the line out then look for something to keep the hose from getting stuck again. You don't want to leave your Halligan, but you find a suitable substitute in the freezer. A smile covers your face as you think about the irony and cram a family-size bag of potato wedges under the toe kick. Your crew has reached the fire, so you follow the line and meet up with them in the back of the store. The fire is in a locked utility room, and the officer's tool they are using just doesn't have enough leverage. You pass forward the Halligan and they force the door with little effort. As they make a hit on the fire, one of the members uses the Halligan to hold the spring-loaded door open. There is no question it's a great tool for firefighting, but it's still just a tool. It's worthless when you're feeding hose, raising a ladder, or relay pumping.

    The trouble with considering a "handie talkie" a tool is that we treat it like a tool. We leave it in our pocket, holstered until we need it. If we're pulling ceiling and we want to call for more poles but there's chatter on the radio, we wait until everybody is done yapping so we can use it. Then when there finally is a break in the squawking, we call in our request. This is a common occurrence on firegrounds and it's a habit that kills firefighters. If the portable radio is "chattering" or "squawking" because everybody wants to hear themselves "yap on the radio," we obviously aren't listening to it. If we aren't listening to it, how do we know if we are missing a message we need to hear? By simply giving the radio a little more attention and priority, we catch more transmissions, cut down on traffic that has to be repeated, and even save lives.

Effective Communications. If we truly desire to improve our radio communications, we should start by listening to effective communication. Investigating how other high-risk professions communicate is a great place to start. Paul Mannion, a career firefighter paramedic and part-time flight instructor from Ohio, makes it clear that the first-time firefighters ride with him. "This isn't like the fire department," he once said. "When the radio talks, stop talking and listen." Mannion said this because pilots don't use the radio as a tool, they use it as one of their most vital instruments needed to fly the plane. Take some time to listen to air traffic control on the Internet. There are numerous websites and airports, so pick a busy one. Now get back out your pad of paper and pen and start scoring. You'll find that air traffic control and airplanes are about 95% effective, but don't be surprised if your last structural fire was closer to 75%. You may attribute their high score to the fact that they talk on the radio all day and have quiet, relatively stress-free environments with excellent equipment and no airpacks. All those facts are true, but after listening for awhile, you'll notice that their communication is effective simply by its structure.

- *Hail first.* Airline and military radio transmissions (and some fire and emergency services) hail the unit being called first, much like you would yell your friend's name when you wanted to get her attention: "Engine One from Command." The theory has some merit, because your call sign tends to catch your attention before the message is delivered. The problem is that so many agencies hail second "Command to Engine One" that there is a sufficient training issue to teach every agency the same procedure. Stress makes you forget what you recently learned, and you tend to revert to your old way. This can cause significant issues on an emergency scene. Initiative 11 strives to establish common policies and procedures, and hailing first may be more effective.

- *Acknowledge and repeat.* Most transmissions are repeated back, especially those regarding heading, elevation, barometer, and frequencies. "American flight 3-6-zero-1 turn left heading 1-4-zero, climb and maintain 21." "Left 1-4-zero and climbing for 21, 3-6-zero-1." This can be accomplished even if your department doesn't hail first. "Roof to interior, we're about to drop a chimney bomb, let us know when you're ready." "Go ahead and drop the bomb, roof."

- *Sandwiched transmission.* Note on the last examples that traffic concerning a specific unit opened and closed with that specific unit number. This allows other units that wish to speak to know when a conversation is completed. Like many radio techniques, it's subtle and makes radio traffic smoother.

- *Short and to the point.* The messages are clear and succinct. They don't ramble or communicate worthless information. You may have heard an officer continue to tell command that there were flames from the roof every five minutes on a defensive fire. Airwaves are too valuable for conversational transmissions with no tactical value. Additionally, if we are advocating more attention to the radio, we obviously have to kill the actual chatter.

- *They are polite.* Listen for a while and you're going to hear "thank you" and "good day" a lot. But does politeness really matter? What happened the last time a dispatcher was snotty to you? Have you ever seen someone throw a microphone? The fact is we are under enough stress without getting an attitude over the radio. We don't have time to wish everyone a nice fire or happy transport to the trauma center, but cutting down on the rudeness is fairly easy and a way to make everybody's job a little easier. *Radio etiquette* is an actual term.

FIGURE 3-8
We need to expect that radio transmissions during a firefighter emergency will be worse than normal.

Stressful Communications. There is no doubt that the stress of emergency scenes will reduce the effectiveness of communication. Worse yet, listen to any recording of a trapped firefighter, and you'll hear radio communications get even worse (Figure 3-8). We know what firefighting stress sounds like, but it's helpful to listen to pilots when they are in a mayday situation. If you took the time to surf for air traffic control, head back to your favorite search engine and look for airline crash audio. If you listen to and score these transmissions as you did before, you will probably find that effective communication has dropped by 30% when pilots are unsure if they will be able to regain control of the plane. By listening closely to the radio traffic in mayday situations, we learn some important lessons in emergency communications.

- *Multiple requests.* In a stressful situation, the mind can only process so many things. As Aloha Flight 243[20] was coming into Maui for an emergency landing, the copilot was asked by the tower how much fuel and how many passengers were on board. Although she was able to answer the correct number of passengers, she forgot that he asked about the amount of fuel on board and **air traffic control (ATC)** never remembered to ask again. A trapped or lost firefighter must be able to answer specific questions to assist RIG in proper location and rescue. Keep the questions simple, and only ask one at a time.
- *Emergency traffic.* When a plane crashed into the Everglades several years ago, the first radio transmission from Value Jet (Critter) 592[21] declaring an emergency was a simple "Flight 592 needs an immediate return to Miami." The controller immediately responded "Critter 592 Roger, turn left heading 2-7-zero descend and

maintain 7000." Rather than asking the pilot why, Miami gave him the headings and an order to descend. Have you ever had to call dispatch to get law enforcement on the scene? If a victim of an apparent medical emergency suddenly pulls a knife and you call for backup, do you want dispatch to ask you why or just send the deputy? When radio traffic sounds urgent, that's the way we should treat it. Respond to the request, then ask questions.

- *The fire is still burning.* A great example applicable to commanding an incident with trapped firefighters comes from the Pacific Ocean off the coast of California. Alaska 261[22] had a catastrophic failure of its controls, causing the plane to nosedive. As the crew fought to gain control and troubleshoot the problem, ATC assisted with whatever he could. The controller stayed calm and cleared other planes from the area. Most importantly, he continued to accept planes into his airspace and pass them off to the next towers at the right altitude and speed. Although you can hear the despair in his voice when he realizes that 88 souls were just lost in the ocean, he continues to lead the others. No greater pain can bestow an incident commander than when the IC can't reach a crew that is in peril, but the fire is still burning. The rescue is clearly the most important function at the scene, but if the IC gets too caught up in it and forgets about the fire, more will be lost.

- *Say it now, say it later.* Another lesson learned from Alaska 261 is what we say on the radio. At one point on the radio when the pilot was trying to gain control, another airplane in the area saw the plane nosedive and reported it to ATC. The controller responded with "very good." Because it was the same controller who had despair in his voice a minute later when they crashed, it's clear he didn't think it was "very good" that the plane was going to crash. The only explanation is that "very good" was his standard response on the radio, and that in times of stress we revert to what we've always done. If we have poor radio practices when we go for fuel, we're going to do it when we are in a stressful situation.

- *Stay here with me.* Our last lesson comes from the controller back in the Everglades. As discussed earlier, planes are normally passed off from tower to tower on different frequencies. Planes are required to physically change channels when that occurs. As Critter 592 was getting closer to Miami, the controller told the pilot to "contact Miami approach on, correction you, you, keep on my frequency." The controller was just doing his job and following policy when he tried to pass off the plane. He then realized that the pilot was a little too busy to try to change radio channels and the policy didn't make any sense. Sometimes when a firefighter becomes trapped, procedures require the crews inside to change channels, which is a difficult task to achieve without the stress of an emergency. On the other side of the continent, Alaska 261 was instructed to change channels. They had to ask for the frequency to be repeated, and ultimately never made the correct change. The worst thing that could happen is for someone inside a burning building to make the switch and end up on some other unmonitored channel.

IMPROVING THE SAFETY OF THE FIREGROUND

Fire scene radio communications are not the same as air traffic control. Emergency scenes are much more distracting, have more background noises, and often use substandard equipment. However, the similarities in airline mayday situations are numerous. If we don't choose to model our radio procedures based on past fire emergencies or the lessons learned from aircraft maydays, we are destined to stay poor communicators. Earlier, we scored air traffic at 95%, but gave them a 30% reduction during a mayday. This results in approximately 65% effective communications when they need it most. Now take firefighting communications, starting with 75%, and knock off the same 30% during a firefighter emergency—45% is simply not acceptable when lives are on the line. The only way to improve emergency communications is to improve fireground communication. With the exception of the "hail first" theory, every other tip listed previously takes no approval from an authority or policy change in a department. It simply takes a conscious effort to improve (Initiative 2). Although it's only one component of risks on the fireground, communications is put to the test when a member needs it the most.

SUMMARY

One of the best ways to reduce injuries and deaths in the fire service is to concentrate on how we make decisions. Risk versus benefit must be the basis for actions. We know that emergency scenes are dynamic situations with inherent risk, but human decisions play an enormous role in the outcome of the event. Strategic, tactical, and task decisions made during the emergency not only dictate the timeline of the mitigation of the event, but also directly affect the chances of injury or death. We are trained to understand that time is our worst enemy when it comes to fire development, building collapse, traumatic injuries, and the environmental exposure of a hazardous material. Emergency workers are therefore ever mindful of the ticking clock. Similar to a football team in a "two-minute drill," we use predetermined plays in an attempt to win the game. It's important to realize that the football team deploying a "hurry-up offense" in the closing seconds of the game is *always* losing, and is getting desperate. Many times the team resorts to high-risk plays such as a deep pass in heavy coverage in an attempt to pull off a win. It's also not uncommon to see a desperate throw result in a game-ending interception. An IC might resort to a similar tactic in a desperate situation, risking much more than originally intended.

Risk management is designed to slow the speed of decision making at an incident. The IC needs to establish specific components of incident management and use situational awareness to make intelligent risk-benefit decisions. It's equally important to be mindful of the incident clock, but not allow it to force a poor decision. Likewise, firefighters need to be *educated* about how to weigh the options of individual tasks and make good risk management decisions at the task level, and be *trained* to react appropriately. Communications is an area where emergency services can make an immediate improvement. This integration of risk management techniques at all levels of emergency management is best accomplished through education, training, and practice.

KEY TERMS

"A" (Alpha) side - The front of a structure.

accountability - A process of tracking the location of firefighters.

air management - The process of maintaining an awareness of the cylinder air level in a SCBA while using it, and ensuring that it will allow time to escape.

air traffic control (ATC) - Air traffic control.

assistant safety officers (ASOs) - Assistant safety officer.

"C" (Charlie) side - The rear of a structure.

"D" (Delta) side - The right side of a structure.

eddy - A dangerous area of a fast-moving body of water.

emergency medical services (EMS) - Emergency medical services.

emergency traffic - The term used to describe a firefighter emergency.

firefighter assist and search team (FAST) - Firefighter assist and search team.

Firefighting Resources of Southern California Organized for Potential Emergencies (FIRESCOPE) - Firefighting Resources of Southern California Organized for Potential Emergencies; one of the early versions of NIMS.

heads-up display (HUD) - A visual indicator of cylinder air level in the mask of a SCBA.

immediately dangerous to life and health (IDLH) - An acronym used to describe an atmosphere that requires special precautions.

incident action plan (IAP) - Incident action plan.

Incident Command System (ICS) - Incident command system.

incident safety officer (ISO) - Incident safety officer.

line-of-duty deaths (LODDs) - Fatalities that are directly attributed to the duties of a firefighter.

mayday - A term used by the aviation and maritime industries to declare an emergency.

N-95 filter mask - A dust mask that is effective for filtering some sizes of particles, but is ineffective at filtering dangerous gasses.

National Fire Protection Association (NFPA) - National Fire Protection Association.

National Incident Management System (NIMS) - National Incident Management System.

Occupational Safety and Health Administration (OSHA) - Occupational Safety and Health Administration.

Operational Risk Management (ORM) - A risk management program utilized by the USCG.

oxidizer - A material that emits oxygen as it burns.

personal alert safety system (PASS) - Personal alert safety system.

personal protective equipment (PPE) - A generic term used to describe the minimum apparel and gear needed to safely perform a specific duty.

personnel accountability report (PAR) - A verbal or visual report to incident command or to the accountability officer regarding the status of operating crews; should occur at specific time intervals or after certain tasks have been completed.

rapid intervention crew (RIC) - Rapid intervention crew.

rapid intervention group (RIG) - Rapid intervention group.

rapid intervention team (RIT) - Rapid intervention team.

recognition-primed decision making (RPD) - A process of using experience to guide decisions at a later date.

rehabilitation (rehab) - The designation of an area where emergency responders can rest and recover.

risk management - Identification and analysis of exposure to hazards, selection of appropriate risk management techniques to handle exposures, implementation of chosen techniques, and monitoring of results, with respect to the health and safety of members.

risk-benefit analysis - The weighing of the facts to determine the advantages and disadvantages of a certain activity.

roentgens - A measurement of radiation.

self-contained breathing apparatus (SCBA) - Self-contained breathing apparatus.

self-contained underwater breathing apparatus (SCUBA) - Self-contained underwater breathing apparatus.

situational awareness - A term used to describe the recognition of an individual's location, the surrounding atmosphere, the equipment being utilized, and the evolution of an incident.

span of control - The ideal number of personnel that a person can effectively manage.

SPE (severity, probability, exposure) model - The USCG's risk management tool for assigning a numeric value to the three components of a risky task.

training-based decision making - A process of using training to guide decisions at a later date.

United States Coast Guard (USCG) - United States Coast Guard.

REVIEW QUESTIONS

1. What is risk management? Why it is an important component of incident management?
2. How does a risk-benefit analysis apply to emergency services?
3. What are the features of recognition-primed decision making?
4. What are the five safety features of the incident management model?
5. What is OSHA's "two-in, two-out" rule and what is its effect on rapid intervention?

FIREFIGHTING WEBSITE RESOURCES

http://everyonegoeshome.com
http://cdc.gov/niosh/fire
http://www.hazmat-news.com
http://www.planecrashinfo.com
http://www.osha.gov/pls/oshaweb/owadisp.show_document?p_table=standards&p_id=12716

NOTES

1. CBS News, 48 Hours, Make no mistake: To err is human. Reported by Susan Spencer; broadcast March 22, 2009.
2. U.S. Department of Transportation, U.S. Coast Guard. COMDTINST 3500.3; 23 NOV 1999. Operational Risk Management.
3. Dorner, D., and F. Pfeifer. 1993. Strategic thinking and stress, *Ergonomics* 36(11): 1345–60.
4. CBS 60 Minutes, interview with Katie Couric. Broadcast February 8, 2010.
5. Kowalski-Trakofler, K.M., and C. Vaught (Pittsburgh Research Laboratory) and T. Scharf (Cincinnati). DATE. *Judgment and decision making under stress: An overview for emergency managers.* National Institute for Occupational Safety and Health (NIOSH).
6. Peniston, B. 2006. *No higher honor; saving the USS* Samuel B. Roberts *in the Persian Gulf.* Annapolis, MD: Naval Institute Press.
7. Klein, G. 1998. *Sources of power; how people make decisions.* Cambridge, MA: MIT Press.
8. Gladwell, M. 2005. *Blink: The power of thinking without thinking.* New York: Little, Brown and Company.
9. http://www.dhs.gov/xabout/laws/gc_1214592333605.shtm
10. National Fire Service Incident Management System. 2000. *Model procedures guide for structural firefighting*, 2nd ed. Oklahoma State University: Fire Protection Publications.
11. Dodson, D. 2007. *Fire department incident safety officer*, 2nd ed. Clifton Park, NY: Delmar Cengage.
12. Fire Department Safety Officer Association; ISO certification. http://www.fdsoa.org/certification/iso_certification.html
13. NFPA 1500, Standard on fire department occupational safety and health program, chap. A.8.1.11. 2002. National Fire Protection Association.
14. Field operations guide ICS 420-1. 2004, June. FIRESCOPE California.
15. http://www.osha.gov/pls/oshaweb/owadisp.show_document?p_table=standards&p_id=12716
16. Emery, M. 2007, March. The ten commandments of intelligent and safe fireground operations. *Firehouse Magazine.*
17. Bledsoe, B.E. 2009. *Rehabilitation and medical monitoring.* Midlothian, TX: Cielo Azul Publications.
18. NFPA 1561, Emergency services incident management system. 2008. National Fire Protection Association.
19. On the theory of war. 1832. From *On war*, chap. 2. http://www.clausewitz.com/readings/VomKriege1832/TOC.htm
20. Aloha 243 transcript and audio recording; Boeing 737. April 28, 1988. http://www.planecrashinfo.com/cvr880428.htm
21. Critter 592 transcript and audio recording; DC-9. May 11, 1996. http://www.hazmat-news.com/archives/valujettranscri.html, http://www.planecrashinfo.com/cvr960511.htm
22. Alaska 261 transcript and audio recording; MD-83. January 31, 2000. http://www.planecrashinfo.com/cvr000131.htm

CHAPTER 4

Eliminating Unsafe Acts

LEARNING OBJECTIVES

- Explain empowerment and how it relates to the fireground.
- Define what constitutes an unsafe act.
- List some examples of energy conversion and how they cause injuries.
- List some of the perceived problems with using PPE.
- List examples for each of the four categories of unsafe acts.
- List examples of using education, engineering, environment, and enforcement to prevent injuries.
- Describe the differences between training fires and real fires, and why it's important to understand them.
- Describe ways to address a safety concern with your supervisor.
- Describe crew resource management and its application to a fireground situation.
- Explain the difference between "bolt-on" safety and "built-in" safety.
- List the four components of human interaction necessary to institute crew resource management.

INITIATIVE 4. All firefighters must be empowered to stop unsafe practices.

Throughout the history of the fire service, tradition has been the backbone of our operations. The paramilitary style that we learned as rookies has proven itself over and over again at countless emergencies. We've become experts at it, and it has proven its worth. Like military operations, successful fire attack relies on quick decisions and a coordinated attack. This type of success is not possible through group discussions or team meetings. It is only effective with swift actions and multiple resources. On top of that, we add unparalleled protective equipment that can withstand the harsh environment of fire. Protecting us from our surroundings, we work in harm's way to accomplish our mission. As with the *Titanic,* the reason we are great is the very reason for our demise. We are told to expect the iceberg, but not to fear or question it. We will work alongside the iceberg but feel that it cannot hurt us. The iceberg is simply no match for the engineering and knowledge that humans have developed, and we will be fine. This "no fear" reasoning starts very early in our careers.

Much like icebergs to ships, emergencies pose real hazards to responders. We are expected to reduce the risk as much as is feasible, but our task is to ultimately save lives and property and then stabilize an incident. Some incidents exceed the capabilities of first-arriving units, but safe actions build the foundation for effective actions and, ultimately, a positive incident outcome. Unfortunately, some aspects of our tradition tend to build the foundation for unsafe practices.

THE FOUNDATION OF UNSAFE PRACTICES

New firefighters are trained to take orders from their first day at the academy. They learn to trust their officer's judgment and follow their lead the first time they enter a burning building. Providing them with the latest protective gear on their back and escorted by an officer with years of experience, the sense of security we give them far outweighs their worries of self-preservation. They are taught that what they had previously perceived as fear is both normal and exciting, and they can't wait to impress their peers and instructors. We push them through dangerous scenarios over and over again, and they emerge unscathed each time. As they get more experience, their armor grows thicker and their confidence builds. Sure, we teach them building construction and lightweight trusses and the dangers of collapse in the classroom, but then it's back to the burn building where they are victorious again and nothing falls down. Eventually, they become firefighters. During their career, they learn to pick up on the visual clues that allow them to make decisions for themselves. Unfortunately, history also proves that they pick up unsafe habits that they learned on the job. Years of firefighting without a serious injury build up a false confidence of safety. The bank of files they rely on to make decisions gets skewed by false alarms and near-misses, allowing complacency to set in. In no time, they choose to let their suspenders hang out of their coat, they drive too fast, or they feed a line to the interior without an airpack. If they don't use a component of their **personal protective equipment (PPE)** and come out "accidently unhurt," the scale gets tipped even more. This spiral of bad habits kills firefighters every year.

To make it even worse, the most realistic training we offer is not very realistic at all. **NFPA** 1403 was a great accomplishment to make our training grounds safer. Besides procedures for building and personnel preparation, it limits fuel to the use of Class A materials in burn buildings. Although this helped eliminate the ludicrous behavior of training officers using flammable liquids to start fires, research shows that fires fueled by pallets and hay are incredibly different than the synthetic materials that are found in real structures today. It's not just the increase in fuel load according to the **National Institute for Standards and Testing (NIST)**,[1] but more importantly it's the way the fire reacts and the rate that it releases heat. Flashover can occur more than twice as quickly in a child's bedroom than a burn building at the academy. Training videos from fire tests by NIST prove flashover is capable in as little as 45 seconds after flame. At the burn building, we are able to lie on the floor with a hoseline to feel the heat and watch the fire grow. When it starts to roll over the student's heads, we hit it (but don't put it out; it's too hard to relight without diesel fuel). In addition to differences in fire growth and flashover, visibility in a real building fire is nothing like the burn building. Flooring materials and furnishings are made of different materials, many of which are petroleum based. Ventilation points and changes in ceiling heights and shape make a fire react differently. NFPA 1403 requires walk-throughs with the attack crews prior to lighting so that room layout, exits, and hazards are located. Fires on Main Street offer little evidence of what lies around the corner or at the bottom of the steps. Finally, crews of three or four on an attack line are common at the academy but many crews in the real world are limited to two.

The point here is not that we should consider going back to the old way of setting training fires, but rather realize that our training has shortcomings (Figure 4-1). Without addressing these deficiencies in other ways, we give the recruits a false sense of security and set them up for failure. Training officers should stress to students in this situation that this is not the "real world" and to expect the unexpected in an

FIGURE 4-1
Changes continue to make live fire training safer, but the differences between training fires and real structure fires must be pointed out to students.

uncontrolled fire environment. The solution is not a foreign concept to emergency responders. We can simulate what we can't do, and talk about what we can't simulate. For instance, some responders will someday have to perform endotracheal intubation on a vomiting patient in a bathroom. Some might have to remove victims from a stalled amusement park ride 150' in the air during a heavy wind. These are not situations we may have trained specifically for, but instead are situations we should consider when drilling (Initiative 5). If realistic fire training is impossible, we need to compensate for that in order to perform effectively at the real fire. If flashovers are the problem, use simulators. If visibility is a problem, cover the mask. If reading the fire is the problem, do more size-up training. This is where creativity makes a training program successful, and actually prepares the student for the real world.

FIREFIGHTER EMPOWERMENT

Initiative 4 declares that all employees should be empowered to stop unsafe acts, but the word *empower* by itself is sometimes confusing and is often misused. **Empowerment** can be described as granting permission to subordinates to exceed their normal authority in an effort to better achieve organizational goals. There are a couple problems with taking this textbook definition and actually implementing it. The first is that there has to be a clear line on what is permitted and what isn't. Emergency services empowerment is not insubordination. It's accurate observations and communication intended to keep the crew safe. If your department hired a new fire chief who announced that he believes in empowering his employees, what exactly are you permitted to do?

- Can you stop by a daycare and show the kids a rescue squad?
- Are you allowed to give a disabled motorist a ride home in a fire engine?
- What about using the aerial to fix the pulley on the top of a 75' flagpole at a car dealer?
- Are you permitted to restart somebody's water heater that lost the pilot light?
- Surely you should be able to pull a police car out of the ditch in a snowstorm.

The fact is, the word *empowerment* means you don't have to ask *anyone* for permission, and are expected to speak up when something doesn't seem right. Departments may have specific orders prohibiting specific situations, such as filling swimming pools, but special orders can't cover every possible scenario. As a result, true carte blanche "empowerment" is not very common. We usually are required to follow the chain of command and **standard operating procedures** or **guidelines (SOPs/SOGs)** for tough decisions. A supervisor holding up the empowerment flag with no specifics should raise both concerns and questions on your part.

The second problem with empowerment in emergency services is that although you may be empowered to solve a specific problem, you may not know how to do it safely or have the proper equipment to do it properly. In his book *Creating an "Open Book" Organization*,[2] Thomas J. McCoy uses the model of a triangle to ensure that true empowerment can occur. Success relies on three equal components: education, enablement, and empowerment. Let's tackle the controversial subject of pumping basements for homeowners during a power outage and see what McCoy's model would look like.

Education

The first component of the model is to educate all personnel on the issue. Pumping water from a basement that is flooding actually falls right into the fire service's goal of protecting property. At face value, it is something we should not hesitate to do. To be even more accurate, most people have homeowners' insurance that covers fires but probably doesn't cover floods. This almost creates a duty to act that is even greater than if there was a fire with no life hazard. Additionally, "pumping" is right up our alley, and we are incredibly efficient in moving water. On the flipside, pumping a basement has risks as well. Firstly, firefighting pumps are too valuable to use to pump flood water. We simply can't afford to damage a fire pump by jamming it up with debris. Secondly, pumping one basement might not be too taxing, but if one needs it there are probably more neighbors in the same situation. It may be that there are just too many homes with the same problem for us to help. Besides, storms aren't exactly slow times for emergency responders. Many times there are much more pressing emergencies than a flooded basement. Electric panels are almost always found in the basement, inviting the electrocution of crews if power comes back on while removing water. Finally, a basement filled with water usually means that the footer drains and ground surrounding the basement are also saturated. By removing the water from inside the wall, we may remove the only thing holding the house up. Basements could collapse as a result of being pumped out, which is a liability we simply can't afford.

If we were to make sure our crews had all the information available to make an educated decision, providing knowledge about all aspects of pumping basements would be vital to its success. **Risk-benefit analysis** models would be used to ensure structural considerations were second only to safety concerns. Firefighters would learn how to differentiate between signs of foundation saturation versus a simple submersible pump failure. Finally, proper education would teach "triage" situations about when to assist a homeowner and when and how to decline. If the goal is to improve service to the citizens, misinformation and damage control must be a part of the educational process.

Enablement

The next step is to enable employees to provide the desired service. In the case of flooded basements, trash pumps or battery-operated submersible pumps and hoses must be available as opposed to using fire pumps and equipment that may be needed during a fire. Although the theory of empowerment doesn't rely on policies and procedures for every situation, general policies may be used to enable crews to use the equipment appropriately. For instance, the department may decide to invest in two battery backup submersible pumps for your station. A policy may allow those pumps to be loaned out to a citizen in need, as long as a release and loaner form is signed and attached to an incident report. The policy may be expanded to cover any loaned equipment from pumps to carbon monoxide detectors, or even child safety seats when the owner's has been compromised in a car crash. Say, for example, you respond to a small electrical fire at a convenience store over the weekend. The most serious threat discovered during salvage operations is the lack of electricity for a freezer that is full of frozen food and will surely be ruined by Monday. One of your crew members is an electrician, and determines that the generator on the reserve engine could supply power to the majority of the frozen stock until the building's electricity can be restored or a rental generator can be obtained by the owner. A release and loaner form may be applicable for a portable generator in a situation such as this.

Empowerment

Once personnel are trained and equipped, empowerment can take place. In the case of pumping basements, a fire department could change its general statement from "We don't pump basements or fill swimming pools" to "We'll assist our citizens whenever possible in the preservation of their property at risk." As far as Initiative 4 is concerned, the exact same process would apply. Firefighters need to be empowered to stop unsafe practices. Without being educated about what exactly constitutes an unsafe act or being enabled with tools to stop it, empowerment is a useless claim.

ACCIDENTS

One of the most misused terms in fire and emergency medical services is **accident,** which is sometimes defined as an occurrence with tragic results that could not be predicted or prevented. One example would be when a meteor falls from the sky and strikes you on the head as you walk across the street. However, if the news reporters had warned you of the meteor shower this afternoon and you chose to walk across the street to get a better look, it's not an accident anymore. In fact, there are very few vehicle *accidents*, tragic *accidents*, or even *accidental* fires. They may be unintentional, but usually they are predictable and preventable. In an effort to predict them, we first have to look at how injuries are caused.

Energy Conversion to Injuries

Firefighter injury epidemiology can be described as the study of the causes of injuries, as well as the interventions applied to reduce those injuries to firefighters. The law of conservation of energy states that energy is constant, and that it can neither be created nor destroyed. Energy can, however, be transferred, and presents itself as either **kinetic energy** (at work) or **potential energy** (at rest) and is found in many forms. For example, when you are dispatched to a working detached garage fire, a large amount of energy is already on the scene. As you arrive and give your on-scene report, kinetic energy is obvious to you. You note the smoke rising from the eaves (thermal) and the adjacent exposure starting to off-gas (radiant). You also recognize some potential energy in the drooping overhead electrical lines, but you could miss the chemical and gravitational energy sources inside the building. Although we cannot create energy, we certainly can bring more with us. As you roll up on scene you bring kinetic motion. Aware of the potential motion, the apparatus driver sets the parking brake and chocks the wheels. Pulling lines and forcing doors are both forms of kinetic mechanical energy. Two attack lines are charged, temporarily converting potential hydraulic energy to kinetic. As one line protects the exposure (hydraulic and thermal), another line is deployed through the door and knocks down the fire (thermal). The force of the stream displaces loose drywall and insulation (mechanical) while the cooling effect converts the fire (thermal) to steam (pressure). Ladders are raised with mechanical energy, but the potential of gravitational energy is an always-present danger. Ventilation saws are fired up, converting potential energy to kinetic, in the forms of mechanical, heat, and sound energy (Figure 4-2).

To say that a lot of energy conversion occurs at an emergency scene is a serious understatement. If it cuts something, moves something, makes noise, or stops something, it takes a transfer of energy. Every time energy is transferred, the chance of injury

FIGURE 4-2
Significant energy conversion occurs at emergency scenes. The key is to keep it from converting to injuries.

to a firefighter is present. The reason why is simple. Energy doesn't care where it transfers, and it can be transferred to you via trauma, burns, or even hearing loss. Simple kinetic energy is an obvious hazard. For instance, it doesn't take much common sense for a firefighter to keep his hand off of a rotating chainsaw chain. The hidden danger is the transfer of energy, especially when kinetic energy unleashes a form of potential energy. We all know thermal kinetic energy can weaken a truss system, which contains a whole lot of gravitational potential energy. Although the odds are that energy will eventually be transferred to emergency responders, there are two separate defenses we can employ to either defer an injury or at least reduce its severity. They are safety equipment and operational redundancies, and are both effective if in place.

Safety Equipment

After the explosion and fire on the Deepwater Horizon wellhead in the Gulf of Mexico in 2010, news crews honed in on the environmental disaster starting to unfold. You couldn't help but snicker at the pictures of hazardous materials crews making their way across beaches in their protective suits, boots, and gloves as they weaved between sunbathers in bikinis. Obviously, the pictures were misleading because vacationers were not swimming in the petroleum and were in no immediate danger, while the workers were coming back from an area that was potentially contaminated. Although it makes an interesting photo, it illustrates some basic concepts of the perceptions and limitations of PPE that warrant examination.

PPE Perceptions. Different people perceive things differently. PPE can sometimes give the illusion of absurd dramatization. People might see the hazmat workers on the beach and say things like, "Is it really necessary to have all that gear or is somebody trying to make a political statement?" Others watching from home may have thought it was an overreaction on the part of the spill cleanup crew: "It's only oil. Heck, I know mechanics that have bathed in that stuff for 30 years and they are fine." Remember the first time you heard of a bicycle helmet? Most kids who grew up before 1990 didn't have one, and very few even knew someone who got a head injury from a bicycle wreck. Abrasions to the palms, pinched skin in the chain, and maybe a broken bone were the common injuries. However, just because you may not have grown up with someone who received a debilitating head injury from a bicycle crash doesn't mean it hasn't happened.

New safety equipment is constantly being introduced in all industries. In an effort to make auto racing safer for drivers, NASCAR started encouraging their teams to install a **head and neck support (HANS) device** in all their cars.[3] Some drivers fought the change because they thought the existing harnesses were sufficient to protect them in a crash. Others claimed their head movement was too limited to see the cars around them. Some even made the argument that this "safety" equipment actually made driving less safe. It wasn't until after February 2001 that sweeping changes took place. Dale Earnhardt, a legend of the sport, was killed when he hit a wall at Daytona. Although it didn't appear to be a spectacular crash—the car simply rolled backwards to the infield afterward—he reportedly was killed instantly with a basilar skull fracture from sudden deceleration. As a result, the sanctioning body made the head and neck restraints mandatory, and the voices that had fought the new equipment faded away.

The fire service is no different. When protective hoods were first introduced as a component of the structural firefighting ensemble, many firefighters didn't see the need for them. Helmets were already available with protective flaps to come down and coats had collars to come up. The neck was already protected, making a hood useless. Some firefighters argued that hoods blocked off their "thermostat" inside a fire. "When my ears start to burn, that's how I know to pull back." How about the first time **EMS** was told they had to wear gloves when starting an **IV line**? "I guess we'll just assume they all have **AIDS**. Besides, how's a paramedic supposed to palpate a suitable vein with a mitt on?" The fact remains that fire and emergency responders too have their "bike helmets." What we need to realize is that when a new piece of safety equipment is introduced, it probably has merit and should be given the benefit of the doubt. Rather than initially fight it, simply take a deep breath and answer some questions.

- What is it designed to do?
- What events lead to the changes?
- Will it make me safer in my duties?
- Can I still do my job with it in place?
- Would I want my son or daughter to use it if he or she had my job?

Notice that "Does it look silly?" is not on the list. The fact is that we, too, are critical of new equipment, especially when it comes to looks. We have to remember that PPE is designed as the last line of defense for emergency workers, not to be stylish. We also have to admit that although we fight change when it comes to new PPE, we usually end up realizing that it doesn't take that long to put on a high-visibility traffic vest on the side of the roadway and that we can still start an IV while wearing gloves. Historically, we have adopted some of the best practices from the health-care industry and other nonemergency organizations that either chose to improve their PPE or were forced to by a regulating agency. They also have proved that they are still able to do their job with it on.

PPE Limitations. On the flipside, PPE is only effective if it is used properly. The cleanup crews working on the beach had one hazard to deal with: petroleum. They didn't need thermal layers, hardhats, or dust masks. Firefighters, on the other hand, have several types of hazards to deal with and must use the appropriate PPE based on the specific situation. Pictures from Ground Zero in New York City showed firefighters searching the dusty rubble with filter masks in an attempt to protect their airways. Now it appears as though they weren't adequate for the size of the microns present in the dust, and not sufficient protection for other chemical hazards and heavy metals they were exposed to. As a result, many have lingering respiratory illnesses.[4] Another unknown hazard at the time was the effect of particles with high alkalinity. The pH of most of the dust particles from the **World Trade Center (WTC)** was extremely high due to the amount of masonry materials pulverized. According to the **Center for Disease Control (CDC),** substances with high alkalinity impair the body's ability to clear itself with cilia and mucus. One of the continuing problems in the fire service regarding **SCBA** use is that premature removal during overhaul and fire cause determination/investigation operations can have severe health effects. For a long time, carbon monoxide levels were used to determine the point when airpacks could safely be removed. Today, studies show that numerous chemicals continue to be released for several hours after extinguishment of the synthetic byproducts of a fire and are more hazardous than originally thought.[5]

The last problem with using PPE is not using it correctly. Lieutenant Jeffrey Neal[6] claimed it wasn't his helmet that saved his life, but rather the chin strap that held it to his head. When he fell head-first through the floor at a structural fire, his helmet broke apart but the inside shell and chin strap still protected his head when he landed. He told Channel 12 News that it was common for firefighters to not use the strap, but he was glad he did. This leads to a question: Why wouldn't a firefighter wear a chin strap? An officer from one of the largest departments in the country once said in a safety officer class that his department's largest safety concerns could be narrowed down to two words: seatbelts and chinstraps. It's hard to believe that chinstraps are uncomfortable or that they take too long to put on and adjust. One plausible answer is that we think it looks cool held on with gravity and the strap clipped over the rear brim. The fact is we still look cool with it clipped under our chin.

Operational Redundancies

Although the Deepwater disaster of 2010 took 11 lives and caused one of the worst environmental disasters in history, the Piper Alpha[7] explosion in 1988 off the coast of Scotland was the most deadly. A series of events lead to the explosion and fire that killed 165 workers who could not escape. The main cause of the disaster was operator error: running a hydrocarbon pump that had been taken out of service earlier by another shift. A relief valve had been removed for repairs, and a cover was put in its place. The crew that removed it put the information in the permit-to-work (PTW) log, but did not lock or tag out the controls for the pump. When a different pump failed on the next shift, the operator switched to the out-of-service line. The cover had not been tightened, and a hydrocarbon spray was released that apparently found an ignition source. The system in place had the potential for redundancy, but the process had several holes in it and an "accident chain" ensued.

Psychologist James Reason once compared an accident chain to Swiss cheese.[8] Each slice has holes, or potential flaws that, alone, cause no problem (Figure 4-3). For instance, taking the relief valve out and putting on a blank flange is both an acceptable and necessary process. The illustration goes on to show that an accident chain only occurs because several holes line up when they are stacked next to each other. Operational redundancy would rotate the slices 90 degrees from each other or shuffle them in an effort to reduce the chance of holes lining up. In the case of the Piper Alpha, a tight cover and proper PTW alongside lock-out and tag-out controls for the pump would have ensured that there were enough failsafe measures to prevent a leak and fire. Although there was a catastrophic failure with a fire and explosion, this specific chain of events did not kill the 165 people. There were actually several more holes in the cheese that exacerbated the incident.

FIGURE 4-3
Operational redundancies are an effective way to prevent undesirable events from occurring.

The Piper Alpha was the hub for other offshore rigs. In other words, they not only pumped from the ground to shore, but they also received an oil and natural gas supply from other rigs. When the explosion occurred, others knew there was a problem but continued to pump oil and natural gas to the fire. Procedure called for an order to shut down, which the Piper Alpha was not permitted to give. After feeding the fire for more than a half hour, they finally shut down. In another chunk of cheese, the crew on the fated rig reported to their living quarters and awaited rescue (shelter in place), as their emergency plan called for. No help ever came, and 81 died from smoke inhalation. Records also showed that fire and life boat drills were scarce, with some survivors stating that they didn't even know how to deploy them and instead jumped into the ocean. The Piper Alpha disaster will always be remembered as having a profound effect on safety practices and systems. Operational redundancies are commonplace industry-wide as a direct result of that July 6th in the North Sea. Even though these measures are in place to stop a catastrophic chain of events, it's more important to identify an unsafe act that starts the cascade in the first place.

RECOGNITION OF UNSAFE ACTS

When identifying unsafe acts, they can be categorized into four different types; visible, invisible, poor decisions, and distraction events. Realize that they often overlap and can include underlying (contributing) factors that heighten the chances of injury.

Unsafe Acts in the Spotlight

These unsafe practices are taught in basic training. From "put down your shield" to "don't step over a ladder," we can recite them all. Simple safety infractions are easy to spot on the cover of fire-related trade magazines, and quickly garnish editorials from "safe" readers complaining about how the editors should choose pictures with no infractions. The fact is they don't just happen to fire departments on the cover of magazines, we all do them. We cut with a power saw without safety glasses and we lean on the step of the aerial ladder when it's in operation. We don't wear hearing protection when exercising extrication tools, and we've been known to work from ground ladders without locking in. Do you know someone who answered a cell phone while driving an apparatus? Do you stop at all red lights and stop signs and drive at safe speeds? Too many years of bad habits catch up to us and somebody ends up getting hurt.

So how do we get firefighters to improve on these safety infractions? In Chapter 2 we talked about no-fault management. The practice is vital to achieving a safety culture, but does not pertain to safety infractions in any way. That's because the basis of the theory considers a fire and emergency services employee a professional (whether you're paid or not), who has a sincere desire to mitigate someone else's bad situation. Split-second decisions must be made, and no-fault management stresses that those decisions are not handled best with discipline. For example, if a "professional" lieutenant chooses to leave the thermal imager on the engine during a structural fire because she feels that taking irons and a second line are more important in this specific instance, she would not be disciplined. Supervisors that feel the imager was dictated in that situation would instead address it in a postincident critique and future trainings. If she simply forgot to take in the thermal imager, she needs to be reminded of its importance, and the subject again needs to be brought up in training. If she believes the thermal imager is a waste of money and refuses to use it, or never even learned to use it, she should be disciplined. Because no-fault management really applies to intent, it additionally does not apply to a "poser" who has no desire to make a bad situation better. A poser is someone who pretends to have noble intentions when he or she joins, but is willing to compromise safety by driving maliciously, performing dangerous tasks for no tactical benefit, or refusing to wear protective equipment properly.

Unsafe Acts in Disguise

Besides the obvious unsafe acts, sometimes hidden dangers sneak onto our scene and catch us when we aren't ready. Although the hazard may have shown up somewhere on earth at some point before, word did not get out. Take, for instance, a construction worker running a hydraulic excavator (trackhoe). Large excavating machines have changeable teeth on the buckets to allow digging in different types of soil. Although this particular operator was not aware of the danger, the replaceable teeth eventually work loose and need to be replaced. That is exactly what happened to his machine. As he scraped the bottom of the ditch, the loose tooth scraped a piece of shale, bending the tooth back. The retainer pin acted as a rubber band and catapulted the tooth back in the direction of the cab at a surprisingly high rate of speed. The operator had absolutely no time to react as the 6-lb tooth exploded through the windshield. The tooth continued through the open side door of the cab and landed 35' behind the machine. The operator said that he remembered feeling resistance as the bucket scraped the rock but he never saw the tooth coming. He had a quick flashback to playing softball at third base when a line drive was hit that he couldn't react to, and then realized he was covered in shattered glass. He still wasn't sure what happened until he noticed the tooth missing and found it behind the machine. A string line was used to reconstruct the flight path of the tooth from its resting point back to its original position on the bucket. The tooth missed the operator's head by less than 2", which probably would have killed him instantly. Other construction workers had heard of teeth being thrown but never had remembered one coming through the cab.

What good does keeping this incident silent do for other construction workers? This is what we refer to as a near-miss. Initiative 9 deals with the problems of near-misses, and ways to incorporate those into a format that will create an awareness of the potential problem. The Firefighter Near-Miss Reporting System was created to address this problem, and is covered in depth in Chapter 9. All backhoe operators should be aware of the hazard, and adopt a rule of "no loose teeth." Emergency responders have their own versions of our loose teeth. Pictures of aerial ladders tipped over continue to be found on the Internet. This event in particular deals with unknown factors for several reasons. An aerial device is basically an industrial crane used to lift personnel and water to an elevated location. Construction crane workers go through extensive training before they are certified to operate one. Large fire departments usually have designated operators who have set up the truck hundreds of times after completing an in-house training program. Partially due to **ISO** giving points for an aerial device as well as sprawl to the suburbs, many smaller departments now have them in service. There is no specific certification for a tower ladder operator, and training is many times limited to the salesperson who delivered the apparatus. It's not uncommon for new employees to get trained on correct setup second- or third-hand from shift members. In fact, it's possible for different shifts to have different opinions on the correct setup for the same piece of equipment. The problem with aerial devices with regard to unsafe acts in disguise is that they can be split into two distinct columns. One column contains the list of all the dates your department's ladder has flopped over as a result of an unsafe act. The other is a list of all the dates when your department set up the ladder and an unsafe practice had no effect. The problem is that the list on the right is usually hidden until you have to enter one on the left.

Poor Risk Management Decisions

As discussed at great length in Chapter 3, "Applying Risk Management Techniques," we know we only risk a little to save a little. Statistics show that in reality we don't follow that rule very well at all. All firefighters need to understand **risk management,** and recognize when an unsafe practice is occurring. Although it's been said that you don't get a "time-out" at a fire, the statement is not entirely true. Officers can make a risk assessment, and decide to pull back to take a second look before committing. For example, your crew has just arrived on the scene of a working fire involving the rear axles of a semi trailer on the interstate. It appears the brakes overheated and started the tires on fire, and is seriously threatening the trailer itself. A law enforcement officer tells you he has the driver in his car under suspicion of **DUI,** but that the trailer is empty. As you advance a line to the rear corner of the trailer and hit the fire, you see a white placard on the trailer with the number 9163. It's also clear that the fire has spread to the trailer floorboards. Your strategic choices are threefold:

1. Offensive—Continue to advance on the fire and extinguish it as quickly as possible.
2. Marginal—Stop advancing, knock down the fire from as far away as possible, and look up the UN number.
3. Defensive—Pull everybody back and evacuate the freeway.

Many firefighters would choose the first one. We've been trained to get in there and put the fire out. Even with the limited information available, there is a very good chance that you can do it safely and have no problems. The cop said the truck was empty, and there are no signs that the fire has spread to the cargo area anyway. This incident might make a paragraph in the fire and police section of the newspaper tomorrow, but probably won't be any big deal. On the other hand, you could have a pretty severe hazmat situation on your hands. The word is that the truck was empty, but then again the cops think the driver is drunk. Maybe the driver was so intoxicated that he continued to drag the burning trailer until someone forced him off the road or the axles seized. Regardless, he's probably not the best witness to the situation. White placards are reserved for special hazards that require special attention, and all placards are usually removed when a trailer is empty. What exactly is the risk benefit in this situation? This may turn out to be a front-page story tomorrow.

With a fire actively burning and bystanders in danger, we have to do something. However, it's not time to sacrifice any firefighters to the trailer gods just yet. Pull back and drown it while you research exactly what it is you're dealing with. Have the police pull bystanders and motorists back and maybe even close part of the freeway. Sometimes backing off is the best choice you can make to reevaluate conditions. A wise poker player once said, "You only lose a little by folding on a good hand but you always lose big by staying in on a bad one." Don't be afraid to toss in your cards. If something doesn't feel right, it's probably because it isn't.

Let's take the past situation involving the fire on the interstate and add a very common contributing factor that most reports don't include. You are the lieutenant in charge of the crew, but you arrive on the scene **burdened.** In other words, at the time of this incident you already have a perceived disadvantage. Maybe you are newly promoted and your crew doesn't trust you just yet. It could be that your boss is not happy with a recent decision you made or that you are accused of crying wolf all the time. What if you are the first minority in your department to get promoted and that every decision you make is under a microscope? To say that a burdened decision maker doesn't bring more baggage to the decision-making table is simply putting our head in the sand. In fact, the one thing that is clearly missing from **NIOSH** reports is the underlying psychological factors that affect the decisions that were made. Regardless of the officer training programs completed or certifications held, there still is a human side to all decisions. Unfortunately, there is no easy way to know what people are thinking. Prejudices are present in everyone in every organization to varying degrees. Fifty years ago we didn't have to guess because people who were biased would just blurt out their opinions. Discrimination has become socially and legally unacceptable, and therefore less measurable. That is why they are referred to as "perceived" disadvantages. The disadvantage may be invisible to some in or out of the organization, but to others it explains exactly why they did what they did.

Distraction Events

These unsafe practices are usually pretty obvious to everyone else, but you just don't notice them. This is like the lifeguard at the pool who is tending to a child with a

FIGURE 4-4 Distractions are everywhere on emergency scenes, and all responders must remain vigilant to not becoming complacent.

nose bleed. As you arrive on the scene, she proudly displays her use of **body substance isolation (BSI)** equipment—as she stands barefooted in the blood. For firefighters, we can sometimes get so wrapped up in what we are doing that we don't notice fire behind us or a roof line sagging. Time is usually distorted in emergency situations, and can easily be overlooked. Sometimes the warning signs for dangerous situations such as flashover and backdraft appear to go unnoticed by some due to tunnel vision. Highway incidents are a common location for distracting events. Emergency workers can be so involved in patient care, fire control, or extrication that traffic is simply forgotten. An ambulance might provide a relatively safe barrier, but stepping to the side compartment to retrieve another backboard may put you into a lane and the path of a moving vehicle. Many departments have SOPs on where and how to place apparatus at vehicle rescue scenes in order to protect the working crews. Regardless, we must all be safety officers, staying vigilant in noticing these unsafe practices and to prevent getting "sucked in" (Figure 4-4).

PREVENTING UNSAFE ACTS BEFORE THEY OCCUR

The best time to catch an unsafe act is before it ever happens. One of the most applicable to the fire service was The President's Conference on Fire Prevention, which established an action program in 1947 to combat fire.[9] President Truman identified them as the three Es: education, engineering, and enforcement. Other sources list additional Es, with environment being one of the most common. The Es of prevention can be utilized as a way to identify prevention control measures in an effort to prevent undesirable events, such as motor vehicle crashes. For example, let's say there is a road by the high school that is considered the "back way" route. Many students choose this route because it doesn't go through town, has less traffic, and has fewer police officers on it. The road is not maintained as well as the main roads, and there is a dangerous curve about 2 miles out of town. "Rollover Curve," as it's known, has claimed the lives of several teens over the past 20 years, and your town sets out to see to it that no teenagers ever die there again. The Es of prevention outline some possible solutions.

Education (Initiative 14)

In an effort to reduce crashes, the police and fire departments join forces with the high school. They schedule an annual assembly where they talk to the junior class about the dangers of Rollover Curve and unsafe driving in general. A survivor of a crash addresses the students and proves that it can happen to them, and the difference a seatbelt can make. Driver's education classes in the town begin to use it as a dangerous example, and billboards are posted in an attempt to convince drivers of the dangers while approaching the curve.

Utilizing the Es of prevention is an effective way to look at the options of preventing unsafe acts in the fire service as well. For example, we all know wearing seatbelts in emergency apparatus is vital to safety, yet some still fail to buckle up. Using this model, we can come up with creative ways to encourage emergency responders to wear them. Preaching to firefighters to wear seatbelts is fairly common. Signing the seatbelt pledge is an encouraging step, as are annual emergency vehicle driving classes. One area that could be stressed more is teaching firefighters to pack up while being belted in. Although it's not an easy task, it is possible. This task should be taught at every fire academy and recruit school, and should be timed and checked off. Surprisingly, many schools still teach the over-the-head method and coat method as their only donning education, even though most firefighters put on the SCBA from a seated position. Keeping on the seatbelt and not tangling it with the airpack straps is a vital skill that must be learned. Finally, some officers have learned to keep their apparatus door open until everyone is belted, whereas some won't leave the station without verbal confirmation from everyone. Just like we train to react to situations, we have to train ourselves to buckle up every time.

Engineering (Initiative 8)

We can use math, physics, and technology to reduce the chance of a crash. Computer simulations could solve some of the issues. The curve could be straightened or banked. Trees and telephone poles could be removed from the outside of the curve, while streetlights are added to the inside radius. Rumble strips could be cut into the pavement. On a broader scale, safety advances to vehicles can be made, such as seatbelts, airbags, antilock brakes, and rollover protection.

Common excuses for not wearing the seatbelt are that "I forgot" or "I can't get on the airpack." Although neither are acceptable excuses, they both can be engineered into compliance. Seatbelt warning lights and loud buzzers can be ordered on vehicles, allowing the officer to see who is restrained and who is not. Some companies have designed airpack brackets that clamp the pack tight enough to use the straps as a seatbelt. The clamp is released upon setting the parking brake. This would allow you to use the seatbelt or the airpack, depending on the emergency. Some departments have simply moved the airpacks outside the cab to eliminate any excuses.

Environment (Initiative 1)

Some organizations list environment as a subset of engineering. Although similar, environment involves getting someone to alter his or her actions based on choice, rather than on need (Figure 4-5). Why do the students choose that route home? Changing the environment makes the decision less desirable, thus having fewer drivers subjected to the danger, whether it means having more police officers patrol that route, or building an ice cream stand or "skate park" on a different road. Routing slower moving busses that way is a sure way to discourage teenage drivers from going home that direction. Changing the environment means hedging their choice and making them want to do something different.

Red or orange seatbelts don't hold much tighter or are engineered with greater strength than black ones, but they sure do stand out when you're buckled. Citizens today call the fire chief when they see firefighters smoking or hear them curse, and they also call to report someone that's not buckled. Creating seatbelts that don't get jammed in the door or get mixed up with the airpack straps make them easier to use,

FIGURE 4-5
Greasing the back of a roadway sign is a good example of changing the environment to prevent theft. The solution isn't engineered, but it certainly deters people from stealing it to hang on their wall.

thus increasing their use. As more and more firefighters enter the field, we saturate the ranks with people who never had the option of not using a seatbelt in a vehicle. Firefighters who learned to drive before seatbelts were standard are becoming less common in the station. Lastly, simply utilizing your seatbelt creates an environment of expectation in which others will use theirs.

Enforcement (Initiatives 9, 11)

The speed limit can be reduced on the road, and radar can be used more often. Warnings can be eliminated through zero tolerance, issuing tickets for all violations. Although enforcement is one of the most effective tools used in behavior modification, its uses are sometimes limited and must be carefully implemented in emergency services.

We don't hear about too many firefighters getting a ticket for not wearing a seatbelt. The public probably thinks police officers should get a ticket if they don't use their seatbelt in a police car so why are firefighters any different? Enforcement, however, doesn't have to be instituted from outside the department. Many organizations have taken the courageous route versus the safe one by instituting procedures (SOPs) or guidelines (SOGs) that lay out the punishment for not using seatbelts. The only way these SOPs work is if they really have bite and are enforced. As with other safety procedures, seatbelt compliance should have its own line in annual employee evaluations. This has the reciprocal effect of ensuring that the officer completing the evaluation also uses his or her seatbelt. It's pretty hard to write up someone for something they are guilty of themselves.

STOPPING UNSAFE ACTS IN PROGRESS

Some books categorize unsafe acts according to the location where they are committed. For instance, they are broken down into groups such as responding to emergencies, fire attack, ventilation, and search and rescue. The way you categorize unsafe acts is much less important than having the ability to identify them. Once that is done, firefighters must be empowered to say something. Even more importantly, they must know how to bring up a concern. Unfortunately, these skills are not taught in most basic training classes. So how does a new firefighter politely suggest to his officer that he is concerned about a safety infraction? Depending on the situation, there are many different ways to handle it, all of which require the use of respect.

Plan for the Problem

Assume you were just hired as a firefighter and it's your first day on the job. Although you have fire experience with another department, this particular fire department, station, and your coworkers are all new to you. You are assisting the seasoned and slightly gruff pump operator, Pat, with his daily engine check sheet when he asks you if you can handle starting the portable generator. You respond with a strong "Yes, sir" and make a quick check of the fluids. They look good, so you find the choke and the run switch. You already have on your helmet, but you look in the compartment for the hearing protection. Recruit school pushed the utilization of safety equipment, and after you find none you contemplate your options. Recruits should be trained that this problem will arise several times in the course of their career and that many firefighters like Pat have never learned to identify unsafe conditions in the same way. They may not expect you to question anything as a rookie, and take it as an insult if you raise a concern. Ideally, the question should be brought up prior to an emergency, such as during this apparatus check. As Pat sarcastically says, "Well—it ain't gonna start itself," you politely turn to him and ask if there are any earplugs carried on the engine. This hopefully prompts him to answer by telling you where you can find them. If he responds in any other way, you are now forced to start the generator without hearing protection or start your new career on the wrong foot. Most rookies tend to avoid upsetting the senior man on the first day, so the best answer might be to start the generator. At the end of the day, thank him for his help training you on the equipment and continue to show your desire to learn.

That doesn't mean it's over. The next step is your lieutenant's office later in the day. Be polite and ask your boss if you can talk to him for a minute. Explain how much you enjoyed your first day and that you hope that you live up to the expectations of the shift. Tell him that Pat was a big help, and that when you were running the equipment you couldn't find any hearing protection. Ask him if it's possible to get some for the engine. Hopefully, your lieutenant is a little more on top of things and explains that he will fill out a request for some to be carried with the tools. However, if the lieutenant answers that "you don't need them," it sheds some light on the safety culture of your new fire department. You are now down to a couple more choices. One might be to contact the person in charge of purchasing equipment. The captain that orders generators is probably a good source to provide all the engines with hearing protection when his budget allows. It could be that instructors at the fire academy might have suggestions, as well as the training officer who usually has a vested interest in safe training. Another might be to contact a member of the safety committee if your department has one. NFPA 1500 requires safety committees to meet often to address concerns such as this. Your last option could be to purchase a pair yourself to carry in your gear. Some departments give a clothing allowance for incidentals, and hearing protection is relatively cheap. Many departments were at one time strapped for funds, and it wasn't uncommon for firefighters to purchase their own gear. This practice can still be found today in certain situations, and might apply to other affordable equipment, such as extrication gloves, traffic vests, or safety goggles. If the object in question is too expensive to purchase or too large to carry yourself, you may have to wait until you are able to influence someone with the authority. Eventually you might get promoted or join a purchasing committee yourself to make the changes that are needed.

If only all safety issues were as easy as ear plugs. Chances are you'll run into safety issues that are far more complicated. Personnel issues and attitudes ranging from unsafe driving practices to someone refusing to wear PPE are far more complex. When a situation arises that puts emergency responders or the general public in danger, it must be exposed. Sometimes supervisors are unaware of the extent of the problem. Other times they aren't sure how to handle it, so they instead ignore it. Regardless of the reason why it is occurring or if it is being covered up, empowerment means you are encouraged to speak up. If your department has embraced the 16 initiatives and empowerment, you now have the responsibility to voice your concerns, and are accountable if you don't.

For instance, let's say that you are a firefighter in a part-time department with two firefighters on duty 24 hours a day. The station is located in the middle of town, and was recently renovated. With a nice locker room and dorm accommodations for up to six, it's become a fairly common practice for some firefighters to go out drinking, then "crash at the station." Sometimes it's because they are on shift the next day and don't want to get up early. Other times it's because they don't want to chance driving home. Although it's annoying to some members that are on shift, it hasn't really been a safety concern until this weekend. Early Sunday morning at about 0415, you're on duty when your station gets dispatched to a fully involved shed fire out in the township. You respond with your captain in the pumper and find the building already collapsed. While extinguishing what's left of the fire, the water tender arrives on scene. You're surprised to see the two guys that were at the station "sleeping it off" walk up the driveway. They appear to be sober, but you notice the smell of alcohol on their breath. Your captain jokes with them and tells them they look "pretty rough." While loading up the equipment, you think about what just happened and the way that the captain responded. You are surprised he's going to let them drive back. There are some specific steps you can take to make the problem known.

Speak Directly to Your Supervisor

Whenever you talk to your supervisor about heavy issues like this, remember that "power" is in play. The proper way to bring up a safety concern is to do it without questioning his authority or power. This involves being polite and explaining how the issue affects you or others as opposed to how he's not doing his job. In this situation, it's clear that your supervisor is aware that they have been drinking so there can only be two reasons he doesn't seem to care. First, he may be aware of the problem but unaware of its potential. You might say *"How about I drive the tender back while you drive the pumper. It doesn't look like those two should be driving."* A simple question if they are in any condition to drive the apparatus home may do the trick. If that doesn't work, try a complex question to make him think about it. *"Hey, if they were to wreck on the way back to the station, couldn't we be held partially responsible?"* By expressing genuine concern for the safety of the crew as well as the potential of the situation, most officers would agree to driving them back.

The other reason could be that he's taking the cowardly way out. He might be trying to fit in with the guys by ignoring the issue. He probably realizes it's a bad situation but isn't sure how to address it and figures that if nothing happens, he won't have to deal with it. The one thing that's true about someone like this who avoids confrontation is that he's not willing to use his "power." Power is still in play, but he doesn't want any part of it. Many times in a situation like this, borrowing his power is an effective tactic. Picture him giving you a dumbfounded look when you suggest driving the tender for them, so you take the power. *"You're kidding, right? Do you honestly think they are in any shape to drive that truck? They couldn't even drive their car home from the bar 3 hours ago but they're good to go now? You are responsible if they crash! I know you don't want to, but you gotta step up and take the lead here. Do the right thing before somebody gets hurt."* By making it crystal clear that you expect him to do something, it forces him to make a decision. He may acknowledge you are right and step in. On the other hand, he may yank his power back and tell you to mind your own business. If that's the case, a good response might be *"Okay. I apologize for coming off insubordinate but I'm really scared something is going to happen. I wish you'd reconsider moving me over to the tender, nobody's gotta die today. If you are ordering me to shut up and drive, I will, but under protest."* If it gets this far and he doesn't stop them from driving, there are very few reasonable things you can do. Because he chooses when to use his power and when to bury it, you have a valid complaint to go over his head back at the station. He clearly is okay with throwing you under the bus in favor of his ego, so you have nothing to lose by going to his boss.

Speak Directly to Your Peers

Another option in some situations is to grab the guys at the fire scene and talk to them without the captain present. You can't just ask them if they're okay to drive. You have to start with a surprised expression to make them feel like they underestimated the seriousness of their action, and then offer them a way out. *"What are you guys doing? If you sleep at the station because you've been drinking, you can't go on a call! How about if you let me drive the tender back so nothing happens. We can tell the cap that I offered and you'd feel more comfortable to not drive."* Many times, talking with your coworker is better than going right to the captain. In no way does this mean that you're covering it up. The captain already knows they've been drinking—you're just trying to improve the present situation.

This tactic is very effective when it comes to personal problems that start to affect safety (Initiative 13). Sometimes attitudes about safety can be changed by simply talking. If your career has paralleled someone else's that makes poor safety choices, who better to institute a change than you? Talking about an incident that occurred or a near-miss can open up lines of communication about avoiding it in the future. Tactical or operational discussions can easily lead to safety concerns and agreements about how to handle them in the future.

Politely Request to Complete a Task

If the problem doesn't become evident until an emergency scene, you have fewer options available to you. An incident is not the time for a discussion about safety procedures or refusing orders. For example, your first shift on a heavy rescue turns out to be a busy one, and you get to try out those extrication tools that very evening. The victims are pinned in a vehicle on its right side, and the two patients need to be extricated as quickly as possible. In the orders that were blurted out by the excited captain, the words *cribbing* or *stabilization* never came out. Your training tells you to stabilize first, but it doesn't appear to be in tonight's program. While pulling the tools requested, politely ask permission to quickly place some cribbing. If the request is denied or ignored, you may have to change your tactic. One effective way might be to relate your question to other firefighters working alongside you. If all else fails, another way to get your point across might be to begin the extrication, then stop momentarily and respond *"Sir, this vehicle is very unstable. I'm afraid if I do any more without blocking it we may lose the whole thing."*

It's not about arguing or proving you are right. It's about reducing the risk of an injury that occurs as a result of an unsafe act. It does no good to follow a catastrophe with "I knew that was going to happen but he just wouldn't listen." It doesn't matter what type of incident it is or even where it's at. Assume you respond to a trench rescue in a neighboring town where a basement water-proofer is almost entirely buried under wet clay. You and your fellow responders have been digging for almost an hour, and inserted several panels and air-operated shores to hold back the dirt. You've managed to expose most of the conscious victim, and are lowering the final air shore to hold back the bottom panel when it slips from its knot and falls into the hole. The IC, who is running the scene from a living room window over the victim, becomes impatient and orders you to climb into the hole and recover the last air shore. You realize that the trench is not fully stabilized, and that it is not safe to enter yet.

It's obvious that the IC is trying to cut corners in an effort to get the victim out of the hole, but it's important to not stray from the **incident action plan (IAP)**. Politely explain that you can't get it, but would be more than happy to retrieve a replacement from the trailer on the street. By offering to obtain the air shore yourself, you make it clear that you are not questioning authority or trying to get out of work, you simply don't have the equipment that you need.

Call a Time-Out

Can a firefighter call a time-out? That depends on your department's commitment to safety. At least one department out there uses a time-out they refer to as "Whiskey

Tango Foxtrot." It not only has an official acronym, but also comes with a cool hand signal (the traditional atlas pose as if you are holding a planet up over your head with your palms up). It's very effective when hand signals are the best form of communication. They use this signal when the applied tactic isn't working. Because everyone on scene understands the intention of the time-out, it is never misconstrued as a slam on the **officer in charge (OIC),** or a jab that a crew doesn't know what they are doing. Instead, it uses common communication to say "it's time to reevaluate what we are doing." The tactic can be used between any crews when something doesn't feel right.

Working with Unsafe Acts

Sometimes an unsafe act is recognized, but simply cannot be avoided. When a risk-benefit analysis has been completed and we are ready to risk a lot to save a lot, we may have to actually perform an unsafe practice rather than avoid it. A good example would be a wind-driven fire on the sixth floor of an apartment building. As the initial attack crew tries to advance on the fire in the center hallway, heavy wind forces the fire back into the window of the apartment of origin, out the apartment door, and in the face of the fire attack. Crews are unable to overcome the blast furnace by themselves and rely on other crews to back them up and the fire to lose some of its fuel. FDNY is far too familiar with this situation and has been hit hard in several incidents. As a result, they partnered with NIST and other fire departments to discover safer ways to combat wind-driven fires.[10] The potential for working with an unsafe act should be identified and planned for whenever possible.

Postincident Critique or Debriefing

One of the most important functions to perform when an emergency is over is the postincident critique (PIC). Also referred to as postincident debriefings (PIDs), postincident evaluations (PIEs), and after-action reviews (AARs), these critiques are used primarily to discuss tactical effectiveness. An active safety culture would utilize the safety officer(s) from the scene to assist the incident commander or operations chief with facilitating the PIC/PID/PIE/AAR. If safety concerns arose that were not handled, this would be the correct forum to address them. Again, respect is vital. If crews had no choice but to perform unsafe acts, it's a time to talk about changes that should be made or equipment that should be purchased to address it the next time it comes up. In the case of the research of wind-driven fires, positive-pressure ventilation, wind-control devices, and high-rise nozzles were all introduced as possible tactics to combat future fires. Firefighters and researchers from different ranks and different organizations have been instrumental in identifying and solving problems after the event is over. But the question is whether or not firefighters can be effective at identifying and solving problems *during* the event. To answer the questions, we need only to look "up."

CREW RESOURCE MANAGEMENT

As discussed earlier, identifying and having resources to combat unsafe practices puts us in the position of being empowered to stop one. There is no better example of empowerment in the name of safety than **crew resource management (CRM),** adopted by the aviation industry in response to senseless airline crashes. Much like the organizational structure of the fire service, airline flight crews were molded after the military. This put the captain in charge, and prevented interference by the flight crew. CRM, however, allows and encourages members of a flight crew to make observations and suggestions when a perceived danger is present. This in no way takes authority from the captain. It simply gives the captain more information to make decisions with. As you would expect, it was not well received by most pilots when it became mandatory.

CRM initially began as a "bolt-on" module to flight school in the early 1980s. Let's say you purchase a new pickup truck, and you later decide you want a heavy-duty trailer hitch to pull a trailer. You can have a bolt-on Class III aftermarket hitch attached to your frame that will probably be sufficient for the job. Mechanics then tap into the truck wiring for the trailer harness lights. Bolt-on accessories have some advantages in the short run. If the hitch you picked is not strong enough, a Class IV hitch can be bolted on in its place. If you don't use the hitch anymore, you can take it off and sell it. There are problems with a bolt-on hitch. One is that many times they don't fit just right. The hitch wasn't made specifically for your Ford, and it may have some spacers or extra holes that decrease its strength. The connection where the mechanics cut into your taillights to hook up the trailer plug may eventually corrode and fail. The point here is that a "bolt-on" hitch or philosophy is a great approach to try something new. If it doesn't work, yank it off and throw it in the shed. However, if it is what you want, you're best off ordering it from the factory and keeping it under warranty (Figure 4-6).

CRM followed that exact path. It started as a bolt-on component, but was reviewed and adjusted until it was finally interwoven throughout the training modules (built-in). One of our biggest problems in society is we never have evolved and had safety "installed at the factory." We still treat safety as a bolt-on approach. The last time you bought a lawnmower or chainsaw did you read the owner's manual? If your department recently bought a new piece of portable power equipment, take a second to page through the instruction book. The first portion of the manual is dedicated to safety. There are a lot of *DANGER!* and *WARNING!* notices on the first several pages, but nothing about how to change the blade or what type of oil it takes. If you're like most people, you blow right by the safety information to page 6 for instructions on how to start it or adjust the tension on the chain. What good does safety information do if nobody reads it? The solution is to take it out of the front of the book and mix it with the "important" information. If we did that, an owner's manual for a ventilation chain saw might contain the following steps for starting it:

1. Be sure that the fuel tank and lubrication oil reservoirs are full (Section 1).
2. Check the chain for proper tension (Section 3).

FIGURE 4-6
Built-in protection is a better long-term solution than bolt-on.

3. Don your eye and hearing protection.
4. Push the priming bulb 4–6 times.
5. Ensure the chain break is set by pushing it toward the end of the bar.
6. Turn the switch to "on." You can turn off the machine at any time by flipping "off."
7. Pull the choke out all the way.
8. Pull the handle until the motor "pops."
9. Push the choke halfway in and pull the handle.

CRM is successful because it progressed. In other words, they don't teach someone how to fly, and then add a couple weeks of CRM training. They instead learn how to fly using CRM. The most important point of CRM is it doesn't teach you how to fly an airplane or extinguish a fire. It instead concentrates on the human side of work, how we communicate and interact. It takes safety from "bolt-on" to "built-in."

Although initially not very popular, attitudes about CRM quickly changed after the first air disaster was averted, especially when the captain identified CRM as the real reason they survived. But can CRM be implemented successfully in the fire service? According to a research project by the **International Association of Fire Chiefs (IAFC)**, entitled "Crew Resource Management,"[11] it is not only possible but vital to improve empowerment in the name of safety for the fire service. The report emphasizes "Mission Safety," which requires involvement from all members and respectful communication. Empowerment is at the root of this involvement and communication.

We've already discussed how much influence human decision and mistakes have on safety, and CRM addresses ways to minimize these unsafe practices. There are four different components of human interaction that are applicable to fire and emergency services: (1) policies and procedures, (2) situational awareness, (3) communication, and (4) problem solving.

Policies and Procedures

The first component of CRM applicable to emergency services is the use of policies and procedures. Instituting CRM forces a hard evaluation of past policies regarding the intent of the policy and whether or not it complies with the intent of CRM. Old policies may have to be discarded or upgraded and new policies may need to be drafted to outline the system. SOPs and SOGs are addressed further in Chapter 11, and influence the success of CRM by addressing authority, the use of checklists, and training.

Authority

The first SOPs or SOGs that need to be looked at are those relating to authority. It's important to realize that nowhere in CRM is authority or the chain of command undermined. It instead contains a subtle nuance that the commander is only human and that mistakes are possible and facts can be overlooked. Policies must reinforce the decisions made by commanders, while focus is placed on utilizing every tool available to make an informed decision. Available tools in this situation include the rest of the crew.

Checklists

Flight crews started using checklists to ensure that no steps were missed during demanding periods, such as take-offs and landings. The fire service has adopted their own checklists, usually in the form of tactical worksheets or **accountability** logs. With **NIMS,** NFPA, and **OSHA** pulling different terminology and operations together, more nationally standardized checklists should be developed. Numerous departments and training academies have identified the NFPA 1403 checklist as a significant improvement to the safety of live fire training, ensuring that nothing is overlooked. Flying a plane in Colorado is basically the same as in Alabama. There are some differences, such as weather, altitude, and terrain, but checklists are very similar. It's really hard to say fighting a fire is much different in one state than in another. Water supply is water supply no matter how you get it or how much equipment it takes.

Training

The last policies that must be checked are with regard to training. Like forcible entry, CRM takes some occasional practice. Forcible entry drills require substantial preparation to be worthwhile. Obtaining a building that is set for demolition or door props with different types of locks takes some planning. CRM is no different. Problem solving in a simulation is the best way to accomplish a worthwhile drill. As with most simulations, the more realistic it is, the better the educational experience. The airline industry uses line-oriented flight training, a simulation involving both distracting and serious problems during a simulated flight. Although tabletop incident command training has been used in the fire service, line-oriented fire training is a relatively new concept but one that would improve the effectiveness of CRM in the fire service. Chapter 8 investigates the increasing use of fireground simulators and the importance of being able to hone decision-making skills. Besides applications for CRM, training can also be effective for situational awareness.

Situational Awareness

Another important component of implementing CRM is improving **situational awareness**. Although covered in more detail in Chapter 3, two significant aspects of situational awareness are highlighted in CRM: task saturation and mistakes.

Task Saturation

Statistics show that many airline emergencies occurred during a hectic period in the flight. Additionally, significant problems occurred when the pilot was preoccupied with a minor problem. The fire service is no different, with errors occurring during task-saturated portions of an emergency. Many NIOSH reports of firefighter fatalities make note of the fact that initial crews missed key signs about the building or fire progression that likely contributed to the incident. Line-oriented fire training is an excellent way to role-play the observations and decisions that must be made immediately after arrival.

Mistakes

As a result of task saturation, the second element of situational awareness to address is making a mistake. Firefighters need to expect mistakes and be flexible to change the game plan. Things go wrong all the time, so we need to expect it. Many times it's not a lethal mistake, just one to increase the stress level a little more. Unfortunately, that sets the table for the next mistake, which is probably a little bigger. Say, for instance, that you respond from home as a volunteer firefighter and hop in the officer's seat of the engine headed for a possible barn fire. It's just you on the pumper with Bill driving, so you go over the plan in your head before you get there. Although you're not sure exactly which barn it is, the address puts it in an area where the potential exists for a long lay and a pretty good-sized horse barn. On scene, you have a long curved driveway, with nothing showing from the **A (Alpha)** or **D (Delta) side** of the barn behind the house. You instruct Bill to not lay in, but instead drive back to investigate. As you round the turn, you find the other barn—a 50' by 100' horse barn well involved. The owners are running in and out of the barn pulling horses out one at a time and screaming. You're now 300' back on a 700' lay and Bill stops. You know you screwed up and should have laid in, and are now faced with three choices:

1. Stop and lay the last 400', then call the next truck to reverse lay out to establish a water supply.

2. Stop and hand-hump 300' to the road, then lay the last 400'.
3. Back up the driveway and fix your mistake.

Any of the three are acceptable, but number one or two are probably better choices depending on the type and **estimated time of arrival (ETA)** of the next-due apparatus. Many of us would choose the last one because it's basically a redo. (We didn't lay in, so let's fix it and do it now.) The real problem is mistakes beget mistakes and eventually one is significant enough to cause a serious problem, an injury, or a death. It turns out that Bill has some experience driving, so you try the last option. He's also under some serious stress as he backs out and is probably going faster than he should. You guide him while half-running backwards with an SCBA on and almost slip into his path. He looks in his other mirror as he rounds the turn, and drops the right side duals off the driveway into soft mud. He feels it start to slide off the drive and punches it to try to get back on. By then it's too late and buries it to the axle. It's clear to you now that hand-jacking was a better option. The best decision you can make when things go wrong is always to stop the downward spiral of mistakes and gain control of the situation.

Communication

Although the "C" in CRM stands for *crew*, it might well have stood for *communication*. Everything covered in your policies, training, and situational awareness has brought you to the point where you have to ensure that everybody in your crew is on the same page. Let's say that you are on scene of a working house fire, and have pulled the attack line to the front door. Your partner has the irons and you both kneel on the sidewalk to mask up. Your acting lieutenant returns from his 360 and tells you it's in the rear of the house in the kitchen and he just vented it. The acting officer continues to yell and push as the three of you move into the foyer of the two-story house where visibility is near zero. As you make your way down the hallway adjacent to the stairs, you see a glow ahead around the wall, but a light catches the corner of your eye in the bathroom on your right. You stop your advance long enough to see flames emitting from the heat duct on the floor. The acting officer grabs you and continues to push you toward the kitchen while yelling at the top of his lungs. Your situational awareness tells you the fire is below you, but you apparently are the only one who knows it (Figure 4-7).

Terminology

The first lesson in communications from our friendly skies is to have common phraseology. It does no good to use words that mean different things or sound like different things. Yelling back to your acting boss that "*it's in the heat duct*" doesn't really help much as he's yelling "*GO, GO, GO.*" Besides, your SCBA doesn't do your words justice and the other two aren't sure if you want them to duck under the heat or if you like to eat duck. Either way, your choice of words isn't the best. "*It's in the basement,*" or

FIGURE 4-7
Effective communication is the key to the success or failure of CRM.

"*Back out*" are probably better options. Another example of a terminology choice is how we say yes and no. *Affirmative* and *negative* are good words to use if you can get all the syllables out. If your transmission gets cut off at the beginning, you end up with "ative" when you ask a question, which does nothing for sharing information. Most firefighters use the term "all clear" to report that there is nobody in a room or the structure. Unfortunately, it's also common to hear someone say that they checked the attic for extension and found it "all clear." If "clear" means people, we need to refer to signs of fire as something else. When a pilot says "flaps down," you can bet it has nothing to do with mud flaps, ear flaps, or flapjacks.

Overcommunication

One tactic they teach in flight school is what they call "overcommunication." The fact is that repetition ensures better communication, especially when different words are chosen. By reporting to your crew in the hallway that "*it's in the basement*," consider following that statement with "*It's under us*" and "*Back out*." If the person behind you understands, he or she responds by saying "*Backing out*." When your friend tries to give you directions to a restaurant downtown, you probably repeat them back at least once, and more if there is a discrepancy. It's also important to note that overcommunication does not mean babbling senselessly (BS) on the radio, but, rather, ensuring that important information is relayed effectively.

Another aspect of overcommunication is maintaining open and honest communications. Concerns must be verbalized between members of the crew, especially when there are physical or emotional problems developing. Picture you and your partner completing a **vent, enter, search (VES)** in a residence late at night. Although there are flames from the living room windows, there is a report of people trapped, so you quickly set a ladder to an A (Alpha) side second-floor window. Jason ascends the ladder, dives through the bedroom window, and scurries to the door to close off the smoke and fire spread. You check on him from the window, and notice him crouched on the floor, with the door still open. You yell out and he says he's fine, but he's not moving. You yell out again, and he tells you to give him a minute. This lack of communication is common for some responders when they are injured or scared. Unfortunately, this isn't basketball where he can take a breather or "walk it off." If he banged his knee going through the window, this has to be communicated immediately.

Finally, we need to use this theory during prolonged periods of low communication. Say you are commanding a fire that is winding down. If you haven't heard from the crew in the basement or the radio has been unusually silent, it's probably time to check in with them. You don't have to wait for a predetermined time for a **personnel accountability report (PAR)** check. Sometimes you can see the crew move past a window or notice an occasional spray of water out the back that tells you that they are still in normal operations, but always err on the side of safety if you're not sure. One successful technique is to ask them a question rather than just ask if they're okay. *"Basement, you still have a lot of steam out of C (Charlie) side. Are you gonna need some relief in there?"* With the proper question, you can check up on them while obtaining additional information.

Briefings

Briefings are a common component of flight plans. Basically, the captain talks through the flight sequence from takeoff to landing to ensure that everyone is on the same page and nothing was inadvertently omitted. This same theory has become standard in NFPA 1403 when dealing with live fire training. By talking through the sequence of events ahead of time, another instructor or a student may ask a question about a subject that had been overlooked. It's similar to reading back a letter or report aloud to check for errors. This process includes everyone involved.

Briefings are not something that can be used on every emergency situation, but can be effective especially for later-arriving crews. For instance, you are the lieutenant in charge of a ladder company being dispatched second alarm to a restaurant fire late in the night. Other units on scene have just pulled back to a defensive position and are working on water supplies. A briefing in the vehicle while responding could sound something like this:

> *"Okay, guys—sounds defensive so nobody get in any harm's way. Rich, if we don't get orders for placement, stay uncommitted in the parking lot on the B (Bravo) side. The hydrants are on the other side of the road so we have to plan on the road being blocked by four inch. Craig and Mark, when we stop, you guys hop out and be the ground crew for Rich so we can get it placed as quickly as possible. If we don't get orders right away,*

> *I'll face-to-face command and find out where they want us and where we'll get water. Keep your packs on for now, and check your radios and make sure they're on Tactical 2, but let's keep our radio traffic to a minimum. Any questions?"*

Once you report back to your crew after receiving orders, your briefing will probably resemble a football huddle (without putting your arms around each other). It's going to be a quick plan of attack that everyone understands completely. Briefings can occur in the hot zone as well. Take, for instance, that you and your partner became disoriented in a large machine shop fire with zero visibility. A briefing will calm both of you down, conserve air, and force you to think. There are many discussions about what specific actions should be taken by a firefighter who gets disoriented. Most suggest remaining calm, using your air wisely, calling for help early, activating your **personal alert safety system (PASS),** staying together, and concentrating on finding an exterior wall.

> *"Tony, looks like we got mixed up in this friggin' place. Let's stay calm and get the hell out of here. How much air you got? I'm at 700. First rule—we stay together. We need to find an exterior wall; we'll know it because it's got that foil paper insulation on it. Let's start by following this seam in the concrete floor. If we hit a wall with no insulation, we follow it to an exterior wall. I'm going to call for help now and once it's acknowledged we'll sound our pass alarms. Be ready to muffle it when we hear somebody on the radio so we don't miss anything. Got it? You and me, buddy."*

Challenge and Response

The last lesson for communication involves getting an important message to the boss or your partner. A challenge and response model is the proper way to address concerns you might have regarding a specific order. In the case of you briefing your ladder crew en route to the restaurant fire, it turns out Rich has a different idea. After you ask for questions, he says:

> *"Hey LT, remember that B (Bravo) side parking lot doesn't give very good access to the restaurant for the truck because of the patio. You might want to consider having me come in from the bank lot off Market. There won't be any apparatus or hoselines there and we can hit either side that they want us. What do you think?"*

There are actually two important components to the challenge and response model. As stated earlier in the chapter, this in no way disrupts the chain of command. The lieutenant is still in charge and has to make the ultimate decision. The challenge is polite and nonconfrontational. It simply gives another viewpoint of the situation in a respectful way. The second aspect worth noting is that it's worded in a way to elicit a response. The best way to find out if our option is being considered is through feedback. *"Well, I think we should go through the bank parking lot"* doesn't show respect or solicit a reaction. An ideal challenge and response atmosphere balances assertiveness on behalf of the firefighters and authority exhibited by the officer. It should then lead into a worthwhile discussion. The only way to become proficient at using the challenge and response system is to practice it.

Problem Solving

The final process of CRM that is vital to the empowerment of fire and emergency services is problem solving. According to Neil Krey,[12] problem solving can be completed based on three different methods: skills, rules, and knowledge.

Skills-Based Problem Solving

The first way to solve a problem is considered skills-based. By learning skills the correct way, humans eventually become proficient in the process without thinking about each individual step. You may see a firefighter on the news put up a ladder by herself quickly and efficiently, but the process is actually a series of steps. By carrying the ladder butt first and slightly lower, she reduces the chance of smacking someone in the head and also is in a position to butt it against the base of a building to assist in raising it by herself. She is taught to look up before raising it to identify hazards, then "walk up" the ladder with her hands until it is tight against the wall. By grabbing a rung high and one low, she keeps the tip held tight to the building while she lifts and pulls the butt away from the wall to the proper climbing angle.

Rules-Based Problem Solving

The second problem-solving technique is more advanced and is used often. Rules-based procedures are in the format of an *if/then* statement. After placing the ladder at the proper climbing angle, look up and make sure that the ladder is vertical. If there are windows, chimneys, or other vertical lines on the building, then you can use those to ensure that it's not leaning to the left or right. If it is leaning, then attempt to straighten it by stomping one spur into the ground and recheck it. If the ladder is on hard pavement, then move the ladder to a more level location. If the ladder must be set up on uneven ground, then consider blocking one side with cribbing or other suitable materials. If you choose to block up one beam, then make sure the ladder is sufficiently butted or tied off to prevent it from slipping.

Knowledge-Based Problem Solving

Knowledge-based problem solving is the most advanced type used. Taking every skill and rule we learned, we "misuse" equipment to complete a task. Let's assume you are an airport firefighter and have been dispatched to stand by for an incoming small aircraft (10 passengers) with a landing gear problem. As the plane lands, the front gear buckles underneath and it skids off the end of the runway into a gully. There is no fire, but other units lay down a layer of foam as a precaution while you and your partner, Christian, are assigned to help evacuate the aircraft with ground ladders.

Setting up ladders against the side of a round fuselage on uneven ground is a little tricky. You discover that the tip of the extension ladder lays nicely on the bottom of the escape door but that the angle is much lower than a proper climbing angle and will be difficult to descend. Additionally, the position of the plane prevents you from putting the spurs level on the ground and it rocks with weight on it. Three feet back, there is a chain-link fence and you get an idea. You extend the ladder two clicks and jam the spurs into the bottom links of the fence. The ladder is still at a bad angle, but is now stable. As an extra step for safety, Christian uses his carabineer to lock the truss beam to the fence. You decide that although it's a bad angle, weight shouldn't be an issue because the extension ladder is still collapsed most of the way. Having passengers climb down the ladder, however, is another issue. You consider using a Stokes basket to ferry them down, but decide it will take too much time. Instead, Christian climbs the ladder carrying a second extension ladder, places it on its beam, and ties it off to the door hinge of the plane. You extend the base into the fence and tie it off on your end. You now have created a handrail that can be used to make the escape easier and safer. You never learned these methods in school, but instead used your knowledge of stabilizing a ground ladder to safely perform an operation.

SUMMARY

Although empowerment is a great way to improve firefighter safety, it must be addressed early on in a firefighters' career. Teaching recruits how to identify and address unsafe acts responsibly is vital. Current firefighters must learn how to accept the application of this empowerment and not misinterpret it as insubordination. Emergency scenes and training grounds are dynamic environments that can easily cause injury or death. Vigilance with empowerment is about stopping unsafe acts or poor practices that would lead to injuries or death; therefore, it is important to remain vigilant to stop them, which then can translate into no injuries.

Two of the most promising ways to stop unsafe acts are prevention methods and crew resource management (CRM). Prevention of unsafe acts can utilize any combination of education, engineering, environment, and enforcement. CRM, on the other hand, uses all personnel to identify and stop unsafe acts as they occur. By using a respectful system of challenge and response, firefighters are empowered to stop unsafe acts and take corrective action as needed regardless of their rank. This also ensures that safety is "built-in" rather than "bolt-on." A system that incorporates both prevention and CRM has the best chance of using empowerment for the elimination of injuries.

KEY TERMS

A (Alpha) side - The front of a structure.

accident - An event that is not predictable, preventable, or avoidable.

accountability - A process of tracking the location of firefighters.

AIDS - Acquired immune deficiency syndrome.

BSI - Body substance isolation.

burdened - A perceived disadvantage of a person which, under certain circumstances, could encourage the person to make poor risk management decisions in an effort to try to prove themselves.

CDC - Center for Disease Control.

CRM - Crew resource management.

D (Delta) side - The right side of a structure.

DUI - Driving under the influence of drugs or alcohol.

empowerment - Granting permission to subordinates to exceed their normal authority in an effort to better achieve organizational goals.

EMS - Emergency Medical Services.

ETA - Estimated time of arrival.

head and neck (HANS) device - A device used by race car drivers.

IAP - Incident action plan.

Insurance Services Office (ISO) - A fire industry insurance organization.

International Association of Fire Chiefs (IAFC) - Organization of fire chiefs from the United States and Canada.

IV - Indicates an intravenous line.

kinetic energy - Energy at work, as in moving, or expanding.

National Institute of Standards and Technology (NIST) - A government laboratory with a fire research division committed to the behavior and control of fire, and providing valuable information to the fire service.

NFPA - National Fire Protection Association.

NIMS - National Incident Management System.

NIOSH - National Institute for Occupational Safety and Health.

OIC - Officer in charge.

OSHA - Occupational Safety and Health Administration.

PASS - Personal alert safety system.

personal protective equipment (PPE) - A generic term used to describe the minimum apparel and gear needed to safely perform a specific duty.

personnel accountability report (PAR) - A verbal or visual report to incident command or to the accountability officer regarding the status of operating crews. It should occur at specific time intervals or after certain tasks have been completed.

potential energy - Energy at rest.

risk-benefit analysis - The weighing of the facts; determining the advantages and disadvantages of a certain activity.

risk management - Identification and analysis of exposure to hazards, selection of appropriate risk management techniques to handle exposures, implementation of chosen techniques, and monitoring of results, with respect to the health and safety of members.

SCBA - Self-contained breathing apparatus.

situational awareness - A term used to describe the recognition of an individual's location, the surrounding atmosphere, the equipment being utilized, and the evolution of an incident.

SOG - Standard operating guideline.

SOP - Standard operating procedure.

VES - Vent, enter, search.

WTC - World Trade Center.

REVIEW QUESTIONS

1. What constitutes an unsafe act?
2. Name some examples of energy conversion and how they cause injuries.
3. What are the four categories of unsafe acts? List examples for each category.
4. What is the difference between "bolt-on" safety and "built-in" safety?
5. What are the four components of human interaction necessary to institute crew resource management?

STUDENT ACTIVITY

For each of the following emergency scene tasks, list three unsafe acts that could occur, and the safest way to complete the task. Follow the example below as a guide.

Example: You are ordered to retrieve and start a gas-powered PPV fan on the porch.

- Lifting and carrying the fan alone; obtain help
- Placing the fan in the path of firefighters; ensure that the fan doesn't obscure exits
- Introducing oxygen to the fire; ensure that command is ready and you are set up at the correct location before starting the fan

1. You are performing overhaul operations in a second-floor bedroom of a single-family home.
2. You need to pull a preconnected attack line to the front door of a working detached garage fire.
3. You must drive the engine to a dumpster fire in a snowstorm.
4. You are drilling with the aerial ladder set up to the fourth-floor window of a drill tower.
5. Your company has a desire to play a prank on the rookie.
6. You begin a new strength-building workout routine.
7. You are preparing to begin the annual hydrant testing program for your department.
8. You respond to the jail as a first responder for an inmate in respiratory arrest.
9. You are ordered to use a ground ladder to access a chimney fire.
10. You are in the jumpseat responding to an automatic fire alarm where false alarms are common.

FIREFIGHTING WEBSITE RESOURCES

http://everyonegoeshome.com
http://cdc.gov/niosh/fire
http://www.hazmat-news.com
http://www.planecrashinfo.com
http://www.osha.gov/pls/oshaweb/owadisp.show_document?p_table=standards&p_id=12716

NOTES

1. Bukowski, R.W. 2006. *Determining design fires for design-level and extreme events.* NIST Building and Fire Research Laboratory.
2. McCoy, T. 1996. *Creating an "open book" organization: Where employees think & act like business partners.* New York: AMACOM; a division of the American Management Association.
3. St. John, A. 2008, February. Anatomy of a NASCAR crash; A bad day at the office. *Popular Mechanics.*
4. http://www.cdc.gov/niosh/blog/nsb090808_wtc.html
5. Firefighter safety during overhaul at the Manhattan Fire Department. EFO NFA.
6. Firefighter Jeff Neal, Cincinnati FD television interview with Channel 12, October 2008.
7. UK, *The public inquiry into the Piper Alpha disaster* (the Lord Cullen report).
8. Reason, J. 2000. Human error: Models and management. *BMJ* 320:768–70 (doi: 10.1136/bmj.320.7237.768).
9. The President's Conference on Fire Prevention. Federal Works Building, Washington, D.C., May 13, 1948.
10. Kerber, S.I. and D. Madrzykowski. 2009, April. Fire fighting tactics under wind-driven fire conditions: 7-story building experiments (58118 K). NIST TN 1629; NIST Technical Note 1629, p. 593.
11. Crew resource management; a positive change for the fire service. 2003. International Association of Fire Chiefs.
12. Neil Krey's CRM Developers Forum. http:/www.crm-devel.org

CHAPTER 5

Implementing Training and Certification Standards

LEARNING OBJECTIVES

- Explain national standards for training, qualifications, and certification.
- Describe the purpose and advantages of credentialing for emergency responders.
- Identify ways to use cognitive learning skills and apply them to an effective psychomotor lesson plan.
- Explain the advantages of the professional qualification standards with regard to safety.
- List the components of a job performance requirement (JPR).
- Describe some of the differences in training requirements between volunteer firefighters and career firefighters, as well as state to state, and the effects on safety.
- Describe how the methods of training should be adjusted for the various risk scenarios you may encounter.
- Describe how the FESHE model combines training and education into one complete package.
- Explain the advantages of a tiered system of certification and how it could improve safety.

CHAPTER 5
IMPLEMENTING TRAINING AND CERTIFICATION STANDARDS

5. Develop and implement national standards for training, qualifications, and certification (including regular recertification) that are equally applicable to all firefighters based on the duties they are expected to perform.

Are you familiar with the process used to hire career firefighters these days? It could be you're taking tests and trying to get your fire science or comparable degree to look just a little bit better than the next applicant. Maybe you've already been on the job for 20 years and see the kids coming in today, wondering if you'd even get hired by today's standards. Or it could be you are one of the lucky ones trying to sift through applications and test results, trying to pick out a diamond in the rough. Regardless of where you are on the ladder, it's obvious that it's not anything like it was in the old days. It used to be that the kid walking in the door looking for a helmet grew up in town but couldn't identify a booster from a boulder. These days, many of them have all their certifications and maybe a college degree. There's a good chance some of them have more **education** than you. They may even come from out of state and stroll in, looking for a career. Society is on the move. Some of our past coworkers or fellow students from school have moved out of state for work in fire and **EMS**. It's actually fairly common for people to move out of state. According to the U.S. Census Bureau, in a 12-month period in 2007 and 2008, an estimated 6.4 million people relocated to a different state.[1]

OPPOSING CONCEPTS

There is a significant impact on emergency services because so many people tend to relocate. Problems sometimes arise when an emergency responder tries to obtain a job in another region. Besides the fact that many **training** certifications are not recognized by other states and different classifications exist, the general public may have expectations of us that we simply cannot meet. For example, if we take a broad look at emergency medical care across the nation, we see two sets of guidelines directly related to the service or care we provide. Although they are not normally compared to each other, they could be viewed to the end user as two opposing approaches to the same need.

Standard of Care

If your parents were two of those who chose to move out of state, you probably know what fire station and what EMS agency protects their house. Fire and emergency service responders are unique in that respect. We want to know what the local organization has to offer, and how well they are going to handle an emergency. For instance, if you get the call that your dad had a heart attack and was rushed to the hospital, it's reassuring to know what kind of care he is getting before you can fly in. As far as EMS is concerned, **standard of care** is defined as the care a person would expect to receive in a similar medical crisis in a similar area. Standard of care differs from region to region across the country. It's not uncommon for an ambulance in some areas to do everything that an emergency room would do for a heart attack in the first 15 minutes. An **S-T elevated myocardial infarction (STEMI)** should be identified and treated the same, whether you live in Montana or Maine. Some ambulances can differentiate between a left- and a right-sided heart attack and administer totally different medications based on its anatomical location, whereas some just drive to the hospital (Figure 5-1).

You may have had a citizen walk into your fire station or EMS office to let you know that his child was on some sort of advanced medical device, such as a ventilator or heart monitor. If so, chances are that that person was probably a firefighter, an **emergency medical technician (EMT),** or other medical professional and wanted you to have as much information as possible if there was an emergency. (He probably was also evaluating your standard of care to see what he's in for if he has to call 911.)

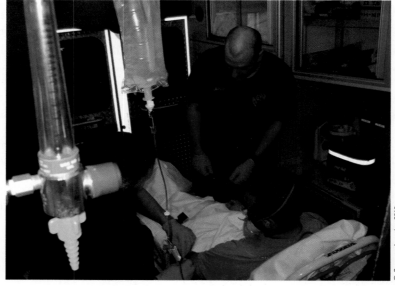

FIGURE 5-1
Standard of care can vary based on your geographic location.

Sometimes the individual wants to know how many firefighters are on duty or if there are paramedics on every ambulance. It could be simply to see if it looks like you know what you're talking about. Although we hate to see it happen and hope it's never said about us, you may actually know somebody in the business who told his or her family to drive themselves to the neighboring community if they need an ambulance. It's a sad state of affairs to see it happen, but it makes you wonder why that is. We have to agree that *what* happens is just as important as *where* it happens. That leads us to the other theory.

Chain of Survival

A casino happens to be a great place to go into cardiac arrest. The reason is because they are very good at the **chain of survival**. The American Heart Association[2] says you need five things to go right to have the best chance to survive sudden cardiac arrest:

1. *Immediate recognition of cardiac arrest and activation of the emergency response system*—Someone has to identify the emergency and be able to call 911. Cameras are everywhere in casinos and they have trained emergency responders ready to roll.
2. *Early cardiopulmonary resuscitation (CPR)*—Someone has to know CPR and be willing to do it. The responders do a phenomenal job at circulating blood in a person with no pulse for the best chance of survival.
3. *Rapid defibrillation*—Someone has to have a defibrillator handy to shock the heart, with the best results in less than three to five minutes. All ambulances can't meet that timeframe, but **automated external defibrillators (AEDs)** hanging on a wall can. Most casinos have them, as do several airports, libraries, and other public places.
4. *Effective advanced life support*—Paramedics are equipped and trained to handle cardiac arrest patients using the latest techniques and advancements to improve the chances of a successful resuscitation.
5. *Integrated postcardiac arrest care*—Getting them to advanced care as soon as possible means quick 911 response and transport to an appropriate facility. Security personnel make that 911 call as soon as they see someone drop.

What's interesting is that the chain of survival doesn't take into account if you're on a cruise ship or a ski trip. It still has to be in place when you need it. Standard of care, on the other hand, is entirely based on geographical location. In a perfect world, the chain of survival and standard of care would be the same for your parents whether they lived in Albuquerque or Amarillo. What about firefighting? Surely, a volunteer firefighter in South Carolina is trained and certified the same as one in South Dakota. Not necessarily. What's surprising is the vast distance in both training and certification for firefighters in the United States.

HOW WE GOT HERE

The root cause of the differences probably goes back to the mid-1800s. Most historians agree that growing pains led to this country's struggle for power, which eventually led to the Civil War. Although most people in the North believed the Civil War was about slavery, much of the South felt it was predominantly to defend states' rights,[3] to be able to make decisions about the way they lived. When President Lincoln tried to pull the states back under one law, states in the South chose to secede from the Union and create the Confederate States of America. This led to the deadliest war for Americans in the history of our great country.

Unfortunately, struggles between states and the federal government continue even today. Speed limits, the death penalty, and gay marriages are just a few of the most heated debates. Many states, such as New Hampshire,[4] have recently passed "state sovereignty resolutions" protesting the federal government's intrusion into state issues with regard to the 10th Amendment to the Constitution. Historically, the carrot the federal government hangs over the states' heads is tax dollars. Highway funds are sometimes withheld (or at least threatened) from states that have differing opinions on maximum speed limits. These present-day political issues directly affect fire and emergency services. Federal grant money for the fire service is dependent on following federal rules and regulations, such as **National Incident Management System (NIMS)** compliance and **National Fire Incident Reporting System (NFIRS)** reporting. Like the state governments, some emergency responders are unhappy with the apparent increasing involvement of the federal government. Others look at it as an opportunity to standardize what we do and how we do it. No one wants more red tape or government "meddling" in their business. However, more cooperation by emergency services on a national level has the potential to improve service, opportunities, and the safety of emergency responders. This is especially true when we have direct input in the standards themselves.

If emergency incidents would be considerate enough to contain themselves within specific municipal, county, or state lines, the system that we have in place for training and certification would be adequate. Unfortunately, history has proven that annexation, response areas, contracts, mutual aid, fire districts, and regional disaster responses blur the geographical lines, if not completely erase them. It would be nice for you to know that when a truckload of emergency responders show up at your next hurricane, ice storm, tornado, earthquake, or terrorist attack with uniforms, tools, and a can-do attitude, they actually have the training and certifications to back up the image that they are projecting.

If we sit down and compare the initial fire training and certification processes of all 50 states, we see a difference as vast as the political opinions of 1863. This time, however, it's not based on the Mason-Dixon line. Differences in both requirements and terminology make direct comparisons extremely difficult. Additionally, states make changes in their requirements fairly often based on decisions by their governing bodies and on suggestions by their constituents. Once firefighters have completed initial training, department drills vary from department to department with no set standard for hours or content. Some departments follow a specific curriculum to ensure that their members are improving their knowledge and skills, whereas others just meet. Mix in the variables of numerous higher-education degrees with different requirements and it's difficult to find two firefighters in two different states who are "equivalent." As a result, complete reciprocity between states is rare. It should be obvious that there is a problem with the system in place. There should be standard responses to emergencies (Initiative 11) and they should be handled by similarly trained professionals (Initiative 5). Our friends, family, and neighbors deserve to receive competent care and attention when they need it, regardless of who hops off the truck.

EVALUATING WHERE WE ARE

Many leaders in fire and emergency services have recognized this deficiency and are working to bring certifications and training to common ground. Organizations such as **Fire and Emergency Services Higher Education (FESHE)**, **International Fire Service Accreditation Congress (IFSAC)**, and the **National Board on Fire Service Professional Qualifications (NBFSPQ, or ProBoard)** have been formed for this specific purpose. IFSAC and ProBoard are working toward the goal of standardizing both accreditation and certification, whereas FESHE promotes higher education to enhance the recognition of fire and emergency services as a profession to reduce the loss of life and property from fire and other hazards. These organizations help provide components of what is becoming a national framework for education and certification. The first step to evaluating where we are is to look at where we start—the fire academy.

Initial Training Requirements

The first piece of common ground is basic training. Although the training is not always the same, almost every emergency responder passes through this turnstile. If the **Transportation Security Administration (TSA)** wants to keep airline passengers from bringing illegal weapons onto an airplane, it's best achieved at a point in the airport where everyone has to pass to get on their plane. If basic training requirements are the same nationwide, we are already well on our way to our goal.

How We Teach

Many educational processes are based on learning domains as a result of research by Benjamin Bloom and his associates. Firefighter training is no different and has followed Bloom's taxonomy[5] for several years. The theory is that people learn through the following domains: cognitive (memorization), psychomotor (physically doing), and affective (feelings and values). For instance, we concentrate on the first two domains when we teach pump operations. The classroom portion teaches the difference between the volute and the impeller (cognitive), whereas the relay pumping practice of maintaining a minimum intake pressure (psychomotor) teaches the skill. Based on how well we teach the cognitive and **psychomotor domains** coupled with the student's interest in the subject, the **affective domain** can follow. This balance ensures the best educational experience based on the instructional design and delivery methodologies matched to the learning styles of the student. In 2001, Anderson and Krathwohl[6] "modernized" Bloom's theory with regard to the cognitive aspect and defined six specific levels of learning: remembering, understanding, applying, analyzing, evaluating, and creating. Although intended to target the cognitive portion, they make a great outline for developing the lesson plans for psychomotor skills. Depending on how the student and instructor interact, there can be some affective domain components as well. These are most likely to take place at the last three levels. The levels use key action verbs to better distinguish which specific skill is being endorsed. Using pump operations as an example, this is what a complete lesson plan might look like.

Remembering. The most basic of learning, *remembering* includes *labeling* the parts of a pump and *stating* each of their functions in the operation. Auxiliary components of the pump must also be *classified*, such as the primer pump and pump shift device. *Identification* of key terms pertaining to pump operations (cavitation) and hydraulics (residual pressure) is also vital.

Understanding. Understanding the process of drafting enables the student to *describe* the theory of removing air from the suction line to lower the pressure in the hose below atmospheric pressure, which in turn fills the suction hose and pump cavity. The student *understands* that centrifugal pumps are not able to "suck" water or pump air, and *comprehends* that no pumping can occur until the air is removed. He or she is trained to *be aware of* the dangers to personnel while pumping water near a roadway, to *identify* safety equipment that should be utilized, and to *recognize* unsafe procedures.

Applying. *Applying* the skill means not only *operating* the pump and *utilizing* a continuous supply from a static water supply in a specific amount of time, but mathematically *proving* the maximum lift of a centrifugal pump and *relating* the same principles to other hydraulic calculations. The student must be able to *solve* friction loss calculations, *complete* water flow tables, and *demonstrate* relay pumping (Figure 5-2).

Analyzing. *Analyzing* means *reviewing* the water flow operations to *ensure* sufficient supply, as well as *scrutinizing* each job function to reduce the chance of an injury. It could include *diagnosing* failures, such as air leaks on the suction side, and *problem solving* as in a plugged intake screen. The student should be able to *stand back and look at* the operation objectively, *distinguishing* between effective and ineffective practices to *evaluate* performance.

FIGURE 5-2
All emergency responders should be trained to the application level as a minimum.

Evaluating. After analyzing the set up, *evaluation* includes *making choices* for improvement, *selecting* the best course of action, and *supporting* the operation. This might be *assessing* the need for another draft engine or *estimating* the value of a parallel supply line from the supply engine. *Decisions* are made to increase efficiency and provide adequate gallons per minute.

Creating. Finally, *creating* or *synthesizing* predicts future events and *generates* new ideas and systems. *Establishing* safer ways of completing a task should always be a priority when *building* a system. The student must be able to look at a problem *innovatively*, such as an overheating motor on the pumper, and *make* a change to keep it in operation for the duration of the emergency. This could be by opening a bypass line, *constructing* a way to increase airflow with a PPV fan, or even *design* a way to use a handline to cool the radiator. This may include a review of the overall process and *establishing* a new and better way of drafting in the future. For example, maybe a great drafting source is identified and a *determination is made* that a dry hydrant installation would make it even better. Creating is very seldom taught, but has a profound effect on future operations and safety.

What We Teach

One common feature of the fire service for initial training is the adoption of **NFPA** 1001, Standard on Firefighter Professional Qualifications, Firefighter I, and Firefighter II. NFPA standards are internationally recognized and adopted, readily available, and reviewed regularly. There are numerous levels of certification offered, created through what they refer to as an "open consensus-based" method, which in turn makes them commonly adopted. They are by far the most commonly used standards by the fire service. The **professional qualification** series of standards are written as **job performance requirements (JPRs)** that describe the performance required to accomplish a specific duty. Duties are similar to a list of broad job requirements. For example, a firefighter may have a duty of fire attack, whereas JPRs explain each task the firefighter would have to be proficient at to perform a successful fire attack. This standardization of the job performance requirements ensures that students learn both the knowledge and skills necessary for the job.

Another advantage of the professional qualifications standards is the different levels of certification within the standard. For example, NFPA 1021, Standard for Fire Officer Professional Qualifications, has four different levels. Although the tiered system has some common duties in each level, such as human resource management, they are covered much more in depth for a Fire Officer IV class than a Fire Officer I class. Some duties, however, are exclusive to a particular level, such as emergency management to Fire Officer III. As a result, the standards have a direct effect on educational materials. Publishers of fire training materials use the NFPA standards to develop texts, curriculums, and supplemental training aids they produce for the fire service.

Knowledge. All training programs start with the knowledge component. For instance, NFPA 1001 Level I[7] requires that specific aspects of forcible entry be taught in regard to the construction and operation of various types of doors and windows. Students learn the normal operation of a properly operating door or window, and the associated hazards with forcing entry. Weaker components are pointed out to ensure that when the student is ready to begin the skills portion, decisions can be made so that the task is more successful. The knowledge component is also known as the **cognitive domain**.

Skills. The skills associated with forcible entry are derived from the JPRs outlined in the NFPA standards. This includes components such as the identification, care, and use of various hand and power tools. Additionally, proper techniques for forcing entry through doors and windows are taught. JPRs usually consist of three parts:
1. A specific task to be completed
2. The equipment and other materials needed to complete the task
3. An assessment parameter used to ensure competency and consistency

This combination of JPRs, equipment, and checklists makes certain that firefighters who are being evaluated are capable of the psychomotor skills needed to force entry into most buildings.

Resources. The final advantage of teaching to a common standard is that the resources generally don't contradict each other. Regardless of the publisher or creator, information is valid as long as the standard is adhered to. Variations in technique and emphasis are common, and even encouraged. For instance, one book might explain several ways of securing the handle of a door before forcing to prevent it from swinging open too fast, whereas another spends a significant amount of time on "through-the-lock" techniques. Perhaps an in-house produced video may concentrate predominantly on a unique type of roll-up door common to the department's community. As long as the resources build on the knowledge and skills of the standard, they are all acceptable.

Certification Requirements

For the purposes of this text, certification means that an individual has been tested by an accredited examining body on clearly identified material and found to meet a minimum standard. By meeting a minimum standard, it ensures that a curriculum can be developed that puts an emphasis on completing tasks using safe practices.

Career Firefighter Basic Training

Career firefighters are normally certified, but the time requirements vary greatly from 200 hours to well over 400 hours. This is predominantly because many add classes to the NFPA 1001 training. Some supplement basic training with hazmat, NIMS, or other technical training. Additionally, large departments sometimes have their own training academy for candidates, whether they are already certified in the state or not. This leads to the problem of reciprocity. Firefighters who wish to relocate can have a very difficult time transferring certifications from one state to another because they simply aren't the same.

Volunteer Firefighter Basic Training

As far as volunteer firefighters are concerned, some states *certify* them with a testing process, whereas others *register* them for participating in a basic-level training class. These classes can range from less than 40 hours in length to well over 100 hours. One state may have a mandatory live burn and another may not even require putting on an airpack. Some chiefs feel that if the initial training is too much of a burden, nobody would volunteer. Other chiefs require the exact same certification for a volunteer that a career member would get because they respond to the same emergencies. Once again, what seems to be missing is a balance.

Let's say your 18-year-old younger brother gets hired by a fast-food chain, working evenings and weekends. He starts his training for bussing tables and mopping the bathroom floor, then progresses to stocking supplies and working a grill. If he does well, he can move up to drive-through and working the counter. If he is committed and learns fast, he could someday become a manager. The question is, how many hours of training does he need before he becomes proficient at his job? Can he be fully trained in one week or ten weeks? How long until he can be a manager and can be responsible for locking up the restaurant or making decisions? How about answering questions on the fly for a health department inspection or customer complaint? It's safe to say that whatever that number, they will prepare him for his job.

Consider instead that your brother joins the local volunteer fire department. He's excited to be the first in your family to serve and has no experience at all. When he enrolls in school, he finds out that the state of Ohio only requires volunteers to get 36 hours[8] of training to be certified. He finishes his class, passes the test, and comes home with a card. One Saturday afternoon he gets a call on his pager for a motor vehicle crash and responds from home. In the next 17 minutes, he will have to make several decisions:

1. Should I drive with lights and siren on or with the flow of traffic without them?
2. Should I go directly to the scene or to the station?
3. Should I put on my gear or take it with me?
4. Which truck should I take?
5. Should I drive the engine by myself or should I wait for someone else?
6. Which route is the best?
7. Where should I park?
8. Am I providing patient care, fire control, hazmat identification, extrication, or traffic control?
9. Should I wear **body substance isolation (BSI),** turnout gear, or a traffic vest?
10. Am I proficient enough to run the pump if there's a fire?
11. Do I know how to safely cut someone out of a hybrid car?
12. Am I in over my head?

Your brother has a lot of decisions with a simple car crash in a matter of 17 minutes. The scary part is that every decision directly affects his life or somebody else's. It's not uncommon for some volunteers to respond by themselves and many times they are forced to make decisions that affect people's lives. Hopefully, the training that he has and the situations he runs into work out well as he continues to learn to incorporate safe practices over the years. The lack of a balance is proven because the burger joint probably spent more time training your brother than we did. The most he had to worry about at the fast-food restaurant was somebody slipping on the wet floor or declaring that they ordered extra pickles and didn't get them. The most he had to face at the wreck was somebody (maybe himself) dying (Figure 5-3).

The question is, how long does it take to train a firefighter to know the basics of what a firefighter should know? New employees in almost any career need more than just a couple weeks of training to be able to learn enough about the operations to make educated decisions. The **National Highway Transportation Safety Administration (NHTSA)** trains firefighters and EMTs to become car seat technicians as a public service. Just a course for installing car seats is 32 hours.[9] The federal government runs a program for **on-the-job training (OJT)** through the Department of Labor that

FIGURE 5-3 Emergency responders can be forced into difficult decisions on their first incident.

will support career retraining for up to six months.[10] Although fire training programs in different states have vast variations in hours, initial training programs have set guidelines to ensure that there is adequate time to meet all course objectives. Many states feel that certification is important for both volunteer and career firefighters.

You could make a convincing argument that if firefighters are being trained to NFPA standards, they should be tested on them. In other words, we should value the professionalism of career and volunteer firefighters by validating their training. Yoga instructors get certified and so do dog trainers. Certification ensures that candidates have met the objectives of the course, and that the course taught the skills necessary to be a firefighter. The training certification process wouldn't have to be long or arduous to be effective. The certification needs to be attainable for volunteers who have other commitments and the process needs to be reviewed often. Whether or not the federal government is in charge of the certification process means nothing. If we feel we are able to police ourselves and set our own national standards, we should prove it. As the late radio commentator Paul Harvey often said, "Self-government won't work without self-discipline." His point was that any leadership that is not accountable is not leadership at all. Nowhere is this truer than with emergency services certifications.

In-Service Training Requirements

In-service training takes place on the job, after a firefighter completes initial training. It assists firefighters in improving their job proficiency, learning new skills, and building on past experiences or theories. A well-defined training system will also concentrate on reiterating safety concerns. In-service training can range from lectures to hands-on training. Simulations are an effective way to drill for situations that are difficult or impossible to replicate. Although in-service training can be required by some states for recertification, most fire departments primarily use it to expand on initial training and improve the professionalism of its members. Many departments have their own requirements for minimum drill hours or special mandatory training. Sometimes those requirements help meet state minimums for recertification or **Insurance Services Office (ISO)** standards for fire department insurance ratings. Drills might be daily for paid firefighters, or weekly for volunteers. Some departments have a dedicated training staff, whereas others rely on company officers or senior members to conduct the drills. Effective trainings allow members to practice the things that they need to know safely, and prepare for the things that they could see in the future. Many times drills are classified by their frequency as well as their risk.

Common Skills (High Frequency, Low Risk)

Some drill subjects cover the basic operations in which every firefighter needs to be proficient. Pump operations, driving, and **SCBA** use are some examples. Occasionally, firefighters become bored with the subjects, but a good instructor can make it more interesting, challenging, or even fun with a little imagination. Take pump operations, for example. You can run a little competition to see who can flow water the fastest from a deck gun supplied by a hydrant. Now add in flushing the hydrant and using a short section of **large-diameter hose (LDH)** to supply the pump before the tank water runs out. Not tough enough? Make it a two-person team to flow from a hydrant and knock down bowling pins (empty **aqueous film-forming foam [AFFF]** jugs) across the parking lot. Schedule it after the garbage man picks up and see who can set up and hit the side of the dumpster the quickest with a master stream from the deck gun. Relay pump with three engines in a loop without overflowing any tanks. Keep it safe, but keep it interesting. These "bread and butter" skills must be commonly practiced, but don't forget the first rule of effective education: Students have to be awake to learn anything. Most instructors also believe that they learn best when they teach someone else, so mentoring in this situation is usually beneficial.

Target Hazards (Low Frequency, High Risk)

We all have our own target hazards in our response areas. Whether it's silo fires, a zoo, a nuclear plant, a quarry, or an amusement park, they all take special considerations with regard to safety. It may just be an occasional festival or other public gathering. Preplanning with walk-throughs is vital to prepare yourself for the hazards you might run into. These could include access problems, dangerous substances, equipment, or even animals. Many times special high-risk facilities have their own team of emergency responders that are more than willing to meet with members and show them around. Setting up joint drills is a great way to ensure that their procedures and actions fit into

FIGURE 5-4
Target hazards may be categorized as low frequency, high risk.

a significant incident of which you could be a part. Differences, such as hose fittings, foam compatibility, or medical control, are best exposed in a drill setting. Learning what specialized equipment is needed and what is available from other agencies is also a good idea (Figure 5-4).

Another type of target hazard is a natural disaster. Drilling for response to a tornado must take into account certain variables, such as emergency equipment that has fallen victim to the storm. Lack of suitable roadways, communication issues, power outages, and mass casualties must all be addressed. Accounting for citizens is never easy when a neighborhood is erased. Shelters, transportation, and fresh water are all longer-term concerns. The only way a proper and safe response can be planned is through practice.

Funding Issues

Although much in-service training can be completed in house, on the computer, and on duty, some simply cannot. Conferences, advanced medical classes, and specialized training all come with a cost. It could be registration, travel, lodging, compensation for time, or even fill for the student's position at the station. Unfortunately, training is usually a pretty easy thing to kill in a budget when times get lean. This is especially true when it comes to special teams—those technicians who play a key role in the outcome of an event when your department is showcased on national news digging in the hole, hanging from the harness, wearing the Level A suit, or diving 50' under the water. Unfortunately, the potential for bad press is increased by poor performances as a result of deficient training. It's extremely difficult to put a price on the training throughout the year until the day you need it.

Perhaps even easier to eliminate is travel to national conferences and fire academies. Political leaders find it hard to justify trips with hefty expenses that appear to be nothing more than job perks. The truth is that the most cutting-edge technology, theories, and practices are exposed at these events, and the only way to learn first-hand, as well as to question, evaluate, and discuss them, is to actually be there. The networking goes a long way to breaking down the geographical or state lines that are sometimes too prominent in our service. Exposure to the way other departments perform their duties is an eye-opening experience for members. Some come back with new ideas to improve safety or service, others come back with a better appreciation of their own department, and almost all of them come back excited and recharged to make changes for the better.

Funding cuts also can hurt safety directly by cutting positions in paid departments. Managers in a retail store might be able to open another register with somebody from the stock room when checkout lines grow, but emergency scenes don't work the same way. Budget cuts may call for the elimination of a training officer or incident safety officer positions, assuming that someone else can pick up the slack. Unfortunately, for some departments, the potential for catastrophe without designated positions is real.

Accreditation

Accreditation is awarded to certifying agencies, educational institutions, and programs that meet the criteria of the accrediting body. The accreditation process normally consists of three components:

1. A self-study is conducted by the entity seeking accreditation.
2. An on-site visit is conducted by representatives of the accrediting body.
3. A report is made to the accrediting body who grants the final accreditation.

One significant advantage to accredited programs in two different geographical locations is that reciprocity is very common. For example, if two different state fire academies taught a certified course for NFPA 1021, Fire Officer I, and both are accredited by the same organization, each state is encouraged to accept the other. There are two major fire service accrediting bodies in the United States, each with its own rules: the International Fire Service Accreditation Congress (IFSAC) and the National Board on Fire Service Professional Qualifications (NBFSPQ, or Pro Board).

International Fire Service Accreditation Congress

The International Fire Service Accreditation Congress (IFSAC)[11] provides accreditation in the United States and internationally. There are two assemblies within IFSAC: the Certificate Assembly and the Degree Assembly. The IFSAC Certificate Assembly accredits certification programs that are based on a specific standard, such as NFPA 1001. After completing a review of an institution's self-study, policies, and procedures, a site visit ensures that the testing process is secure and valid and follows IFSAC's criteria for accreditation. If the certifying program meets the criteria, it can then claim itself as being IFSAC-accredited.

The Degree Assembly of IFSAC accredits educational institutions for postsecondary degree programs. Colleges and universities that offer degree programs in fire and emergency services often seek accreditation with IFSAC for their programs of study. A site visit is also conducted for degree accreditation. A unique characteristic of IFSAC is that it is both peer-driven and governed by its own members. According to IFSAC, the benefit is that there are no outside influences on the accreditation process.

National Board on Fire Service Professional Qualifications

Another accreditation committee is the National Board on Fire Service Professional Qualifications (NBFSPQ, or Pro Board).[12] In 2010, they offered accreditation for 72 levels of 16 NFPA standards. Certifying organizations that are interested in accreditation complete an application that is reviewed by the **Committee on Accreditation (COA)**. A site visit is followed by COA's decision on the institution's accreditation status.

Additionally, Pro Board issues certificates to individuals who have completed the certification process from one of their accredited institutions, and enters the information into their certification database. Certifications from a Pro Board institution in one state are normally accepted in another.

Higher-Educational Requirements

For many firefighters and officers, the certification and training offered isn't enough and they seek educational degrees. Some departments give incentives to further one's education, or require it for promotion to administrative positions. Different community colleges and universities offer associates, bachelors, masters, and even PhDs in subjects that are related to our field, including:

- Fire science
- Fire protection systems
- Emergency management
- Public administration

Several of the degrees are becoming obtainable online. As we will discuss later in this chapter, many of the higher-education institutions have joined forces to ensure that the degrees they offer are comparable to each other and meet safety guidelines. This has a direct benefit to fire and emergency services when comparing schools. Like certifications, most colleges and universities are accredited. Regional accreditation is critical for the colleges and universities that offer degrees related to fire and emergency services. It is also important for you as the student to know that the college or university you select to complete your degree is regionally accredited. The United States Fire Administration/National Fire Academy offers the Executive Fire Officer Program (EFOP). The EFOP is designed to provide senior fire officers with a broad perspective on various facets of fire and emergency services administration. The applicants to the program must have attained a bachelor's degree from a regionally accredited institution of higher learning to be accepted. There are many other reasons that accreditation is a critical component but being able to apply and be accepted to an EFOP is important to you and your career.

Business Associations

Business associations are organizations used to advance the occupations of their members. Many times they offer things such as training and **continuing education (CE)**. For instance, the **American Institute of Certified Public Accountants (AICPA)** not only offers but requires continuing education to remain a member.[13] Many of the business associations today are Internet-based, providing a forum for discussions and information sharing at little or no cost. Emergency services have a wide variety of such sites, including:

- Periodicals (magazine based)
- Union (International Association of Firefighters)
- Associations (National Volunteer Fire Council)
- Organizations (Everyone Goes Home)
- Social networking (specific sites or groups in social sites)

The advantages of Internet-based business associations are numerous. Information can be accessed any time of the day from anywhere with internet access. Email can be used to reach experts you normally would have a difficult time talking to directly. Videos, pictures, sketches, and opinions can be bounced off several people from different areas at one time. Many new theories pertaining to safe practices evolve from brainstorming sessions on the Internet.

Physical Ability Testing

Another commonality in the fire service is the movement to establish a standard physical ability testing procedure. One of the most popular nationally is the **Candidate Physical Ability Test (CPAT)**.[14] The test consists of a series of activities closely tied to actual firefighting duties to simulate the actions of a firefighter at a fire scene. The test is pass/fail, with specific criteria for each station of the test and an overall maximum time. Other departments and institutions have their own version of physical agility testing. As with initial training, these rarely are accepted outside their respective geographical areas. Initiative 6 gets into greater detail about the national standards for fitness capabilities.

ESTABLISHING WHERE WE SHOULD BE

If you've never been on a cruise ship before, it might surprise you that everyone on board has to join in on an emergency drill before the ship even unties from the dock. Imagine just unpacking from a long flight and cab ride and having your eyes set on the swimming pool, a cold drink with a great view, or simply touring the ship. Instead, you find yourself being herded like cattle to life boats with your life jacket on. A vacation of a lifetime delayed for a silly muster drill! U.S. Coast Guard regulations and other international rules require that this low-frequency, high-risk recreational activity be as safe as possible. When the ship docks at certain ports and most of the passengers disembark, you might see your waiter transform into a firefighter, pulling hoselines from closets on the ship in full gear and SCBA simulating a fire attack. Lifeboats are lowered into the water and medical teams are assembled. It doesn't matter what country's flag is on the stern or what language is spoken on the bridge; the rules for safety training apply to everyone.

Fire and emergency services are responsible for far more lives than the cruise ship industry. We've already discussed the value of common initial training and certification for all levels in each state, utilizing NFPA standards. Additionally, there are several other components of an ideal system that have proven successful in emergency services or other industries that should be used as models to improve our national system.

Recertification Requirements

Many agencies that certify individuals have moved to recertification. Although common in the past for EMT certifications, in some states, the process is fairly new to the fire service. Some of the advantages of recertification include:

- The ability to track who is currently active
- Ensuring that continuing education or in-service training is occurring
- Tracking the number of active individuals for statistical planning

It's not uncommon for states to require other emergency responders, such as police officers, to attend continuing education in order to recertify. Although in 2010, Tennessee had no training requirements for full-time firefighters to recertify, peace officers in an equivalent position are mandated to complete at least 40 hours of training annually.[15] At one time, NFPA 1500 specified the minimum hours of annual in-service training for firefighters involved in both structural and nonstructural firefighting[16] that could be applied to a recertification standard. The 2007 edition instead references individual department standards for fire training and federal regulations for hazardous materials. It also addresses specific skills, such as respiratory protection training, drilling on new technology or procedures, and the use of annual skills checklists to "prevent skill degradation." Some EMTs not only have an obligation for a set number of continuing education hours to recertify, but they must complete a set amount in special

FIGURE 5-5
Continuing education requirements can sometimes encourage the addition of new classes, such as hybrid automobile emergencies.

subjects, such as pediatrics. In the future, standard recertification requirements could require all emergency responders to obtain a specific number of hours in specific safety subjects, such as driver's education or risk management (Figure 5-5).

Some feel that continuing education requirements are just another unfunded mandate placed on individuals and departments. Others feel it's an effective way of validating their certifications through a "use it or lose it" process. Let's say you complete fire school with a guy who gets certified, then goes into another career for 20 years. With no recertification requirements, he can walk right back in and get hired next to you. Mind you, he may have learned the job with a John Bean High-Pressure Pump and an SCBA with a belt-mounted regulator riding on the rear step of the pumper. There's a good chance he has no idea what a PASS alarm is or what a thermal imaging camera does. As a result, most people agree that without some sort of a recertification process, our certifications aren't worth quite as much.

The FESHE Model

One of the most inclusive models for the future of training and education to come out of the fire service is the Fire and Emergency Services Higher Education (FESHE) model. Last adopted in 2009, the model lays out the path for a firefighter to progress to the level of fire chief. Although most firefighters won't ever become a fire chief, the blueprint provides the information to work through the ranks to a desired position. Earlier we talked about how more and more firefighters are obtaining college degrees. Other firefighters with years of experience are unimpressed, and feel that "you can't put out a fire with a piece of paper." They believe that having a college degree doesn't

FIGURE 5-6
The FESHE model successfully blends educational and training/certification requirements into a roadmap that can be used for career advancement.

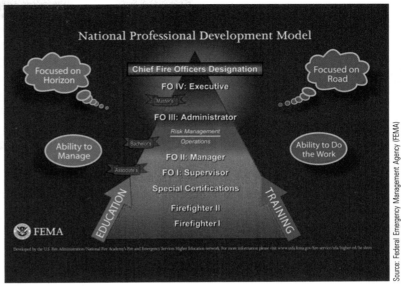

necessarily make a firefighter. On the other hand, many municipal leaders and citizens are surprised that many fire officials didn't go to college. Under our present system, there could be a fire chief in charge of millions of dollars of taxpayer money managing 100 employees and making a six-figure salary with only a high school education. Neither side is entirely wrong but, once again, we need to find a balance.

This is where the FESHE model shines. Originally concerned with ensuring that colleges and universities were on par with each other and that degrees were comparable, the organization realized that the training and certification side was equally important. The model takes into account the total package of what training and education should include to create an effective leader. Although there is a long way to go to include all aspects and goals of training organizations, the model is well on its way to becoming an industry standard. One of the best attributes is that the model is peer driven. Although the **United States Fire Administration (USFA)** facilitates the meetings and annual conference, fire and emergency services instructors are the driving force for the direction of the model (Figure 5-6).

Tiered Training

When it comes to special topics, the proven method for training is the tiered system of training. NFPA uses the same format with many of their standards by offering specific levels within a certification. Special topics are subjects that are usually covered in basic training but generally only touch the surface. Some individuals with an interest in the subject continue to further their education with additional levels of certification. Earlier, we discussed the advantage of using the tiered system when it comes to safety. Everyone should learn the basics, whereas some should become experts based on their position. A great example of implementing the tiered system in an existing certification is car seat installation.

As previously stated, NHTSA has a 32-hour certification for car seat technicians. However, they have no awareness- or operations-level certifications. It would probably be safe to say that although most emergency responders would like to learn how to inspect a child safety seat, they simply can't devote a week to becoming a technician. As a result, departments that offer free checkups are seriously understaffed, and some departments can't offer them at all. Say a new mom pulls into your fire station looking for advice on a car seat that is obviously unsecured, endangering her child. You may have been told that because you are not a certified technician, you are liable if you adjust the seat and there is a crash. What about the liability if you send her off knowing that it's in wrong? Surely, if you make it a little safer, there is a lower probability that the child would be hurt in a crash than if you did nothing. Firefighters are not electricians, but we still occasionally tape off wires and turn off breakers. We then call a certified electrician. There's no reason why this wouldn't work for car seats. If a tiered system for car seat inspections was initiated, it would consist of awareness, operations, and technician level training.

- *Awareness level*—Introductory level that identifies basic principles and risks. Awareness classes usually target new employees and other agencies involved with first responder activities, such as law enforcement. The types of car seats, the ages and sizes of children, state laws, and car seat installation tips would be taught.
- *Operations level*—The level that is the standard by which all firefighters and EMTs should be trained. This actually gets into the theories behind the practices and introduces the hands-on portion. Identifying an improperly installed seat would be the primary responsibility, but individuals would also be trained to "make emergency repairs." They would advise the parent where to go for a complete inspection.
- *Technician level*—The level at which individuals have been trained to the highest level, and have become proficient in all phases of installation. They would be familiar with problems associated with specific vehicle installations, information about recalls, and parent education. They would take the lead for community inspection events, using operations-level personnel to assist.

Credentialing

After the September 11 attack on the Pentagon, hundreds of emergency responders converged on the scene, as they would for any fire in their jurisdiction. Due to the fact that the Pentagon was a high-level security military building that had just been victim of a major terrorist attack, security was heightened to an unprecedented level. Many responders had difficulty entering the scene, trying to prove that they were who they said they were.[17] As federal investigators began to arrive on the scene, the emergency responders were then forced into the role of trying to verify who they were and who had the authority to be there. It became apparent that some sort of **credentialing** would be necessary to prevent security issues in the future.

Presidential Directive 12

In 2004, President George W. Bush ordered the Department of Homeland Security to establish a system of **personal identification verification (PIV)** credentials for the state department. PIVs were developed as a means of identification and validation of federal employees and contractors. They are essentially a smart card, containing an electronically embedded chip holding all pertinent information. A hologram is sometimes included with the issuing logo for a lower security visual check. Four components to the system ensure its security.[18]

1. During application, an individual's identity and credentials are verified through several means. Original documents and multiple sources of information are used.
2. The system itself is resistant to tampering or counterfeiting.
3. The credential can be quickly verified through interoperable electronic devices.
4. Credentials are issued only by accredited organizations.

Additionally, the PIV card is issued with a **personal identification number (PIN)** which ensures validation by the user for more secure access. Next time you're at the airport, watch the airline and TSA employees slide their access card, and then punch in a PIN to obtain access to secure areas. Besides the extra security, it allows tracking of when a card and an individual opened a door.

First Responder Authentication Credentials

Credentials have always been a part of emergency workers gaining access to certain locations. It started with a uniform and flashing a badge, then evolved into photo ID cards. For first responders, credentialing has begun to take form as **first responder authentication credentials (FRACs)**. The order that created the rules for credentialing was the Federal Information Processing Standards Publication 201 (FIPS 201). All federal PIVs and FRACs would be required to comply with PIPS 201. Some counties in Virginia, Maryland, and Colorado were among the first to pilot the national model of FRACs. Other states, such as Pennsylvania, have set goals to achieve credentialing for their state in an effort to determine qualifications of responders that arrive on emergency scenes.[19]

Besides using them for entry into emergency scenes, these credentials have the potential to make several aspects of emergency services safer and easier. Incident commanders can quickly ascertain whether responders have the training to complete a task, making the scene safer. The possibilities of secure access into some buildings in our jurisdictions or accountability at a fire scene are not out of the question. Airport security screening for travel should be easier as would crossing the border back into the country. Since FRACs contain encrypted personal information, agency representation, and certifications held, it's obvious that certifications (security clearances) must be the same from state to state.

SUMMARY

One of the challenges of improving safety in the fire service is to establish minimum training and certification standards. Emergency responders have evolved in a large part due to the process of the teaching and learning components of successful incident mitigation. Although many occupations use certification processes to ensure that their professionals continue to evolve through standard instruction, emergency services have aspects that are sporadic at best. A critical step to improving safety is to ensure that everyone learns what we already know, and is then encouraged to build upon it. Firefighters who receive no standard training may not have access to this vital information, unless they search on their own on the Internet.

There are successful models from educational systems that we can use to produce effective and safe lesson plans. We have access to job performance requirements and professional qualifications. A career ladder has been developed that encompasses both standard certification and the integration of higher education. What we lack is a coordinated effort to create minimum professional standards for volunteer firefighters, part-time firefighters, and career firefighters, regardless of the locations where they respond. Standard credentialing, certification, and recertification would provide the opportunity for safety to be interwoven into all aspects of training, making the entire fire service safer.

KEY TERMS

AED - Automated external defibrillator.

affective domain - A learning area defined by Benjamin Bloom involving feelings and values.

AFFF - Aqueous film-forming foam.

AICPA - American Institute of Certified Public Accountants.

BSI - Body substance isolation.

candidate physical abilities test (CPAT) - A type of criterion task test involving several firefighting activities with a focus on agility testing.

CE - Continuing education.

chain of survival - A series of steps identified by the American Heart Association that is needed to successfully resuscitate someone in cardiac arrest.

COA - Committee on Accreditation.

cognitive domain - A learning area defined by Benjamin Bloom involving memorization.

credentialing - An official way of designating the qualifications of an individual by a governing body.

education - The process of learning through higher education.

EMS - Emergency medical services.

EMT - Emergency medical technician.

FESHE - Fire and Emergency Services Higher Education.

First responder authentication credentials (FRACs) - A proposed national system of credentialing emergency responders.

IFSAC - International Fire Service Accreditation Congress.

Insurance Services Office (ISO) - A fire industry insurance organization.

JPR - Job performance requirements.

LDH - Large-diameter hose.

NBFSPQ - National Board on Fire Service Professional Qualifications, or ProBoard.

NFIRS - National Fire Incident Reporting System.

NFPA - National Fire Protection Association.

NHTSA - National Highway Transportation Safety Administration.

NIMS - National Incident Management System.

OJT - On-the-job training.

personal identification verification (PIV) - Credentialing system used by some federal employees and contractors, usually in the form of a smart card containing an electronically embedded chip.

PIN - Personal identification number.

ProBoard - National Board on Fire Service Professional Qualifications (NBFSPQ).

professional qualifications - A series of NFPA standards identifying the requirements needed to attain specific certification levels.

psychomotor domain - A learning area defined by Benjamin Bloom involving mastery of skills.

SCBA - Self-contained breathing apparatus.

standard of care - The care a person would expect to receive in a similar medical crisis in a similar area.

STEMI - S-T elevated myocardial infarction.

training - The process of learning a skill by practicing.

TSA - Transportation Security Administration.

USFA - United States Fire Administration.

REVIEW QUESTIONS

1. What is the purpose of credentialing for emergency responders? Name some of the advantages of credentialing.
2. What are the components of a job performance requirement (JPR)?
3. What are some of the differences in training requirements between volunteer firefighters and career firefighters, as well as from state to state? What are the effects on safety?
4. How does the FESHE model combine training and education into one complete package?
5. What are the advantages of a tiered system of certification? How could it improve safety?

FIREFIGHTING WEBSITE RESOURCES

http://everyonegoeshome.com
http://cdc.gov/niosh/fire
http://theproboard.org/
http://www.ifsac.org/
http://www.usfa.dhs.gov/nfa/higher_ed/index.shtm

NOTES

1. Faber, C.S. 2000, January. Current Population Reports P20-520; U.S. Department of Commerce Economics and Statistics Administration.
2. Highlights of the 2010 American Heart Association Guidelines for CPR and ECC; AHA ECC Adult Chain of Survival 2010.
3. McKay, J. 2003. *Civil War sites of the southern states.* Guilford, CT: The Globe Pequot Press.
4. http://www.gencourt.state.nh.us/legislation/2009/HCR0006.html
5. Bloom, B.S. 1956. *Taxonomy of educational objectives: The classification of educational goals; Handbook I, cognitive domain.* New York: [publisher].
6. Anderson, L.W., and C. Krathwohl. 2001. *A taxonomy for learning, teaching, and assessing: A revision of Bloom's taxonomy of educational objectives.* New York: [publisher].
7. NFPA 1001: Standard for Fire Fighter Professional Qualifications. Level I and II. Forcible Entry. 2008.
8. http://www.publicsafety.ohio.gov/links/ems_fire_minutes0309.pdf
9. http://cert.safekids.org/certification.html
10. US Department of Labor, Job Training Partnership Act; Job Training Reform Amendment of 1992. Section 141(g)(2).
11. International Fire Service Accreditation Congress (IFSAC), Certificate Assembly Orientation Committee on Site Teams, April 2001.
12. http://theproboard.org/
13. www.AICPA.org
14. "Developing the CPAT." The IAFF/IAFC Fire Service Joint Labor Management Candidate Physical Ability Test Program Summary. http://www.iafc.org/displaycommon.cfm?an=1&subarticlenbr=389. 2010
15. Rules of the Tennessee Peace Officer Standards and Training Commission; 1110-4.01(1); revised February 2007.
16. NFPA 1500, 1992 edition; section 3-4.
17. NIMS Standards Case Study: Responder Authentication, National Preparedness Directorate, August 2008.
18. http://csrc.nist.gov/publications/fips/fips201-1/FIPS-201-1-chng1.pdf
19. The KEMA Current. 2010, March 16. Keystone Emergency Management Association 1(1).

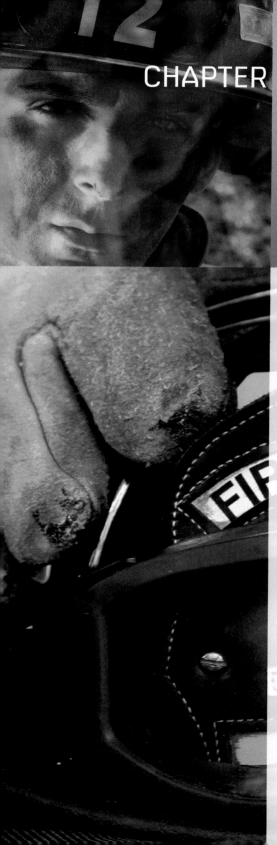

CHAPTER 6

Developing Medical and Fitness Standards

LEARNING OBJECTIVES

- Discuss what national medical and physical fitness standards are and the relevance to fire and emergency responders.
- Explain why improving the health, wellness, and fitness of emergency responders is important.
- List the reasons heat dissipation is difficult for firefighters.
- Explain maximum heart rate and the influence that emergency responses have on it.
- Describe the importance of a responder's ability to use oxygen.
- List the components of a firefighter's criterion task test, such as the CPAT.
- Describe the difference between wellness and fit-for-duty.
- Identify the NFPA standards developed for medical and fitness application.
- Explain the application of maintenance versus repairs when discussing health.

CHAPTER 6
DEVELOPING MEDICAL AND FITNESS STANDARDS

 6. Develop and implement national medical and physical fitness standards that are equally applicable to all firefighters, based on the duties they are expected to perform.

In Chapter 1, "Defining a Cultural Change," we discussed the value of adopting a safety culture for the fire service. A culture change reaches into every aspect of one's organization, influencing its values. No aspect of a cultural change for safety is as daunting as implementing a national medical and physical fitness standard for the fire service. One reason this is so challenging is the fact that there aren't many medical or fitness standards in other industries to compare to. Additionally, opinions of health and physical fitness vary widely and emergency responders have shift schedules that place them in a "disadvantaged" position compared to the general public.

For instance, let's say you have a neighbor who appears to be a health and fitness guru. Willie is 54 years old, but looks much younger. If you see him out of his house, he's usually jogging down the street or working in his garden. He grows most of his own "organic" food during the summer. He works an 8-5 shift at the bank and helped you refinance your house a couple years ago. He and his family like to hike and camp for vacation, and this year plan to finish the last leg of the 469-mile Blue Ridge Parkway. He's not really a marathon runner but he and his wife Mindy enter a lot of the charity run/walk events all year. Although they invite you to join them sometimes, your schedule never seems to allow it. Instead you pledge some money and "buy your way out" of it. He denies that he is a "health nut" as you joke, but says he just wants to live longer than his dad did.

You, on the other hand, are quite different. People guess you are a couple of years older than you really are. Your neighbors know you as Mr. Yardwork, and your weekend barbecues are the place to be in the summer. This year you've been saving up for a cruise and plan to take the whole family. You're pretty happy with your life; you have a good job, decent pay, and 48 off. Of course you'd like to lose 20 pounds and your doctor keeps yelling about your cholesterol, but that's no big deal. You do pretty well getting exercise on calls, you coach little league, and you even golf a couple times a month. You're not "unhealthy" by any means.

When we say that emergency responders are disadvantaged, what we mean is that many aspects of a firefighter's life are not conducive to living like Bill. It's hard to eat right when our schedule is so mixed up and we eat with different people who have different tastes. The stress of shift work and serious calls affects our blood pressure and the way we digest our food. It's hard to keep an exercise regimen if we don't know when the next call will come in or if we have to hold over. How can we plan to do a cancer run an hour after we get off shift? We might have to be up all night. We work many weekends, and sometimes have to attend extra classes, such as ACLS, or teach at the fire academy. It's hard enough for a person our age who's not a firefighter to exercise and eat healthy. It's almost impossible for us.

This brings us to the cultural change that is before us. We've seen it coming for quite some time and we hoped it would pass, but here it is. If we need to lose 20 pounds and our doctor says our cholesterol is too high, we *are* unhealthy (Figure 6-1). If we rely on emergency runs, standing in a dugout, and carrying our golf clubs to the cart as exercise, we *are* unhealthy. If we rely on cigarettes or alcohol to cope with our stress or nerves, we *are* unhealthy. The worst part is that none of this is news to us. It's not like we live on a mountain in Tibet and know nothing about health, fitness, heart attacks, obesity, and strokes. We see people every day who deny the obvious. From patients who didn't want to call 911 until 2 a.m., to a store owner who never knew a rear exit had to be clear of stock. Denial is a part of our job, but shouldn't be part of our job description. This chapter is not going to define how many grams of protein you should have or how many reps you need to be able to do at a certain weight load. There are numerous books and reference materials from experts in nutrition and fitness, with many of them designed specifically for emergency responders to use when designing their specific diet and fitness program. Instead, this chapter will focus on what is being done to improve health and fitness and how our body uses energy and reacts to firefighting. Finally, it will draw conclusions about what is necessary for us to be able to do our job safely and efficiently.

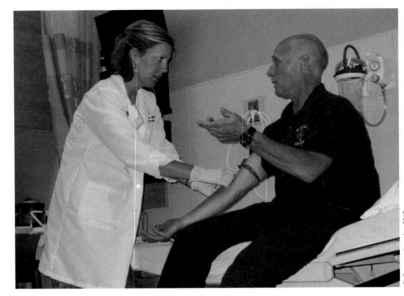

FIGURE 6-1
Numbers don't lie. Cholesterol levels can easily be obtained with a simple blood test.

A COMMON PUSH FOR HEALTH AND FITNESS

Although talk of improving health and fitness is nothing new to emergency response providers, there are signs that it has now been moved to the front burner. As a result of the unacceptable number of line-of-duty deaths every year, movements have come from different directions to ensure that health and fitness standards are developed that apply to what we really do. Both the **National Volunteer Fire Council (NVFC)** and a partnership between the **International Association of Fire Fighters (IAFF)** and the **International Association of Fire Chiefs (IAFC)** have taken steps to fix the problem, by first admitting that there is one. The Fire Service Joint Labor Management **Wellness-Fitness Initiative (WFI)** was created by the IAFF and IAFC.[1] Although the partnership for WFI represents career firefighters and chiefs, many components of the initiative are applicable to part-time firefighters as well as volunteers. Volunteer and career firefighters have relied on each other for decades to come up with common solutions that all of the fire service can use. Usually it's been with technology or service, but it's also been effectively used in grant programs and with the adoption of NFPA standards. For example, both 1710 (Standard for the Deployment of Fire Suppression Operations, Emergency Medical Operations, and Special Operations to the Public by Career Fire Departments) and 1720 (Standard for the Organization and Deployment of Fire Suppression, Emergency Medical Operations, and Special Operations to the Public by Volunteer Fire Departments) were adopted together. Although questions and debates continue about the application of specific aspects of each standard, they will continue to be improved through revisions in the years to come. If the fire service can pass parallel standards for something as globally different as staffing and response for career and volunteers, there is no reason why we can't do the same for health and fitness.

Career firefighters are not alone. In 2003, the National Volunteer Fire Council (NVFC) created the Heart-Healthy Firefighter Program.[2] The goals of the program are to improve the health, fitness, and nutrition of all emergency responders, both volunteer and career. They provide free "train the trainer" type workshops for health and wellness advocates to promote everything from developing a safe exercise program to weight management in their department. In his book, *The Fire Fighter Conditioning Workout Plan*, Stew Smith explains that both volunteer and career firefighters in the country cover the complete spectrum from fit to unfit.[3] As a trainer in New York who specializes in firefighter, police, and military fitness preparation, Smith believes that although programs for career and volunteer firefighters might differ, they must both have the same end result. The physical demands and stresses are the same for a similar fire in a similar building, regardless of where it is or who is compensated. Regarding WFI, essentially the "blue shirts" and the "white shirts" agreed that too many firefighters were dying as a result of poor physical fitness and health issues that were either not identified or not corrected. What's comforting about WFI is that two organizations with normally opposing views have created a common group with a common goal and voice.

The first thought that sometimes comes to mind when you bring up a fitness program in your department is panic. Many people who had to pass a medical physical or fitness evaluation when they got hired or joined the fire department may not have had one since. They might fear the possibility of not being permitted to continue fighting fires if they don't pass, or argue that the fitness test doesn't replicate what the job really entails. They may think that a physical should be done by their primary care physician, and that one hired by the department has ulterior motives. Firefighters may feel that fitness time and equipment should be made available, but shouldn't be mandatory. Administrators may be primarily concerned with the bottom line. They might not want fitness in the station because it costs money and can actually cause on-duty injuries or even death. Besides, there certainly are more important things to accomplish between calls than working out, such as prefire plans, hydrant inspections, and training.

Firefighter Statistics

Before we panic or jump to conclusions about what the other side is really up to, let's start with some of the indisputable facts about why management and labor both agreed to sit down and formulate a solution to the problem. It's hard to argue with the statistics. According to the NFPA,[4] approximately half of the firefighters that died in the line of duty in 2009 died as a result of stress or **overexertion** (usually heart attacks or strokes). If half of the 11-year-olds in the United States who died every year from trauma were killed by skateboard injuries, you can bet we'd outlaw skateboards. We know that heart attacks are the enemy for firefighters, and we even know the major risk factors. We can't eliminate them all, but with some evasive maneuvers, we could eradicate or reduce the frequency and severity of the rest. Some argue that firefighting is a strenuous job, and that sudden cardiac arrest is just the cost of doing business. If that were really true, we'd see the same trend of fatalities in other physical occupations, such as professional sports. Although professional athletes have to try out each year and firefighters tend to have a longer career, not many active athletes die from overexertion. When a professional team invests a significant amount of money in a player, they insure their investment through proper diet, physical training, and medical observation. Although we may not get as much attention or money as an NFL star, we still have a physically demanding job. As the NFPA 1583 Standard on Health-Related Fitness Programs for Fire Fighters[5] explains, firefighters need "to be medically and physically fit in order to perform the required tasks."

Until recently, a serious problem for firefighter health and fitness has been that we were working off theories instead of hard facts. Although there were some research projects completed to study the effects of firefighting on the human body, many were from small groups under simulated circumstances. We knew that more firefighters died from **sudden cardiac arrest (SCA)** than other occupations but weren't sure exactly why. Two major research projects involving the physical stresses of firefighting have been conducted. In 2008, Indiana University and the Indianapolis Fire Department released the results of a thorough and technologically advanced health research project performed on firefighters while they worked. According to the final report, members wore data-recording equipment that recorded "every heartbeat, breath, and footstep"[6] of the 56 firefighters they monitored. The results were obvious to the researchers who spent six months responding with the firefighters. In their conclusion, researchers explained that if a person has heart disease, statistics show they probably will die as a result of it. However, firefighters with the same condition will die from a heart attack or stroke much sooner. A few years before that, the University of Maryland partnered with the Maryland Fire and Rescue Institute to conduct a study on the effects of live fire training. Together these two research teams have proven what many professionals had argued for years: Firefighting takes a significant toll on the human body, and health and fitness is the best way to prepare for it.

BODY METABOLISM

The first thing we need to look at is how our body works. By understanding what is actually taking place inside our gear as we ventilate a roof, we can see how the information gathered in research has an application at our next emergency call. Rather than getting into the physiological systems in place or discussions using medical terminology, we can instead look at how we generate the energy to work as a machine. Gasoline and diesel engines use a combination of fuel and oxygen to produce energy. Regulating the amount of fuel or oxygen to the engine directly affects the energy output. For example, let's say you respond to a motor vehicle crash involving a couple of cars and a dump truck. When you arrive, the dump truck motor is still running, with the engine racing. The driver is frantically trying to pull the engine stop, but the body damage is preventing the linkage from moving. The police officer looks baffled about how to shut down the screaming engine. Your options are to cut the fuel or the oxygen. The primary way to cut the oxygen is by accessing the air damper on the emergency shut-down valve. If you can't do that, you can simply discharge a CO_2 extinguisher into the air intake. Taking away the oxygen takes away the fire, with no residue or damage to the engine. NASCAR uses a similar principle when they want to slow down race cars for safety reasons at some of their faster tracks. A restrictor plate is placed on the carburetor to restrict the amount of air permitted in. Less oxygen = less power = less speed.

Oxygen

Like dump trucks and race cars, humans need oxygen to run and the same rules apply: Insufficient oxygen means insufficient endurance. That's why one of the first signs of cardiovascular problems is shortness of breath. The respiratory and cardiovascular systems work hand in hand, and have a direct effect on each other. The cellular process of using oxygen and glucose (sugar) together to create energy is referred to as metabolism. When we relax on a lawn chair and watch the clouds go by, we are at a very low metabolic rate. When we drag a 2-1/2" handline to the rear of the structure with a portable deck gun in full gear, we are operating at a high metabolic rate. Our body needs more oxygen to do the work, so our heart rate and breathing increase to meet the demand. The three components of this equation that are responsible for your ability to place that portable monitor behind the building are:

1. A healthy pump to get the oxygen and sugar to the cells
2. Adequately sized pipes to get the blood where it is needed
3. The ability to load and unload oxygen into and out of your blood

The Pump

Your heart is one of the strongest muscles in your body. Endurance wise, it's unbeatable. It beats from before you're born until the day you die. Its job is to pump blood around the body, but because it's also a significant muscle, it needs its own fresh supply of blood. Blood that is rich with oxygen and nutrients is pumped from the aorta into large arteries that feed the heart muscle. The harder the heart works, the more oxygen it requires. A person who has high blood pressure forces his or her heart to work harder than it normally would, increasing the need for oxygen. The oxygen demand is even higher for someone who is overweight. Eventually, the additional workload takes a toll on the heart muscle, affecting its performance. **Congestive heart failure (CHF)** can be caused by hypertension, and can lead to symptoms such as shortness of breath and fluid in the lungs. Your heart has several components and safety systems that make it pump efficiently, making it very similar to that Hale or Waterous fire pump you have on your fire engine.

Throttle (Heart Rate). Much like a centrifugal fire pump, your heart has a throttle that directly relates to discharge pressure. The pressure is usually adjusted by a pressure governor on the throttle (neurological input), but can also be controlled by a manual throttle (chemical input). Our brain is the main neurological computer that controls the speed of our heart. This is generally a result of a physiological response. You work harder, your heart rate increases. It could also be a result of a psychological signal, like when you step on a ground hornet's nest. Either way, the brain tries to anticipate the needs of the body. The manual throttle, on the other hand, can be modified by humans. Through certain cardiac medications, a doctor can manually adjust the throttle of the heart, thus affecting blood pressure. Some illegal drugs or unintentional electrolyte changes can make the heart race out of control or crawl to a stop.

Your fire department pumper is equipped with a governor to prevent the throttle from over-revving, thus damaging the diesel engine. Your body also has a mechanism for regulating heart rate, but sometimes simply is overrun. Many sources refer to this as **cardiovascular collapse,** or overexertion. The maximum heart rate (MHR) is a theoretical number based on your age. One of the most common ways to calculate your MHR is to subtract your age from 220. Most research, fitness, and weight loss information sources use this number to prescribe workout routines at a specific heart rate. For instance, a source might recommend exercising at 60% to 75% of your MHR for the best fat-burning results, while they believe that 70% to 85% is the best percentage for overall calorie consumption and aerobic training. A pulse of more than 100% of your MHR is generally considered unsafe, with unknown consequences. This is especially true for patients who are older, overweight, or who have cardiovascular disease.

As stated earlier, all muscles need oxygen as they work. However, one significant difference between your heart and the other muscles in your body impacts ascending heart rates. Most muscles receive their oxygen during the pumping stage (**systole**). When you feel a pulse, cells are receiving oxygen. Therefore, an increased heart rate means increased oxygen to the tissues (perfusion). But because your heart muscle is squeezing during systole, the only way it can receive fresh blood itself is when it rests (**diastole**) between contractions. Most people are familiar with systole and diastole as the two numbers used for measuring blood pressure (e.g., 116/76). Because diastole is the lower number, you can see that there is less pressure to get fresh blood into the heart muscle. Worse yet, as your heart rate climbs, less time is spent in diastole. When your heart approaches its MHR, it simply doesn't have time to get enough oxygen, even though it's consuming oxygen at an alarming rate. If this scenario continues, the heart tissue starves for oxygen and can be injured or even die, known as a **myocardial infarction.**

Finally, the very fact that we are excited has a huge effect on our heart rate. Our nervous system has two different branches, referred to as the **sympathetic** and the **parasympathetic.** Sympathetic is described as our "fight or flight" side, which is what results when you realize that a Doberman Pincer just broke loose from his owner and is running at you from across the street. Some organs, such as the heart and lungs, as well as eyesight and hearing, improve in an effort to jump a fence and run or to kick and fight. One study in Indianapolis showed that just the sympathetic response from the alarm and dispatch of an incident put firefighters at as high as 85% of their MHR[7] before they even donned gear. Research also showed that experience levels had a significant effect on controlling this response. Chief Billy Goldfeder believes some firefighters get too excited going to calls.[8] He illustrates the point by explaining what a friend of his said years ago, that garbage men don't turn the corner onto a side street, see trash at the side of the road, and run, yelling "Trash!" His point is that we are firefighters and we need to accept that we will see fires and need to learn to reduce our sympathetic response by staying calm. Being mentally prepared for duty (Initiative 13) is as important as being physically prepared (Initiative 6), and the sympathetic nervous system has the two stitched together.

Heart rate is one of the most interesting components to come out of the study in Indianapolis. Let's look at the components that affect your heart rate in the first 20 minutes of a structural fire.

1. When the alarm sounds or the pager alerts, it creates a sympathetic response. Sometimes the alert volume is set too high, or lights come on at night, which both add to the adrenaline response. If you were asleep, you see a drastic change in heart rate and blood pressure immediately.
2. As you get dressed and make your way to the trucks, or volunteers run to their car to drive to the station, your physical activity creates an oxygen demand that increases your heart rate even further. Couple that with the psychological stress of looking for your keys or responding in heavy traffic and your pulse continues to rise (Figure 6-2).
3. Dispatch may give updates that law enforcement is on the scene and reports a working fire in the basement or that the mother called back and said her child was trapped in the fire. When you get on scene, you may see fire spreading to the exposure next door or hear propane tanks exploding. Neighbors might be yelling at you when you get off the truck. All this throws more adrenaline into your bloodstream.
4. Now drag the hoseline and put up ladders. Firefighting is a strenuous job, and the adrenaline is pushing you to move faster. Most athletes, such as runners, pace themselves for a strong finish, but our race is at the onset of an incident. The metabolism that is occurring may be consuming oxygen faster than you can replace it. Your heart rate climbs even more.
5. Time to mask up and get in there. It's hot in your gear already, but now we're going someplace hotter. We all know dogs pant as they get hot, and people breathe faster too. The heat from our breath gets trapped in our mask. Heart rates climb due to temperature, in an effort to get the hot blood from the core muscles and to

FIGURE 6-2
Emergency response can significantly influence the stress level of responders.

the surface of the skin. Sweating improves the heat transfer from the skin to the atmosphere. The trouble is that convection, radiation, and evaporation don't work too well in full protective gear. More heat buildup equals a faster heart rate. Depending on your geographical location and season, ambient temperature may be working against you. The Maryland study on the effects of live burn training[9] came out with a recommendation to consider cancelling drills in full gear when the heat index approaches 110°F (43.3°C). Unfortunately, we can't cancel fires due to weather.

6. Although the sweating doesn't do much for cooling the body in a fire, it does contribute significantly to dehydration. Fire scene dehydration is exacerbated when a firefighter arrives dehydrated prior to the fire. Football players know to prehydrate prior to a football game on a hot day, but firefighters aren't aware of what time the game starts, or even if there is one at all. Now add in dehydration from sweating, and we get a drop in total blood volume. Less blood volume leads to lower blood pressure, which in turn causes the heart to pump faster.

You may remember incidents in which you noticed some things from this list and had to stop to catch your breath. As already discussed, MHR is a number we'd rather not approach but it happens fairly often. Researchers proved that some Indianapolis firefighters worked *over* their estimated MHR at fire scenes for as long as 40 minutes at a time. Even after returning to the station, some still had a heart rate over 100 beats per minute two hours later.

The other branch of the nervous system is the parasympathetic. This "rest and restore" mode has the opposite effect as the sympathetic system on many organ systems in the body. As hormones that were dumped into the system from the sympathetic response are metabolized or excreted, the body can take time to complete vital tasks such as digesting food, healing, and resting. Problems can occur when the body is in a heightened state for prolonged or frequent periods during a shift.

Transfer Valve (Volume or Pressure). Another similarity to the heart is that centrifugal pumps can either pump more pressure or more volume. Some pumps are designed as a two-stage pump, allowing the operator to pick pressure or volume at the onset of a fire. The transfer valve is positioned to send the water either in series (pressure) or parallel (volume). Most pumps today are single-stage pumps, which use the throttle rather than a transfer valve to make changes. The rules of pressure versus volume still apply. For example, if your fire pump is capable of supplying 1000 **gallons per minute (gpm)** at 150 **pounds per square inch (psi)**, it would only be able to pump 700 gpm at 200 psi, and 500 gpm if you were to crank it up to 250 psi. Your heart behaves in the same fashion. As the rate approaches 200 beats per minute, it becomes less efficient. This reduces the amount of blood pumped per contraction (stroke volume), thus lowering the blood pressure. Unfortunately, your heart sometimes recognizes this inefficiency and tries to compensate by pumping even faster.

Discharge Pressure (Blood Pressure). When you relay pump between apparatus, your intake pressure has a direct effect on discharge pressure. For instance, it's hard to keep up with a master stream and a couple handlines from only one 3" supply line. If you suddenly lose your water supply at a fire, the RPMs take off in an attempt to compensate, exactly the same way a pulse rate climbs for a trauma patient who is bleeding internally. On the other hand, if your fire pump can't handle the amount of intake pressure it's getting, it blows out the pressure relief valve. When your body has a high intake pressure, it tends to back up in your lungs (**pulmonary edema**).

Auxiliary Cooler (Sweating). Finally, a pump that is running builds up heat that must be dissipated. A fire pump generally cools itself easily because the water flowing through it takes heat with it. Problems arise when the pump is not flowing a sufficient amount of water, and the heat instead accumulates. The solution is to circulate water through the tank or to open another line. The motor also can start to overheat without enough airflow through the engine compartment. Some vehicles are equipped with auxiliary cooling lines to assist in the process. All energy consumption ends in heat, from a television that's on to tires on the turnpike (**thermodynamics**) (Figure 6-3). Body metabolism is no different. Sit in a class all day and you get cold; climb a flight of stairs and you warm up. Through metabolism and increased work, our bodies continue to build heat when working in the confines of **personal protective equipment (PPE)**. Sweating usually helps cool the body, but encapsulating gear prevents this from occurring. With no immediate way to cool and slow down the pump, we must rely on prehydration and rehab. Rehab should be mandatory at bottle changes when the weather and

FIGURE 6-3
The use of energy many times results in heat buildup, as seen through a thermal imaging camera.

workload dictate it. Every rehab should include vital sign checks to ensure that firefighters are not working too close to their MHR.

Many sources cite acclimation as a way to assist in combating heat stress. When a vacationer from Chicago visits Atlanta in February, there's a good chance he won't even wear a coat. He's already acclimated to 10°F (−12.22°C) for the winter, so 45°F (7.22°C) feels balmy. The same thing occurs on the fireground. If you're like many firefighters, you work out in shorts and a navy blue t-shirt in the air conditioning with a fan on. Although it does a fair job of keeping you cool as your body builds up heat, it doesn't do much for acclimation training. Part of a diversified workout regimen should include an occasional session in full gear and an airpack. If spare, clean gear is not available to use, consider bulking up with a weight vest and sweatshirts to simulate the diminished ability to lose heat as you exercise. If you spend some time shooting hoops at the station, try a game of "pig" in full gear and on air. You can get knocked out by getting the "g" or getting a low air bell. Besides heat acclimation, you learn how to conserve your air and make complex muscle movements in restricted gear while getting some laughs. Many cities have a firefighter stair climb where members can climb a high-rise building in full gear to benefit a charity. Getting your department to compete against another is a great way to get fit, build camaraderie, and help a good cause.

The Pipes

The next component to being physically able to complete a task at a fire scene is having adequately sized pipes to move the blood. Blood pressure is an easy way to gauge cardiac workload by measuring how hard the blood is pushing on an artery's wall. This is affected by the size and number of discharges as well as the **viscosity** of the blood.

The Size and Number of Discharges (Peripheral Vascular Resistance). How
much water flows at a fire scene is directly affected by how many discharges are being used, and what size diameter they are. As more water is pushed through handlines, pressure and friction loss both increase. **Friction loss** is a term used to describe the effects of friction and turbulence on the overall pressure. For humans, **peripheral vascular resistance (PVR)** can be compared to friction loss, and dictates how high your blood pressure reads. During a sympathetic response, blood is shunted away from veins near the surface of the body to prevent bleeding if wounded and to keep oxygen in the muscles and vital organs. **Vasoconstriction,** or shrinking of the pipes, can also be caused by smoking, cold temperature, or medications. As a result, PVR is increased. On the other hand, exercise training lowers blood pressure by reducing PVR.

Viscosity (Elements of Blood). Another component to the pipes is the consistency
and components of the blood. Viscosity describes the thickness of a fluid. If blood were as pure as the water we pump and our arteries were as smooth as the inside jacket of a fire hose, we wouldn't have nearly as many problems. Unfortunately, blood is relatively thick and contains cholesterol, fat, and platelets that can start to plug up arteries. Blockages can cause heart attacks and strokes by restricting the amount of oxygen that can reach the heart muscle or brain tissue, respectively. They also tend to block up arteries in the rest of the body, leading to further PVR.

Loading and Unloading Oxygen (Oxygen Uptake)

The final component of moving oxygen is how efficiently you can get it into your blood and back out where you need it. If you're out of shape and find yourself huffing just to make it up the stairs, you probably have an oxygen transport problem. Smoking and past exposure to smoke and other chemicals can contribute to such problems. Anatomically, there are different places problems can occur. For instance, the lungs could have problems moving oxygen to the blood stream. Asthma, emphysema, and other forms of **chronic obstructive pulmonary disease (COPD)** affect how fast oxygen can be loaded from the air sacks (**alveoli**) in the lungs to the bloodstream. The second problem could be the amount of red blood cells in the blood. Red blood cells carry oxygen and affect how much can be transported at a time. The final component is how well oxygen can be off-loaded at the cells that need them. Some diseases and poisons affect oxygen use at the cellular level.

Oxygen uptake is the most commonly used term to describe how much oxygen a person can take in and use. Commonly expressed as **VO$_2$ max** in ml/kg/min, it can be measured several different ways. Most measures don't actually measure "the maximum" because of the risk involved, but instead measure a percentage of oxygen-carrying capacity, and then estimate what the maximum would be. Sometimes the testing uses changes in the heart rate during a specific workout to estimate VO_2, such as in the **Queens College Step Test,** whereas others use **spirometry** and measure the chemical difference between the oxygen concentration present in exhaled air compared to that in inspired air. The type of test completed is less important than consistently using the same type of test for more comparable results. Past studies have suggested that firefighters should have a minimum VO_2 max of 40, although some have placed the number as low as 33.5. The validity of the actual number has been contested in court.[10] The Indianapolis study noted that the mean VO_2 max of the firefighters monitored was 46, which is well above the general population of similarly aged people, as well as that of many firefighters. As a comparison, scientists have placed Lance Armstrong's VO_2 max at 85.[11]

Another way of expressing the amount of energy being exerted, and thus oxygen being consumed, is by the **metabolic equivalent (MET)**. Using METs is less confusing than using chemical values to describe exertion. When it comes to hazardous materials, firefighters commonly use the terms **specific gravity** and **vapor density** to describe the physical characteristics of a material compared to water or air. Water has a specific gravity of 1, meaning that a liquid chemical spilled into a waterway will sink if it has a specific gravity of more than water. The same concept applies to vapor density, because normal air has a vapor density of 1. Propane (1.52) will settle to the basement, whereas natural gas (0.65) will rise to the second floor. Similarly, you have a resting MET of 1. Riding a bicycle can range anywhere from 3 to 12 METs, depending on your speed and resistance. Oxygen uptake is directly proportional to the METs.

Fuel

The second component necessary for metabolism is fuel. Although complex, it's important to understand the basics of how your body supplies fuel to the cells for muscles to perform work. By understanding what is occurring inside as you complete physical tasks at a fire scene, you can better prepare yourself for those duties. Let's assume that you need to carry an apartment pack and other tools to the sixteenth floor of an apartment building. When your foot hits the first step, the muscles in your left leg get the message to start climbing.

Adenosine Triphosphate (ATP)

The cells in your leg muscles contract, using **adenosine triphosphate (ATP)** as the fuel source. ATP is commonly used for short-duration tasks that require a burst of energy, such as lifting a patient or the initial pull of a crosslay that's wedged between the dividers. If you wanted to, you could probably run up a couple flights of the stairs carrying the apartment pack before you would start to run out of ATP. Because ATP is not replenished nearly as quickly as it can be used, instead, you pace yourself. As you hit the third-floor landing, your stored ATP is just about depleted and you notice your heart rate and breathing has increased. Oxygen uptake is part of the reason, but an increased heart rate will also shuttle nutrients that are needed to replenish ATP to the muscles. ATP is created from four specific body sources:[12] carbohydrates, fat, protein, and creatine phosphate.

Carbohydrates (Sugars, Starches, and Dietary Fibers). As blood is delivered to the muscles, it carries glucose (blood sugar), which can be used to form ATP. As the blood glucose level drops, the body immediately replaces it from the muscles. As you continue to climb, it looks in places such as your digestive system. If you ate an hour before the apartment fire, that glucose will be used to get you up another 13 floors. Sugar that is absorbed from the digestive system is carried into the cells by insulin that is secreted from the pancreas. If there isn't any glucose in the digestive system, it looks for glycogen. Leftover glucose from past meals is converted to glycogen and stored for later use in the liver and in muscles.

Carbohydrates can either be complex or simple. Complex carbohydrates are considered healthier than simple carbohydrates, and consist of whole-wheat foods and certain beans and vegetables. Simple sugars are found in most junk food and candy. Those sugars are fast-acting, but short-lived. If you get called to assist a diabetic person that has a blood sugar reading of 23, EMS might start an IV and push D50 (dextrose/simple sugar) into the bloodstream. This immediately raises the individual's blood sugar and she becomes responsive again. Unfortunately, this is only a quick fix. She needs to sit down and eat some complex carbohydrates to maintain an adequate blood sugar. Simple sugars give a boost of fuel, like starting fluid in a carburetor, but don't last very long. If your breakfast prior to the apartment fire was just a donut, it gave you a quick fix and you felt pretty satisfied. However, the carbohydrates won't last in the stairway and you will feel hungry again soon. Oftentimes, simple carbohydrates contain extra calories you can't use right away, so they are stored as fat. Physical training increases the body's ability to store and burn carbohydrates.

Fat. Fat is stored throughout the body and is a relatively abundant source of energy. With adequate oxygen, fat can actually produce more than 10 times the amount of ATP than the same amount of carbohydrates can.[13] Anyone who has tried to lose weight can attest to how slow weight loss occurs due to this fact. Although fat is a plentiful source of energy, it's not nearly as accessible for use by muscles due to the fact that it takes longer to convert to ATP. As a result, fat is not going to help you get to the sixteenth floor when your donut runs out. Recently, some fad diets have removed carbohydrates from the menu, forcing the body instead to consume fats. Unfortunately, fat burns most efficiently in the presence of some carbohydrates, so lowering your daily carbohydrate intake below 100 grams is not typically recommended. Physical training also improves the ability to burn fat during exercise.

Proteins. Proteins, like fats, are also a source of energy in the absence of carbohydrates. They are first broken down into amino acids, some of which are ready to release energy. Although proteins can be used to supply energy, they do so at the expense of breaking down muscle tissue. Without physical training, muscle tissue will continue to be reduced (**atrophy**).

Creatine Phospate. The last source for ATP generation is **creatine phosphate.** Generally considered a muscle's energy reservoir, creatine is stored in the muscles and easily produces ATP. Creatine supplements are common in sports, and are not considered illegal substances because they are found in normal muscle. Researchers working on the Indianapolis study recommended that more training be devoted to creatine phosphate metabolic capabilities.

Aerobic Versus Anaerobic Metabolism

There has been much debate regarding firefighter fitness with regard to physical training. The body works in two distinct modes, aerobic and anaerobic. An example of **aerobic** (with oxygen) training is running on a treadmill for 30 to 40 minutes (Figure 6-4). An example of **anaerobic** (without oxygen) training is heavy resistance training (Figure 6-5). The two types of training are different from each other and have profoundly different affects on the body. In his book *Aerobics Program for Total Well Being*, Dr. Kenneth H. Cooper[14] tells a story of evaluating an Olympic champion from Trinidad who won the gold medal in the Montreal Olympics for the 100-meter dash. As an experiment, he placed Hasely Crawford on a treadmill, walking at a steady pace. The champion reached a heart rate of 187 and complete exhaustion after less than 17 minutes! This shouldn't be a surprise because he trained only for sprinting, a predominantly

FIGURE 6-4
Firefighters can benefit from participating in aerobic training.

FIGURE 6-5
Firefighters can also improve their fitness level by completing anaerobic training.

anaerobic activity. His aerobic capacity was extremely limited, even for walking. A similar example would be a marathon runner who can't even bench press his weight. So which type of training does a firefighter need?

According to firefighter Michael Stefano, the author of *The Firefighter's Workout Book*,[15] firefighters must be able to maintain muscle activity over long periods of time. He refers to this as "cardio-respiratory endurance." Aerobic activities that significantly elevate the cardiac and respiratory rates for at least 30 minutes are the best ways to improve this type of endurance. Aerobic training also helps lower the resting heart rate. A lower resting heart rate means a lower working heart rate, which maintains a safer distance to the **maximum heart rate.** Dr. Cooper adds that without an aerobic base, anaerobic capabilities suffer. He explains that statistics show that athletes with limited aerobic capacity become fatigued quicker, which in turn brings on higher instances of injuries.

However, it's more than a stretch to say that aerobic exercise is the only key to firefighter fitness, because the majority of firefighting activities are anaerobic. It doesn't matter how far you can run if you can't drag your partner out of a house or lift the hydraulic tools into the compartment. In fact, a firefighter fitness reading supplement published by Gymflesh School™ of New York refers to the "fixation" of the fire service on aerobic exercise as a myth.[16] They instead make a convincing argument that a concentration of anaerobic training is much more effective based on the specific duties of a firefighter, as aerobic training takes into account only 25% of the body systems. There is sufficient evidence from professionals both in and out of the fire service that anaerobic training on an aerobic base has definite benefits for firefighters and other emergency responders. Stew Smith explains another benefit of "the ideal combination of exercise"—aerobic exercise and resistance training working together to burn fat. As an example, compare the body shape of a construction worker who performs both aerobic and anaerobic functions daily with a bodybuilding firefighter who spends several hours per day lifting weights in a gym. The firefighter might appear to be in better shape, but which one tires first from physical labor? Because the best workout regimen includes both types of exercise, it would follow that any fitness assessments of physical capabilities should involve each component as well.

PHYSICAL AGILITY TESTING

For quite some time, the fire service has struggled to develop the perfect physical capability test for new hires. Officially known by various professions as a **criterion task test (CTT),** it attempts to mimic actual fireground tasks in a controlled environment (Figure 6-6). There are dozens of agility tests across the country, with many validated for the skills being tested. Many of them include both aerobic and anaerobic components, with the majority of the test under specific time constraints.

The Combat Challenge

The **combat challenge** started as a physical agility test in Maryland in 1976. Although it continues on as a CTT, it evolved into a popular competition for existing firefighters in 1991. There are variations of relays, but the most popular is a "head to head" race consisting of five sequential segments.[17]

FIGURE 6-6
Criterion task tests consist of skills that can be expected as a firefighter.

High-Rise Pack Carry

The high-rise carry starts at the base of a five-story tower. The candidate wears complete gear and an airpack while carrying a 42-pound, 3" hose pack to the top floor and placing it in a container. Handrails can be used and steps can be skipped on the way up, but the hose cannot touch the tower or fall.

Hose Hoist

Once the hose has been placed in the container, the second event begins from the top of the tower. A piece of rope is attached to another 42-pound donut roll of hose, which is hoisted from the ground. Control must be maintained without dropping the hose or hitting the tower. Once the hose is hoisted and placed in the container on the top deck, the candidate descends the tower without missing any steps.

Forcible Entry

At the bottom of the steps lies the Keiser Force Machine. The Keiser Sled is a chopping simulator used to simulate forcible entry. Swinging a 9-pound sledgehammer between his or her feet, a competitor drives the 160-pound sled in a horizontal beam a distance of 5'. Accuracy is vital, as points are taken off for striking the beam of the sled or the handle of the mallet.

Hose Advance

Candidates then navigate an obstacle course 140' long, and simulate a fire attack with a charged section of 1-3/4". The hoseline is advanced 75', where it breaches swinging doors. The nozzle is then opened to strike a target. Once the target is hit, the nozzle is shut down and dropped to the ground.

Victim Rescue

Once the fire attack is completed, the simulated victim rescue takes place with a 175-pound mannequin. The dummy is lifted (not carried) and dragged 100' backwards while the firefighter maintains position in the firefighter's respective lane. The victim cannot be grasped by its clothes, arms, or legs, and timing stops once the finish line is crossed.

Candidate Physical Abilities Test© (CPAT)

One of the most inclusive forms of physical ability testing is the **Candidate Physical Abilities Test©**, or **CPAT**. Developed through the WFI, a task force was assigned to create the CPAT using "job analysis and job task surveys."[18] They started with 31 fireground tasks to consider in the test. Surveys were completed by random firefighters and were validated. The final list of skills associated with the tasks was placed in an order that could be expected at a fire scene. Used by several fire departments in the

United States, it's slowly becoming one of the most popular because of its similarities to actual firefighting tasks. Even nonfire service fitness experts like Stew Smith have lauded the CPAT for simulating the physical demands of firefighting so well.[19] The CPAT consists of eight events,[20] completed successively.

Stair Climb

The stair climb begins the test, requiring the candidate to use a stairway climbing machine for three minutes at a set pace of 60 steps per minute. The firefighter wears a 25-pound simulated hose load for this part, and is not permitted to stop or grab the handrail.

Hose Drag

The next component of the test simulates stretching an uncharged 1-3/4" hoseline in preparation for a fire attack. The hose is dragged around obstacles to a designated point, where the firefighter then kneels down and pulls the line a set distance.

Equipment Carry

During the equipment carry element, the firefighter approaches a cabinet and removes two power saws, one at a time, and sets them on the ground. The candidate then picks up both of them carries them a total of 150', and returns them to the cabinet.

Ladder Raise and Extension

This portion of the test ensures that the firefighter is capable of standing up an extension ladder, as well as extending it. Utilizing two different ladders that are partially secured ensures safety during this event.

Forcible Entry

Unlike the combat challenge, this simulation of forcible entry more closely resembles the actual breaching of a wall or door. Rather than swing a hammer between their feet, candidates use the sledgehammer in a controlled horizontal swing at doorknob height. The device measures each blow and sounds a buzzer when a predetermined amount of total force is achieved.

Search

The next segment of the test involves crawling in a dark, confined space to simulate searching under zero visibility. Besides the space's diminishing in size, it includes sharp turns and obstacles.

Rescue Drag

Simulating a victim or firefighter rescue, this part requires the firefighter to drag a 165-pound mannequin a total of 70' with one sharp turn around a drum.

Ceiling Pull

One of the most unique components of the CPAT is the final one, the ceiling pull. A common overhaul function at a structural fire is to pull ceiling. The pushing and pulling overhead requires looking up while performing a resistance exercise with arms extended. Similar to a fire scene, this function is critical to firefighting and is completed when firefighters are fatigued.

Fit for Duty Versus Wellness

Historically, the big dilemma with regard to fitness is the term *fit for duty* versus *wellness*. Most firefighters are more than willing to partake in a **wellness** program, one designed as a voluntary program to help make healthy choices in diet and exercise. The term **fit for duty** generally refers to being able to complete specific skills or meet set criteria to be permitted to work. Some career fire departments have phased in a fit for duty clause in their union contract. However, similar rules for volunteers appear to be less common. One of the biggest problems with fit for duty is defining exactly what is "fit." According to the state of Nevada, firefighting tasks have been estimated to require between 12 and 16 METs.[21]

Let's assume for a minute that those numbers are proven in several studies and are determined to be valid in your state. One of your fellow firefighters is 10 years from retirement and is out of shape. For his mandatory fitness evaluation, he proves on a treadmill that his body is only capable of 5.5 METs in a controlled environment. It would be a hard sell to say that he is capable of doing his job safely, especially after factoring in the sympathetic effects of a fire scene. Medical research essentially has proven that he is unable to meet the physical demands of his job. If he is physically unable to do his job, it's hard to justify him staying in that particular job. This discussion infuriates some firefighters. The trouble really isn't with the facts, it's with our culture. For instance, if he instead was found to have a seizure problem that would not pass medical clearance, he would be removed from his position. He wouldn't be happy about it, but most firefighters understand that the doctor has to clear you to work. Both examples involve removal from duty as a result of being medically unfit for duty. The big difference between them is that he may never be able to eliminate seizures, but a change in habits could improve his METs to a level to return to duty.

This debate is far from over. WFI third edition does not support a minimum standard for fit for duty. It suggests using data to establish "norms" that are nonpunitive in nature, but instead promote individual improvement. The system appears to be effective. Some emergency responders believe that fit for duty will eventually become a component of firefighting. Besides the probable reduction in line-of-duty heart attacks and strokes, financial incentives will perhaps drive the change. Some fire departments already get discounts on insurance or workers' compensation by instituting programs that reduce risks, such as a drug-free workplace and emergency vehicle driving.

We can expect that if more fire departments see improved health and financial savings by requiring fit for duty, others will follow. Both career and volunteer firefighters shouldn't see this potential change as a *career-ending situation*, but, rather, an opportunity to *prevent* a *life-ending situation*. Whether they're compensated for firefighting or not, all firefighters deserve the opportunity to enjoy their retirement. Variations of grandfathering and phasing-in are proven techniques to bring about change as painlessly as possible and may be effective in this situation.

Regardless of what the specific rules will be, it's advisable to be proactive in establishing our own standards. Participating in writing our own fitness standards is similar to writing a will—no one wants to do it, but it's still better than someone else making the decisions for us. Many professionals in the fire service feel the same way, and have initiated two NFPA standards to do just that: NFPA 1583 and NFPA 1582.

NFPA 1583

NFPA 1583, the Standard on Health-Related Fitness Programs for Firefighters, describes the necessary components of a health-related fitness program. These include a fitness coordinator, a fitness assessment, an exercise training program, education and counseling, and data collection.

Fitness Coordinator

The first step is the assignment of a qualified **health and fitness coordinator (HFC).** WFI refers to this person as a **peer fitness trainer (PFT).** The fitness coordinator administers all components of the program, and has received appropriate training.

Fitness Assessment

Both NFPA and WFI require periodic fitness assessment for all members, including annual assessment in NFPA 1583. The five components of a fitness evaluation include:

- Body composition
- Aerobic capacity
- Muscular strength
- Muscular endurance
- Flexibility

Exercise Training Program

A regular exercise training program must be available to all members. WFI suggests different ways to obtain equipment and ensure time for workouts. Because the goal is improved health, members with medical problems must be cleared medically before beginning a program.

Education and Counseling

One of the best ways to exercise efficiently is by having education and counseling available. Normally the HFC or PFT would be available for the education, but utilizing outside resources through referrals is also a possibility.

Data Collection

The final component includes a process for collecting and maintaining data collected from the **Health-Related Fitness Program (HRFP).** Overall departmental statistics can be reported as long as individual results are not released. Individual information is private, and kept in each member's permanent HRFP file, and includes the following categories of information.

- Personal information
- Preassessment questionnaire
- Fitness assessment
- Program participation data

NFPA 1582

NFPA 1582, the Standard on Comprehensive Occupational Medical Program for Fire Departments, outlines the medical components necessary to ensure the health of personnel. Although not specifically broken down by category, some of the main components of the standard include the fire department physician, annual medical physicals, and special testing situations.

The Fire Department Physician

The fire department physician must be familiar with the duties of firefighting and the associated risks. Through thorough exams, the doctor screens the member for injuries or disease likely to interfere with firefighting duties. The physician identifies disqualifying conditions that must be acted on immediately, and ensures proper referral to the appropriate specialists. By discussing the results of the examination and answering questions, the physician also can make individual recommendations for the member to improve his or her health. Although specific findings are confidential, the physician reports back to the department whether the member is cleared for firefighting duties.

Annual Medical Physicals

NFPA 1582 outlines what specific body systems should be evaluated and what diagnostic tests should be completed. The purpose of these tests is to find a medical problem before it becomes debilitating, as well as to identify **trending.** For instance, a pulmonary function test (spirometry) is one recommended component of the annual diagnostics. If a firefighter has continually done well on the test for the past seven years but now cannot pass, it's clear that something has occurred. Perhaps the member was involved in an emergency incident that caused injury to her pulmonary system that hadn't yet been identified. Without annual testing, it could go unnoticed and a direct cause could be more difficult to pinpoint down the road. Sometimes firefighters argue that they should be permitted to see their own doctor rather than the department physician. The biggest problem is that some primary care physicians aren't capable of administering some tests, such as spirometry, and may not understand their value.

Another important component of the testing is blood work analysis. Next time you get a blood draw for a physical, take your results to a computer and pull up the American Heart Association's website (www.heart.org). Punch your readings and personal information into the heart attack calculator to see what your chances are of having a heart attack in the next year. Of course, the calculator doesn't take into account the increased cardiac stress of your job, so it might actually be on the low side. But by trending these results each year, you can monitor the progress of your improvement.

Special Testing Situations

The last component involves the specific needs of members. Hazardous materials team members might go through more extensive diagnostic testing, such as chest x-rays. Staff members who do not wear SCBA may not receive the same amount of pulmonary tests as a line firefighter, whereas members over a certain age or with a specific medical history will probably get additional tests. Sometimes tests are necessary due to a specific hazard in their response area. A good example of this is a fire brigade for an industrial fire department working around specific chemicals.

MAINTENANCE VERSUS REPAIRS

Many times, discussions about health draw the analogy of fire trucks to personnel. The WFI references a study in Oregon regarding the way we treat our two valuable resources: equipment and personnel. Both take a significant investment to acquire and put into service. They both fight fires and respond to emergencies, and they each have the department's logo on the side of them. They both start off looking pretty sharp but eventually need maintenance and repairs to keep them going. Unfortunately, that's where the comparison fades away. We know that personnel are more valuable, and that many have a duty cycle twice as long as apparatus. Common sense would dictate that, as a result, we would take better care of our personnel. Statistics show that this is not the case.

When it comes to apparatus, we tend to invest a lot of money in **preventative maintenance (PM)** to prevent the big, expensive repairs (Figure 6-7). A good PM program also reduces the chance of a catastrophic failure at an emergency incident. Occasionally, these failures result in the apparatus being scrapped out. Personnel, on the other hand, are run until they're broken and then are sent out for "repair." Very little

FIGURE 6-7
Emergency services agencies feel strongly about preventative maintenance for apparatus, but what about preventative maintenance for the personnel?

is invested in PM compared to the cost spent on repairs. If you're not sure, check the numbers at your fire department. Compare money spent on workers' compensation and lost time with annual physicals, workout equipment, workout injuries, and annual fitness evaluations. Although we invest heavily to prevent the catastrophic failure of apparatus at incidents, a lack of PM for personnel leads to far too many failures. WFI studies have shown that by increasing the *proactive* costs of personnel maintenance, we reduce the *reactive* cost of personnel "repairs."

Our discussion usually uses statistics from **line-of-duty deaths (LODDs).** Initiative 9 deals with near-misses, the times during which an LODD could have easily occurred. When we're talking about catastrophic failures of personnel, it has to include near-misses. The Indianapolis report cited that an estimated 765 firefighters had heart attacks in the line of duty in 2005. Just because they survived doesn't mean they weren't catastrophic failures. It would be safe to say that a good many of those were not able to return to work. Preventative maintenance for the engine includes an oil change, lubrication, and inspection. Personnel need a proper diet, exercise, and inspection. Because we already do a great job with apparatus maintenance, why reinvent the wheel? There are several steps we take throughout the year to ensure that our apparatus is ready to respond. Some of these steps can be compared to personnel.

Truck Checks

Whether your vehicle checks are daily in the morning for a staffed fire station or weekly for volunteers, most apparatus go through a fairly quick evaluation to make sure there are no obvious violations that need to be repaired. Fluid levels, emergency warning devices, and equipment are checked, and the vehicle might even get washed. Most firefighters, if they're scheduled to work at the station, go through a similar routine with breakfast and a shower. Maybe this daily check should involve a full set of vital signs and weight that are logged into a personal journal. If your EMS jump kit has a glucometer, an occasional blood sugar check before breakfast might be a good idea. If you don't work out in the morning, a stretching regimen can get you ready and possibly prevent an injury. Daily truck checks look for what is broken, but they also look for trending. If you have to put coolant in the pickup truck every other day, there probably is a leak. If the oil level in the heavy rescue apparatus slowly gets higher each day and the viscosity gets thinner, you've got big problems. Trending vital signs can also alert you to changes that should be addressed at your next physical. If you found serious changes with the engine, you'd voluntarily take it out of service. The same should occur with you.

Some departments require a vehicle check after returning from an alarm. When your pumper returns from a fire, you probably do a quick inventory of equipment and replace supplies that were used. You might check the fluids and top off what's low, including the water tank. After the engine is ready to respond, you might head off for a shower and a change of clothes yourself. Now would be a great time to do a check of your own. Get a set of vitals and compare them to what your measures were in the morning. "Top off your fluids" by rehydrating completely. As Brian Fass, author of *The Fit Responder*[22] explains, "By the time you feel thirsty, you are already dehydrated." Basically, ensure that you too are ready to respond.

Annual Pump Testing

Most fire departments cycle their pumping apparatus through a testing process annually per NFPA 1911 standard. The pump testing process is designed to push the engine and pump to their limits while being closely evaluated for any signs of potential failure. Actually a series of several tests, pump testing results are compared against previous years and ensure that it is operating as efficiently as it was the last time it was tested. Although wear to the impeller can cause a reduction in gallons per minute, there is a minimum criteria that must be met to recertify. At that point, the pump needs to be either taken out of service and overhauled or recertified at a lower capacity. Because pump testing is rough on the motor, many departments send the apparatus through preventative maintenance just prior to pump testing to find and fix any potential problems. Much like pump testing, personnel should be evaluated in a similar method. Both NFPA and WFI recommend annual fitness evaluations. Like fire apparatus, there is a definite advantage to scheduling the annual medical physical just prior to annual fitness testing.

EFFICIENCY FACTORS OF FIREFIGHTING

Although we can't restrict the types of work that firefighters complete, there are ways to improve the efficiency of how we complete those tasks. In the *Essentials of Exercise*

Physiology,[23] McArdle, Katch, and Katch point to specific factors that increase physical efficiency, thus decreasing the energy needed to complete a job. These exercise guidelines are equally applicable to firefighting. For instance, let's look at the example of breeching a masonry wall with a sledgehammer in full protective gear. It's a physically demanding function that can be completed more easily and safely following specific criteria, including the work rate, the speed of movement, and **extrinsic** factors.

Work Rate

The pace that a firefighter sets while completing a task affects overall performance. Although two swings to the same spot with the same force create the same amount of damage to the wall, they don't necessarily take the same amount of effort. Rather than a linear relationship (1 swing = 1 damage), faster rates create a **curvilinear** relationship (5 swings = 4 damage). Besides accuracy suffering and injury risks increasing with faster hits, energy usage outpaces the work completed, resulting in a reduced efficiency.

Movement Speed

The speed of the swing also affects efficiency. For every action, there is an optimal speed for the best results. When breeching a cement block wall, the goal is to break out large pieces that can be removed quickly and efficiently. Swinging too softly wastes energy by not creating any damage for the effort put forth. On the other hand, swinging too hard has a surplus of energy evident by pulverized block. A hard swing also increases the chance of injury, sends dust and shrapnel airborne, and can reduce the stability of the wall above, creating a collapse hazard. Utilizing the exact speed of labor-intensive jobs takes experience. Adrenaline pushes a firefighter to swing as hard as possible, but overkill is possible.

Extrinsic Factors

Extrinsic factors are those having to do with the equipment available. Some aspects of the job, such as PPE, fall into this category. It's harder to swing in full gear, but the first concern is making sure you have the minimum PPE in place. Depending on the situation, you may have to wear full gear but could drop your airpack. Other times, you may be able to swap out your firefighting gloves for extrication gloves. Other extrinsic factors are the tools you are using. A flat-head axe can breech a brick wall, but a heavy sledgehammer is much more efficient. Sometimes you are forced to use the tool you have, but realize that sometimes calling for the right tool can save time and energy in the long run, especially if one firefighter continues to use the flat-head axe while waiting for the sledgehammer.

Additional extrinsic factors include muscle fiber composition, individual fitness level, an individual's body composition, and technique.

Muscle Fiber Composition

Almost every muscle in the body is made up of two different types of muscle fibers: type I (**slow twitch**) and type II (**fast twitch**). Exercising and physical training can change the percentage of each fiber in a specific muscle. For example, the muscles that control your eyes are predominantly fast twitch, and complete movements with very little oxygen. Fast-twitch fibers are generally anaerobic, and use ATP for fuel. Slow-twitch fibers require less ATP and more oxygen for the same amount of work. Generally speaking, higher percentages of slow-twitch muscle fibers are more efficient. Picking workouts that include both types are beneficial; however, short bouts of anaerobic increase the fibers most used by emergency responders.

Fitness Level

Obviously, the more fit you are, the more efficiently you will perform a duty. Let's say you have a resting heart rate of 65 and it doubles while swinging a sledgehammer for four minutes. You are far ahead of your partner, who has a resting heart rate of 80. She will soon find the gap between her oxygen demand and her maximum heart rate closing fast. A fit firefighter is more efficient because less energy is required for normal body functions, such as circulation and waste removal. This allows a higher percentage of the available energy to swing the sledgehammer.

Body Composition

Similar to fitness level, a firefighter with a heavier load to carry will have to deduct the energy needed to move his or her body from the energy available for work. If your arms weigh twice as much as your partner's, you need twice as much effort to make them swing. Your muscles might be stronger underneath the fat, but they will tire quickly. This not only takes away power that is available, but brings on fatigue quicker and reduces the energy delivered to the block wall. Did you ever see an offensive lineman pick up a fumble and try to run it back for a touchdown? This player tends to be pretty good at throwing off tackles for the first 10 yards, but overall efficiency starts to suffer the longer he runs. If he does end up breaking away, many times he falls from exhaustion. Firefighting is usually more than a 10-yard run.

Technique

The final component to ensuring maximum efficiency is the technique involved. It's not always the biggest guy at the carnival who can ring the bell. Sometimes it's the 16-year-old kid who breaks down semi tires at the truck stop after school. Unfortunately, many of the physical movements firefighters must complete are different than activities performed outside emergency services. Splitting firewood and breaking beads on truck tires is not the same as breeching a wall or forcing entry on a door. It is similar to roof ventilation with an axe, but when was the last time you did that at a real fire?

Training is the only way to become proficient at the skills firefighters need using the tools firefighters have. Besides the technique of swinging the sledgehammer, there's also a specific place to hit the cement block. Knowledge of block construction allows you to hit between the webs and maximize your effort.

SUMMARY

The physical demands of emergency response and firefighting on the human body have been well documented by two specific studies in Maryland and Indiana. The results prove what we already expected—the physical and mental stresses of firefighting tax our bodies severely. As a result, it's not uncommon for some emergency responders to exceed their physical limitations. Research uses data such as maximum heart rate to expose the dangers, and statistics display it as overexertion. Responders must prepare for the anticipated stresses well in advance of an emergency.

If you were to drive cross-country with your family, you certainly would check the vehicle thoroughly before leaving. Similarly, emergency responders must ensure that they are ready to respond. This requires a personal accountability to safety, which consists of continuous medical and physical maintenance followed by annual testing. NFPA has provided the standards, and organizations such as the NVFC and WFI have provided the endorsements. Emergency responders have a responsibility to those they protect, their fellow responders, their family, and themselves.

KEY TERMS

adenosine triphosphate (ATP) - A predominant source of energy for cells.

aerobic - Term used to describe metabolism in the presence of adequate oxygen; also used to describe sustained exercise requiring sufficient stamina.

alveoli - The air sacs of the lung tissue which transfer gasses between the lungs and the bloodstream.

anaerobic - Term used to describe metabolism without the presence of adequate oxygen; also used to describe short bursts of exercise requiring sufficient muscle exertion.

atrophy - The loss of muscle, generally from lack of use.

Candidate Physical Abilities Test (CPAT) - A type of criterion task test involving several firefighting activities with a focus on agility testing.

cardiovascular collapse - Overexertion, or the sudden inability of the heart and vascular system to meet the demands of the muscles.

chronic obstructive pulmonary disease (COPD) - A disease affecting the lungs and the ability to utilize oxygen. Some examples include emphysema and chronic bronchitis.

combat challenge - A type of criterion task test involving several firefighting activities, many times in head-to-head format as a race.

congestive heart failure (CHF) - Medical condition in which the muscle of the heart is unable to meet normal demands of the body, often resulting in fluid buildup in the ankles (edema) and possibly in the lungs (pulmonary edema).

creatine phosphate - A molecule that is stored in muscles and can readily be converted to ATP to provide energy.

criterion task test (CTT) - A type of test used to determine the physical capabilities of a person based on specific job functions. A firefighting CTT would likely include tasks such as wearing PPE while dragging fire hose, rescuing a simulated victim, and carrying equipment.

curvilinear - A term used to describe curved lines on a chart, especially helpful in estimating results. In this text, it refers to estimating the results of an action based on other influences which could affect the outcome in an unanticipated way.

diastole - The resting stage of the heart, also the lower number (denominator) of a blood pressure.

extrinsic - Affected by an outside or controllable source, such as smoking's effect on a disease.

fast twitch (Type II) muscles - A type of muscle that predominantly operates quickly using little oxygen.

fit for duty - A term used to describe a rule or policy that requires firefighters to accomplish the duties that they could be expected to perform in a specific amount of time. This could be accomplished through periodic medical physicals or the use of a CTT.

friction loss - The conversion of useful energy into nonuseful energy due to friction.

gallons per minute (gpm) - A common form of identifying a specific amount of water being pumped in a specific amount of time.

health and fitness coordinator (HFC) - The person who, under the supervision of the fire department physician, has been designated by the department to coordinate and be responsible for the health and fitness programs of the department.

Health Related Fitness Program (HRFP) - A comprehensive program designed to promote the member's ability to perform occupational activities with vigor, and to assist the member in the attainment and maintenance of the premature traits or capacities normally associated with premature development of injury, morbidity, and mortality.

International Association of Fire Chiefs (IAFC) - Organization of fire chiefs from the United States and Canada.

International Association of Fire Fighters (IAFF) - Labor organization that represents the majority of organized firefighters in the United States and Canada.

line-of-duty death (LODD) - Fatalities that are directly attributed to the duties of a firefighter.

maximum heart rate (MHR) - A theoretical number that an individual's heart can rise to before uncertain health risks can occur. It can be calculated by subtracting a person's age from 220.

metabolic equivalent (MET) - A measurement of work, assuming 1 MET is equal to a resting metabolic rate.

myocardial infarction - An event where heart muscle dies; a heart attack.

National Volunteer Fire Council (NVFC) - A professional organization for volunteer firefighters.

overexertion - Cardiovascular collapse, or the sudden inability of the heart and vascular system to meet the demands of the muscles.

oxygen uptake - The ability of an individual to obtain and utilize oxygen during periods of high demand.

parasympathetic - A component of the autonomic nervous system in which the body rests and restores.

peer fitness trainer (PFT) - An individual certified to help develop a fitness program and assist the members in achieving their goals. Similar to a health and fitness coordinator defined by NFPA.

peripheral vascular resistance (PVR) - A term used to describe the amount of pressure the heart must overcome to pump blood throughout the body. Results are commonly referred to as blood pressure.

personal protective equipment (PPE) - A generic term used to describe the minimum apparel and gear needed to safely perform a specific duty.

PM - Preventative maintenance.

psi - Pounds per square inch (psi).

pulmonary edema - A medical emergency where fluid accumulates in the lungs, often a result of congestive heart failure.

Queens College Step Test - A test that is capable of measuring oxygen uptake.

SCA - Sudden cardiac arrest.

slow twitch (Type I) muscle - A type of muscle that predominantly operates slowly using oxygen.

specific gravity - The weight of a liquid as it is compared to water.

spirometry - A pulmonary function test that can be used to estimate oxygen uptake.

sympathetic - A component of the autonomic nervous system in which the body uses adrenaline to react to a stressful situation.

systole - The pumping stage of the heart, also the higher number (numerator) of a blood pressure.

thermodynamics - The release of heat that occurs as energy is converted from one form to another.

trending - The identification of patterns in an effort to predict future occurrences.

vapor density - The weight of a gas as it is compared to air.

vasoconstriction - A body's ability to shrink the blood vessels to increase peripheral vascular resistance and blood pressure.

viscosity - The thickness of a liquid.

VO$_2$ max - Oxygen uptake.

wellness - A term used to describe the encouragement and promotion of a healthy diet, exercise, and life.

Wellness-Fitness Initiative (WFI) - A fire service management initiative.

REVIEW QUESTIONS

1. Why is improving the health, wellness, and fitness of emergency responders important?
2. What are some of the reasons why heat dissipation is difficult for firefighters?
3. What is the maximum heart rate and what influence do emergency responses have on it?
4. What is the difference between wellness and fit for duty?
5. What are the NFPA standards that are developed for medical and fitness application?

FIREFIGHTER WEBSITE RESOURCES

http://www.iafc.org/displaycommon.cfm?an=1&subarticlenbr=382
http://www.iaff.org/HS/Well/index.htm
http://firefighter-challenge.com/

NOTES

1. The Fire Service Joint Labor Management Wellness-Fitness Initiative; WFI 3rd Edition; 2008.
2. http://www.healthy-firefighter.org/atp/about.php
3. Smith, S. 2007. *The fire fighter conditioning workout plan.* [City: Pub]
4. Fahy, Ph.D., R.F., P.R. LeBlanc, and J.L. Molis. 2010, June. Firefighter Fatalities in the United States, 2009, and U.S. Fire Service Fatalities in Structure Fires, 1977–2009. NFPA Fire Analysis and Research Division.
5. NFPA 1583, Standard on Health-Related Fitness Programs for Fire Fighters, 2000.
6. Physiological Stress Associated with Structural Firefighting Observed in Professional Firefighters; Indiana University Firefighter Health & Safety Research; School of Health, Physical Education & Recreation. Department of Kinesiology; Bloomington, Indiana 2008.
7. Table 4.1; Stress Associated with Structural Firefighting Observed in Professional Firefighters; Indiana University Firefighter Health & Safety Research; School of Health, Physical Education & Recreation. Department of Kinesiology; Bloomington, Indiana, 2008.
8. Chief Billy Goldfeder; FDIC speech, 2009.
9. *Health and safety guidelines for firefighter training.* 2006. College Park, MD: Center for Firefighter Safety Research and Development, Maryland Fire & Rescue Institute, University of Maryland.
10. United States Court of Appeals, Eighth Circuit. *Smith v. City of Des Moines, IA.* No. 95-3802. November 12, 1996.
11. Kolata, G. 2005. Super, sure, but not much more than human. *New York Times,* July 24.
12. Davies, P. Energy Sources to Replace ATP. http://www.sport-fitness-advisor.com/energysystems.html
13. McArdle, W.D., F.I. Katch, and V.L. Katch. 2006. *Essentials of exercise physiology,* 3rd ed. Baltimore, MD: Lippincott Williams & Wilkins.
14. Cooper, MD, K.H. 1982. *Aerobics program for total well-being: Exercise, diet, and emotional balance.* New York: Bantam Books.
15. Stefano, M. 2000. *The firefighter's workout book.* New York: Harper Collins Publishing.
16. Gymflesh School Library; Firefighters' Edition Stop Loss Preparedness Protocol. Required Program Reading, Supplement MEHA-FF 06.4 The Firefighter Aerobic-Fixation Myth Exposed.
17. Firefighter Combat Challenge Rules (© OTC, 2009). http://www.firefighterchallenge.com/
18. Developing the CPAT. 2010. The IAFF/IAFC Fire Service Joint Labor Management Candidate Physical Ability Test Program Summary. http://www.iafc.org/displaycommon.cfm?an=1&subarticlenbr=389
19. Smith, S. The fire fighters workout. http://store.stewsmithptclub.com/newfifiwo.html
20. Fire Service Joint Labor Management Wellness/Fitness Initiative Candidate Physical Ability Test© Orientation Guide.
21. Peak Performance; State of Nevada Risk Management Division, Carson City, April 2008.
22. Fass, B. 2008. Fit responder. Precision Fitness, Personal Fitness Inc.
23. McArdle, *Essentials.*

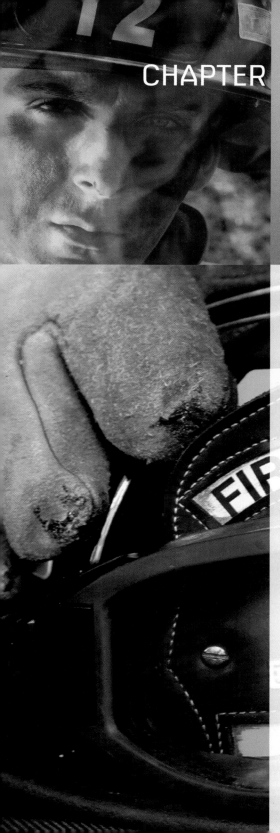

CHAPTER 7

Creating a Research Agenda

LEARNING OBJECTIVES

- Explain what a data collection system is and how it relates to the 16 initiatives.
- Explain how research and development could be used by emergency services to add another element of safety.
- Describe some of the ways research can benefit risk management.
- List some of the dangers of implementing changes without research.
- Explain the uses of statistical data from incident reports and the importance of accuracy.
- Discuss the differences between qualitative and quantitative data.
- Identify the effective components of hypothesis, and how it can be used in research.
- Discuss the unique problem of "exposure" as related to injury and LODD data.
- Explain how data collection, analysis, and utilization are used in conjunction with each other.

INITIATIVE 7. Create a national research agenda and data collection system that relates to the initiatives.

It's been said that one of the greatest problems with the advancement of the fire service is the lack of hard data available. Many feel law enforcement has been more successful at collecting and compiling data in regard to their needs, and thus have been able to secure more grants and resources. It's not uncommon for a police department to have extra personnel on duty for traffic enforcement or **school resource officers (SROs)** as a direct result of those grants. It's not that law enforcement is more important—it's just hard to justify what we do when we can't prove what it is that we do.

As a result, the fire service has worked hard over the past several years to try to collect data and get grants to the forefront. For example, utilizing information gained from NFIRS and converting that into legislation for Fire Act grants has proven successful. Unfortunately, statistical reports are only as good as the data that is recorded. Let's say you are the officer in charge of an engine company, and have been dispatched to a vehicle fire on the interstate. As you arrive on scene, you find the hood up on a late-model SUV with nothing showing. The occupants are in the median and meet you as you exit the engine. They report that there was smoke under the hood, so they pulled over and found a small fire on the motor, which they extinguished with a cup of coffee. You ensure the fire is out and that a tow truck is dispatched. While gathering information for the report, you get sent to another alarm, which evolves into a pretty busy morning. After lunch, you finally get a chance to sit down and try to catch up on reports. You pull out your notepad with the vehicle and contact information from the car fire, and find that you are missing some key information for the report.

As you scroll through the choices of "Incident Type," you weigh the options of picking "vehicle fire" or something more along the line of a "good intent" call. You were dispatched to a car fire. The evidence you found in the engine compartment and the statements by the owner both confirm that indeed there was a fire. However, the vehicle fire report is fairly lengthy compared to picking a simple good intent. There may have actually been just a short circuit with no actual "fire," and the fact that someone called you to help the owner could justify good intent. Besides, if you choose vehicle fire, you'll be missing some key information that you're pretty sure will get kicked back as incomplete. As a result, you choose to complete the much shorter "good intent" report with no actual fire.

This is the problem with any data. Computer programmers sometimes refer to this as "garbage in, garbage out." Inaccurate data that is entered cannot produce accurate reports (Figure 7-1). Although the error of referring to a car fire as something else is not in itself devastating, the overall effect of misrepresenting fires hurts the department and the fire service in the long run. The incidences of fires will always be "down" if we never report them. Retired Captain Tom Brennan, FDNY, noted the problem and made his opinion clear in a speech in 2002 when he said, "What the hell's a good intent call? It's a fire!"[1] Accuracy is important because information obtained directly from incident reports represents one of the most commonly used types of data in the fire service. As a result, fire departments that submit no data provide little help to the fire service in general.

Once data is obtained, it is of little use until it is researched, analyzed, and utilized. Initiative 7 calls for a **national research agenda,** utilizing a complete data collection system. A research agenda would clarify what data needs to be collected, and identify the problems with the way information is recorded today. A research agenda would likely utilize statistics gained from incident reports, but would also include components of

FIGURE 7-1
Accuracy of our data is vital to research.

the fire service that would benefit from more research: specifically, the content of the other 15 initiatives. For example, Initiative 9 has the goal of improving the way we learn from **line-of-duty deaths (LODDs),** injuries, and near-misses. An effective research system would establish how the data are submitted, organized, and scrutinized so that positive changes could be instituted.

RESEARCH AS A TOOL

Research can be used to benefit the safety of emergency responders. Some components of research used by other industries that could assist us in our goal include different types of research, the categories of data, utilizing a research **hypothesis,** and using research and development.

Types of Research

Although there are many different types of research, most can be categorized as basic or applied. Basic research is conducted out of curiosity, which leads to no immediate product or treatment. Let's imagine a scientist discovers that certain fluorescent dyes, when introduced into someone's bloodstream, tend to stick to hormones secreted by people who are under extraordinary stress. On the surface, it may not appear that it was a significant scientific breakthrough, but the finding allows others to imagine what effect it might have on firefighter safety. This basic, or pure, research allows applied research to step in and take over. Imagine that another scientist develops a harmless additive to a sports drink that contains this fluorescent dye and another creates a skin electrode capable of identifying the dye in the bloodstream when it is attached to the stress hormone. Essentially, firefighters who consume the sports drink in the morning are now able to have their stress levels monitored through a sticker on their arms throughout the day. In this example, applied research has developed a system to monitor how much (chemical) stress an individual is experiencing using information gained from basic research.

Categories of Data

Data also can take different forms, falling into the broad categories of qualitative or quantitative. **Quantitative data,** or empirical data, consists of hard numbers. A "black box" installed in a fire engine can record data of a quantitative nature to use in analyzing the speed of the vehicle and braking techniques to use in crash reconstruction or driver evaluations. **Qualitative data** takes into account the feelings and opinions of people involved. Other firefighters in the apparatus are good sources of information and can provide qualitative data as to the performance of a driver. Witnesses to the crash could provide insight as to whether the driver/operator appeared to maintain **due regard** for the safety of other vehicles and pedestrians. Many states have laws that allow drivers of emergency vehicles to operate outside the scope of normal traffic laws, yet still require that they operate them safely, protecting the public from injury with due regard.

Using a Research Hypothesis

One of the most common research processes outside emergency services involves the introduction of a hypothesis. A hypothesis is a prediction of why something occurs or will occur based on scientific evidence. Theories are more general in nature, and don't follow the same format. For example, suppose your department wanted to improve risk management on the fireground. Several key components have been identified, and your group is currently working on a process to limit the risk to first-arriving firefighters at residential structure fires. You have already been provided with quantitative data, including:

- Past incident data that includes incident times, the number of personnel, duties, etc.
- National statistical data on firefighter close calls, injuries, and deaths
- NFPA Standards, including 1500, 1710, and 1720

Qualitative data has also been received, consisting of:

- Officer surveys regarding standard priorities and concerns on the fireground
- Firefighter surveys regarding personnel required for specific duties
- Suggestions from members on apparatus response, staffing, and automatic aid
- Opinions from members on the most serious scene threats and close calls

Several hypotheses may be developed for different safety aspects of the first-arriving units. One such hypothesis might be: "If improper or incomplete size-up is a contributing factor in many close calls, injuries, and LODDs, then increasing training in size-up will make our firefighters safer." (See Figure 7-2.) Notice that the first half of the statement is based on fact, and the second half draws a conclusion from it. A good hypothesis must be testable, and thus supported or disproven. In this case, your team identifies several examples that support the hypothesis from information in the qualitative and quantitative research, supporting it. A training system is developed to improve the size-up skills of firefighters, testing the hypothesis. By evaluating the firefighters' knowledge before and after the drills, the hypothesis may be supported even more. Other hypotheses your group may come up with might include the following:

- If standards and rules require firefighters to wait until adequate members are assembled to safely launch an interior attack in most situations, then firefighters should be trained about what an adequate number is, and what exceptions there are to the rule.

FIGURE 7-2 Size-up has such an effect on firefighter safety that training should continually emphasize situational awareness.

- If we determined that an adequate number of firefighters needed to safely launch an interior attack is seven, then a system of ensuring a minimum number of seven firefighters are dispatched on the initial alarm is required.

In the second example, the hypothesis is assuming that the group had already identified seven firefighters as their minimum for launching an interior attack under normal circumstances. You might also note that the hypothesis doesn't conclude in training like the others, but instead an SOP/SOG change for dispatching apparatus. If all your hypotheses conclude the same way, you may be missing some other important components of the system.

An interesting fact about a hypothesis is that it cannot be proven, only supported. It's impossible to prove that training firefighters will make them safe, and that a minimum of seven is the magic number for fireground safety. However, if they are supported by evidence and facts, they are safe assumptions to make, and when assembled into a system, they increase fireground safety. The more hypotheses that are created and tested, the better the chances for success in achieving the goals of your department,

Research and Development

It's not too often that research and development (R&D) is addressed in the fire service unless we're talking about new equipment or technology being introduced by a manufacturer. Let's consider for a minute that you are plant manager for a lawn mower manufacturer in charge of new products. An employee asks why propellers and fans have at least three blades but lawnmowers only have two. You contemplate the idea and decide it might be worth trying.

Your first step would be to create plans in the engineering department. Next would involve constructing your prototypes. You'd make sure you had your best fabricators on the project to ensure it was built to specs and balanced perfectly. Tests would have to be initiated to see if existing motors had enough horsepower to run the blades and what the optimal RPM was. Performance of the blade would be checked to ensure that the cut was superior to normal models. Research would also include the feasibility of converting existing machinery to build the new blades and cost projections for manufacturing. Market analysis would identify the applications of the blade and establish projections on profit. Assuming it was beneficial, corporate would give you the "green light" and begin a marketing strategy to ensure success.

If we were to apply this everyday scenario from the manufacturing world to researching line-of-duty deaths (LODDs), we would have a proven template for creating solutions. Initiative 7 identifies the need for better research, and research is worthless unless development follows. R&D is a vital component to business strategies and contributes to the success of the company. Sometimes firefighters look at the causes of LODDs and see a complex problem with no easy answer and relegate it to the back burner. The fact is that businesses run into the same situations, but instead face them head-on. If they don't, they're out of business. The authors of *Fundamental Issues in Strategy, A Research Agenda* compare insurmountable business problems with no visible solution to treating a person with cancer.[2] They ask, "Isn't the promise of an answer worth a try?" The person who has terminal cancer has nothing to lose by trying new treatments. The person who has yet to get cancer is the one who loses if no attempt is made. If a firefighter is shot and killed by a patient as he approaches at a motor vehicle crash, some might say it wasn't preventable. However, there is no harm in researching exactly what happened to him. Identifying any mistakes that were made or contributing factors can lead to engineering plans to prevent it from happening again. They can be passed on to our best fabricators who convert the plan into a training program that is tested, compared, and adjusted. The real harm occurs when nothing is learned and another firefighter is killed in the same fashion.

NATIONAL RESEARCH AGENDA

A coordinated National Research Agenda would organize and prioritize what research is needed to save lives in the fire service. Firefighters have always used research to make firefighting safer. For instance, Lloyd Layman did numerous tests on shipboard fires in the 1940s and proved that the fog nozzle, used in conjunction with an indirect attack, can be a safer and more effective way to extinguish fires in a location with limited ventilation. His testing continued utilizing the procedure in structural firefighting.[3] Research was then picked up by a group at Iowa State University who calculated required water for extinguishment along with developing the combination method of attack. Although we still teach these tactics and they have proven effective, there are still questions about how modern building materials and methods affect attack, and how much fire is actually "pushed" with a fog or combination nozzle. Most firefighters that learned structural fire attack in the last 25 years learned that pushing a fire the wrong way can actually burn down the rest of the house. As a result, hitting a bedroom fire from an outside window is a poor tactic that will not work. Some fire researchers suggest now that pushing a fire might be more dependent on the size and layout of the house rather than whether you're attacking from the front yard or the hallway.

Let's say you are returning from fueling the engine by yourself when a woman and her children in their front yard flag you down. You notice there is smoke emitting from the open front door of their small ranch. As you call dispatch to report the alarm, you see that the fire is a fully engulfed couch in the living room. The mother says there is nobody in the house. As you put on your gear, you weigh the options for your next move. Risk management tells you that in this situation, you need to risk a little to save a little. Therefore, setting the pump and entering the structure by yourself is not an option. Your training says that if you hit it from the open front door, you'll push it into the kitchen. Your instinct and the homeowner's yelling say put out the fire. The real problem here is that what you've been taught often flies directly in the face of safety. That's perfectly acceptable as long as what you've learned is based on facts. Facts allow you to weigh the options and make an educated choice. In other words, if you pack up and hit the fire from outside the doorway or window, do you have a 20% chance of knockdown and an 80% chance of fire spread? Is it more like 80% knockdown and 20% fire spread? What signs should you look for to hedge your bet one way or the other? What about using a smooth bore nozzle at 50 psi versus an automatic combination nozzle at 100 psi set on straight stream? It's impossible to make an educated decision without facts, and facts only come from sound research.

In 2005, a group of fire service professionals with the goal of creating a national research agenda met on the grounds of the **National Fire Academy (NFA).** In an effort to lay the groundwork for Initiative 7, the National Fire Service Research Agenda Symposium was held. The group of participants created and ranked specific areas of the fire service that could benefit the most from additional research. The Executive Summary[4] from the symposium outlines some of the key components to the research agenda by severity or importance, whereas this text has arranged them by the initiative.

Defining a Cultural Change (Initiative 1)

In Chapter 1 we discussed the advantages and difficulties of establishing a safety culture. Because an existing culture is based on beliefs and attitudes, research specifically devoted to the values and resistance to change would be beneficial to ensure a successful change. As with accountability, cultural change must take place both individually and organizationally. For example, let's consider the fact that some firefighters are photographed with chin straps unfastened or strapped around the back of the helmet at emergency scenes. Individually, researchers might look into the reasons why some firefighters fail to buckle them. They might find several explanations:

- "The helmet will still protect my head without being fastened."
- "The emergency didn't dictate that the helmet be fastened."

- "The strap takes too long to fasten in emergency situations."
- "The helmet looks better without a strap."
- "The SOP/SOG states that a helmet will be worn. It says nothing about a strap."

By identifying the reason for the resistance, a number of factors could be applied: education (training and practice), engineering (researching better straps), environment (making straps look cool), or even enforcement (punishment for not using them). Similarly, organizational research could expose that some leaders don't believe straps are worth their attention or that loose straps even pose a safety risk. If the organization doesn't see a problem with an unsafe situation, it's not likely to improve.

Researching the existing culture and how it affects safety would involve much more than just the usage of helmet straps. A bulk of the research would likely examine **risk management** and how the culture of emergency responders affects decisions. Change, however, doesn't have to be negative. Rather, the identification of motivators can be utilized to achieve job satisfaction while increasing safety. **Maslow's hierarchy of needs**[5] theorizes that the only way to achieve self-esteem or self-actualization is to first believe that you are safe. Our existing fire service culture has, in a way, circumvented this natural order by convincing us that we are *safe enough* because our job is inherently dangerous and that we are protected from the effects of fire with our training and **personal protective equipment (PPE).** This allows some of us to progress through our careers with perceived invincibility. Researching ways to instill safety into our culture without upsetting tradition is a great way to start.

Enhancing Accountability (Initiative 2)

Initiative 2 is probably one of the easiest initiatives to understand and agree with in theory, but is one of the most difficult to achieve. Accountability is really nothing more than a decision, and it starts with **personal accountability.** A conscious decision to accept accountability for safety doesn't take an act of Congress or even a policy from the fire chief. It's a grassroots effort that spreads easily to your coworkers, your company, and eventually the department. If you are accountable for safety, you refuse to move the engine until everyone is buckled. Others see it, and tend to follow your lead. More research involving ways to improve personal accountability could shed light on new systems for teaching accountability.

Organizational accountability, on the other hand, is generally a corporate decision. It's not a matter of just adding it into the department's mission statement, but involves accountability for safety in every policy, procedure, decision, and change. An organizational change to accountability for safety is a formal adoption, which starts at the top and leads the department into change.

A simple example of a decision for personal accountability was suggested in Chapter 3, when discussing radio communications. By making a decision to actively listen to *all* radio traffic and give it a priority on scene, you increase the chances of receiving all information that is broadcast. Others who work with you on scenes can see the emphasis you put on radio traffic, and may pick up the good habits. Before long, the radio communication of the entire department could improve. Organizational accountability might address researching the upgrade or total redesign of existing radios, a culture of poor radio habits, providing communication training to all personnel, and holding them accountable for their performance.

Applying Risk Management Techniques (Initiative 3)

Many firefighters feel that risk management is an area in which continuous improvement has the best opportunity to increase safety. As a result, a reader might question why risk management is not addressed in the first two initiatives. One of the best explanations is that without a cultural change and personal accountability for safety, risk management is less likely to succeed. For instance, suppose your department had an experienced officer that led crews into some dangerous situations until his eventual retirement. In one specific instance, he forced entry into a well-involved, abandoned bowling alley without considering any risk or benefit. Although nobody was killed, the building soon collapsed as firefighters barely escaped through several different exits. Risk management training would have likely been scoffed at by him as a result of his cultural beliefs. He may have truly believed that his job was to get in the building and put out the fire, and that his experience is what prevented him from getting killed. He essentially had no personal accountability for his safety or the safety of his crew. Risk management can't coexist with this type of an antiquated belief system.

Today, risk management is at the forefront of some of the biggest discussions in the fire service. The advantage of hot topics is that they tend to invoke a lot of passion and drive that evolves into actions. Unfortunately, without facts they are prone to misconceptions and mistakes. Research can pay great dividends in the early stages of these new ideas. Take air management, for example. For a long time, the fire service referred to air bottles as "a time" (e.g., a 30-minute bottle). It's easy to jump to conclusions or do limited studies at your burn building about how long air bottles last and make changes accordingly in policy. However, whether the policy is correct or not is really dependant on how accurate the research was. In the Indianapolis study discussed at length in Chapter 6, "Developing Medical and Fitness Standards," the average time for a "45-minute" bottle at a structure fire was only 28 minutes.[6] If the average of these *fit* firefighters was only 28 minutes, imagine what you can expect out of an unfit member of your department. Without solid research, any new topics under the umbrella of risk management will be less effective.

Scientific Community Risk Assessment

Statistics and experience often lead decisions about what type of equipment to purchase and what emergencies to train for. Sometimes departments invest resources, instead, on opinions and feelings. For instance, you may know of a department that wanted a heavy rescue truck so badly that other needed equipment could not be funded. The rescue truck may actually get more use at parades and community functions than on emergency calls. Risk assessment systems have been in place for some time, but may have lacked scientific methods. Software packages have been developed to address some of these concerns, but more scientific research and validation is needed in the area.

Let's say you are considering the purchase of the rescue truck. An advanced software package might allow you to enter information such as the miles of different types of highway you cover, weather patterns, traffic makeup and flow, hazardous materials being transported, and emergency vehicle access. Other information could consist of local crash history, adjacent or regional resources, and manpower available. The software may utilize national statistics for similar components, factor in trends, and produce a statistical analysis of the risks associated, which could be used to make decisions based on need rather than opinion. The obvious benefit is when preparing budgetary needs.

Flashover and Collapse Prediction

There has been much discussion about the way that buildings burn. From the construction materials to the methods, we aren't fighting our grandfather's structure fire. The **National Institute of Standards and Technology (NIST)** has done tests that show how buildings burn and how our estimates about time till flashover are outdated.[7] Additional research would be beneficial for the collapse potential of modern building systems. Results of the research, coupled with flashover statistics, could give firefighters and officers clues in order to evacuate before a catastrophe occurs (Figure 7-3). As building fires become less frequent, experience has less of an effect on the decision-making skills of incident commanders. As a direct result, situational awareness suffers.

Situational Awareness

In 2008, 40% of all construction worker deaths on nonresidential job sites occurred as a result of falls.[8] Like firefighters, construction workers have to concentrate on other

FIGURE 7-3 Research should be based on statistical data of near-misses, injuries, and fatalities, such as building collapse and flashover.

tasks while working at heights, and tend to let their guard down. It would probably be safe to say that many of the workers who fell simply lost track of where they were as they concentrated on their job and became complacent. Lack of situational awareness has been identified in many fireground fatality reports as well, specifically by the **National Institute for Occupational Safety and Health (NIOSH).** Theses reports include when firefighters apparently lost track of where they were, how much air they had left, or what the fire was doing. The **Occupational Safety and Health Administration (OSHA)** dictates both proactive protection (railings) and reactive protection (tethers and cables) for some industries. Unfortunately, our PPE just can't hold up a collapsing building or protect us from a flashover. Reactive responses have limited success in our field. Military operations utilize a "cover me" attitude (proactive) when faced with similar situations. Once the decision has been made to make entry, some of the situational awareness load has to be carried by personnel outside of the hot zone. Researching ways to monitor personnel, the building, and environmental conditions through technology can pay huge dividends. Similar types of technology are already used in other fields, but need to be adapted to withstand the rigors of structural firefighting.

Wildland Firefighting Operations

Risk management is not limited to structural firefighting. Besides decisions on tactics (reactive), research needs to be supported in the planning of wildland interface zones (proactive). Sometimes firefighter fatalities are a result of protecting a vacant home in a subdivision that shouldn't have been built. Regulations and the enforcement of codes consistently save firefighter lives, but aren't always adopted. Researching what practices have been successful and enforceable in the past would benefit the fire service in general.

Another potential area of research for interface problems is improved real-time mapping. Besides simply locating vegetation, advanced methods might identify types of organic material and their moisture content to allow better risk management decisions for specific areas. Computer modeling of slope and wind-driven fires applicable to specific areas would also be beneficial in forecasting the direction and possible threats of a moving wildfire.

Risk Management for Incident Management

Managing risk has been identified as a component of incident command, but that might not be the best way to categorize it. Because every decision in incident management is based on risk, the hierarchy might be more accurately demonstrated by placing incident command as a critical element of risk management. In other words, incident command exists to support risk management. Researching ways to define how risk management can be implemented in every decision at every incident should be a priority. The system must be applicable to all risks at all incidents and, like incident management, should be expandable. It should be easy to implement by officers initially on scene. The system would involve training, application, and evaluation phases.

Eliminating Unsafe Acts (Initiative 4)

At the heart of **crew resource management (CRM)** is the empowerment of individuals to identify and prevent unsafe acts. As discussed in Chapter 4, the aviation industry has successfully used CRM to improve safety and performance. The maritime industry has taken the aviation version and adapted it to their specific needs. Fire service research, as well, could investigate components applicable to emergency services and formulate an effective system to implement.

For example, consider a crew using CRM in an emergency situation. The resources available to an airplane that is unable to fly or a boat that is sinking in the middle of the ocean are limited to the personnel on board. Although willing help might be on the other end of the radio, a rescue by another aircraft or vessel could be out of the question. As a result, CRM maximizes the use of all personnel on board to improve the chances of a favorable outcome. Now consider a truck company crew that is trapped completing a search in the basement of a pet store. In this case, resources are available on the scene and are capable of performing a rescue. CRM as used in other industries is of little help in assisting rapid intervention. Researching ways to implement the key concepts of CRM into rapid intervention would be beneficial in making CRM our own (Figure 7-4).

FIGURE 7-4
If CRM is adopted by the fire service, research needs to examine what changes must be made to existing programs to be effective in emergency services.

Implementing Training and Certification Standards (Initiative 5)

Recruit-Level Training

Certification starts with recruit training, and research needs to begin there as well. Fire candidate training has historically followed paramilitary parameters. Success on the fire scene is crucial to learned skills such as discipline and control. The problem is that we expect the new breed of firefighter to also know how to incorporate risk management and situational awareness. One way to make a difference would be the addition of the 16 initiatives into all recruit training programs. Although safety proficiencies are not necessarily in contrast with order and duty, they are not presently taught the same. One of the areas of research with the most potential for improved safety in the fire academy is to look at variations of the paramilitary organization for application in the fire service. An example would be one that builds empowerment into the organization and expects input from members of the crew when safety is involved. This would likely utilize the challenge and response model as a component of CRM, as explained in Chapter 4.

Another area that could be looked into is the disparities between certification levels and the duration of initial training programs. Most instructors across the country realize that similar certifications should involve teaching the same cognitive curriculum and skill sets, and include minimum contact hours. The real dilemma is which state has the most applicable program for our needs to use as a model. One of the greatest benefits of university-type research is that it gives a third-party view that is most effective when sensitive issues such as this are evaluated. Third-party research would eliminate most of the politics involved in the process of deciding which initial training program is the most beneficial for the fire service. After development, the program could be adopted by an organization that operates the testing and certification system much like the National Registry for EMS certifications. States could then adopt it in a similar way.

Instructor Development

It's difficult to discuss changing basic training without researching the best way to improve our instructor skills as well. Most of us have had a high school or college instructor that refused to change with the times. Students continue to evolve in regard to their norms and values, as well as the way they learn. We've identified and labeled these groups with titles like **baby boomers** and **Generation Xers**. An instructor who cannot adjust the way he or she teaches to different groups over the years has a difficult time being effective.

Additionally, many seasoned instructors may have difficulties implementing risk management and empowerment—or even safety in general. Research could help identify some of the cultural issues and opinions that are allowing outdated views of heroism to be cultivated by instructors. If a fire instructor won't wear a **self-contained breathing apparatus (SCBA)** while "teaching" a crew how to properly extinguish a car fire, it has the same effect as a drivers' education instructor refusing to wear a seatbelt in the passenger seat. "Do as I say, not as I do" doesn't work on teenage drivers and it doesn't work on firefighters.

Live Fire Training

Although live fire training is valuable to firefighters' education, it also carries many of the inherent risks associated with genuine fires. Much data exists in the reports of near-misses, injuries, and deaths that can be researched to identify common critical or contributing factors. By analyzing the components, conclusions can be incorporated into training materials, instructor updates, and future revisions of the **National Fire Protection Association (NFPA)** 1403 Standard on Live Fire Training Evolutions. By instituting changes in the way we train, in effect, we change the way we act in actual emergencies.

Training Simulators

One of the most interesting components of future training programs incorporates the use of training simulators. Although prevalent in the military and aviation industries, firefighting simulation generally lags behind. Video games have evolved to a point where football players have the same physical characteristics and skills of the actual players they depict. Auto racing games are so realistic that sod torn up from a car sliding through the grass on lap 78 is still evident in every camera view for each of the remaining laps of the race. Technology is available to simulate actual firefighting experiences in a virtual reality realm. Some of the benefits of simulation include:

- Training in a safe and controlled environment
- Training in skills difficult or impossible to re-create
- Training in human issues, such as risk management and situational awareness

Research can investigate the latest capabilities of simulators, and how other industries utilize them. Care must be taken to program success only when proper tactics and safe procedures are followed. If you are playing a video game and know that a specific batter will hit a fastball down the middle for a homerun, you won't throw it. Similarly, if you know the building will collapse on you when the truss system is involved, you won't go in.

Professional Development

Leadership and management skill training is very limited at the entry officer level. Although we need officers who can organize a successful rescue, we can't afford to let them learn how to manage "on the job." Although some promotional testing takes into account leadership abilities, it's questionable how accurate the results are after only one afternoon of interviews. You may know of some officer candidates that did well on the test but can't make a decision at emergency calls. Others may be great leaders on scene, but just can't pass the test. Additionally, there aren't well-defined courses available to help a promising candidate succeed. Researching the way business or the military train their leaders could establish plans for officer schools or preparation courses. For instance, NFPA 1021, Standard for Fire Officer Professional Qualifications, has four specific levels, but many regions don't even offer a way for an interested candidate to get certified.

Computer-Based Training and Education

Postsecondary education has established a trend toward distance learning, most commonly utilizing computer-based education. Some colleges and universities have devoted a substantial amount of resources to the backbone of the systems, allowing fire and emergency service classes to offer their curriculum as well. The advantages of Internet-based courses for emergency responders include flexible scheduling around shift work as well as the ability to work at a preferred pace. Technology also adds greater dimensions to adult learning. For instance, let's say you are taking an online course for fire hydraulics, and the instructor added short videos while performing friction loss calculations on the board. If you are having trouble understanding a specific calculation, you may be too embarrassed to ask for help in class. At home, you can play the videos over and over until you understand the formula completely. Emailing a question or discussing it in a forum with other students is far less intimidating than speaking up in class (Figure 7-5). Investigating colleges that offer online versions of classes and their success rates would help other cognitive materials move to the computer.

FIGURE 7-5
Internet-based training and education should be utilized when possible to increase safety.

Developing Medical and Fitness Standards (Initiative 6)

Health Maintenance

The Fire Service Joint Labor Management Wellness-Fitness Initiative (WFI) has laid the groundwork for proving the value of a wellness and fitness initiative in career departments through cost-benefit analysis. By comparing prevention program costs with injury and time lost expenses, initial results show a substantial savings for health maintenance. By expanding the program to willing departments and creating a database for all results, data would be produced to allow further research for the individual components of an effective system. Eventually, research can help create a way for part-time and volunteer departments to participate.

Cardiovascular Disease. Cardiovascular disease among emergency responders also needs to be researched. Very little research has been conducted on firefighters with heart disease who are exposed to the rigors of their job. By identifying the influences of cardiovascular aggravation in hostile environments, corrective measures or diversionary tactics could be employed. For instance, some heart attacks occur when an area of plaque in an artery ruptures. If the body mistakes this injury as bleeding, it works to try to clot the blood, causing the heart attack. If a firefighter is identified as having these susceptible areas, research may show ways to reduce the chance of a rupture during times of high exertion or ways to prevent the clotting mechanism from engaging if a break occurs.

Cancer. Along with cardiovascular disease, cancer should receive significant attention when it comes to research. Carcinogens are cancer-causing agents, and they are abundantly supplied at fire scenes. However, little is known about the events that can lead to cancer. A firefighter who collapses on scene and is revived en route to the hospital can be diagnosed with a massive heart attack that can be attributed to reloading the hose on the engine. Another firefighter assisting may be fine for years until a lump develops on her neck that turns out to be cancerous. It's sometimes difficult to prove (presumptive) that the cancer was related to firefighting. It's even more difficult to blame it on one specific incident, while it's simply impossible to determine the specific event. Was it completing overhaul in the garage when she removed her glove and toxins were absorbed into the skin or when she removed her airpack and took a breath? There are four routes of entry for poison, but do we know if absorption of a specific carcinogen is worse than inhalation of it? As with cardiac events, studies have been limited when it comes exclusively to emergency responders.

Reproductive Health Issues. Finally, reproductive health issues should also be researched. Rumors of anomalies specifically affecting firefighters in regard to infertility and birth defects would benefit from research. Studies might show that emergency responders have higher instances of reproductive problems based on specific incidents or job classifications. The effects of chemicals that create birth defects (**taratogens**) for male or female responders is limited. Research might show links or trends, suggesting changes in assignment for at-risk responders.

Physiological Response to Emergencies

Both the Maryland and Indiana research projects discussed in Chapter 6 explored the effects of firefighting on the human body. Now that many stressors have been identified, future research might build off the results and look into ways to condition the body to be better prepared. **Functional capacity evaluations (FCEs)** are a series of tests commonly used to monitor the rehabilitation of workers. Other projects might involve studying technological advances that either reduce workload or monitor the physiological responses of crews. Results could be used to influence incident rehabilitation procedures.

Candidate Physical Assessment

As discussed in Chapter 6, the **Candidate Physical Abilities Test© (CPAT)** was created to evaluate the physiological capabilities of a candidate. Because it's a relatively new test, the program could benefit from research that evaluates the present program and considers applicable changes. Further adaptation of the program can ensure a safe and reliable measure of physical agility and explore its use for ongoing physical training.

Creating a Research Agenda (Initiative 7)

Data Requirements

The first step in ensuring that any research conducted is accurate is to establish the requirements for data. Besides ascertaining what specific data are needed, formats, parameters, and rules must be created. For instance, one of the most preventable LODD causes in the fire service is motor vehicle crashes. Data collected for crash research must include:

- Fire department versus **personally owned vehicles (POVs)**
- Vehicle type, size, weight, and year
- Maintenance and inspection records
- Road and weather conditions
- Time of day; day of week
- Hours into shift
- Driver impairment
- Emergency versus nonemergency
- Compliance with laws and procedures

- Safety equipment use
- Driver qualifications and experience
- Investigation results

Data Related to Exposure

One of the unique problems for fire service data is exposure. Let's say you are at a party and are introduced by the host to a guy who used to play baseball for the Mariners. You don't recognize his name, but he says he played for seven years. You might guess that he's a pretty good hitter to be around that long, but don't really know without his stats. In baseball, a player's batting average takes into account the number of "at bats" compared to hits. A batter with an average of 1000 could be phenomenal, or may have only been up to bat once and got lucky. Over the course of a season or career, "at bats" continue to mount, making the batting average more accurate. Unfortunately, a firefighter NIOSH LODD report can only use years of service, and classify someone as career or volunteer. Generally speaking, more years means more experience, but we all know that's not true. You might rationalize that a career firefighter has more experience than a volunteer of the same years, but we know that's not necessarily true either. Regrettably, we lack the means to track statistics for experience. Some departments are busier than others, some members tend to always be on the nozzle, and some seem to disappear on scene.

Another issue with exposure is hours spent on fire prevention activities, both fire safety inspections and safety education. It's difficult enough to track success with prevention without factoring in that many times it's not documented. If you get sent to a block party where a discussion about fire extinguishers evolves into a class for a dozen adults, do you record it? Even if your department tracks the information, what becomes of it? Let's assume that one of those adults actually has a fire in her kitchen and she extinguishes it the way you taught her. Unless she tells the engine crew that she learned her skill from you, you have no way to connect the dots. At the end of the year, if our fire loss went down, we generally assume it's a result of good fire prevention, but have a difficult time proving it. Worse yet, if she never calls the fire department after extinguishing the fire, we know even less.

Data Sharing and Access

Research can pay dividends by identifying other organizations or agencies that are presently collecting data related to our needs. Hospitals, insurance companies, and other governmental agencies may already have systems in place for their needs (Figure 7-6). By investigating what's already out there and how to partner with those groups, we can improve our chances for success. A **geographic information system (GIS)** such as ArcInfo® uses digital photos and map data to create 3D maps with different layers for different needs. If access is possible, overlaying water mains and hydrants on district maps allows an incident commander to make strategic decisions from the command

FIGURE 7-6 Emergency response agencies are not alone in data collection and research.

vehicle. Traffic cameras might not have been developed for emergency response, but access to real-time video can be effective if convenient. There may be other agencies that could benefit from a joint venture that haven't figured out how to make it work. Working together can improve the speed, accuracy, and quality of the information by ensuring different layers of safeguards.

Utilizing Available Technology (Initiative 8)

Respiratory Protection

Since the early 1980s, SCBAs have become standard equipment for firefighting. Unfortunately, they aren't drastically better than they were 30 years ago. Of course, there have been improvements, but respiratory protection is not yet where it needs to be. Regulators have been moved from the belt to the mask and now provide positive pressure, while bottles have become lighter and now carry more air. Technology has been added in the form of redundancy alarms, integrated PASS, and heads-up display. But running out of air and mechanical malfunctions still occur. Compare a 1981 Toyota with a 2011, a Rand McNally folding state map with your GPS, or a Commodore 64 computer with the processing power of your new cell phone. Time and technology has made many things we used in 1982 obsolete.

A new generation of respiratory protection has recently begun to take shape. For the first time in 40 years, an SCBA has been developed that doesn't even resemble an SCBA. Research is needed to ensure that as new equipment is phased into the fire service, it meets the needs but also increases safety. Lightweight and durable materials, such as carbon fiber, are common in other industries and could be used to create PPE that reduces the work of firefighting tasks while increasing protection. Developing hybrid models that utilize air purifying or rebreathing technology in conjunction with SCBA should be examined. Lastly, technology must undergo more comprehensive testing in hazardous atmospheres. Rumors and reports of electronic failures on existing equipment due to high heat or moisture must be researched to ensure that the next generation is more resilient.

Fire Detection and Extinguishing Agents

We know that fire is a chemical process, and that detection and extinguishing agents are designed to recognize or stop its progression. Unfortunately, most current systems aren't involved in the extinguishment process until the open flame stage (reactive). Many times in this book we have identified the advantages of being proactive versus reactive. Detecting and extinguishing fires at the incipient stage would save lives and property. This area of research has numerous possibilities.

Although water is still the extinguishing agent of choice for most fires, combustible metals have become much more prevalent in vehicles and other mobile equipment, challenging both firefighting tactics and safety. Although Class D fire research hasn't led to many significant changes, there has been an increased production of new additives and foams for Class A and B fires. **Compressed air foam (CAF)** systems and stronger concentrations of **aqueous film-forming foam (AFFF)** are some examples. Environmental concerns have influenced research for alternative "green" chemicals, with extinguishing agents such as **Halon** being phased out. Although production of Halon ceased in 1994, present supplies are expected to outlast their usefulness.[9] Research continues to find a replacement agent suitable to make Halon obsolete. Many feel that fine mist (aerosol technology) may provide some of the answers.

Investigating Fatalities, Injuries, and Near-Misses (Initiative 9)

Resources such as the **Everyone Goes Home (EGH)** campaign and the **National Firefighter Near-Miss Reporting System** have created ways to share statistics and information gained from the fire service for the fire service. Continuous research is needed to look for future trends as well as the results of mitigation programs. For instance, **supplemental restraint systems (SRS)** were added as standard equipment on vehicles in an effort to save lives predominantly in front impact crashes. As a result, surgeons began seeing patients survive the crash with devastating knee, hip, and pelvis injuries that hadn't been seen before. It was presumed that victims who previously wouldn't have survived a crash of that magnitude now could survive and receive new injuries that needed to be prevented.[10] This led to research into side curtain and lower dash airbags. Similarly, changes in safety practices as a result of the initiatives will create new problems. For example, more firefighters working out (Initiative 6) will increase the number of exercise injuries. Monitoring what the injuries are and how they occur can lead to new training programs.

Blending Grants and Safety (Initiative 10)

Anyone familiar with grant writing knows that one of the basic requirements of grants is to establish a need. Research is the most effective way to validate the necessity. Initiative 10 looks into requiring specific safe practices of applicants, as well as a preference for the awarding of grants to programs that specifically increase safety. Studying the effectiveness of such programs can result in stronger and more successful future agendas.

Establishing Response Standards (Initiative 11)

Strategic Response to Risk

One of the most interesting potential undertakings when it comes to research is with Initiative 11. To establish national response standards requires utilizing data gathered from incidents, and comparing that to how departments operate. If a smoke detector alarms on the fourth floor of an occupied hotel with no other supporting information, how often does it turn out to be a malfunction versus a fire? Based on construction and fire protection systems, how many times are occupants injured or killed? Does the **authority having jurisdiction (AHJ)** allow security to check it out first, deploy a single unit to respond, or dispatch a full first alarm assignment? What exactly comprises a first alarm assignment? Similar to researching common certification programs (Initiative 5), Initiative 11 is one of the most emotionally charged. The only way to succeed is to use a third party with vast experience in scientific research to evaluate how we respond to incidents.

Communications Operability and Interoperability

One of the vital components of response standards is the ability to communicate. Some firefighters can remember how the addition of **citizen band (CB)** radios in some fire trucks 50 years ago brought communications to a whole new level. Although not many CBs remain as primary communication devices between responding emergency vehicles, they did offer operability (and interoperability) capabilities that are envied by some systems in use today. The term *interoperability* refers to a system's ability to allow communications among different agencies. Communication issues are commonly exposed as critical contributing factors in LODD reports. One reason could be that many radio systems aren't specifically designed for the fire service, but are the best option we have.[11] It's not uncommon to read an owners manual for a new portable radio that states it may malfunction if subjected to "high heat or moisture." Most also suggest there could be "dead" spots in many buildings, rendering them ineffective. There is no guarantee that a portable radio will transmit when you are crawling in a basement and spraying water at a fire that is ready to flash, but generally these radios work surprisingly well. Developments in the last 20 years with 800-MHz frequencies and repeater systems have provided some improvements, yet system problems still remain. Research could provide possibilities for more effective systems by looking at past failures and complaints. New technology that brings a radio system, complete with repeaters, into a building or on a scene should be researched. Communications systems that aren't simply dropped into a pocket on a turnout coat (bolt-on) but instead are integrated into gear in a protective fashion may prove effective. Designing a comprehensive radio system that permits a ladder truck from Boston to respond to Washington, D.C., and communicate with the incident commander is no longer a desire but a necessity.

Performance Measures

In Chapter 2 we covered ways to increase organizational responsibility, specifically in regard to safety. Further research could be used to create a standard set of performance measures that track the service delivery of the department in general. The advantage of performance indicators is for negotiating budget decisions and resource allocations with local government by using facts to determine what level of service they are interested in funding. Although some departments may not like the idea of being compared to others, they should understand that they already are. The main difference is that the comparison would be based on facts rather than opinions. Some of the indicators presently utilized to rate fire departments are not true measures of performance.

Examining Response to Violent Incidents (Initiative 12)

Extensive research is needed to ensure that our response to violent incidents is not only smart but also effective. Roles need to be established for what fire and rescue services will do in a hostage situation and what won't be done. Without standard procedures, it leaves too many questions for interpretation. For instance, you get called to respond with the **strategic weapons and tactic (SWAT)** team. If the commander makes one of the following requests, what is your answer?

- "We want to use your fire engine to provide protection for your medics to retrieve a victim from the parking lot."
- "We want to use your fire engine to transport and hide SWAT team members as we respond to an imaginary fire next door to the drug bust."
- "We want to use your fire engine to blast the gunman off the roof with a deck gun."
- "We want to use your fire engine to spray down and disperse the angry crowd."

Many firefighters would struggle with decisions such as these on the scene. Although we want to support law enforcement, there are numerous ramifications for misusing our equipment and injuring the public or ourselves. We aren't covered by the same laws, and we don't have the same policies and procedures. Without standards developed ahead of time, we will continue to put ourselves in tough situations. Unlike Initiative 11, in which most departments have response policies, many don't have specific policies for potentially violent events. Research in this area is vital to developing these types of procedures.

Providing Emotional Support (Initiative 13)

Substance Abuse

Substance abuse can result from the way that many humans choose to cope with stress. Historically, the use of alcohol and tobacco products has been the self-treatment of choice, yet ineffective. Researching how alcohol and tobacco use can turn into abuse for some emergency responders would be beneficial to preventing it in the future. Additionally, studying the ways that the human brain operates under stress and designing alternative ways to train the mind (proactive) or debrief after stressful events (reactive) would provide alternatives to substance abuse for emergency responders.

Post-Incident Stress Management

The military has designated significant resources to study the effects of **post-traumatic stress disorder (PTSD)** and ways to "erase" the significance of an event. Researching alternative methods of counseling and rehabilitation need to be examined, including the effectiveness of critical incident stress management (CISM) and chaplaincy programs for emergency responders. Recent studies have introduced psychological first aid as treatment choice due to its success outside emergency services.

Enabling Public Education (Initiative 14)

Targeted Problems and Evaluation

Public education presentations sometimes look more like "open mic night" when it comes to subject and delivery. An engine company that is assigned to visit a daycare to show the equipment normally isn't given a script or a lesson plan, but instead talks about whatever they think is important. Unfortunately, this leads to substandard education and missed opportunities. Departments that have assigned personnel for education may use curriculum from NFPA (Learn Not to Burn® or Risk Watch®), but still may teach lessons with no research or purpose behind them. Identifying appropriate audiences, messages, and programs that are effective in reducing the number or severity of fires needs to be a priority. Programs for specific social groups also need to be examined to ensure that economic and cultural barriers are eliminated. Public education has saved lives, but more research into the effectiveness is vital to target effective practices.

Advocating Residential Fire Sprinklers (Initiative 15)

Public Awareness

The very first step in making residential fire sprinklers commonplace is to convince everyone that they are not only effective, but vital to life safety. Forty years ago we debated the benefits of residential smoke alarms, which now have become commonplace. Now we find ourselves down a similar road with residential sprinklers, which are not yet mandatory in all areas. Researching ways to display the effects of fire on the building materials and methods used today both with and without sprinklers to the insurance industry, builders, and the general public is important to its success. Ways to change the image of sprinklers from unfunded mandates that create water damage to effective life-saving equipment must be a priority.

Active Versus Passive Fire Protection

Investigations need to be conducted that compare the options for fire protection and their associated costs. Active fire protection takes steps to mitigate a situation, such as extinguishing the fire. Passive fire protection simply announces that there is a problem, such as a strobe light or audible alarm. By completing a cost-effectiveness analysis using installation, maintenance, and repair costs against effective financial and life savings, standard methodology can be used to make educated decisions for specific levels of protection.

Engineering Safety into Equipment (Initiative 16)

Much of the raw data to ensure that safety is engineered into all new equipment already exists in the form of injury, fatality, and near-miss reports. By categorizing the injuries by "equipment being used," engineers can identify ways to build equipment that is safer. Manufacturers could contribute to this initiative and firefighter safety by increasing research during the design phase about past injuries.

DATA COLLECTION

Once a research agenda or project has been identified, data must be collected, analyzed, and then utilized to provide the information needed to make decisions for instituting programs. As stated earlier in the chapter, standard research suggests starting with the creation of a hypothesis.

Creating a Hypothesis

Let's say you decide that you want to research the effectiveness of one of your unique public education programs, Camp 911. This summer camp for kids ages 11–14 meets for one week at a city park with a recreation hall. Each day is mixed with lessons, games, and outdoor activities to introduce the students to emergency services:

- Day 1—Introduction: safety equipment, 911 dispatch, overview of emergency services
- Day 2—EMS Day: certification in first aid and CPR, visit by an ambulance
- Day 3—Police Day: self-defense, Internet safety, abuse (drug, physical, sexual), police car
- Day 4—Fire Day: extinguisher training (with a simulator), SCBA search and rescue, fire equipment
- Day 5—Graduation Day: simulated rescue, simulated fire and police scenarios

You believe that the camp is successful based on interest, but have specific questions about the effect on the teens. For one, you would like to think that Camp 911 helps the graduates make safer adolescent decisions, but the only way to know this is to do research. As a result, you develop the following hypothesis:

- If students attend Camp 911, then they are more likely to make safer choices in everyday decisions throughout adolescence.

Data will have to be assembled as a form of evidence. There may be evidence that both supports and discounts your hypothesis, yet the evidence will either largely prove or largely disprove your hypothesis. By accepting and acting on the evidence of the stronger side, similar program results can be expected.

Data Collection Plan

Once your hypotheses have been identified and recorded, it's time to start collecting the data needed for research. A data collection plan will ensure that a thorough process is used that allows the data to either support or oppose a specific hypothesis. It will list the tasks to accomplish, who is responsible for them, and the timeline associated with each. Without strict adherence to the plan, research can stray toward opinions and feelings. There are several components of a data collection plan, many of which are applicable to this example.

Identifying the Form of Data

The form of data can be either *hard* data, such as statistics, or *soft* data, such as opinions that utilize surveys or questionnaires. It would be very difficult to track future statistics of the graduates in regard to their behavior. School disciplinary reports, police records, and medical records for minors are not public record. As a result, surveys are a better choice. How the data are collected and recorded are also important components of this step.

Information recorded as a yes/no choice is far more limiting than **scalar** (1–10) references. Variations of "strongly agree" to "strongly disagree" are usually more effective.

Another consideration to the form of data is the ethical and legal implications of the data. Anytime children are involved, another layer of risk must be factored in. Permission from parents or guardians should be obtained, and the entire process should be evaluated by the fire department's legal advisor. Schools may already have some safeguards in place if surveys are conducted anonymously in school with no personal information provided. Many times the real problem lies with how the data are stored or used. Procedures should be in place to dictate how the data is used, shared, and ultimately destroyed. For our example, written parental permission may be adequate because they will be involved in providing much of the information for the survey. A full scope and detail of the research project should be supplied to them.

Identifying the Type of Data

Selecting what type of data is needed to support the hypothesis is sometimes the most difficult step. In the case of making safer decisions, you might identify that the student does the following:

- Utilizes safety equipment during recreational activities and sports
- Is considered "safety conscious" by coaches, family, and friends

In this case, you might formulate the types of data into qualitative questions that could be used in a survey. It's important to include questions of preprogram compliance as well. For instance, if Camp 911 spent sufficient time on the importance of bike helmets, complete with activities such as dropping cantaloupe from the aerial ladder both secured in a helmet and then by itself, we would hope to see a change in attitude. The change in attitude should lead to a change in actions. A 10-question survey that you develop from the two data points above would include these components.

Identifying the Data Collection Point

The next component of the plan involves where the data will be collected and how many surveys need to be completed. It's important to choose subjects to survey that can give honest information. Ideal subjects would be willing to assist you, are familiar with the data needed, and are easy to access. Let's say that the survey you developed was designed for people who are friends or family to one of the graduates of Camp 911. Parents have supplied names of the student's baseball coach, friends, and teachers. Surveys with a cover letter might be mailed to subjects with an anonymous return envelope postage paid.

Analyzing the Data

Once the data has been received, it needs to be analyzed. Interpreting what the data actually means is as important as the data received. For instance, let's say one of your questions asks how likely the graduate is to wear a bike helmet. Although the survey results might show that he is not very likely to wear a bike helmet today, it's not clear why that is. Does he even have a bike helmet? Were most of the surveys completed by friends who don't wear a helmet themselves? Does his mom or the baseball coach really know how likely he is to wear one? Distorted data can occur as a result of misunderstanding or bias (Figure 7-7). This is where flaws in some of the questions on the survey may become evident and may indicate that the survey isn't valid.

Utilizing the Data

After completing the data collection, results can be used to further the research agenda. An interesting example of this format occurred at a summit in 2000 to reduce medical errors in the United States. The report, "Research Agenda: Medical Errors and Patient Safety," which resulted from the meeting, assembled data into three specific project timelines: short, medium, and long term.[12] Using this model is a good way to improve the overall success of your program. For instance, your survey results might show that 36% of the graduates of Camp 911 showed an increase in bicycle helmet use after completion of the course, whereas none saw a decrease. Although you are happy with the rise in safety compliance, you believe you can improve your success and convince at least half of the students to wear their helmets.

Short-Term Goals

Short-term goals are high-priority items that need to be initiated immediately. They have a high likelihood of success and require little investment. Short-term goals might be to increase the class time spent on bike helmets and add new activities to prove the

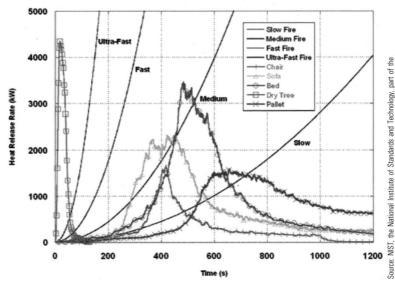

FIGURE 7-7
Careful interpretation of data is many times a difficult component of research.

lesson. Perhaps more time invested will show up in the statistics as more children wearing their helmets.

Medium-Term Goals

Medium-term goals tend to build on data and successes gained from short-term projects. It may involve using the short-term goals as an experiment, while the medium term tweaks it to make it more permanent and effective. Medium-term projects could involve creating a safety video that can be used at camp, or include raising funds to purchase all students their own helmets. Medium-term projects have the opportunity to be the most effective because of their investment.

Long-Term Goals

Long-term goals include items not identified or that are too complex for short and medium completion. For instance, long-term projects might include bike rodeos or a school program to expand your audience outside of Camp 911. Just as short-term projects serve as testing grounds for medium- and long-term projects, Camp 911 can be an experiment for programs that reach many more students.

SUMMARY

Emergency responders are well aware of how much science is involved in firefighting and EMS. From the fire tetrahedron and BTUs to milligrams per kilogram and electrophysiology, it's evident that we wouldn't be where we are today without science. It's also true that research is responsible for the science that responds with us on every emergency call. What we haven't learned to do globally is harness the energy of research for our other needs. Some leaders might research where to put the next fire station or which aerial ladder to buy, but is the research scientifically sound? What data are being used and how accurate are they?

Research is an important tool, but we haven't yet learned to use it to its fullest potential when it comes to creating a safer work environment. Initiative 7 calls for a national research agenda, an industry-wide team effort to design a system to collect, evaluate, and utilize data to strengthen each initiative. The fire research conducted by NIST has benefited the fire service in an unprecedented way, and gives us a glimpse of what research could do for us when used correctly. By assembling the experts in scientific research and channeling them to improve safety on every level, we can make positive changes in reducing injuries and death.

KEY TERMS

active fire protection - A form of fire protection that extinguishes fire.

AFFF - Aqueous film-forming foam.

AHJ - Authority having jurisdiction.

baby boomers - A term used to describe a generation of people born between 1946 and 1964.

CAF - Compressed air foam (system).

Candidate Physical Abilities Test (CPAT) - A type of criterion task test involving several firefighting activities with a focus on agility testing.

citizens band (CB) - A public mobile radio system.

crew resource management (CRM) - A procedure adopted from the aviation and maritime industries that recognizes the value of all personnel.

due regard - A legal term used to describe the professional and safe driving attitude of emergency responders as a standard of care.

Everyone Goes Home (EGH) - A prevention program created by the National Fallen Firefighters Foundation in an effort to reduce future line-of-duty deaths. One of the major accomplishments was the creation of the 16 initiatives, the basis of this text.

functional capacity evaluations (FCE) - A series of tests commonly used to monitor the rehabilitation of workers to facilitate return to work.

Generation Xer - A term used to describe an individual from the generation of people born between 1965 and 1985.

geographic information system (GIS) - A computer system such as ArcInfo® that uses aerial photographs and maps in conjunction to form layers of information.

halon - A chemical extinguishing agent that is no longer being produced due to its adverse effects on the environment.

hypothesis - A scientific deduction used to predict or explain an outcome.

line-of-duty death (LODD) - Fatalities that are directly attributed to the duties of a firefighter.

Maslow's hierarchy of needs - A graphical depiction of the order of human needs.

National Firefighter Near-Miss Reporting System - A voluntary, nonpunitive, fire service reporting system for near-misses.

National Institute of Standards and Technology (NIST) - A government laboratory with a fire research division committed to the behavior and control of fire, and providing valuable information to the fire service.

National Research Agenda - An official plan that would organize and prioritize what research is needed to save lives in the fire service.

NFA - National Fire Academy.

NFPA - National Fire Protection Association.

NIOSH - National Institute for Occupational Safety and Health.

organizational accountability - A term used to describe the ability of an association to be held to certain standards.

OSHA - Occupational Safety and Health Administration.

passive fire protection - A form of fire protection that alerts occupants or a monitoring station of a fire.

personal accountability - A term used to describe the ability of a person to be held to certain standards.

personal protective equipment (PPE) - A generic term used to describe the minimum apparel and gear needed to safely perform a specific duty.

POV - Personally owned vehicle.

PTSD - Post-traumatic stress disorder.

qualitative data - Data or research utilizing feelings or opinions.

quantitative data - Data or research utilizing numbers or percentages.

risk management - Identification and analysis of exposure to hazards, selection of appropriate risk management techniques to handle exposures, implementation of chosen techniques, and monitoring of results, with respect to the health and safety of members.

scalar - A scale of ranking, such as 1–10.

SCBA - Self-contained breathing apparatus.

school resource officer (SRO) - An individual assigned to an educational institution as a liaison between students, faculty, and law enforcement.

SRS - Supplemental restraint system.

SWAT - Strategic weapons and tactics.

taratogens - Chemicals that can create future birth defects.

REVIEW QUESTIONS

1. What is a data collection system? How does it relate to the 16 initiatives?
2. What are the uses of statistical data from incident reports and why is accuracy important?
3. What are the differences between qualitative and quantitative data?
4. What are the effective components of a hypothesis? How can a hypothesis be used in research?
5. What are some aspects of the unique problem of "exposure" as related to injury and LODD data?

STUDENT ACTIVITY

Use the fictitious news headlines below to create at least three hypotheses for each to increase safety at your incidents.

Example: "Firefighter struck on roadway while clearing crash debris"

- If roadway incidents are dangerous scenes, then minimizing time on scene will reduce the chance of injury.
- If emergency responders are preoccupied with duties on roadways, then a spotter should be assigned to watch traffic.
- If firefighters are working on a highway, then they should be as visible as possible to traffic, utilizing high-visibility clothing and scene lighting.

1. "Fundraiser Saturday for paramedic who succumbed to disease from ill patient."
2. "Fire department radio tape shows calls for help from firefighter trapped in basement."
3. "Latest statistics show our state leads the nation in firefighter heart attacks."
4. "Not the first time the axle came off one of the town's fire trucks."
5. "VFD to burn down elementary school for training on Sunday."

FIREFIGHTER WEBSITE RESOURCES

http://everyonegoeshome.com
http://firefighterclosecalls.com
http://firefighternearmiss.com
http://fire.nist.gov
http://cdc.gov/niosh/fire

NOTES

1. Tom Brennan, FDIC Speech, Random thoughts, Indianapolis, 2002.
2. Rumelt, R.P., D.E. Schendel, and D.J. Teece. 1994. *Fundamental issues in strategy; a research agenda.* Boston, MA: Harvard Business School Press.
3. Layman, L. 1955. Attacking and extinguishing interior fires. NFPA.
4. Report of the National Fire Service Research Agenda Symposium; Executive Summary. Emmitsburg, Maryland, June 2005.
5. Maslow, A. 1954. *Motivation and personality,* 3rd ed. New York: Harper and Row Publishers.
6. Physiological stress associated with structural firefighting observed in professional firefighters. 2008. Indiana University Firefighter Health & Safety Research; School of Health, Physical Education & Recreation. Department of Kinesiology; Bloomington, Indiana.
7. How fast does fire spread? 2001, September. National Institute of Standards and Technology. NIST Tech Beat. http://www.nist.gov/public_affairs/techbeat/tb2001_09.htm
8. U.S. Bureau of Labor Statistics; Fatal occupational injuries by industry and event or exposure, Table A-1. All United States, 2008.
9. Status of industry efforts to replace halon fire extinguishing agents. 2002, March. Robert T. Wickham, P.E. Wickham Associates.
10. Kuppa, S. An overview of knee-hip injuries in frontal crashes in the United States. National Highway Traffic Safety Administration.
11. Voice radio communications guide for the fire service. 2008, October. U.S. Fire Administration; FEMA.
12. *Research agenda: Medical errors and patient safety.* 2000, October. National Summit on Medical Errors and Patient Safety Research.

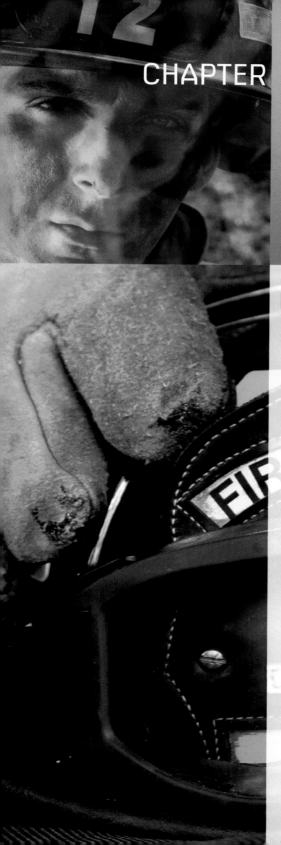

CHAPTER 8

Utilizing Available Technology

LEARNING OBJECTIVES

- Describe how technology can produce higher levels of health and safety.
- Describe some safety improvements that have evolved since the introduction of newer technologies in the fire service.
- Explain the problems that can arise by over-relying on technology.
- Discuss the use of technology in improving the medical monitoring of personnel.
- List the ways technology can be adopted from other industries to reduce injuries and deaths from motor vehicle crashes.
- Explain the theory of using time, distance, and shielding to prevent injuries.
- Discuss the design aspects of simulators that are necessary to improve training systems.
- Describe the application of robotics to improve safety.

INITIATIVE 8. Utilize available technology wherever it can produce higher levels of health and safety.

Earlier in this book we looked at how difficult it is to change. Nowhere is this as obvious as when we are introduced to new technology. We all know someone who won't text on a cell phone or resists using a computer at work. Sometimes it's a matter of being familiar and comfortable with the old way of doing things; other times, it's a matter of trust. When it comes to firefighting, some make the argument that you can't put lives on the line with technology because it can fail. To a certain extent, they are correct. Technology is flawed primarily because it is designed by humans who are flawed. Then again, ignoring that technology can be responsible for saving lives in the form of vehicle airbags, enhanced 911, and carbon monoxide detectors isn't fair, either. It does no good to leave your newest technology in the charger on the truck simply because you don't feel comfortable with it. Technology is most appropriately used for safety in conjunction with training and common sense. For instance, a team of firefighters searching a residence for a trapped victim may use a **thermal imaging camera (TIC)** to assist them in searching more quickly. But just because they have a camera doesn't mean they can ignore sound search procedures. If the battery goes dead halfway through the search, they should still be able to complete their search and know how to get out. Technology doesn't compensate for bad decisions. When we rely exclusively on technology, we set ourselves up for failure.

OVERRELIANCE ON TECHNOLOGY

In October 1995, a school bus loaded with teenagers was struck by a commuter train in Fox River Grove, Illinois. The bus driver had crossed a set of railroad tracks and stopped for a red light at an intersection adjacent to the crossing. Although the driver pulled completely up to the white line for the signal, the back of the bus still hung out in the path of an approaching express train. The resulting crash killed 7 students and injured 24. In the final investigation report, the **National Transportation Safety Board (NTSB)** put a significant portion of the blame on the lack of "signal system integration" between the highway and the railway systems.[1] Investigators felt that the warning system for an approaching train should have been capable of overriding the traffic signal to ensure the grade crossing was clear. The potential existed for conflicting signals, which could have led to a crash at either the train tracks or the highway. As you would expect, the report also cited as a contributing factor that the bus driver did not ensure that she was clear of the grade crossing. In the report, the driver stated, "It never entered my mind that there wasn't enough room for that bus to fit." Although she clearly left the bus in jeopardy by her actions, it's hard to argue that the control system wasn't flawed.

A similar hazard exists in many areas of the country for emergency responders. Technology has allowed many towns to obtain **traffic preemption systems,** allowing an emergency vehicle to change traffic signals to green as they approach (Figure 8-1). This usually allows traffic at the intersection to clear ahead of responders, thus reducing the chance of a crash and allowing the emergency vehicle to maintain its lane. Unfortunately, the technology intended to increase safety can actually increase risk in two specific situations. First, let's say you are approaching an intersection with your lights and siren on. If your speed is faster than engineers had factored or traffic is heavier than normal, traffic can't clear your lane before your vehicle is in the intersection. You may then be forced to go left of center. A vehicle in a turn lane in front of you may see a green arrow for them and make a left turn into your path. Likewise, a vehicle from your left could make a right turn on red, causing a head-on collision. The worst mistake you could make is to put your faith in technology and drive through the intersection without stopping, *even* with a green light. Another potential problem is the response of multiple emergency vehicles. Most preemption systems have some sort of indicator to announce that an emergency vehicle has taken control of the signal, such as a white

FIGURE 8-1
Putting too much faith in technology can allow responders to let their guard down.

spotlight or blinking blue light. However, rarely do they distinguish which direction they are acknowledging. As a result, collisions between responding units can occur while responding to the same emergency from different directions.

As you can see, technology itself can be a contributing factor in a crash. If the bus driver mentioned previously was forced to make a grade crossing with no warning lights and then proceed across a highway with just a stop sign, she probably would have ensured both intersections were clear and assumed full responsibility for crossing safely. An engine company responding to a fire would normally slow to a stop at a red light before entering a hazardous intersection, but the engine driver may become complacent with a preemption system in place. Technological advances designed to increase safety rarely fail due to a malfunction (such as a traffic signal giving both directions a green light) that causes a crash, but humans tend to forget what they know and put complete trust in them. It's not uncommon to hear drivers argue over who had the green light after a crash, rather than acknowledge that they should have looked for traffic before entering the intersection.

UTILIZATION OF TECHNOLOGY

Initiative 8 has the goal of consolidating all available technology in any application that can raise the level of health and safety. Although no one can predict the inventions we will see in the future, there are four specific ways we can get a glimpse of the direction in which we are heading. These include present needs, current trends, military applications, and statistics. Each have already had an influence on technological advances in emergency response and will continue to do so.

Present Needs

Most companies that design new technology for a profit base their research on a specific need. For instance, tire companies are spending little, if any, research money on a new type of fire apparatus tire compound that can reduce the number of flat tires per year. Flat tires simply don't affect too many emergency responses. If you received a marketing survey from a tire company asking if your department would pay twice the cost of a tire to cut the chance of a flat by 75%, you probably wouldn't think it was a wise investment. If the survey instead asked about double the cost for reducing the stopping distance of your fire apparatus by 75%, you might consider that the risk benefit is worth it. Consumers have to be convinced there is a need for a product and that it's a fair price to make a purchase.

It's also important to remember that a primary goal of advertising is to convince you of a need, whether you really have one or not. For instance, a luxury car commercial during a football game has a specific demographic in mind, and targets the advertisement to hit the "triggers" of who is watching the game and likely to purchase one. A viewer might actually picture himself in the car, imagining what it would be like. Not many viewers *need* one, but some purchase one. Fire and **emergency medical services (EMS)** equipment advertisers are no different. Pictures and words in magazines and on the Internet are clearly intent on playing into the emotion of emergency responders. Photographs of firefighters commonly show carbon smudges on their faces while they carry uninjured children from the scene of a fire. Slogans infer that you are brave and professional, and that you choose their product because, like you, it is the best. As with a luxury car ad, it's pretty easy to get pulled in and imagine yourself as the hero. Advertising works for soap and it works for fire helmets. Unfortunately, the ads sometimes display unsafe acts as heroic, further eroding a safety culture.

Current Trends

Another way technology is introduced is to address current trends. One such example in our society is "going green." If it's environmentally friendly, it's got a potential for the future. This has a trickledown effect on emergency responders in the form of responding to emergencies for hybrid or alternative-fueled cars, homes with standby generators, solar and wind power, and buildings with engineered construction (trusses). A current trend specifically in the fire service is a push to improve safety. Technologies that will improve safety, such as better air management or improved accountability, are always being considered. One example that has become standard was the addition of the **personal alert safety system (PASS)** device to our **personal protective equipment (PPE)**. Officially referred to as an **automatic distress signal unit (ADSU)**, the device has absolutely no tactical value while extinguishing a fire in a rubber factory. However, the benefit to firefighter safety makes it mandatory equipment for every person in the hazard zone.

Military Applications

Military applications will always have a significant influence on technology that the fire service uses. The **National Aeronautics and Space Administration (NASA)** and the Department of Defense have bigger budgets and the ability to research new technology at a level not attainable by the fire service. One of the more recent crossover applications from the military is thermal imaging. In the past several years, creating standards for TICs has been a priority for the **National Fire Protection Association (NFPA)** and the **National Institute of Standards and Technology (NIST)**. Suppose your department was in the market for a new TIC, and you review the choices of technology available. What we refer to as thermal imaging is really three different types of technology, with the most common being **vanadium oxide (VOx), amorphous silicon (ASi)**, and **barium strontium titanate (BST)**.[2] Each type has its own advantages and disadvantages, but you may not understand the differences. As a result, it's vital for technology adopted from other agencies to be tested in our environment and have standards for use.

Other technology adopted from the military includes field trauma treatment, protective clothing, and satellite photography. "Hand-me-downs" from the armed forces will continue to benefit the fire service. Looking ahead at what the military is using gives us a good idea what we might see in the future. For example, military drones (unmanned aircraft) have been used increasingly over the past several years. As they become more reliable and safe, they will probably make their way into public safety for observing disaster scenes. It's not unreasonable to suggest that someday they could be used for reconnaissance missions at wildfires, or even to fly firefighting tankers.

Statistics

The last way to introduce technology is through the evaluation of statistics. Unlike needs or trends, statistics don't take into account opinions. If statistics showed that an average of eight firefighters died annually on duty from food poisoning at the fire station, it wouldn't matter if you thought it was a problem or not. Statistics would prove that some sort of disinfection or radiation is needed in firehouse kitchens. A much more realistic example is the technological research being conducted at the **Worcester Polytechnic Institute (WPI).** As a direct result of the **line-of-duty deaths (LODDs)** of six Worcester firefighters in a warehouse building in 1999 in their hometown, WPI has taken on the goal of the "development of a prototype Precision Indoor/Outdoor Positioning System"[3] to prevent firefighters from getting lost in a building. The research has been conducted in a competitive format for the past several years and has uncovered promising possibilities. The final system will likely be able to plot firefighters' locations in a building on a 3D map in real time.

TYPES OF LINE-OF-DUTY DEATHS

Tracking firefighters' location inside a building is a technology that has been difficult to develop. Most successful tracking devices utilize satellites, which don't work well inside buildings. If WPI were to take the approach that the problem is too difficult to solve, they would never have solved it. Similarly, we can't look at *any* statistic and concur that it's "the cost of doing business." Doing so causes missed opportunities that could save lives in the future. The balance of this chapter takes broad classifications of LODDs and matches them with areas of promising or potential technological devices.

Stress and Overexertion

Technological improvements are a key strategy in reducing the leading causes of LODD, stress, and overexertion. Although we understand more about what a firefighter's body goes through now than we did 20 years ago, this type of research uncovers more facts—and questions—with every study. Chapter 6 identified two research projects that evaluated the physical stresses of firefighting, and Chapter 7 explained the value of and implementation procedures for potential research projects, along with collecting the data to support them. New technology will be used to reduce these types of injury and death in several ways.

Medical Monitoring

Researchers in Indianapolis and Maryland observed firefighters at work while wearing medical monitoring LifeShirt® vests. As components of **remote patient monitoring (RPM)**, these vests have numerous applications in the medical field. The technology also proved its ability to monitor firefighters in hazardous environments. Although it's likely more research studies will utilize similar monitoring apparel, the opportunities for widespread medical monitoring is possible. They may be used to observe high-risk personnel with a pertinent medical history. However, based on the success of early models and the current price, it's plausible that they could someday be integrated into all PPE ensembles. Patient monitoring electrodes that are both flexible and stretchable have been developed that can be inserted into washable textile material. These sensors and electrodes could easily be installed in duty uniforms, allowing monitoring of the entire tour of duty.

Although the vests might be able to record the medical data from a firefighter, it doesn't do much good after the fact. Real-time monitoring would rely on a separate component, such as telemetry, which is already a part of structural firefighting in some areas of Europe. Some **self-contained breathing apparatus (SCBA)** have the ability to communicate with base units at the entry points of a building, transferring information about cylinder air level, ambient temperature, and PASS status. Future technology could integrate medical monitoring to communicate vital signs and other firefighter information, allowing the base computer to monitor the capability of crews. In the Indianapolis study, some members volunteered for saliva tests in which **cortisol**, a hormone used to measure stress indicators, was evaluated and compared against an "emotional state" mood survey they completed. Special mouthpieces could be developed that monitor cortisol in the saliva, providing an accurate scale of firefighter stress in real time. A computer could essentially factor in environmental data (ambient temperature), equipment status (bottle level), vital signs (heart and respiratory rates), and emotional stress (cortisol level) into a complex algorithm against a firefighter's baseline and give a scientific evaluation of whether the firefighter should be pulled out and rehabbed, or even transported to the hospital for evaluation. This "rehab criteria" would take out many of the personal issues involved with rehab by requiring that a member return to a certain baseline before going back in (Figure 8-2).

Medical monitoring doesn't have to wait for an ignition source and fire response. Better diagnostics at the annual physical and other doctor visits might help hone in on potential problems. Analyzing historical data with future computer programs might spot trending before any doctor would. For instance, there are many methods to check for an increased chance of colon cancer, but they may not be noticed right away. A comprehensive database of medical records of the future might link a family history from a form filled out years ago, a stray marker gene in a blood sample from a work physical last year, and a complaint of weakness and weight loss today to order a "virtual" colonoscopy. Hindsight often reveals what might have been noticed by simply assembling the sporadic information we already have available.

FIGURE 8-2
Technology has the capability to monitor our personnel, to better plan for rotation on the scene.

Incident Integration

The potential exists for monitoring personnel at the scene of an incident to an extent not possible before. Many departments use an accountability officer to assist in tracking interior crews and rapid intervention groups, but 3D tracking systems, such as the ones WPI are working on, would make the position a mandatory assignment. There may be a time when a **supervisor of tracking employee resources (SOTER)** would monitor the location and medical status of interior crews. In Greek mythology, a Soter was considered the spirit of preservation and deliverance from harm. Much like an air traffic controller, a Soter might watch blips on a computer screen and communicate orders on a radio. An air traffic controller doesn't put much thought into flying the plane or who's on board, but does track the speed, elevation, and location of numerous aircraft at one time. The Soter's responsibilities would be similar, tracking numerous personnel and monitoring each member's time on air, the crew rotation to rehab, and environmental conditions. Computer software could be able to forecast crew rotations, much like a **global positioning satellite (GPS)** in your car forecasts the time of arrival to a destination. A GPS adjusts the time of arrival based on your travel speed, and in the same way crew rotation could be forecasted based on changes in firefighters' vital signs or the rate that they are using their air supply. This would allow the Soter to communicate to command how soon crews would need to be replaced. The same software would be monitoring crews at rehab, giving forecasts of when specific personnel will have vital sign levels that permit them to return to work and how many are out for the remainder of the incident. Rather than simply counting bodies at rehab and asking who's ready to go back to work, medical evidence will be coupled with how personnel feel to provide better planning and the use of personnel.

Motor Vehicle Crashes

Historically, technology has had a significant effect on motor vehicle safety. Traffic preemption systems, airbags, and GPS are just the beginning of what we might see in the future that could make responding to emergencies safer. Specific technology should be researched and implemented to improve vehicle safety, emergency response, and crash investigation.

Vehicle Safety

Compared to automobiles, fire apparatus and ambulances don't provide the amount of protection you would expect. Much of the engineering designated to automobile design is related to either fuel efficiency or meeting federal crash standards. In contrast, custom emergency vehicles are engineered more to function and aesthetics. For instance, a modern **sports utility vehicle (SUV)** with three rows of seats might safely transport seven people. Lap and shoulder seatbelts are in every riding position, and side curtain airbags protect the driver and passengers in an intersection crash. Padded seat backs and cushioned door liners assist in containing passengers in the event of a rollover. Compare that to a Type III ambulance with the same wheelbase. It may have lap and shoulder belts for the two front seats and the captain's chair, yet only provide lap belts for the other bench positions, which are positioned perpendicular to the direction of travel. Although it may be capable of transporting seven occupants, very little restraint protection is provided. Now factor in that some of the occupants in the back will be providing patient care and might not be restrained at all. By the 1980s, most "impaling hazards" had been removed from the dashboards of automobiles, replaced with flush mount knobs and rounded corners on the dash. Take a look in the back of the ambulance. How many sharp corners, aftermarket interior modifications, or potential missiles are present during patient transport? Some ambulances do a good job of securing the sharps container for contaminated needles yet have an open top or nonlatching lid. A rollover could pepper the back of the ambulance with *poisoned* darts.

A glimpse into the future of emergency vehicles will undoubtedly include modern technology and improved safety. We can expect more stringent requirements down the road, including crash worthiness testing and standards on aftermarket additions that, for the most part, are nonexistent today. A safer version of attendant tethers or restraint systems may become commonplace during EMS transport, and the addition of an EMS helmet, much like a medical flight crew's, could become standard. This would protect the head of an attendant in a crash as well as improve hospital and crew-to-crew communications. The possibility exists for quick-change pods, allowing a vehicle to transform from a dive rescue unit to a breathing air supply truck in just minutes. Current trends for chassis design might allow it to run more like a diesel electric locomotive than an ambulance or pumper. Diesel electric hybrids are in various stages of testing and development. Besides increased fuel economy and reductions in emissions, fire apparatus could use their electricity on scene for water pumping, ventilation, extrication equipment, emergency lighting, or other long-term disaster needs. Medical vehicles could have the electrical capability to run advanced x-ray or CT scan systems. Electric motors driving the front wheels would eliminate a rear axle. This could allow mounting operators' controls much lower to the ground. For EMS calls, patients wouldn't have to be lifted up 30" to an existing ambulance height but, rather, could be rolled up a short ramp similar to the back of a military transport plane.

Emergency Response

It's been quite a while since emergency vehicles introduced a new type of warning device. Sure, we evolved from flashing lights to rotating lights to strobes to **light-emitting diodes (LEDs),** but it's still just a light. Studies are under way to bring a new type of alerting method to approaching emergency vehicles that have less to do with visual or audible warning, and more to do with vehicles communicating with each other. In the past several years, GPS systems have been more readily available and affordable as a bolt-on accessory to automobiles. Today, many new vehicles are being produced with GPS systems from the factory **built in,** and there's a good chance they will become standard on all vehicles. Some technological forecasters suggest that someday vehicles will communicate with each other through the GPS. For example, let's assume you are traveling out of state to visit family, and punch the address into your GPS. As your vehicle tracks its location against satellites, it's not much of a stretch to suggest that someday similar routes of vehicles could be noticed by the GPS of a car in front of you. The advantage for highway traffic flow is that communication can allow groups of vehicles with similar destinations to be assembled into "trains" that can travel at a faster pace, closer together, and with greater fuel efficiency. The bonus for emergency vehicles is the ability to warn others. If an ambulance was able to signal vehicles ahead through a GPS map or some other type of display, traffic would know from which direction the ladder truck was approaching, how close it was, and even in which lane it was. Although not every vehicle we overtake will have a GPS, the ones that do will have more information available to them. Applications could include alerting other emergency vehicles as they converge in order to avoid crashes at intersections. The same type of alerting mechanism might change the message when the vehicle sets its parking brake to alert approaching motorists of an emergency vehicle stopped ahead or a crash on the highway. It may even be capable of providing alternate routes to avoid the area, or could direct which lane for other motorists to choose.

Advanced safety systems within emergency vehicles are likely to include a new generation of airbags, increased rollover protection, and better braking systems. Vehicle operators may have greater access to driving simulators, better preparing them for crash avoidance. The future will probably make use of autonomous or semiautonomous vehicles that either take control of a vehicle when a crash is likely, or simply follow GPS coordinates and a traffic monitoring system to drive the vehicle to the scene by the quickest route. If this ever happened, responders would be free to review prefire building plans, prehydrate, or even provide medical treatment en route to the hospital.

Many intersections have installed cameras to signal traffic lights or monitor traffic flow. It's probable that cameras will serve an even greater function for emergency responders in the future. Panoramic-view cameras have now been developed, allowing a 360-degree view from the roof of a vehicle. Imagine a camera mounted to the roof of an emergency vehicle that is utilized to make tactical decisions or critique an incident. Installed on bridges, cameras could be used to make decisions about lane closures or detours several exits back from an incident. They could also be mounted to buildings or towers to assist in the dispatch of emergency vehicles. A **panoramic detection system (PDS)** could utilize the images to automatically change traffic flows or announcement boards. Finally, these cameras and systems could be used to investigate details about a crash scene for investigation, which would be valuable in clearing the scene sooner.

Crash Investigation

One of the future applications of existing technology is the addition of **event data recorders (EDRs),** or black boxes, to all emergency vehicles to collect data for crash investigations. Although most domestic emergency vehicles manufactured after September 2010 have systems in place to download information after a crash, most emergency apparatus lags behind and is exempt.[4] Existing technology is available that records events such as speed, braking, and acceleration. Increasing its usage on emergency vehicles would assist in driver training as well as in crash investigations. The addition of dashboard cameras has helped, but because they are not required, few fire trucks or ambulances utilize them.

Falls

Another significant portion of LODDs is due to falls. Many times these are the result of falling from a building or a piece of apparatus. Advances in safety will likely concentrate on two areas: fall protection and eliminating the need to be in a position in which the firefighter can fall.

Falls from Apparatus

It's probably a safe assumption that most, if not all, falls from apparatus are preventable. Most emergency vehicles are equipped with seatbelts, and their proper use virtually eliminates falls from the cab. Let's say you are reloading 1000' of **large-diameter hose (LDH)** on an engine from a long day of hose testing. It's hot out, you're tired, and when it's reloaded, you're done. In this situation, some of the rules you normally follow on a fire scene might be ignored. Most departments require the use of a helmet during the pressurization phase of hose testing, but it's common to see them come off when the water is drained.

FIGURE 8-3
In many ways, loading hose 60 years ago was much safer than it is today.

Policy may prohibit you from riding on top of the truck when it's moving, but many do when loading hose. Unfortunately, a sudden stop or start of the vehicle could force you to lose your balance. Your first reaction when you lose your balance is to take a step, which is impossible while standing in rows of flat-loaded fire hose. If your boot gets caught, you fall headfirst off the side or rear of the engine. If you do survive, landing on the asphalt parking lot headfirst with no helmet on likely leads to serious injuries.

The trouble is that the hose still has to be loaded. In a situation like this, other industries might use fall protection with harnesses and tethers. It's an exhausting job, which leads to further mistakes and injuries. LDH storage is one example in fire service that is begging for a better solution. Our hose bed has been in a time capsule for 60 years. We still store 5" rubber hose like we stored our 2-½" single-jacket hose (Figure 8-3). Our water flowing capability increased, but our storage system remained the same, except for removing a couple of hose bed dividers. Some manufacturers have worked toward a better solution with hose beds that lower to the ground. Future engineering might find a better way of hose storage, possibly involving reels closer to the ground that deploy and pick up LDH with less chance of an injury. European apparatus tends to be smaller, lower to the ground, and utilizes more reels for hose storage and deployment.

Falls from Ladders and Aerial Devices

Many LODDs occur while working from elevated heights. Like hose beds, ground ladders haven't undergone many safety improvements in your lifetime. In fact, many contributing factors to falls have increased. Aluminum ladders are lighter than the wooden types we used in the past, but they are more slippery to climb when cold or wet. Additionally, the residential building surfaces we lean them against have changed predominantly from brick to vinyl siding. A wood ladder tends to stick to masonry better than an aluminum ladder does to vinyl. Another contributing factor to falls is the

FIGURE 8-4
Aerial devices have undergone major alterations for safety, yet still have room to improve.

way we use ground ladders. Better gear and changes in building construction have allowed firefighters to get caught in rapidly deteriorating fire conditions, possibly using ladders more hastily. Pictures have depicted firefighters climbing out of second-story windows headfirst in an attempt to rapidly exit a flashover. As a result, many training programs have added emergency ladder egress to their "Saving Your Own" segment. Changes in ground ladder design might concentrate on the way we use ladders today, and better ways to keep them in position once they are raised.

Aerial devices have seen significant improvements to safety as a result of better technology. Unfortunately, firefighters still fall from them. As with motor vehicle crashes, very rarely is it a catastrophic event that causes the ladder to fail, but human error. Technology could devise new ways to secure firefighters to a ladder, especially while ascending or descending. Smoother operator controls could prevent sudden movements that could throw someone in a platform off balance, especially when inadvertently contacting a building. Tip detectors might stop movement momentarily and sound an alarm. Improved technology could introduce electrical hazard detectors or better grounding systems in the event of contact with an overhead electrical line. Other improvements could include real-time monitoring of the actual weight load on the tip, and better ways to stabilize the unit (Figure 8-4).

Falls from Buildings

Suppose you and your friends want to ice fish, but the thickness of the ice is questionable. You would probably be nervous about walking out in a group where the water is 30' deep, carrying extra gear, including a chainsaw and an axe. If you bent over and started cutting an observation hole, your partner's hand on your shoulder wouldn't really make you feel much safer. Instead, you would probably prefer a lifejacket and a ladder or boat to distribute your weight. Finally, you would limit your time on the ice until you found out exactly how thick it was. Roof ventilation takes the risks of thin ice and adds two more significant risks: (1) As time passes, the fire burns below and the roof gets thinner, and (2) some roofs are sloped, adding the chance of *falling off* as well as *falling in*. Unfortunately, some videos show firefighters spending time on the roof more like it's a staging area and less like they are on thin ice. When it comes to weight distribution, it might be pretty easy to make the first cut in a roof from a roof ladder, but the next two or three aren't quite so easy. Rooftop vertical ventilation is still one of the most dangerous fireground functions. The job puts you high and over the seat of the fire, which is exactly where the fire is heading. Sometimes the need for ventilation is immediate, but by the time crews and equipment are assembled, the scene has deteriorated to an unsafe level.

Future technology could benefit this tactic by creating a faster and more efficient method to roof ventilation. Rotary spiral bit saws (ROTOZIP®) are effective tools on construction sites that cut through nails, shingles, and metal, yet they haven't been successfully adapted for the fire service. A rechargeable electric motor could be developed to power something similar, eliminating the need to start the motor on the roof. Not only would this type of tool run in oxygen-deficient atmospheres, but it would be lighter than a gas-powered saw and the spinning bit would be far less dangerous than a 14" rotating blade or a 16" bar. Finally, a roof ladder template or guide might be developed to allow one person to cut a D-shaped hole in the roof from the safety of the ladder in far less time. Reducing the time spent on the roof will likely decrease the number of falls.

In an effort to prevent falls, we are likely to see numerous low-tech solutions. They may be as simple as better lighting, such as integrated LEDs in the toes of boots to an integrated harness system for fall protection. Many turnout coats being used today include an **integrated rescue device (IRD).** This rescue strap is looped between the layers of a turnout coat and cinches down around the shoulders as a rescue harness when pulled. Taking the integrated device one step further might allow rappelling and fall protection. The harness strap could easily be used in place of a ladder belt and possibly integrated into a more commonly used and easily deployed system for rooftop operations.

Being Struck by or in Contact with Objects

One broad classification of firefighter injuries is the "struck by" category. Dynamic emergency scenes involve the movement of sharp, fast, heavy, and hot pieces of equipment. Technology could help prevent or reduce the injuries caused by being struck by vehicles, equipment, or debris.

Being Struck by Vehicles

Earlier, we discussed the possibility of adding alerts to emergency vehicles. This also opens up the opportunity for pedestrian alerting. Automobile manufacturers are currently improving a pedestrian avoidance system in which a vehicle computer compares stored images of pedestrians and small animals with images of objects in the path ahead. High-visibility traffic vests could someday give off warning signals that vehicles recognize as special hazards, protecting roadside workers at crash scenes. Sometimes it's our own apparatus that we have to watch out for. Much like the technology used as backup warning devices on family cars, sensors mounted to the running boards of fire trucks could be capable of sensing an object in its path, setting off alarms. Recently, a new tone of safety green was introduced that is extremely noticeable. Traffic vests have adopted the color, and they've also been seen on road signs. Unfortunately, this color will likely show up on billboards, real estate signs, and car dealership flags in the future. The more the color is seen, the more it becomes normal and therefore invisible. Efforts to reserve a certain color exclusively for pedestrians would be a great step toward reducing pedestrians being struck.

Taking lessons from construction crews on the highway has already affected emergency services in the form of high-visibility vests, arrow sticks, and chevron striping. Construction zones utilize two more methods to reduce or minimize the effects of crashes by deflecting the energy exerted during a crash. Concrete "jersey walls" allow a wayward vehicle to stay in a specific lane with vehicles traveling in the same direction, and truck-mounted **impact attenuators** (crash cushions) help absorb some of the energy from a crash into a parked vehicle. Although we aren't quite ready to carry a flatbed full of concrete walls or mount a 10' yellow rubber bumper on the back of a pumper, there may be ways to apply the same theories. Many times construction companies sacrifice old vehicles and mount attenuators to them. Grand Rapids fire department in Michigan put their own spin on the theory and bolted one to an old city dump truck. Now a standard response for a crash on the interstate includes a red dump truck. Other creative ideas could build off what we already do. By parking apparatus on an angle, we essentially deflect a vehicle out of our lane. Future technologies may allow the side of some emergency vehicles to be designed to take a glancing hit with less damage, or the ability to deploy a temporary steel cable guardrail on a winch that can prevent a car from hitting the back of the engine. There's a possibility that using inflatable structures that could be assembled quickly and filled with compressed air or water might be practical and effective in some situations.

Being Struck by Equipment

This category includes tools and equipment that fall, or an unexpected movement that sometimes occurs. It could be a window air conditioner unit that falls from an upstairs window or an axe that inadvertently slides off a roof. Many of these injuries will not be prevented by technology, but proper training, situational awareness, and effective command and control. Technology, however, can assist in some areas of equipment design. Although a fire hose is an effective way to move water at a fire scene, it carries potential energy just waiting to be unleashed. Sometimes it's in the form of water hammer that develops when a hoseline suddenly jumps, a handline that gets loose from a person operating the nozzle, or even a hose that bursts without warning during annual testing. Technology has come a long way in the form of pressure relief valves, but we still see failures in hoselines and appliances. Although hose construction materials and methods have improved over the past 25 years, we still test hose in a destructive method every year. We basically crank up the pressure and wait for it to fail. If it doesn't fail, we might put it back on the rack and wait another year until we do it again. Alternative construction methods and materials have allowed SCBA cylinders to last longer between hydrostatic tests. Similar technologies for the construction or testing of hose and appliances could work toward eliminating the requirement for annual pressure testing, reducing injuries and fatalities.

Being Struck by Debris

Hoselines aren't the only hazard on the scene. We might look at ways to reduce our risk of being struck by debris the way a nuclear plant operator might look to reduce the risk of radiation—through time, distance, and shielding.

Time. Similar to time spent on roof ventilation, the sooner we can get out of danger, the better the risk-benefit ratio. Let's say you are operating at a motor vehicle crash on the interstate. The quicker you can get off the road, the better off you are. Many times at emergency scenes, we find ourselves looking at the wreckage, the ashes, or the water and we realize we probably aren't needed any more. Any technology that reduces our time in a vulnerable position should be embraced. Faster hydraulic rescue tools, more efficient roof ventilation systems, or advanced warning systems will increase safety by decreasing time.

Distance. Earlier in this book, we examined how different types of energy can cause injuries. Anytime we recognize a risk of potential energy, we should work from the furthest possible distance possible. It's been said the most soothing sound to hear when a propane truck is on fire and hissing is your own backup alarm. Improved technologies with infrared cameras, firefighting robots, and thermal imaging can help. Video cameras mounted to bridges or signs might provide information that can be remotely viewed. Reverse 911® allows a dispatch center to call the houses in a neighborhood that needs to evacuate. Future notification technologies might be able to alert all cell phones within a specific distance of a cell tower with the same message.

Shielding. The last way to reduce risk from being struck by debris is by protecting yourself. The first shield we think of is our gear, yet it's really our *last* line of defense. New technological advances in gear would be welcomed. Take the hazards of a modern-day vehicle fire, for instance. Increasing reports of flying shocks, struts, and airbag inflators have been reported. One way for technology to address this concern is to develop some sort of a shield that is easy to deploy and use in conjunction with a hoseline and nozzle, yet is effective at deflecting projectiles launched at an advancing attack crew. All baseball players can get hit with a ball, but we don't send in a catcher to home plate without specialized gear. We might consider vehicle fires as evolving into a specialized situation in which "shin guards, chest protectors, and a cup" are standard issue.

Some fire apparatus manufacturers have accepted the challenge of building safer equipment, yet a pump panel of today still looks much like it did on a 1954 Peter Pirsch. The pump operator for most vehicles still stands for hours at a time within inches of numerous hoselines full of potential energy. Technology might uncover new ways to shield the operator, or might drastically reduce the exposure of personnel close to the apparatus. Many departments utilize a short 10' section of hose for the crosslays to make reloading easier. New designs might provide these first sections constructed of a special material or method, making them virtually indestructible and also eliminating the chance of bursting near the vehicle and reducing the sudden movements associated with water hammer.

Flashover

The precise definition of flashover has been disputed in the past several years, but for the purposes of this text we will define it as the point where a fire spreads very rapidly due to the simultaneous ignition of room contents. This phenomenon has contributed to numerous close calls, injuries, and deaths. Because flashover is a "scientific event" that involves a chemical chain reaction that still is not fully understood, research and technology can be of great benefit. Two of the areas in which technology is most likely to promote safety in regard to flashover include warning devices on gear and simulators.

Warning Devices on Gear

Flashover is generally considered the point in time when all combustibles in a room reach ignition temperature at the same time. When this occurs, the fire is so intense from floor to ceiling that even occupants with the best of PPE have only seconds to exit. Earlier detection of impending flashover conditions saves lives. NIST has already completed numerous tests on fire growth and flashover, and has proven that some of the outdated myths about forecasting time are not realistic.[5] Future research projects could take more environmental clues from sensors on advancing crews to forecast a dangerous event.

Simulators

One of the most promising uses of technology may be all fun and games. If we establish a zero tolerance for friendly fire casualties (training fatalities), we need to develop a whole new theory about training. It's got to be realistic, it's got to prepare someone for real conditions, and it's got to be safe. The military and aviation industries have chosen simulators to fit their needs, and the fire service is well on their way to following in their footsteps. As an industry, firefighting should be relatively easy to design simulators

FIGURE 8-5
Structural firefighting ensembles and equipment provide numerous hiding spots for virtual reality components.

due to the amount of gear and tools we carry (Figure 8-5). For example, a tennis simulator wouldn't be realistic with a virtual reality helmet on, but a fire helmet, hood, and mask can hide some sophisticated equipment. Additionally, the video quality would be far less than that of a fighter pilot simulation. What do you really see on an interior fire attack? The graphics would display a lot of smoke, an orange glow, some intermittent light, and occasionally a blurry piece of furniture. Although the video aspect wouldn't be that great, simulations must be able to accomplish four things: They must (1) use realistic gear, (2) reflect realistic movement, (3) reflect a realistic firefighting environment, and (4) portray a realistic response.

Realistic Gear. The military uses sensors mounted in faux weapons to simulate real equipment. By installing wireless sensors on a prop that feels and reacts like a charged hoseline, you could have the sensation of dragging weight, complete with nozzle reaction, pressure changes, and snagging furniture. Much of your standard gear would be part of the simulation, including a real air pack. The helmet could be specialized, holding video display equipment, speakers, and movable weights. The weights might be thrown suddenly forward to give the illusion of ceiling debris striking the helmet or they might slowly offset to mimic balance changes with position. The hood might contain heating and cooling elements that could simulate increasing heat or a sudden flashover. There could be tiny water lines to inject a small amount of hot or cold water onto the neck. Gloves might have components that assist the nozzle prop to simulate nozzle reaction. Finally, a TIC could be a prop in itself, showing video of heat that matches the simulation monitors on the helmet.

Realistic Movement. Walking and crawling must be components of the simulation. Other simulators use hydraulics to simulate gravity and tipping. Escalator-type treads can appear for stairways. Newer simulators utilize omni-directional treadmills that can rotate for any direction of movement. One such simulator, the **RAVE II**™ at the Army Research Center, was developed to allow soldiers to maneuver through a simulated combat zone.[6] The technology utilizes motion tracking sensors to create a dynamic scenario with different outcomes based on decisions of the personnel.

Realistic Environment. The simulator area might consist of six video displays creating a "room" around the participant. Furnaces could discharge blasts of heat through small ducts aimed at the participant. Smoke would obstruct the video screens as appropriate, and could be used in conjunction with smoke on the video for special effects. Although the hoseline wouldn't discharge water, if you direct a straight stream at a ceiling, it should spray back hot water on you. Plastic balls might shower down to mimic water droplets or blast you like a water spray. The balls could be collected under the simulator and piped back up, much like baseballs in a batting cage. Simulated rollover would emit both heat from the furnace ducts and light from the overhead video screen. Simulated collapses or entanglement forcing a wall breech or other emergency procedure would be standard. Dropping lightweight debris would be easily achieved.

Much like the military, multiple users in the same simulation add affect. There aren't too many firefighters that conduct interior attacks by themselves, and many times their decisions are based on the actions of their partners. Training with your normal crew, in the way they think and react, is a significant component of the training. Simulations involving more than one player also add the element of communication. By placing the individual simulators close enough to each other, voices could carry in a similar fashion. By turning your head or stopping movement, you'll be able to hear others better. Background sounds of the fireground could be used through speakers in the helmets.

Other props must be part of the simulator. If the actual simulator was a rotating stage, other stages could line up, based on the scenario. For instance, a cylindrical wall section might be placed to the right for a right-handed search and rotate as the firefighter moves, creating the illusion of a hallway. Other "stages" could rotate in, containing furniture, a wall section for breaching, or possibly a window for ventilating. The secret would be planning different components that could be added in making endless scenarios, with no two the same.

Realistic Response. The simulation should begin outside of the structure, forcing crews to look at visual signs from the outside before committing. The simulator could allow a 360-degree view if the participant chooses to complete a walk-around of the building before entering. Beginning a simulation by being "dropped" into the front door, the firefighter misses a significant training opportunity. Besides fire location, construction types and building layouts should contribute to the scenarios in order to change the way fires are fought. For example, lightweight residential modern construction should react differently to fire than a Type-3 construction auto repair shop.

The simulations must be accurate in their depiction of the science of firefighting. Fire dynamics should follow proven growth patterns, utilizing research provided by NIST. Poor tactics must give poor results. Proper techniques and water application should give appropriate results, though not all of the time. A significant portion of simulation should be devoted to teaching when efforts are not having the desired effect and therefore when to pull out. Decisions made by the crew to call for vertical ventilation, positive-pressure ventilation, or simply breaking a window and using hydraulic ventilation should change the dynamics of the fire appropriately. Finding a victim should include reversing the path and tracing steps backward while pulling a victim. In the Indianapolis research project, members of the department were weighed with full gear and equipment, like you might find in a rapid intervention situation. Firefighter weights ranged from 203.7 to 434.0 pounds. If the scenario involves a firefighter down, a 175-pound mannequin is less than realistic. If we are rescuing grandma in a nightgown, there can't be handholds or convenient belts to grab. The simulation should be designed for realism, and the props can't be substandard or unrealistic. Every component has to be realistic, and the participants must believe they are in the fire.

Being Caught, Trapped, or Lost

Technological improvements don't necessarily have to be in the form of new equipment, but they could also concentrate on improving existing technology. A good example of such potential improvements is with PASS alarms. Although we know that PASS alarms can be effective in certain situations, there are many questions to be answered about their effectiveness. For instance, if a PASS alarm is most commonly an audible warning device, it requires someone to hear the alarm, recognize it, and find it. Questions arose regarding the effectiveness of the alarm if it became muffled due to firefighter positioning. NFPA added muffling tests to the 2007 edition of NFPA 1982, Standard on Personal Alert Safety Systems (PASS), but designated the test be done in a sound chamber for consistent results. NIST conducted tests in 2009 in various firefighter positions and directional decibels (dBA), but also added normal fireground noises in realistic environments to establish more realistic results. Those results in the final report suggest that "the 95 dBA PASS device alarm level may be seriously challenged by other sound exposures on the fireground."[7] Effective research such as this can only lead to greater improvements in the design of existing equipment.

Collapse

Several of the technological possibilities in this chapter have concentrated on sensors worn by firefighters. What if some technology was a part of building, as a component of a fire detection or suppression system? One of the problems with a deteriorating building is identifying specifically when the building is unsafe. Chicago Fire Department partnered with NASA to research the belief that buildings undergo "a signature change" before they collapse. If sensors could detect the change and give ample warning, incident commanders could have the information they need to make informed decisions. A less expensive system could be nothing more than stand-alone detectors with a unique audible sound installed in the truss systems of new homes. An alarm from these detectors would alert firefighters of fire in the truss system, which would serve to warn them to evacuate. Although sprinkler systems (Initiative 15) would be far more beneficial, the price and relative ease of retrofitting an existing building with attic detectors could make it an effective alternative, with an immediate increase in firefighter safety.

Robots

Bomb squads utilize robots to traverse rough terrain both inside and outside buildings. They can open doors, climb steps, and drag equipment. Many have attachments for specific purposes, such as cameras, microphones, atmospheric metering, x-rays, and even a shotgun. It's only a matter of time before robots are introduced as another tool for firefighting. Although a fleet of robots rolling off an engine and pulling an attack line is not likely, limited-use robots for specialized and dangerous tasks may be a reasonable prediction. For instance, searching the first floor of a residence with a working basement fire is a risky assignment. A robot that would be easy to deploy from the front door might be designed to distribute weight better than a firefighter, reducing the chance of a collapse. The robot could maneuver between pieces of furniture, much like a robotic vacuum cleaner does, scanning for victims. Offshore pipeline inspection systems currently utilize sonar and shape recognition software to inspect pipe. Bodies might be identified by shape or density (we are mostly water) through the use of radar. This type of victim identification system might not even need to access the structure from the floor, but instead could be suspended from a pole, much like a boom microphone on the set of a movie.

Another obvious application for robotics is in the pump panel of fire apparatus. We've already seen electronics and computers debut on pump panels and be utilized as pressure governors, foam inductors, and flow meters. However, it's only the beginning of what might be possible. Picture your engine pumping near capacity for several hours with numerous lines for multiple uses. Adjusting pressures, monitoring pump temperature, and maximizing your water usage can require constant attention. Technological advances to pump controls could better evaluate the overall picture of where water is going and identify potential problems before they arise. Systems could be semiautomatic (sending the operator warnings and suggestions prior to a problem) or automatic (gating back valves, circulating water, or adjusting the throttle accordingly). A pump could recognize that an exposure line has not flowed for quite some time, and gate the valve back reducing the chance of injury. The potential advantages would be improvements to safety, the overall efficiency, the reduction of water hammer, better cooling of pump components, and possibly freeing up the pump operator for other duties.

Communication

Technology will always be a part of communication improvements. One aspect being looked at is the need for intrasquad communications. Headsets in fire apparatus allow person-to-person communications in the cab of a truck, but systems allowing the communication inside a building are extremely limited. With greater advancements in wireless communications, improvements will likely include **push to talk (PTT)** for radio transmissions with voice activation for intrasquad communications. Better communication between crews in the building will reduce both the time trying to yell to one another and the amount of air used. Intrasquad communication would also decrease the amount of overall radio traffic on scene.

One of the most pressing components of communication is interoperability. "Interoperability refers to the ability of emergency responders to work seamlessly with other systems or products without any special effort."[8] Interoperability can further be described as either intradiscipline (fire units to fire units), or interdiscipline (fire units to police, etc.). One promising technological solution utilizes satellites. The Technology Council of the **International Association of Fire Chiefs (IAFC)** completed a study on the future of satellites for the interoperability of radio systems. The advantage of using satellites as a component of future communications systems is the coverage area. Two existing satellites have the capability of covering not only the United States, but "all of North America and its coastal waters."[9] This opens up the possibility of interoperability with Mexico and Canada. The primary goal of this type of communications system would not be interior structural firefighting, so the fact that satellite communications do not work well in buildings would not be an issue.

The military is presently developing a new type of communications system called **software defined radio (SDR).** Present-day radio systems normally occupy a specific wavelength on the radio spectrum based on the hardware installed in the radio. For military operations, this requires several different radios for each type of resource (e.g., aircraft, boat, shore). Computer software is now able to change the frequency of the radio to accommodate different types of communication, creating improved interoperability. The next generation could involve the computer actually picking the best frequency for a given transmission. **Cognitive radio,** as it's referred to, would allow improved communications during disasters or times of peak radio traffic by allowing the radio to choose its own route. With a system such as this, the possibility exists for emergency responders to communicate with truckers on CB channel 19, with HAM radio operators, on cell phones, or even with the coast guard with a **personal digital assistant (PDA).** It's only with the security and ranking provided by cognitive radio "that could make such sharing practical."[10]

SUMMARY

Technology should be viewed as a vital layer of protection, but it does have limitations. If you're planning a ski trip to Girdwood, Alaska, you might shop for the best in an insulated ski jacket. The new material might provide a better barrier than your old one, be more resistant to wind and moisture, be more visible to other skiers, and make you look stylish and professional. Yet without the proper layering underneath, it's unlikely to keep you warm. It won't help you navigate a trail or protect you if you hit a tree. The new apparel will not make you a better skier. The same rule applies to technological advances for emergency services. Thermal imaging cameras don't *replace* good search and rescue techniques, they *enhance* them. Having a good GPS in your apparatus doesn't mean you don't have to learn the streets.

For most of us it's hard to imagine a time when emergency vehicles responded to emergencies and had no communication with dispatch or other apparatus once en route to an incident. Technology will continue to be an asset to emergency service, adding safety and providing more effective operations. Although it should be welcomed, we must continue to insist on rigorous testing and consider its application to emergency services. The new technology must be trained on and used in conjunction with the other tools we have to provide an effective system.

KEY TERMS

amorphous silicon (ASi) - A form of technology used in TICs.

automatic distress signal unit (ADSU) - A piece of equipment that sends a signal when a predetermined criterion is met. A PASS alarm is a form of ADSU.

barium strontium titanate (BST) - A form of technology used in TICs.

built in - The process of building safety into a product or process. It's generally more effective than a bolt-on feature, and ensures that safety is a primary function and consideration.

cognitive radio - A radio that chooses its own frequency based on importance, quality, and availability.

cortisol - A hormone found in human saliva which may be an accurate indicator of emotional stress.

EMS - Emergency medical services.

event data recorder (EDR) - A "black box" used to record information prior to and during a crash to assist in the investigation.

GPS - Global positioning satellite.

impact attenuators - Safety devices attached to immovable roadside structures or construction vehicles designed to disperse energy in the event of a crash.

integrated rescue device (IRD) - A rescue strap built into structural firefighting gear used to drag a firefighter in case of emergency.

International Association of Fire Chiefs (IAFC) - Organization of fire chiefs from the United States and Canada.

LDH - Large-diameter hose.

LED - Light-emitting diode.

line-of-duty death (LODD) - Fatalities that are directly attributed to the duties of a firefighter.

NASA - National Aeronautics and Space Administration.

National Institute of Standards and Technology (NIST) - A government laboratory with a fire research division committed to the behavior and control of fire, and providing valuable information to the fire service.

NFPA - National Fire Protection Association.

NTSB - National Transportation Safety Board.

panoramic detection system (PDS) - A system of cameras used to identify traffic congestion and adjust patterns accordingly.

PASS - Personal alert safety system.

PDA - Personal digital assistant.

personal protective equipment (PPE) - A generic term used to describe the minimum apparel and gear needed to safely perform a specific duty.

PTT - Push to talk.

RAVE II - An advanced military simulator used for combat training.

remote patient monitoring (RPM) - A method of using technology to monitor the vital signs or other physical findings of a patient outside a medical facility.

SCBA - Self-contained breathing apparatus.

software defined radio (SDR) - A radio that uses software to change radio frequencies rather than being restricted to the hardware installed, allowing the user more choices for communication.

SOTER (supervisor of tracking employee resources) - A person used on scene as technology improves who would be responsible for tracking the location, time on air, vital signs, and stress level of emergency responders.

SUV - Sport utility vehicle.

TIC - Thermal imaging camera.

traffic preemption system - A system that overrides traffic light timing devices to give emergency vehicles the right-of-way. Most use a strobe light or siren to trigger them, but some now use GPS.

vanadium oxide (VOx) - A form of technology used in TICs.

WPI - Worcester Polytechnic Institute.

REVIEW QUESTIONS

1. What are some of the safety improvements that have evolved since the introduction of newer technologies in the fire service?
2. What are some of the uses of technology in improving the medical monitoring of personnel?
3. Explain the theory of using time, distance, and shielding to prevent injuries.
4. What are some of the design aspects of simulators that are necessary to improve training systems?
5. How are robotics used to improve safety?

FIREFIGHTER WEBSITE RESOURCES

http://everyonegoeshome.com
http://firefighterclosecalls.com
http://firefighternearmiss.com
http://fire.nist.gov
http://cdc.gov/niosh/fire

ADDITIONAL RESOURCES

http://www.army-technology.com/features/feature92744/
http://www.aes.itt.com/ar/safe.asp
NASA Johnson Space Center: http://er.jsc.nasa.gov/seh/pg53s95.html
http://www.ntsb.gov/speeches/s980401.htm
Wireless Electronic Patches Improve Medical Data and Patient Mobility Monday, 6/30/08: http://www.embeddedtechmag.com/component/content/article/6174
http://www.prnewswire.com/news-releases/vivometricsr-introduces-next-generation-lifeshirt-prototype-a-smart-garment-that-delivers-real-time-vital-signs-remotely-to-healthcare-providers-62053387.html
http://www.theoi.com/Daimon/Soter.html
National Fire Protection Association, NFPA 1982 Standard on Personal Alert Safety Systems (PASS), 1982 Edition, Quincy, MA.
http://www.techbriefs.com/component/content/article/2297
Development of Proposed Crash Test Procedures for Ambulance Vehicles, Nadine Levick, Objective Safety LLC USA, Raphael Grzebieta, Monash University, Australia, Paper Number 07-0074.
http://www.crashdataservices.net/Vehicles.html
Future Drive, *Popular Science Magazine,* May 2010.
WASCO Pipe Coating: http://ppsc.com.my/
http://www.dailydispatch.com/NationalNews/2011/August/04/Modified.dump.truck.is.Michigan.citys.new.weapon.at.highway.crash.scenes.aspx

NOTES

1. National Transportation Safety Board, Report PB96-916202; School Bus at Railroad/Highway Grade Crossing in Fox River Grove, Illinois, on October 25, 1995 NTSB/HAR-96/02.
2. Madrzykowski, D., and S. Kerber, Firefighting Technology Research at NIST. http://fire.nist.gov/bfrlpubs/fire08/PDF/f08025.pdf
3. Precision Indoor/Outdoor Personnel Location Project, Worcester Polytechnic Institute. http://www.wpi.edu/Images/CMS/PPL/PPL_Flier_Apr08.pdf
4. Department of Transportation; National Highway Traffic Safety Administration 49 CFR Part 563 [Docket No. NHTSA-2006-25466] RIN 2127-AI72. Event Data Recorders
5. How fast does fire spread? 2001, September. National Institute of Standards and Technology. NIST Tech Beat. http://www.nist.gov/public_affairs/techbeat/tb2001_09.htm
6. Rave II™ simulator, Army Research Laboratory. InterSense Partnership; Fakespace Systems, Inc. http://www.embeddedtechmag.com/component/content/article/1572
7. Randall, J. 2009, July. PASS sound muffle tests using a structural firefighter protective ensemble method. Lawson Fire Research Division, Building and Fire Research Laboratory. National Institute of Standards and Technology: Technical Note 1641.

8. Voice radio communication guide for the fire service. 2008, October. U.S. Fire Administration; FEMA.
9. A SMART™ Model for Interoperable Communications; Satellite Mutual Aid Radio Talkgroup Program Information Paper. Charles Werner, Chair, IAFC Technology Council, SkyTerra LP. February 2009, Updated September 2009.
10. Mitola III, J. 2000, May. Cognitive radio an integrated agent architecture for software defined radio. Dissertation paper. http://web.it.kth.se/~maguire/jmitola/Mitola_Dissertation8_Integrated.pdf

CHAPTER 9

Investigating Fatalities, Injuries, and Near-Misses

LEARNING OBJECTIVES

- Explain the reasons for thoroughly investigating firefighter fatalities, injuries, and near-misses.
- List the two reasons for thorough investigations.
- Describe the causes for injuries and the different injury pattern theories.
- Define hindsight bias, and explain how it affects events that have already occurred.
- Explain the advantages of indemnity in the near-miss reporting process.
- Explain the disadvantages of voluntary reporting as it applies to near-miss reporting.
- Describe the process of near-miss reporting, and the information needed.
- List the five leverage points that can be used when implementing a safety system.
- Explain how sharing information can reduce future injuries and LODDs.

INITIATIVE   9. Thoroughly investigate all firefighter fatalities, injuries, and near-misses.

Few words invoke more fear than the word *investigation*. We never want to be involved in an **Internal Revenue Service (IRS)** investigation, a murder investigation, or a **NIOSH** investigation. Humans have natural defense mechanisms that resist being probed, examined, or scrutinized by any authority that has the power to discipline, sanction, or confiscate aspects of our life. The fear is that an investigation could conclude with the allocation of blame to something we did or were a part of. Although it is true that a component of **investigations** is to find fault where fault exists, it's not limited to liability. Another valuable component is to search for and identify the actions and **contributing factor** that allowed an event to occur. By identifying these aspects, actions can be taken to prevent future occurrences. Take, for instance, a firefighter who falls from an extension ladder at a structural fire. Although seriously injured, he survives the fall. Which of the following is a thorough investigation likely to lead to?

1. Disciplinary action for the incident commander and the firefighter holding the ladder
2. Liability for the ladder manufacturer, resulting in a lawsuit
3. Recommendations to eliminate contributing factors in order to prevent a similar incident

Obviously, the last choice is the most probable. Fault and culpable liability are always possible, but recommendations and changes in policy are almost guaranteed. To paraphrase the Spanish philosopher and novelist George Santayana, *if we don't learn from our mistakes, we are bound to repeat them.* This phrase applies to many aspects of life. With regard to reducing fireground fatalities, if we fail to learn from injuries, LODDs, or **near-misses,** then we will continue to have more. The only way to learn is to conduct systematic investigations.

Initiative 9 calls for the thorough investigations of all injuries, deaths, and near-misses in an effort to reduce the chances of the next one. As a result, specific components of an effective system include reporting, investigating, evaluating, and implementing changes.

Fire and emergency services have several individual components in place, yet continue to work toward linking information into one effective system.

INVESTIGATING FATALITIES

Suppose you turn on the television and hear the news anchor report that a 30-year-old mine worker was killed in an explosion in West Virginia. You may think about it for a second, and then consider it an acceptable LODD for a miner. Certainly there are coal mines in West Virginia and the mines sometimes have explosions. You may even imagine what happened, assuming that a dangerous atmosphere of coal dust or methane gas probably found an ignition source. The family and coworkers, however, would want to know exactly what happened. The mining and insurance companies have a vested interest in uncovering every factor that led to the explosion, and ensuring that it doesn't happen again at another mine. The **Mine Safety and Health Administration (MSHA)** investigates all mining fatalities and has investigation records going back well into the 1800s (Figure 9-1).

Thorough investigations don't assume anything. When this explosion actually took place in July 2006, it didn't occur underground. It involved no methane or coal dust, and had little to do with mining at all. The worker had been a welder for 10 years, and was welding and grinding a truck rim inside a building when it suddenly exploded, killing

FIGURE 9-1
The Mine Safety and Health Administration uses investigations to help prevent future events.

him instantly.[1] Now assume that you are the investigator for this incident. You might cite numerous contributing factors, but ultimately place the blame on the worker for attempting to make repairs on a pressurized truck tire. However, you would be remiss if you stopped right there. Too many questions still remain.

- He was an experienced welder for 10 years. Wouldn't he know that grinding and welding on a metal rim could weaken it to the point of a catastrophic failure?
- What about heat conduction? Surely he would know that the metal rim would transfer heat to the rubber tire, allowing combustible gasses to build up inside, possibly catching on fire or allowing the bead to fail.
- Why was it still pressurized? It only takes a couple seconds to remove the valve core and eliminate any chance of an explosion.
- What kind of training did he receive? Is there a chance he had no idea of the potential of compromising the structure of a pressurized vessel? Maybe he just welded rebar on bridges for the past 10 years.
- Is welding a cracked rim an acceptable repair, or must it instead be replaced?
- Was he ordered to weld on the inflated tire by a supervisor?
- What experience or certifications did the supervisor have?
- Did the company have a policy on welding rims?

What we're referring to is the **root cause** of the event, a concept that was introduced in Chapter 1. The actual investigation report identified several root causes of the event (none of which actually blamed the worker). If an investigation doesn't expose root causes with contributing factors, it may not prevent the next event. Root causes must then be eliminated in future situations. If the investigation concluded that welder training and certification programs had no mandatory section specifically on the hazards of working on pressurized vessels, it's sure to happen again.

Injury Patterns

In the 1930s, safety pioneer H.W. Heinrich introduced a theory linking **unsafe acts** to fatalities in the workplace.[2] Utilizing our example of the welder, the theory would explain that hundreds of pressurized vessels might be repaired before one failure occurred, and that it might take several failures to lead to one severe injury or death. Other researchers have shown a direct correlation between unsafe acts and deaths, although the predicted numbers vary. We frequently hear the argument in our business that *"I've been lighting these pilot lights for 20 years and never had one do that!"* The pyramid theory suggests what we already assumed: The more unsafe practices that occur, the more likely we are to have an injury or LODD. It follows that if we can significantly reduce unsafe practices (Initiative 4), we will in turn reduce deaths.

More recent theories suggest that there isn't an absolute relation between unsafe acts and deaths, because any unsafe act in itself can result in a death. This argument suggests that it only takes one lottery ticket to win, and that purchasing thousands of them doesn't ensure winning. Therefore, as long as there are *any* unsafe acts occurring, the potential for a death is equally present. Other researchers find fault in the theory of concentrating on the worker's involvement (the unsafe act) rather than focusing on operating systems (risk management). Regardless of which theory you believe in (buy one lottery ticket for an even shot at the cash, purchase several to increase your chances, or don't waste your money because the odds are against you), you still have to buy a ticket to win. In other words, they all suggest that unsafe practices must be identified and eliminated, whether it is the action of an individual or in the overall incident management plan.

NEAR-MISS REPORTING

The concept of a near-miss affecting behavioral change is nothing new to humans. You may have first learned from a near-miss when you were a child and talked back to your parents, thinking they were out of earshot. After you got your driver's license, you may have run your father's car off the road, but no damage occurred and he never found out. It could be that you responded to a fire just last week and forgot your gloves, but didn't get burned. On a personal basis, we tend to learn from our near-misses. In fact, the closer to catastrophe, the more likely we will take note and avoid the situation in the future. However, the intent of Initiative 9 isn't just to learn from an accident or a near-miss, but to expose it so that *everyone* learns from it. Although injuries and fatalities aren't secrets, near-misses are seldom identified by people who were not directly involved. There is often fear and embarrassment in claiming a near-miss, fear of retribution for breaking a rule, and embarrassment for making a bad decision or foolish mistake.

Advantages of Near-Miss Reporting

Numerous industries have identified the advantages of examining a near-miss. Some refer to the event as an **incident with potential (IWP).** In an article addressing medical mishaps and near-misses, *British Medical Journal* authors suggest one distinct advantage to using near-miss data as a learning tool.[3] Rather than concentrating on unfavorable events (injury or death), a successful near-miss program commits all its resources on recovery.

Plentiful Data

There are numerous examples of near-misses. Author Dave Dodson once said the "next NIOSH report is probably already written."[4] In other words, a similar event has probably already happened and *almost* occurred several times. The real question is whether we learn from it. Although no firm numbers exist, it is presumed that near-miss data far outnumber actual injury incidents. This suggests that they have the potential to be applicable in more situations, linked to more contributing factors, and could eliminate a significant number of future events.

Minimization of Hindsight Bias

A simple example of **hindsight bias** is *"I knew that was going to happen."* Sometimes it's used after the other football team intercepts a bad pass in the end zone. Other times it happens when a firefighter gets injured: *"I've been saying all along that those guys were going to get somebody hurt."* It's a natural human reaction in which someone believes that they could have predicted a specific event. A NIOSH fatality report tends to entice more hindsight bias than a near-miss, because a near-miss didn't actually happen.

Proactive Reporting

The advantages of being proactive rather than reactive are well documented in this text. Utilizing near-miss information to prevent an injury or a death is a great example of being proactive. The key is to assemble the lessons of near-misses into a format that we can learn from. (Note the root word "active" in *proactive*).

Cost Savings of Timely Reporting

One of the most significant advantages of being proactive is that there is little or no actual damage or cost as a result of an incident (Figure 9-2). We should consider a near-miss a free pass, and assign value to it. We should save it until we can take advantage of it, then share it with as many people as possible as a learning experience.

FIGURE 9-2
Near-misses should be considered as a free pass, and be exposed rather than hidden.

Indemnity

Successful near-miss reporting systems give immunity in most cases. In an event that leads to an injury, discipline is directly related to the cause. When small infractions such as horseplay lead to an injury, little discipline would likely result. Severe violations, such as running a red light at an unsafe speed with a resulting crash, would elicit more severe discipline. Now take those two examples and make them near-misses, where no damage occurred and nobody was injured. It's likely that, if a supervisor witnessed the event, the most severe discipline would be just a warning in both cases. Warning a firefighter might change the behavior of the few involved, but a properly used near-miss program has the potential of impacting the behavior of numerous firefighters and countless future events. This makes immunity a valuable element of near-miss reporting systems.

Components of Near-Miss Reporting

Several other industries have identified both the value of learning from near-misses and the factors affecting the acknowledgment of an incident. The **Aviation Safety Reporting System (ASRS)** is one such example, and has data going back to incidents in 1988.[5] It's funded by the **Federal Aviation Administration (FAA)** but administered by the **National Aeronautics and Safety Administration (NASA),** which assists in ensuring that no retribution for reporting an incident will occur. The data are primarily used to identify differences between the aviation systems and the human factors involved in delivering the system. It could be said that it compares what the governing body believes is occurring with what is really happening. In 1995, the **National Fire Fighter Near-Miss Reporting System** was created with the support of the **International Association of Fire Chiefs (IAFC).** With a goal of increasing firefighter safety, it followed the lead of other reporting systems, such as the ASRS, to ensure a comprehensive and effective system. The firefighter system is made up of several important components, including voluntary reporting, confidential reporting, and nonpunitive reporting.

Voluntary Reporting

For a near-miss reporting system to be effective, many believe it has to be voluntary. Besides the fact that mandatory reporting is impossible for events that are unknown, voluntary reports tend to be more accurate. They usually identify problems in the system, rather than just problems with individuals, providing information that could prevent future incidents. The **Institute for Safe Medication Practices (ISMP)** believes voluntary reporting is the best strategy when dealing with medication errors. In a mandatory reporting system, "critical information may be lost and proactive error prevention strategies are less likely."[6] Mandatory reporting requirements sometimes contain loopholes or use confusing terminology, allowing incidents to be reported inaccurately. Take mandatory tax reporting, for instance. Most agree that IRS forms can be confusing and easy to misinterpret. Those that thoroughly understand them can identify and make the most of the loopholes. Mandatory systems require an enforcement side as well, which have the responsibility to investigate and discipline those who make mistakes. Narrative-type voluntary reporting systems, such as the firefighter near-miss form, allow the reporting individual to describe exactly what happened and what could have happened in his or her own words.

It is important to note that because reporting is voluntary, some statistics cannot be reliably extracted from the data. If 43% of the firefighter near-miss reports involve nonemergency activities (which tops the list), can it be inferred that nonemergency functions are the biggest problem in the fire service? Because not all near-misses are reported, 43% is not an accurate statistic. Valid data are best achieved through mandatory reporting because every incident is reported rather than just the ones that someone chooses to report. Another component involving the accuracy of voluntary reporting is terminology. If more than half of the incidents reported involve a contributing factor of situational awareness, what exactly does that say? It could mean that we really are not paying attention to what we are doing, that **situational awareness** is a vague term not understood by the person doing the reporting but is the latest expression in vogue; therefore, checking that box seems like the right thing to do.

On the other hand, certain types of voluntary data can produce interesting conclusions. The types of incidents being submitted can be used to gain insights into what firefighters expect and don't expect. Firefighters are more likely to report a good story, one that they didn't see coming rather than one they half expected. For example, let's say there is a surprising abundance of near-miss reports that involve hose bursts or bursts in water flow appliances, yet there are very few reports for slips and falls. Actual injury statistics might show exactly the opposite, that falls occur far more often than injuries that result from being struck by pressurized appliances. We could use that information to conclude that firefighters are more surprised by burst events and that training might have to be adjusted so that they might expect it more often. Likewise, the fact that falls are not being reported in relation to their occurrence suggests that firefighters may have become complacent when it comes to falls. Possibly a new form of training is required to help identify and eliminate hazards. There is a lot of useful information to gain from voluntary reporting, but statistics and trends must be used carefully.

Confidential Reporting

Confidential is the first word to describe near-miss reporting for the railway system in Scotland. Operators of the **Confidential Incident Reporting and Analysis System (CIRAS)** feel strongly that, because of their strict confidentiality, they receive information they wouldn't receive any other way.[7] In most reporting systems, any information that identifies personnel or the organization is removed before the report is stored. Although many reports request a contact person in the event that more information is needed, they assure secrecy in the sources of their reports. The relative anonymity of the Internet has proven especially beneficial for the reporting of near-misses, and many systems use entry forms online to make submissions simple.

On the other hand, there are some disadvantages as a result of the privacy used in the systems. Accuracy is not guaranteed because no verification of the reports is conducted, allowing "creative writing" on the part of the person submitting a report. We all know there are usually two sides to every story, but the reports submitted come from only one view. Additionally, personal accountability may be eliminated from the incident. Some that may have been at fault for a fairly significant mistake could feel as though the reporting in itself eliminates their responsibility in the incident. However, the advantages of confidentiality far exceed the disadvantages and allow the system to provide a safe area to report what firefighters see as potential problems that affect safety. In a worst-case scenario, even if someone embellishes a story or outright makes one up, it still has the potential to save lives in the future, which is the goal of any near-miss reporting program.

Nonpunitive Reporting

One of the best examples of the nonpunitive nature of reporting is found in the ASRS. According to the ASRS rules and regulations, limited liability is awarded to reporters that file a near-miss report within 10 days. As stated earlier, NASA accepts the reports and maintains the records, preventing the regulatory agency, the FAA, from even knowing the specifics of the incident. The confidential nature of the firefighter near-miss reporting system ensures that reporting an incident is not tied back to disciplinary action.

The nonpunitive nature of near-miss reporting could be compared to plea bargaining in our justice system. By admitting a specific level of guilt in a given situation, more information is disclosed than would have normally been discovered in an investigation. Although victims usually feel the theory of a plea bargain is unfair, many professionals with careers in the justice system agree that, when used properly, they can actually benefit society. If they didn't, they would push to eliminate plea bargains altogether. For our purposes of near-miss reporting, there is no victim because there was no harm. The benefit of information that could be used to prevent future incidents far outweighs the fact that we bypass a component of the disciplinary procedure.

The Reporting Process

Reporting to the firefighter near-miss database is achieved by completing a handwritten form or an online form (Figure 9-3). There are five specific components to submitting a close call: the reporter information, the incident information, the event description, the lessons learned, and contact information.

Reporter Information

The first section requires specific information about the reporter and the department where the incident occurred. These fields include department information, such as career or volunteer, what hours of staffing they provide, and in which **Federal Emergency Management Agency (FEMA)** region it is located. Personal information includes the submitter's job or rank, age, and experience.

FIGURE 9-3
All firefighters should be trained and encouraged to complete near-miss reports.

Incident Information

The next section identifies the contributing and environmental factors involved in the event. The weather, time of day, and how far into the shift the participants were are all parts of this section. Also included is the reporter's involvement in the incident and a broad classification of where it occurred (e.g., on an emergency call, during training). Up to five contributing factors can be picked from a list, along with the potential for loss. The submitter can also decide whether he or she thinks this incident will happen again.

Event Description

This open narrative section allows the reporter to put the event into his or her own words. Suggestions for a complete report include human factors such as the decision making involved, how sleep played a factor in the incident, and the communication process utilized. The report also questions how the system played a role through equipment, incident command, and the policies in place. During the review process, identifying names or other department indicators are removed in an effort to keep the event confidential.

Lessons Learned

Immediately following the event description, the person reporting it is encouraged to make suggestions to ensure that it won't happen again. It explores whether human factors, organizational factors, or just "bad luck" led to the event. One benefit of the lessons learned section is to force the submitter to look at contributing factors and consider ways to improve the system in his or her department.

Contact Information

As explained earlier, confidentiality is vital to the system. However, the final component is an option for the reporting party to enter his or her name, telephone number, and email address. This section is voluntary, but it allows an avenue for staff to follow up if there are further questions. For example, a report may have stated, "My partner and I were removing the victim when she suddenly became entangled in debris from a collapsed drop ceiling." In reading the story, it might be unclear whether the firefighter or the victim became stuck. At face value, it might not make a difference to the story, but when trying to establish contributing factors involved in victim rescue and firefighter safety, it's vital. If research was being conducted on the number of near-misses reported that involve firefighter entanglement, more information would be needed. In order to improve the applicability of training materials, the project might include narrowing down specifically what was causing entanglement issues and what part of the body was entangled. Therefore, an obvious benefit of providing contact information exists.

BUILDING A SUCCESSFUL SYSTEM

Now that we've taken a look at investigations and near-misses, we are halfway to establishing a system that uses them to reduce injuries in the future. Let's look at the problem of motor vehicle crashes that involve large trucks and passenger vehicles to see what we should do next. A study in 1998 used statistics to suggest that although truck crashes (versus car crashes) were a relatively low percentage of all crashes, fatalities were higher than average. The analysis also examined factors related to driver actions that led to the crash and concluded that the truck drivers were responsible for 16% of the fatal crashes, as opposed to 70% for passenger vehicle drivers.[8] The remaining 14% identified driver actions on the part of both. Later studies suggested that blame could be assessed equally between drivers when all contributing factors were reviewed and that, rarely, was only one driver to blame. Regardless of fault, most statistics identified blind spots and sudden lane changes as two common occurrences that led to the crashes. Truck drivers and law enforcement also attributed the two factors in countless near-misses. The trucking industry and **National Highway Traffic Safety Administration (NHTSA)** both worked to improve education on the part of all drivers. Suddenly, stickers on the back of semitrailers explained potential blind spots and mud flaps warned of wide turns (Figure 9-4). Interstate road signs suggested gradual lane changes

FIGURE 9-4
The trucking industry used crash data to design new systems to prevent future events.

and maintaining a greater distance in front of large trucks. In an attempt to eliminate unsafe acts that were identified as contributing factors of fatalities, a complete system of crash investigation and the compilation of near-miss information was turned into an educational program. The sections that follow present an example of how several initiatives working together in investigating a near-miss might look.

1. The Event

One morning you open up the newspaper and see a picture of your department's old Crown Victoria crushed under the back of a semitruck. Car 4 is a retired chief's car that members use for attending classes and running department errands while on shift. The paper doesn't say much more than the department vehicle slammed into the back of a beverage truck and no one was injured. You call the station and find out that Cheryl was using the car to go to the hardware store. As she approached the stoplight at Main Street, the brake pedal went to the floor and she slammed into the back of a truck stopped at the light. Luckily, nobody was injured, but the car was most likely totaled.

As your day progresses, you remember that your shift used it to go grocery shopping earlier in the week. While it was out, you noticed some oil in the bay where it was parked. You got a rag and cleaned it up so no one would slip on it. Suddenly you realize that the *oil* you cleaned up wasn't as black as you had expected, and fear that maybe it was brake fluid after all. You now recognize that the crash might have been prevented if you had investigated it a little more.

2. Personal Accountability (Initiative 2)

Human reaction is to downplay your role in the event, even though you are the only one who is aware of it. It's hard to stop the excuses from erupting. It wasn't your job to inspect the car. It was a piece of junk and should have been replaced years ago. Cheryl probably was driving too close; everyone knows she's easily distracted. It probably wasn't even brake fluid that you cleaned up. There's no reason to get involved. Once you calm down and realize that you need to remain personally accountable, you decide you have an obligation to come forward with what you know.

3. Investigations and Near-Misses (Initiative 9)

Sometimes we react to an event without a thorough investigation, but without it you cannot prevent the next incident. When you call the chief to offer your information, you may be surprised to learn that the state police are investigating the crash. Sometimes fire departments don't rely on the local police department to conduct an investigation. Instead, they call in the sheriff's department or state officials. Many might think the reason is to show impartiality. Although that is true, an outside investigation is much more likely to uncover organizational contributing factors than an internal investigation. In other words, an outside agency will probably be more thorough and not take anything for granted. A department that is willing to cooperate with any outside investigation shows organizational accountability (Initiative 2). This is the same reason why state arson investigators or the **Bureau of Alcohol, Tobacco, Firearms and Explosives (ATF)** are sometimes called in to assist with investigations of significant fires that have a high-dollar loss or a loss of life.

4. Data Collection and Research (Initiative 7)

During the investigation, statements are completed by Cheryl and the truck driver. With only a property damage crash, it's likely that the law enforcement investigation will end at this point. An additional city investigation might be limited to looking at maintenance records of the Crown Vic and possibly a postincident mandatory drug test for Cheryl. There's a good chance the case will be closed right there and life will go on without even using your statement. Here's where the intent of Initiative 9 shines. The crash of Car 4 wasn't an accident—it was a crash and a near-miss. It was a near-miss because nobody was injured. It wasn't even in emergency response mode when the brakes failed. The brake failure of an emergency vehicle that is responding to an emergency has been cited as a contributing factor in several injury and LODD reports, as well as in the lives of the public. The department needs to start its own investigation into all contributing factors and root causes. Your statement can be used to assist in finding other potential problems with the maintenance procedures of the vehicles. Researching similar incidents from NIOSH and the National Fire Fighter Near-Miss

Reporting System's database allows data to be collected and assembled into a form of information that can be used in the next step, implementation.

5. Implementation (Initiatives 5 and 11)

Each of the 16 fire life safety initiatives (FLSIs) are intended to bring about actions. Research, accountability, and empowerment are worthless unless they bring about change. The implementation stage is where this occurs. Whether an event actually occurred or not (accident investigation versus near-miss), the root cause and contributing factors must be assembled into a system to reduce the chances of its happening again. With regard to the chemical industry, some professionals recommend assembling the information gained from investigations and near-misses into "leverage points," from which corrective actions can be instituted.[9] Utilizing a system such as this would require results from all near-miss reporting and accident investigation to be fed into a common system.

Establishing Leverage Points

Let's assume for a moment that you are following your department's policy of exiting the vehicle to back up the pumper into the station. It's a clear, cold winter night, and as you round the back of the apparatus, you slip on the ice and fall on your back. You are in pain, but not injured. Before you know it, the pumper is in reverse and heading for your legs. You roll out of the way in time just as the driver notices he doesn't see you in the mirrors and stops. At that point you realize it could have been a really bad night for you. Following the chemical industry's leverage points example, a result of the near-miss report would provide different layers of protection in the future. Leverage points designate different levels of protection, including reducing the chance of future occurrences, providing a barrier, protecting against contributing factors, creating a barrier for contributing factors, and developing an emergency plan.

Level 1: Reduce the Chance of Future Occurrence. Level 1 takes the description of the event and attempts to eliminate the obvious cause. Changing the policy to *not* putting backers behind apparatus is an obvious option, yet the only reason for backers to be in danger in the first place is to reduce backing incidents. Therefore, the policy is probably sound but the method is flawed. Many times, Level 1 actions are directed at the primary offending party—in this case, the driver. When multiple solutions are identified, they are lettered accordingly.

- 1A. When the vehicle is moving, ensure that the backer is visible at all times to the driver.
- 1B. If multiple backers are utilized, all of them must be visible when the vehicle is in motion.
- 1C. The driver should turn on emergency lights, open windows, and eliminate distractions while backing.

- 1D. If the driver loses sight of a backer, the vehicle must be immediately stopped until the backer is located.

Level 2: Provide a Barrier. The next step involves providing a safety net so that if an incident occurs, an injury is prevented (Figure 9-5). Unfortunately, this level sometimes requires obtaining additional equipment, which takes time. You may have to stop extrication efforts to obtain a blanket to protect the patient or fire-line tape to warn of a hole in the floor during overhaul. In our example, it's relatively easy to accomplish by simply changing the way we do things.

- 2A. Backers must remain outside the path of the vehicle and visible to the driver.
- 2B. If the backer must cross the path to switch sides, a stop signal must first be given.

Level 3: Protect Against Contributing Factors. In our example, contributing factors could be related to your actions as the backer and the environmental aspects. It's possible that you were running to the back of the vehicle because it was cold outside and you weren't wearing your coat. It could be that ice is a normal problem in front of the station due to a drainage problem.

- 3A. Backers must remain vigilant when directing an emergency vehicle.
- 3B. Hazards at the fire station must be identified and eliminated.

FIGURE 9-5
Providing a barrier is a common way to reduce the severity of an incident. However, they are only effective if the incident can be foreseen.

Level 4: Provide a Barrier for Contributing Factors. Much like Level 2, using a **barrier** assumes that the event will occur again and that protective measures must be taken. Referencing Level 3 actions guides us in developing plans for this level.

- 4A. Backers should be in proper PPE, including shoes with grip and highly visible clothing.
- 4B. Hazards, such as ice on the pad, should be treated with chemicals regularly to prevent slips and falls.

Level 5: Create an Emergency Response Plan. The final level ensures that a contingency plan is in place in the event that all other layers fail. Many times NIOSH reports contain this component, even if the factor wasn't present in the specific incident. Common references include training more on your accountability system or ensuring that emergency traffic procedures are known by all members. Many fire departments have a policy that an EMS unit is dispatched to all fires for this reason. For our example, the following might be added.

- 5A. All drivers should be trained annually in emergency vehicle driving.
- 5B. All members should maintain a current CPR certification.
- 5C. All non-EMT members should maintain first aid certification.

Distribution of Information

The final step is assembling leverage points into a format to distribute and actually institute change. Most near-miss systems are lacking coordination at this point. For example, if we determine that a significant number of incidents involve firefighters becoming trapped in structural fires with many investigations and near-miss reports pointing to the potential of serious injury or death, what is actually done about it? A thorough system must use the information gathered to make changes to training materials, policies, and procedures. This is precisely why common certifications (Initiative 5) and common policies and procedures (Initiative 11) are so vital to our system. Meanwhile, many avenues are used to get the information out in specialized formats. Some are devoted to investigations, whereas others concentrate on near-misses.

NIOSH Alerts. **NIOSH Alerts** go back more than 30 years and rely on investigation statistics. NIOSH Alerts disclose new observations in regard to "occupational illnesses, injuries, and deaths." Alerts urgently request assistance in preventing, solving, and controlling newly identified occupational hazards (Figure 9-6).[10] For example, a NIOSH Alert was released in July 2010 in regard to implementing risk management principles at structural fires (Initiative 3). The Alert used information from four separate incidents that resulted in five firefighter fatalities and ten injuries. Using examples from each

FIGURE 9-6
NIOSH Alerts can be used to increase firefighter safety by exposing dangers that are creating an impact.

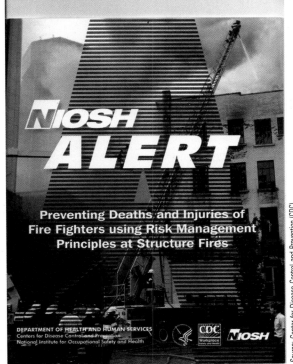

case, the report concluded that "firefighters may not fully consider information related to building occupancy, structural integrity, and fire involvement before entering structures to initiate interior operations and while performing offensive operations."[11] It then provides steps to follow and tips to use in an effort to increase the use of a risk-benefit model.

NIST Research Projects. Much like NIOSH Alerts, **National Institute of Standards and Technology (NIST)** research projects sometimes evolve from a question or idea that develops due to statistics. With its strong research capabilities, NIST uses scientific research methods to prove or dismiss what firefighters believe to be true. For example, firefighters might believe that truss roofs fail quicker than framed roofs, but NIST has the ability to build side-by-side life-size models and conduct numerous tests to use specific construction techniques and materials to show how they actually react to fire (Figure 9-7). Many NIST research projects that involve fire behavior and testing are available in an electronic format that displays scientific data and charts along with side-by-side video footage.

FIGURE 9-7 Researchers at the NIST Fire Research Lab have the ability to prove or disprove what firefighters know.

Firefighting Magazines. Firefighting periodicals have long been an avenue for leaders to spread their knowledge and new ideas. Known experts continue to write monthly columns on basic skills. Essentially anyone with an idea can write an article that may be published for the entire fire service to read. Many of the tactics and concepts we use today began with an article in a magazine. A new concept tends to start "spot fires" across the country, where others pick up the idea and alter or add to it.

For instance, let's say your department has trained on several components of the Saving Your Own curriculum, including the "Denver Drill." The Denver Drill is a technique that is used to remove a downed firefighter from the tight confines of a dormer window. The skill was created after crews had great difficulty removing a trapped firefighter from a window. During a basement fire, your department experiences a similar near-miss when one of your members falls through the floor into the basement. Unhurt, he makes his way to a basement window, where he is pulled to safety (predominantly because it's twice as big as most basement windows). During the critique of the incident, you realize that your department had never considered this possibility, especially if the person was conscious but unable to fit through a normal basement window. You decide to practice the technique and come up with a plan if a firefighter ever presents at a normal basement window.

- Call "Emergency Traffic" on the radio and remove everyone else from the house. Obtain a **personnel accountability report (PAR)**.
- Remove the window glass and frame, and then grab hold of the firefighter by the airpack straps. Secure the firefighter by tying off through the shoulder straps to a tree or tool so that if he or she becomes unconscious, the firefighter cannot fall into the basement.

- Ensure that the firefighter is okay and has sufficient air. If not, remove his or her mask. If heat and smoke are beginning to vent around the firefighter, use **positive-pressure ventilation (PPV)** or a hose on fog stream at the window and vent on the other side or up the stairway into the house.
- Remove cement block above the window to the sill plate while protecting the victim from falling debris. Be aware of anchor bolts that could attach the top course of block to the sill plate. If a bolt obscures the area, remove it with bolt cutters. If possible, remove one block on either side of the window. A sledgehammer works best, but a flathead axe can also be effective.
- If the firefighter is unable to climb out, place a ground ladder over the victim and attach more straps or ropes for lifting. Put one or two members on the ladder and members on either side to facilitate lifting a heavy victim out.
- Total extrication should be completed in less than five minutes.

After practicing the evolution several times, you become proficient at it, taking pictures to document the technique. It's advisable to pick a magazine based on its format. Some are best suited for apparatus ideas, some are geared toward training, and others excel in procedures and management. A submission like this would probably be run by a "hands-on" magazine and practiced around the county. This is a great way to take near-miss results that hit close to home and disseminate the information quickly to a large audience.

Firefighter Websites. If any form of media has the ability to replace magazines, the Internet has the best shot at it. The relative ease of posting ideas to the Internet makes it a quick way to spread information and solicit opinions. One disadvantage is the relative lack of accountability. Some people use the anonymity of the Internet as a way to provoke arguments or be destructive as a way of entertainment. Another is that audiences that access the information are somewhat limited. Although statistics might show a significant number of visitors to the sites, the amount of information on the sites is daunting and readers are likely to step right over good information without knowing it. Many magazines have recognized the abilities of the Internet and have provided additional information there, whereas other organizations stick strictly to the web. Besides the Firefighter Near-Miss Reporting System website, three others are directly related to Initiative 9.

- The **Everyone Goes Home**® **(EGH)** website (www.everyonegoeshome.com) supplies a vast amount of information about the 16 initiatives as well as unique ways to institute changes in your department. Provided by the **National Fallen Firefighters Foundation (NFFF)**, the site provides safety materials, such as printable posters and computer wallpapers. It contains lesson plans to use while conducting department drills, and has an extensive video library, including the Life Safety Resource Kit series. You can also search the site's extensive database for research materials related to the initiatives as well as LODDs.

- The **Firefighter Close Calls** website (www.firefighterclosecalls.com) is devoted to disseminating valuable information through short articles and scene photos. One popular component, The Secret List, is an electronic newsletter that provides firefighting news, along with early reports of firefighter injuries and LODDs. The website provides numerous training scenarios for common emergency responses, complete with lessons learned.
- The **Wildland Fire Lessons Learned Center** website (www.wildfirelessons.net) promotes safety for the wildland firefighting community. The website is devoted to facilitating networking and contains numerous training opportunities, including case studies and lessons learned through the "Facilitated Learning Analysis (FLA)" feature.

Textbooks. This book is another example of disseminating information as a result of research. Although printed textbooks are slower than other avenues in providing new information, these tools and higher education have a profound effect on the long-term practices of firefighting. A textbook utilizes cognitive input and knowledge gained through experience in firefighting or other industries and applies it to specific subjects.

Training Seminars. One of the most popular ways of distributing lessons learned is through training seminars or lectures. Several national conferences allow students to come face to face with authors and others who have excelled in a specific subject. Networking occurs for groups of people with similar interests. Many times, new equipment is also showcased by manufacturers at a conference.

Videos. Training videos are a great way to demonstrate firefighting techniques or display fire growth. Near-misses on video make a significant impression on anyone who watches them. Fire training videos have historically been edited into a professional format to teach a specific subject, such as fire attack or forcible entry. The Internet has opened up opportunities for anyone with a camera and access to place short video clips on public websites. These videos can be utilized in training situations for everything from scene size-up and situational awareness to firefighting hydraulic calculations. Videos can readily be found for virtually any near-miss event, but unfortunately are probably used more for entertainment than educational purposes. Assembling these videos into one location would enable viewers to utilize them for learning.

SUMMARY

When we stand back and objectively look at the causes of firefighter injuries and deaths, we have access to three equally important sources of information. The first is the data provided in LODD reports. These comprehensive accounts from NIOSH supply us with a thorough breakdown of contributing factors, conclusions, and recommendations. The complete reports are easily obtainable on the Internet. The second area we can look at is injury data. Although less inclusive than LODD reports, they do provide some information about the occurrences and causes of injuries. The final source of information is near-miss reporting. It's important to use all three sources of data for the most accurate snapshot rather than concentrate exclusively on LODD information. This is key because *any* injury or near-miss could have been a fatality under slightly different circumstances. Additionally, *injuries and near-misses greatly exceed LODDs.*

Unfortunately, much of the valuable information gained from injuries and near-misses remains locked up somewhere in a warehouse. We need to do a better job of collecting the data, assembling it, sorting it, and reviewing it for distribution. The shipments should be packaged with conclusions and recommendations into usable forms such as uniform SOPs/SOGs (Initiative 11) and training systems (Initiative 5), to be supplied back into the workforce where it can be applied. Other batches of information could be distributed to engineers, manufacturers, and standards committees where technological changes (Initiative 8) might be considered for the development of new equipment (Initiative 16). Without comprehensive investigations of all components followed by an equally effective distribution system, it will be difficult to learn from our past experiences.

KEY TERMS

ATF - Bureau of Alcohol, Tobacco, Firearms and Explosives.

Aviation Safety Reporting System (ASRS) - A voluntary near-miss reporting system for airline incidents.

barrier - Any form of protection in place to reduce the chance of injury if an event occurs.

Confidential Incident Reporting and Analysis System (CIRUS) - A voluntary near-miss reporting system for the rail industry in many areas of the United Kingdom.

contributing factor - A tertiary component of an event that may have encouraged an event to occur or worsened the outcome.

Everyone Goes Home (EGH) - A prevention program created by the National Fallen Firefighters Foundation in an effort to reduce future line-of-duty deaths. One of the major accomplishments was the creation of the 16 initiatives, the basis of this text.

Federal Aviation Administration (FAA) - A division of the United States Department of Transportation, responsible for civilian aviation oversight and safety.

FEMA - Federal Emergency Management Agency.

Firefighter Close Calls - A website devoted to ensuring that near-misses and other important information are shared in an effort to prevent future events.

hindsight bias - A psychological reaction to an event in which a person falsely believes he or she has predicted or forecasted the event, such as "I knew that was going to happen."

incident with potential (IWP) - Another name for a near-miss.

Institute for Safe Medication Practices (ISMP) - A nonprofit organization providing information to patients and the healthcare community in regard to safe medication administration.

International Association of Fire Chiefs (IAFC) - Organization of fire chiefs from the United States and Canada.

investigation - A review of an event in which fact finding provides insight as to the root cause and contributing factors with the intent of preventing future events.

IRS - Internal Revenue Service.

MSHA - Mine Safety and Health Administration.

NASA - National Aeronautics and Safety Administration.

National Fallen Firefighters Foundation (NFFF) - A nonprofit organization created to honor and assist families of firefighters who die in the line of duty, and create programs to prevent future events.

National Fire Fighter Near-Miss Reporting System - A voluntary, nonpunitive, fire service reporting system for near-misses.

National Institute of Standards and Technology (NIST) - A government laboratory with a fire research division committed to the behavior and control of fire, and providing valuable information to the fire service.

near-miss - An event that had the potential for serious consequences but avoided catastrophe. If the event is studied and used as a learning tool, future events could be prevented. Also called an incident with potential (IWP).

NHTSA - National Highway Traffic Safety Administration.

NIOSH - National Institute for Occupational Safety and Health.

NIOSH Alerts - Safety bulletins periodically released by NIOSH to the fire service with a focus on similar events that are causing injuries and deaths, and providing prevention suggestions.

personnel accountability report (PAR) - A verbal or visual report to incident command or to the accountability officer regarding the status of operating crews. Should occur at specific time intervals or after certain tasks have been completed.

positive-pressure ventilation (PPV) - A technique of forcing pressurized air into a structure or enclosed space in an effort to clear the area of smoke or gasses. It can also be used in conjunction with a fire attack in certain situations.

root cause - The primary cause of an event; without it, the event would likely have not occurred.

situational awareness - A term used to describe the recognition of an individual's location, the surrounding atmosphere, the equipment being utilized, and the evolution of an incident.

unsafe act - An action by an individual that is performed in a way that could result in an injury. Many times it involves an individual not using proper PPE, not following safe procedures, or rushing a task.

Wildland Fire Lessons Learned Center - An educational website devoted to the continued improvement of wildland firefighting, using near-misses and lessons learned.

REVIEW QUESTIONS

1. What are the two reasons why thorough investigations are conducted?
2. What is hindsight bias, and how does it affect events that have already occurred?
3. What are the advantages of indemnity in the near-miss reporting process?
4. Describe the process of near-miss reporting, and the information needed.
5. What are the five leverage points that can be used when implementing a safety system?

STUDENT ACTIVITY

For each of the following activities, build safety into planning by creating leverage points. Use each of the five levels and provide at least one supporting point for each.

Example: You have been asked to create a safety plan for firefighters working above ground that will be preparing an acquired structure for future training.

- Level 1 Reduce the chance of a fall
  - 1A All firefighters must use firefighting ladders that have been tested annually.
  - 1B All firefighters must wear harnesses and be secured while working on the roof.
- Level 2: Provide a barrier
  - 2A Full turnout gear must be worn while working above ground.
- Level 3: Protect against contributing factors
  - 3A Weather conditions must be adequate for work being completed.
  - 3B Building must continually be monitored for stability as work is being performed.
- Level 4: Provide a barrier for contributing factors
  - 4A Check weather before planning work.
  - 4B Provide adequate bracing as needed to ensure stability.
- Level 5: Plan for emergency response
  - 5A All firefighters must have a FD portable radio with them to summon help if needed.
  - 5B An ambulance should be available for standby during work from ladders.

1. You are assigned to create a safety guideline to protect members from the dangers of traffic for hydrant flushing that will begin next week.
2. You have been asked to develop a safety plan for your water rescue team who will be providing safety for a fundraising event. Volunteers will take turns jumping into a lake in subzero temperatures, and your members will need to wear cold water rescue suits and be ready to assist anyone who needs help.
3. Your fire department has been asked to provide fire protection for a demolition derby at the county fair. You have been asked to prepare a safety plan for two teams that will be assigned to hoselines in the event a car catches on fire.

FIREFIGHTING WEBSITE RESOURCES

http://wildfirelessons.net
http://everyonegoeshome.com
http://firefighterclosecalls.com
http://firefighternearmiss.com
http://fire.nist.gov
http://cdc.gov/niosh/fire

ADDITIONAL RESOURCES

http://concreteproducts.com/mag/concrete_reevaluating_incident_pyramid/
http://emeetingplace.com/safetyblog/2008/07/22/the-accident-pyramid/
http://www.iafc.org/displaycommon.cfm?an=1&subarticlenbr=204
http://www.fire.state.mn.us/Response/NearMiss.htm
http://www.cdc.gov/niosh/fire/
http://www.ciras.org.uk/whatisciras.html

NOTES

1. United States Department of Labor; Mine Safety and Health Administration. Coal mine safety and health report of investigation, surface coal facility. Fatal exploding vessel under pressure accident, July 30, 2006.
2. Brauer, R.L. 2006. *Safety and health for engineers*, 2nd ed. Hoboken, NJ: John Wiley and Sons, Wiley-Interscience.
3. Barach, P., and S.D. Small. 2000, March 18. Reporting and preventing medical mishaps; lessons learned from non-medical near miss reporting systems. *British Medical Journal* (Clinical Review) 320.
4. Dodson, D. Incident safety officer. FDIC Conference, Indianapolis, 2009.
5. ASRS. http://asrs.arc.nasa.gov/
6. Voluntary error reporting is best public policy. 1999, May 19. Institute for Safe Medication Practices. http://www.ismp.org/newsletters/acutecare/articles/19990519.asp
7. Confidential Incident Reporting and Analysis System. http://www.ciras.org.uk/whatisciras.html
8. Blower, D. 1998, June. The relative contribution of truck drivers and passenger vehicle drivers to truck-passenger vehicle traffic crashes. Publication No. UMTRI-98-25. University of Michigan; Transportation Research Institute.
9. Phimister, J.R., U. Oktem, P.R. Kleindorfer, and H. Kunreuther. 2003. Near-miss incident management in the chemical process industry. *Risk Analysis 23(3)*.
10. http://www.cdc.gov/niosh/docs/99-146/
11. Preventing deaths and injuries of fire fighters using risk management principles at structure fires. 2010, July. NIOSH Alert 2010-153.

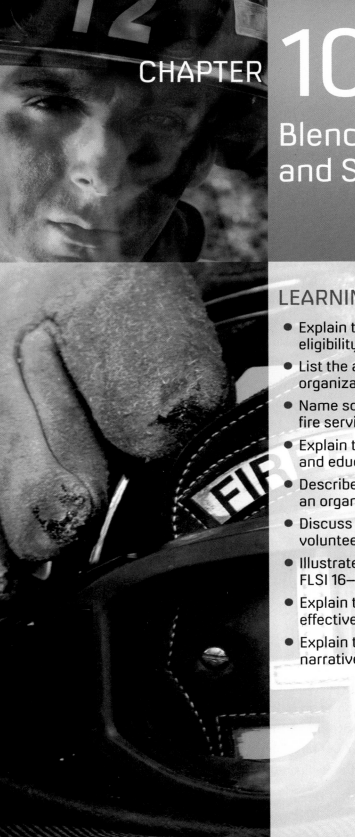

CHAPTER 10

Blending Grants and Safety

LEARNING OBJECTIVES

- Explain the importance of safe practices as an eligibility requirement for obtaining a grant.
- List the advantages of using community organizations for funding opportunities.
- Name some of the companies that support the fire service on the national level.
- Explain the benefits gained by using research and education grants to firefighter safety.
- Describe how a program seed grant can benefit an organization.
- Discuss the benefits of using a SAFER grant for volunteer recruitment.
- Illustrate the importance of compiling a list of FLSI 16–registered organizations.
- Explain the relationship between developing effective grant requests and grant priorities.
- Explain the influence of an effective grant narrative in being awarded a grant.

CHAPTER 10
BLENDING GRANTS AND SAFETY

 INITIATIVE 10. Grant programs should support the implementation of safe practices and/or mandate safe practices as an eligibility requirement.

One of the unique aspects of emergency services is making do with what is available. Firefighters are great at adapting to and overcoming the present situation. There's nothing wrong with being creative as long as risk management and safe procedures are utilized. For instance, say you and your pump operator, Allison, get dispatched to assist with a man trapped under a vehicle in his garage. On your arrival, you find that he was working under his car when it rolled backwards off the ramps, pinning him under it. Worse yet, the garage is extremely cluttered and the car rolled up against the overhead garage door, preventing it from being opened. Access at the rear of the car by the entry and overhead door is limited, but the area at the front of the vehicle is unobstructed. The car is on the slope of the ramps, with the bowed-out garage door keeping it from rolling back any farther. The ambulance has been dispatched, but is still 15 minutes out. You call for additional resources, including a mutual aid rescue truck while your driver checks the victim for injuries. She reports that the car is resting on his chest, but he is conscious and alert with rib pain and significant trouble breathing.

The most appropriate tool to lift the car would be air bags, but it's not something your department can afford. You assign Allison to get what little cribbing you carry on the engine while you grab the victim's floor jack. He tells you it's low on oil and won't lift, and that's why he was forced to use the ramps. The only thing else you see that looks capable of lifting some weight is a "cherry picker." This hydraulic engine hoist is normally used to pull engines and transmissions and has a hydraulic jack attached to a long arm. You see that the hoist has a lift capacity of 3 tons, which is sufficient for the front end of the car, but you know that the farther out on the arm the load is, the less it can lift. You decide that if you can chain it short, close to the jack, it should be able to lift the front of the vehicle. Allison cribs the vehicle to prevent movement and puts the vehicle in park, setting the parking brake. You place the chain around the front bumper by the frame and begin to lift. Although Allison's out of cribbing, she finds several bundles of asphalt shingles and uses them to take up the slack on the frame as you lift. With each pump of the jack, she slides a couple more shingles between the floor and the bottom of the car to prevent it from dropping back down on the victim. The victim is freed just as you hear the approaching siren of the ambulance.

In a perfect world, we wouldn't have to use shingles to crib a vehicle or an engine lift to free a victim. Unfortunately, most of us can't afford every tool for every possible situation and we can't always wait around for someone to bring it. Looking back in the history of the fire service, you can't help but question how they did it 200 years ago. The technology we have today allows us to respond faster, respond more safely, and be better prepared than in the past. Unfortunately, technology comes with a price, and that affects who can use it. For example, the **self-contained breathing apparatus (SCBA)** has been used by firefighters in the United States since the late 1800s.[1] In 1975, the **National Fire Protection Association (NFPA)** began work on a standard for SCBAs and ultimately released NFPA 1981, "Standard on Open-Circuit Self-Contained Breathing Apparatus (SCBA) for Emergency Services."[2] At the time, SCBAs were readily available from several manufacturers, but many departments didn't have enough for all the firefighters on scene. Some departments may have only had one or two in suitcases, on standby for special situations such as basement fires or hazardous materials incidents. Departments that could provide enough for crews rarely had breathing apparatus that was up to the current NFPA standards. Even if they desired the newest and safest models, many departments simply could not afford to upgrade. That was the case until January 2001, when fire equipment grants became a viable option for fire departments to improve safety through the **Assistance to Firefighters Grant (AFG)** program offered by the **United States Fire Administration (USFA)**. Grants are a form of financial assistance used to promote a specific **goal**. They aren't loans or financial aid, and organizations are not entitled to grants. Many times grants are intended to support the operation of the organization applying for the grant (the **grantee**). Usually these are for a one-time purchase, such as for high-pressure airbags or even a fire engine. Other grants have a goal of furthering the cause of the organization that is offering the grant (the **grantor**). Such a grant would be intended to stimulate an organizational change in the grantee. For example, the **Emergency Medical Services for Children (EMSC)** offers grants to states that increase injury prevention activities for children at risk. These funds may then be passed on to fire departments that agree to add injury prevention, such as NFPA's **Risk Watch**® curriculum, to existing fire prevention activities. Although some of these types of grants are a one-time payment, some provide assistance over several years. Emergency responders who are interested in obtaining grants need to know who offers grants, what types of grants are available, and the guidelines for obtaining a grant.

WHO OFFERS GRANTS? THE GRANTORS

In the past several years, the federal government has been one of the leading agencies that offers grants to fire departments. In fact, the federal government has more than 1000 grants available through federal agencies for many different needs.[3] In October 2001, the **Federal Emergency Management Agency (FEMA)** released the final report for its first year of their AFG program. For the first time in history, a federal grant program was available that was capable of replacing the complete SCBA inventories of numerous fire departments across the country. However, grants are not a new concept, and the grantors aren't limited to the federal government. Many state governments offer grants to emergency services organizations within the state. Other common sources of support include community organizations, businesses, and private foundations.

Community Organizations

One of the easiest ways to obtain small grants is through community organizations. They might be locally operated, like a homeowners' association, chamber of commerce, or a women's league (Figure 10-1). Community organizations can also be a local chapter

FIGURE 10-1 Community organizations are a great source of local grants that often aren't advertised.

of a larger, possibly international organization, such as Lions Club International™, Kiwanis International™, the Fraternal Order of Eagles™, or Rotary International™. Regardless of how large the organization is, the key is the tie to your community. There are several advantages to using community organizations, as a result of their purpose, connections, and location.

Their Purpose
Many community organizations have a **mission statement** that includes identifying and supporting the needs of their community. When thermal imaging cameras became readily available in the 1990s, many fire departments relied on community groups to help raise money for their unanticipated purchase. Like community organizations around the country, the Mission Beach Women's Club in San Diego, California, took on the cause and raised the funds necessary to purchase a camera for Fire Station 21.[4] Sometimes groups learn of a need by word of mouth or news reports, whereas others are approached directly by an emergency agency with a request.

Their Connections
Many times, members of community organizations have direct ties to emergency services, either through family or friends. Often, civic-minded individuals are members of both the fire department and community group. Sometimes they have a direct influence on starting or helping strengthen emergency services organizations. For instance, the Black Forest Community Club in Colorado Springs, Colorado, was instrumental in helping Black Forest Fire Rescue grow in the 1960s.[5] Although, in this instance, the help was temporary, some fire departments actually were created when a men's club altered their purpose, bought a fire truck, and changed their name.

Their Location
Besides the fact that many community organizations call the same town or neighborhood home, many times they even share the same space. Fire stations are many times used by other organizations for meetings or fundraisers. As a result, the organization feels a tie to the fire department. It's not uncommon for a Cub Scout troop to receive their training for a first aid merit badge from the local firefighters, only to turn around and use some of their proceeds from popcorn sales to purchase needed equipment.

Businesses
Companies are a great resource for grants and other special funding, but should be matched with a specific need. For instance, it's unlikely that a local pet store would assist in financing a new gas detector, yet it might jump at the chance to help cover the costs of training a new arson dog. The pet store might "adopt" the dog and even have a contest to name the new addition to the department. The expenses associated with

FIGURE 10-2
Some grants are specialized, such as field trip grants.

training the dog would be nominal compared to the advertising and media attention the store would reap from participating. In hard financial times, it's usually much easier for a welding supply company to assist with materials to refurbish a piece of apparatus than to write a check. Some national chains have grants that can be utilized by emergency services, such as Lowes® Heroes program and Target™ Field Trip Grants (Figure 10-2). Many times, these chains have money set aside for each branch or store to use for local needs. Even small amounts are sufficient for smoke detector or fire prevention activities and require little paperwork. Due to their similar goals with the fire service, some fire insurance companies continue to partner with the fire service. Although many local insurance companies have community involvement programs, three of the largest national programs are offered by Fireman's Fund℠, FM Global®, and State Farm®. These companies have offered grants for everything from coloring books to funding the firefighter safety initiatives.

Fireman's Fund Heritage Program℠

The Heritage Program is one of the most beneficial programs to the fire service and has provided numerous types of equipment to fire departments. As of June 2010, the fund had distributed more than $26 million to the fire service, including more than $1 million for national efforts, such as the **International Association of Fire Chiefs (IAFC) National Fire Fighter Near-Miss Reporting System,** the **National Fallen Firefighter's Foundation's (NFFF),** the **Everyone Goes Home (EGH)** program, and the **National Volunteer Fire Council's (NVFC)** Heart-Healthy Firefighter Program[6].

FM Global® Fire Prevention Grant

FM Global® believes that the majority of fires can be prevented, and therefore offers funds to organizations that fight fire. Grants are available for fire prevention programs, such as smoke detector installations or fire safety education, as well as prefire planning software and computers.

State Farm® Insurance Safety Grants

State Farm® has grants available for fire prevention, but puts a significant focus on motor vehicle safety and educational programs. Emergency services agencies interested in obtaining funds for seatbelt compliance or teen driving safety programs such as the Prom Promise may find funding.

Foundations

Foundations are generally nonprofit organizations established to fund specific needs or assist in certain geographical areas. The Hartford Foundation is an example of one such private foundation, limited to grantees in the 29 towns surrounding Harford, Connecticut.[7] Private foundations are a great resource for grants that affect a significant number of people. Although some foundations will offer grants to governmental units, some are available only to nonprofit organizations. As a result, some fire departments have recognized the advantages of creating their own nonprofit organization to assist in obtaining funds for specific needs. A fire department that operates a "safety village" could open up the number of grants available as well as reduce the share cost by obtaining a charitable status. For example, in past years, the AFG program has required a 5% to 20% **match** based on a fire department's size, whereas nonprofit organizations have no **cost-sharing** responsibilities. Creating and operating a nonprofit organization is a serious undertaking, but it can make use of grant funds unattainable by other means. Additionally, businesses and individuals who contribute may be able to write off their donation on their taxes, which they may not be able to do for an emergency services organization.

CLASSIFICATIONS OF GRANTS

Many grants are used to encourage grantees to further the cause of the grantor. For example, the AFG program continues to focus "on firefighter life and safety as well as the safety of the public the firefighters serve."[8] Applicants who demonstrate that their request will have a direct effect on life and safety will receive the highest consideration. Let's say you are interested in obtaining a grant for new fire hose. You might be disappointed to find that a substantial number of the grants available seem to be intended for prevention and program activities rather than to simply purchase needed equipment. Although the AFG program funds equipment, hose may not rank as high as a request for gear or portable radios. However, hose is much more likely to be funded by an equipment grant, such as AFG, than any other type of grant. It's important to understand some of the classifications of grants that are available. They include operating, research and education, quick response, donor-advised, program seed, equipment, and staffing grants.

Operating Grants

Operating grants are intended to provide funding for the everyday costs of nonprofit organizations, but seldom are offered to emergency services. Permitted expenses of an operating grant include anything from salaries to heating bills. Many law enforcement agencies utilize grants to fund a **school resource officer (SRO)**. Safe Schools Initiative grants are available to provide salary funding in an effort to reduce delinquency.[9] Organizations for public health or awareness, such as cancer research, are likely to receive operating grants for their continued existence. Some grantors want to ensure that organizations they have supported in the past continue with their mission. For example, the Jackson Foundation in Richmond, Virginia, funds "capacity building grants," which are intended to assist with organizational growth.[10] These grants are offered to past recipients who show a need and desire to grow. The foundation understands that if an organization is not successful, the money invested in it was essentially wasted. Therefore, funding components such as leadership, legal advice, and human resource management are good investments. Many times they include specific goals or benchmarks to achieve with the funding.

Research and Education Grants

The medical and technological worlds have utilized research grants for quite some time now in an effort to find medical cures and solve industry-wide problems. Although fire and emergency service agencies are sometimes called to assist in live-fire testing (as in Indianapolis and the Maryland Fire Rescue Institute), large-scale projects are normally awarded to educational institutions or technological companies. The time requirements for scientific methods associated with the research are usually out of the reach of an individual fire department. Other industries take a different approach to research and education grants, which could have applications in emergency services. Take the **Sustainability Agriculture Research and Education (SARE)** grant, for instance. These grants are funded by the federal government with a goal of assisting farmers and ranchers in developing and experimenting with new ideas to improve farming on a small scale. The grant not only requires research and application of their ideas, but that education of the industry as a whole takes place as a result. Like most grants, their **call for proposals (CFP)** lists components of projects that they encourage, and thus makes the request more likely to be awarded. Say your department has a unique idea to improve firefighting water supply to a section of your response area. Because your department can't afford the upgrades and the idea could benefit other fire departments as well, you might locate and apply for a research and education grant. The Western SARE Administration Council specifically identifies three components of impacts and outcomes that are equally applicable to the fire service for this type of grant: identification, measurement, and dissemination.[11]

Identified Impacts

Say your volunteer fire department covers approximately 50 square miles, and has no pressurized hydrants. You've been slowly adding **dry hydrants,** essentially pipes installed in lakes and ponds to allow quick hook-up and drafting as a water source for fires (Figure 10-3). One response area is of specific concern, a recreational area around the cove of a 200-acre lake. This resort area once housed small summer cottages, but lately has been replaced by full-size homes with wells and septic systems. You've been successful at getting the homeowners' association to install two dry hydrants on the docks, but the neighborhood requires long lays of supply lines or water shuttles. The time of day and the season not only affect the number of volunteers you can expect to respond, but also how many residents you are protecting. Your members brainstorm to solve the water and manpower problems, and come up with a plausible idea. If you were able to lay 6" pipe in the ground with fire department connections from the two draft sites 1500' uphill to a main road and attached a normal fire hydrant to them, you could essentially protect the center of the village with a water supply. These parallel underground "standpipes" would allow one firefighter to set up a draft site and charge the

FIGURE 10-3
Dry hydrants can benefit fire protection for areas with static water supplies.

water main in no time at all. The attack engine could forward-lay from the closer of two strategically placed hydrants to a fire. Even if there were not enough firefighters to make a safe interior attack, water would not be an issue and a defensive attack with a deck gun might contain the fire to the property of origin with only a handful of firefighters and two engines.

Measured Impacts

Meetings with the homeowners' association to unveil the plan bring up the concern of cost. As you explain the potential of a grant that picks up the majority of the expenses, they become much more interested. Planning involves plotting the best location of the two hydrants, and considering the addition of more. Results of mapping the distance to all structures in the development are compared with the number of driveway and street crossings needed to ensure that the most cost-effective route is taken. Underground utilities are located and bids are solicited. It's determined that the total cost of construction would be under $20,000 and that the 10% match would cost each homeowner about $20. Firefighters and the association decide to send letters to all the homeowners explaining the project and requesting the money in the event the grant is awarded. As a backup, the association agrees to devote the proceeds from their annual pig roast to cover the cost of vacant homes and those who do not pay.

Meanwhile, the fire department conducts drills in conjunction with annual hose testing to get accurate estimates of the time involved in obtaining a water supply without the proposed water line. The grant application includes timetables based on the number of personnel, along with estimates of the anticipated reduction in time and personnel with the water line. Past fire statistics and the number of building permits are used to justify an existing need.

Disseminated Impacts

The educational component of the grant is to share the results with the rest of the industry. Both the successes and lessons learned are equally important. Besides the final report, which will likely become public record to the industry, grantors are looking for end users who share their experience in several ways. It's possible we might present the project with the firefighting world through the use of magazine articles or websites, as discussed in Chapter 9. It could be that future textbooks might reference the project or water shuttle equipment companies could start selling underground standpipe kits to make the purchase of the supplies easier.

Mini or Quick-Response Grants

Quick-response or mini grants are sometimes offered periodically throughout the year, intended to supply funds for immediate or unexpected needs. Generally easy to apply for with little or no wait, these grants fund small programs. For instance, it could be that your department has a sudden influx of emergency responses to "bottle bombs," a dangerous mix of household chemicals contained in a plastic drink bottle that can detonate with incredible force. As a result, you decide to team up with EMS and law enforcement for presentations at the middle school and high school in an effort to show the dangers of mixing chemicals and the effects of unexpected explosions. The police explain the legal implications of creating explosive devices and the specific laws and penalties for those who are arrested while an EMS officer explains the medical implications. A quick-response grant could provide the immediate funding necessary to make a stand and possibly reverse the trend before anyone else gets hurt.

Donor-Advised Grants

Another type of grant is considered donor-advised, or donor-driven. As the name implies, donors may have a specific need in mind when making a charitable donation to a foundation. The foundation is thereby required to use the funds for the specific needs. It could be that your department has recently responded to several **sudden infant death syndrome (SIDS)** emergencies. You wonder if they could be linked, and investigate ways to spread the word about ways to help prevent it. You start a local coalition to improve education to the public utilizing the "Back to Sleep" materials from the **National Institute of Child Health and Human Development (NICHD)**. It would be helpful to find that perhaps a donor in your state made a contribution earmarked for "infant health" to a private foundation that is having trouble locating a suitable program. Simply matching a need to the right donor is many times the key to getting a successful grant.

Program Seed Grants—Fire Prevention and Safety

The **Fire Prevention and Safety (FP&S)** grants are a component of the AFG program and are broken down into two general categories: fire prevention (Initiative 14) and research (Initiative 7). As discussed earlier, research grants are not that common for emergency services. Fire prevention, however, is popular. The main reason is that many departments plan for the purchase of replacement equipment, such as fire hose, but rarely have money in the budget to obtain a fire safety trailer or coloring books. Likewise, grantors prefer to fund cutting-edge and exciting programs rather than simply replace old nozzles. Outside of emergency services, these are often referred to as program seed grants. As the name implies, a seed grant provides the startup costs for a new program.

Many times, the grant provides funding for more than one year on a decreasing scale. The intent is to allow the organization to focus its effort on making the new program successful rather than spending its time trying to fund it. For example, let's say your department is interested in piloting a new defensive driving class intended for commercial truck drivers. Your department's insurance company is offering grants to 10 departments that agree to make the five-year program mandatory. In the first year, the grant funds 100% of the costs incurred for training all 46 members, as well as a 15% reduction in insurance rates. The second year it covers half the costs of retraining, as well as the reduction in rates. For the remaining three years, the department takes over the costs of training and the insurance company locks in the same reduced rates. Seed programs slowly "wean" the grantee into supporting the cost of the operations after the initial funding cycle.

Equipment Grants (AFG)

When many firefighters think of grants, they probably picture equipment grants. Using a grant to purchase a piece of equipment that they otherwise might not be able to afford is a viable option as a result of grants such as the AFG program. To convince a grantor to supply the money needed to obtain equipment, the grantee must demonstrate both a substantial need and proof that "the organization will have the resources to manage and maintain it."[12] No grantor is interested in funding a piece of equipment that is unable to be stored or used. A similar capital grant is the AFG **Fire Station Construction Grants (SCG)** program. With goals of improving the capabilities of the fire protection agency and better serving the members of a community, this grant provides funding for the construction of a new fire station or the remodeling of an existing one.

Staffing Grants (SAFER)

For departments that have the facilities but have been unable to provide the personnel to protect the community, the **Staffing for Adequate Fire and Emergency Response (SAFER)** grant is available. The goal of the SAFER grant is to raise the number of trained firefighters available to meet the standards of NFPA 1710, Standard for the Organization and Deployment of Fire Suppression Operations, Emergency Medical Operations, and Special Operations to the Public by Career Fire Departments; or NFPA 1720, Standard for the Organization and Deployment of Fire Suppression Operations, Emergency Medical Operations, and Special Operations to the Public by Volunteer Fire Departments. Similar to a program seed grant, the SAFER grant is intended to provide initial funding that tapers off over a specific period of time.

SAFER grants can be used for rehiring laid-off firefighters or hiring new firefighters, as well as volunteer recruitment and retention. The intent of the grant is to place more front-line firefighters in a position to improve a department's compliance with NFPA deployment standards. Paid fire departments that recently laid off firefighters can apply for funding to reinstate members. The grant can also be used to hire additional personnel, but the fire department may be required to maintain the added positions for a specific amount of time after funding ends. The volunteer recruitment and retention programs are intended to attract or retain additional members through an incentive program; for example, attracting members could be in the form of marketing or explorer/cadet programs, and retention programs might include insurance or education reimbursements. New ideas for recruitment and retention programs are welcome, but must show evidence of success to be considered.

CHOOSING A GRANT

Once you have identified possible sources for funding and a need has been identified, it's time to research the available grants to pursue the best possible match. A fisherman who knows exactly what type of fish are in the water, what bait they prefer, and where they are swimming is much more likely to be successful than someone who haphazardly casts from shore with whatever lure is available in the tackle box. According to professional grant writer Lynne Marie Paeno, many grants are unsuccessful due to lack of planning and "simply because the applicant did not follow instructions."[13] In her book, *Professional Grant Writer: The Definitive Guide to Grant Writing Success,* Paeno implores grant writers to read the instructions several times and use a checklist to ensure that all the requirements are met. There are several aspects to picking the best match, including eligibility, grant priorities, and sustainability.

Eligibility of Grant Recipients

One of the first aspects of the applicability of grants is the eligibility of the grantee. It's not uncommon for grantors to require specific certifications or to allow specific organizations to be eligible. Many federal grants require applicants to register through Dun and Bradstreet's (D&B) **Data Universal Numbering System (DUNS).** The DUNS number is a nine-digit number assigned to organizations and businesses. Federal grants may also require **Central Contractor Registration (CCR)** as a federal contractor to be awarded funds. Sometimes grant applications require attendance at mandatory meetings or award ceremonies to be considered. Fire departments may have to sign a statement that their personnel have been trained in **National Incident Management System (NIMS)** and that the system will be used for incidents in their jurisdiction. Others may have to report injury or incident statistics to their state to be eligible for state grants. Some grants are available to emergency service providers, whereas others are only available to charitable nonprofit organizations.

Initiative 10 calls for all grant programs in the fire service to require safe practices by the organization to even be eligible. Unfortunately, no mechanism exists for the organizational accountability required by the initiative. If eligibility was as simple as checking a box on the grant application declaring that the organization was safe, there are few who wouldn't do it. Without a prearranged list of criteria, it's difficult to define exactly what "safe" is. In Chapter 2, we discussed the need for a process and a way of recognizing the organizational responsibility of a department that achieves a minimum benchmark of safety and holds its members to higher levels of safe practices. The fire service needs to commit to the creation of a web-based "FLSI 16" registry with specific safe practice requirements for emergency services organizations to comply with Initiative 10 and be eligible to apply for certain grants.

Grant Priorities

When selecting grants for a specific need, it's important to carefully read about the grant priorities. Many grantors put a priority on regional agreements that promote consolidation, whereas others prefer to fund grants that encourage environmental responsibility. Sometimes the grant is intended for a specific end user or risk group, like disadvantaged children. Other times it's specifically for fire department equipment. Even if you have the appropriate end user, the priorities still have to fall in place. For instance, let's say you are interested in purchasing an all-terrain vehicle (ATV) for your department, and the fire chief wants you to look into getting a federal grant for it. Successfully winning the award is as much based on your department's operations as on your need. The AFG Guidance and Application kit for 2010 lists the priorities used to award funds for equipment with regard to replacement equipment, additional equipment, and new equipment.

Replacement Equipment

Highest priorities are given to replacing existing equipment, such as old SCBAs. According to the AFG application, you are much more likely to receive the funding if you are replacing an unsafe or obsolete piece of equipment. The older and more unsafe the equipment is, the better the chance of replacement. If your department already owns a small trailer and an old three-wheeler that is used to support your dive team, the ATV has a good chance of being replaced (Figure 10-4). You already have a substantial investment in your program, and obviously have the resources to support it. Additionally, three-wheelers are not manufactured anymore and have been replaced by four-wheelers, which are much more stable.

Additional Equipment

The next priority for the grantor is to assist you in obtaining additional equipment to supplement an existing program or "functional capabilities." Let's assume for a moment that you don't have an ATV, but have specific locations where your pickup truck can't reach brush fires or technical rescues. Your existing program lacks the necessary equipment, resulting in further risk of firefighter injury and to public safety. The request for additional equipment will enhance a current mission, making it safer for the responders and providing better service to the citizens.

FIGURE 10-4 Replacement equipment has a good chance of being replaced, especially if it is unsafe.

New Equipment

The lowest priority is assigned to backup/reserve equipment or new programs. If your need for an ATV is the result of creating a *new* swift-water rescue team, it will be a tough sell. You may be able to prove that the swift-water team is needed and that the equipment can't be carried to the river, but there is little guarantee to the AFG evaluators that the new program will succeed. Although they may not put a preference on this type of grant, others may. Some encourage new programs, and prefer to supply the seed money to get a new service started. Picking the appropriate grant is one of the best ways to win one.

Sustainability

Whether the grant is supporting an existing program or starting a new one, almost all want to ensure that the program is sustainable. It's not uncommon for a specific question on a grant application to ask how you plan on sustaining the program once the initial funding is complete. This section is especially vital for a seed-type grant. Some organizations, such as the Everyone Goes Home (EGH) campaign, receive operating money in the form of grants year after year, but these types of grants are relatively rare in the fire service. Most grantors want to see the benefits of the award for years to come by planting *perennials* rather than *annuals*.

COMPONENTS OF A GRANT

When applying for a grant, there are several specific components of the application that should be considered, including research and evaluation, the narrative, and the budget.

Research and Evaluation

One of the easiest ways to scare off amateur grant writers is for a grant application to question the methodology involved in the project. This could relate to either the research used to identify a need for your request or to evaluate its effectiveness once the program is complete. Not all grants require this information, but some do. If so, proposals are more likely to be funded if they can justify a need and provide assurance that the proposed program is able to achieve the intended results. This is where the two different types of research discussed in Chapter 7 come in—quantitative and qualitative.

Quantitative Research

The root word of *quantitative* is *quantity*, referring to a specific number. This scientific type of research uses an hypothesis and then backs it up with hard numbers to prove its validity. Let's say your volunteer fire department is investigating the possibility of using a grant to pay for equipping a "duty officer/paramedic" to respond at night. The grant would fund a small four-wheel-drive vehicle equipped with some incident command, firefighting, and paramedic equipment. Qualified personnel could sign up for 12-hour shifts and would be paid a small stipend to respond from home directly to the scene. Your hypothesis or theory might suggest that the department would see a reduction in response times. Quantitative research could prove it by comparing response times before (7.89 minutes) and after (4.65 minutes) the program. This could eventually provide enough data to show increased life-saving and improved incident management capabilities.

Qualitative Research

If **quantitative research** looks at the science or numbers involved with a proposal, **qualitative research** looks at the quality, or perception, involved. Using the same example as above, qualitative research would use interviews and opinions as a basis to prove a need. Surveys of residents might show that response times is a primary concern to them, and that having a guaranteed paramedic responding to EMS calls would make them feel safer. Personnel on the department might appreciate the fact that someone is always responding, and that reduced response times equate to better service by giving prearrival information and calling for additional resources early. Properly completed qualitative research can make a strong case for need, based on feelings and opinions. It's important to note that sometimes grant writers choose qualitative methods over quantitative because they believe it's easier than coming up with hard facts. Unfortunately, many sources suggest that statistics are more reliable than opinions, and that it may actually be easier to use quantitative methods to stay focused on a specific need.

The Narrative

Many grants available for emergency services organizations require a needs or problem statement as a component of the narrative section. This is generally followed by a description of your proposed solution. It's important to utilize this opportunity to clearly convey the reason why it's essential that you receive the funding you seek with excellent writing skills. Remember that almost everyone who is applying for the same money has a similar need. As a result, your best proposal will follow a specific outline or system to communicate your aspiration and convince the grantor to pick you. Firefighters commonly understand the three tiers to describe the operational events at a structural fire. Fire attacks are generally considered to be either offensive or defensive (strategic level), may employ the use of a master stream device (tactical level), and require personnel to provide a supplemental water supply to make it all work (task level). Arranging your thoughts in this manner can effectively follow this same system for grant writing.

Let's assume that you want to conduct three different two-day training seminars for all 60 line officers in your department. The classes would be held off-duty at a local hotel conference center, and concentrate on the practical applications of risk management and NIMS integration for all incidents. You calculate the total cost of the proposal to be just under $40,000, including overtime, instructors, educational materials, food, and facility rental. You also locate a grant that could cover the entire cost. In "Proposals That Work: A Guide for Planning Dissertations and Grant Proposals," the authors suggest breaking down the narrative section into three distinct classifications; significance, objectives, and procedures that we can compare to the specific levels of an effective fire attack.

Significance (Strategic Level)

The strategic-level information should be confined to one or two paragraphs, and should introduce a problem or needs statement. The rationale for your argument might include near-miss or injury reports to set the foundation for better risk management training. You may need to point out the potential catastrophic risks that exist in your county, and why NIMS compliance needs to be instituted. Your angle could be that although initial risk management and NIMS training was completed by all members, it was superficial in nature. Continuous improvement and recently developed lesson plans with group projects using concrete skills are now available and can effectively take the officer corps to the next level of compliance. As a result, there will be improvements to both safety and service.

Objectives (Tactical)

Most of us recognize **objectives** as working components used to achieve a specific goal. In this case, the authors compare the word *goal* with significance. There are several rules to keep in mind when it comes to laying out the objectives of a proposal.

Limit the Objectives. Rather than explaining every possible objective with your proposal, select only two or three to drive home the real purpose of the plan. For our example, you might limit it to two specific objectives:
1. The fire officer will learn to identify risk, understand the process of calculating risk versus benefit, and be able to apply risk management in all emergency and non-emergency operations.
2. The fire officer will identify the positions of the National Incident Management System, and demonstrate the expansion of any incident to involve interagency and federal response teams.

Objectives Are Objectives. It's also important to remember that no procedures or tasks are addressed in the objective section. Grants don't fund objectives, so read your objectives carefully and ensure that they don't list anything you need money for. Likewise, don't bring up the problems or needs. Rehashing or begging doesn't help get funding—it merely shows that you are desperate. Grantors want to know what you expect to accomplish with your objectives.

Be Confident in Your Objectives. Be sure your objectives are direct, absolute, and measurable. Don't use phrases such as "hope for success" or "try to change" (related to a behavior). If you have to leave yourself an emergency exit, eliminate it and change the objective to something you can stand by. In an earlier section, "Limit the Objectives," notice that the two examples of well-written objectives use action verbs such as *identify*, *calculate*, *apply*, *differentiate*, and *demonstrate*. By using strong words that escalate with each phrase, they convey the message that this is going to be an effective action-oriented program with guaranteed success.

Procedures (Task)

Also referred to as the program description, this section explains exactly how the objectives will be met. Because it can be lengthy, it's sometimes a good idea to break it into digestible parts with headers, using the objectives as a guide. Many grant proposals allow the use of graphics, charts, or photographs to clarify the tasks involved. Sometimes vendors can be beneficial in supplying information that can be used in this section, but it's important to read the specific rules on each grant. Some grants allow input from vendors, whereas others allow the vendor to supply the entire grant template. Others refuse to consider grants that have any vendor involvement. Most grant requests for equipment limit the use of specifications that could limit or otherwise restrict competing vendors from bidding. Therefore, it's vital to understand what is permissible and what is not.

For our training seminar example, it would be important to clearly define the instructional materials being used, the instructional methods and activities planned, as well as the instructors' qualifications. If the specific grant gives preference for NIMS, that section should be emphasized. If the grant gives the most consideration to safe practices, push the risk management side more and tie in the importance of NIMS compliance toward safety. Also realize that a cutting-edge proposal such as this will draw sufficient attention from evaluators and professionals who will be watching closely to see if it has possibilities to use elsewhere.

Many times, grantors judging proposals use a mathematical scoring system to better compare one proposal against another. Sometimes they explain their system in advance, but many times they don't. It's important to pick up on statements such as "a stronger consideration will be given for" when explaining your grant proposal. One way to write a proposal is to think of yourself as a contractor bidding on a job. Grantors are essentially hiring someone to complete a task they either can't do or don't want to do. Let's say you want to remodel the kitchen in your home. As you meet with contractors, there are several reasons you might pick one over another. The list could include professional appearance, qualifications, timelines, price, and even a guarantee.

Professional Appearance. First impressions are very important. You wouldn't want the contractor to stomp into your house with muddy boots on, toss the clipboard on the counter, and start measuring. You probably want someone who appears orderly and trustworthy. You expect a polite handshake and an introduction followed by a lot of listening and taking notes while you explain your wants and needs. If a bid comes in that involves removing the window over the sink when you made it clear that you wanted a bigger window for more light, there's no way you'll choose it. Likewise, if a grant has rules, they must be followed. Care must be taken to ensure that your proposal fits in the parameters the grantor established. Imagine how you would feel if the contractor called you back to argue with your requirements or to try to change your opinion to make the job easier. It's acceptable to contact the grantor for clarification and questions, but realize you probably won't change their mind about how the process will work. There are usually more requests than available funds and they simply don't need your application if you aggravate them.

It's also important to note that many times the only image a grantor will have of you is your application. The words you choose and the way you formulate thoughts will be reviewed and picked apart thoroughly. Proper grammar, correct spelling, the correct use of references, and neatness all count. Consider your contact information as well. If they need to contact you by phone, make sure your voicemail message doesn't make them second-guess your professionalism. Email addresses shouldn't include "bigdaddylover@" on a grant application, and an Internet search of your name shouldn't reveal anything you wouldn't want a grantor to know about you or your organization.

Qualifications. Obviously a plumber has a better shot at landing a job if he or she is qualified. You need to be convinced that a prospective contractor has experience in this type of work, and you would expect references. Similarly, grantors need to be assured that you are capable of achieving both your goals and theirs. Past successes need to be documented and linked to the present proposal. For instance, let's say you obtained a fire prevention grant in the past for the purchase of a children's fire safety trailer. You presently use the department's pickup truck to pull it to the schools, but want to expand its use at community events. Your proposal might be to refurbish an old ambulance into a new tow vehicle that could be set up with the trailer at events such as the county fair. The lockable storage space would allow you to carry extra fire equipment and gear, a generator for your own power, as well as a tent to cover the entire display. This would allow you to stay set up for the entire weekend, tripling the amount of people using the smoke trailer each year. The past success in obtaining a grant and carrying it through to an effective program are key assets to use on a proposal. This puts you in a good position to convince the grantor that you are not trying to pitch an outlandish scheme, but are simply asking for assistance in obtaining the next logical step in reaching more people with your existing program while making the most of your past grant.

Timelines. You probably want your kitchen done by a specific date. It could be that the family Christmas party is at your house this year. All grants have some sort of a timeline, whether following their fiscal period or a self-imposed timeline by the grantee. Usually it starts with an application period. They won't accept any applications before a certain date, and they must be postmarked by a certain date. Many times, grantors are actually grantees from another organization, and have deadlines themselves. Therefore, there are numerous deadlines from application deadlines to final reports. If your floor tile installer misses his or her deadline, it's unlikely you will ever use this contractor again. Likewise, it's tough to get a grant if you've failed to meet timelines in the past.

If any requirements of a grant are negotiable, timelines may be the most likely. If you've proven a need and your program scores well overall but you have trouble with the grant timeline, you still have a shot. Many grant programs run on the fiscal year of the grantor. Your proposal, on the other hand, might be a fire safety program in the school system. Say the grant cycle provides funds in February, and a final report is due in December. You application needs to clearly state that your program is restricted to run from October to May due to the school year. Many times, grantors will allow the alteration, as long as a partial report and proposed financial statement is turned in by the December deadline. This ensures that they can meet their reporting deadlines.

Price. The contractors you meet will all compete for price, but you also probably realize that the cheapest bid might not always be the best deal. You will have to look at the materials they use, the timeline they can meet, and the guarantee they give to make your best choice. Grantors will compare your proposal to others (Figure 10-5). Contrary to what many people think, grants aren't just free money to use as you wish. Many grantors compare the cost of a proposal to the value of what they are receiving in an effort to pick one that is most likely to meet their goals. Therefore, accurate and competitive pricing is vital. Sometimes grantors have specific rules about seeking bids. For instance, the FEMA Fire Act grants include a statement on "procurement integrity" with regard to fair pricing.

FIGURE 10-5
Grants are usually highly competitive, and your proposal will likely be compared against others like it.

Sometimes funds can be reallocated within a program, but not always. Say you obtained a grant to purchase 20 sets of turnout gear at $1500 per set. When you order the gear, you find that the supplier is giving a price break for that quantity, and that you will actually be able to purchase 22 sets for the same amount. It's likely that the grantor will deny the purchase because your needs statement and proposal stated you only needed 20. However, the grantor may allow you to upgrade the 20 sets of gear with more reflective striping or radio pockets bringing the cost back up to $1500 per set.

Guarantee. The final component in choosing a contractor is a guarantee. An electrician that gives a 100% unconditional guarantee on the work completed is likely to impress you. Grantors need to feel certain that you can meet your end of the deal. Sometimes references help, but a guarantee can seal the deal. Say your department wants to increase residential sprinkler awareness in an attempt to raise the number of new homes that are equipped with them. Your city holds an annual home improvement show with "contractor day" every year at the convention center and your fire prevention bureau wants to expand on its normal booth by setting up a display in the parking lot. Your department develops a plan to build a sprinkler demonstration prop, complete with a propane-fed fire and plumbing. The grant proposal is complete with drawings and schematics by professional engineers. Pictures of test burns and sprinkler extinguishment are included to show that sufficient time was devoted to research and testing with an assurance that it will work. Grantors still might be concerned with funding something new and unproven. A guarantee might offer a specific amount of capital investment set aside by the department to cover any unexpected expenses that arise. It could be the difference to sway a judging committee to approve your request.

Budget

Because grants are predominantly requests for funds, virtually all grant applications require some sort of a budget. The financial component of your proposal will be scrutinized as much as the narrative, so it's important to read the application carefully and be sure your budget meets the requirements. Sometimes grant applications require both a budget narrative (a description of the expenses) and a line-item budget (laid out in spreadsheet style). Budget narratives are useful when trying to justify a specific purchase. There are several parts of a budget that may be required and should be addressed, including direct expenses, in-kind contributions, and the total amount requested.

Direct Expenses

Direct expenses are the actual cost of purchases, rentals, or other forms of compensation as components of the grant. Depending on the cost of the items, quotes may be required. Care should be taken to ensure that estimates are as accurate as possible. This is sometimes difficult because prices occasionally change in the period between the application and the awarding of the grant. Quotes obtained from vendors should take into account the award date to ensure accuracy of pricing. If a surplus or deficit occurs, some grantors allow you to transfer excess money from one line item to another by a budget adjustment.

In-Kind Contributions

Some grantors require a specific commitment by the grantee that they will invest in the success of a proposed program. Usually in the form of cash, equipment, or materials, these contributions are referred to as **in-kind contributions**. In-kind contributions may also include administrative costs such as salaries, benefits, office supplies, and transportation. Once again, reading the grant rules closely is vital to ensure your in-kind portion is acceptable. For example, some governmental grants state that the equipment listed as in-kind contributions become property of the state for a designated period of time. Therefore, a commitment to purchase a video/data projector as an in-kind contribution for a state grant could result in a state property tag being affixed to it. Rules may also require that the projector would have to be purchased after the award date of the grant in order to qualify as an in-kind contribution, preventing a grantee from claiming existing equipment. When a grant application specifically requires matching funds, the grantee is expected to split the expenses. Unless otherwise stated in the application, matching funds are 50%. A foundation grant request for matching funds to purchase a 12-lead **electrocardiogram (EKG)** machine would require the fire department to pledge half the cost.

Total Amount Requested

Finally, the application will include the total requested amount. Researching past data in advance can help you understand if your request is realistic or not. The grant will be analyzed, and therefore evaluated both objectively and subjectively.

Objective Analysis. Much like the way you look at the price at the bottom of a sticker on a new car, the amount you request will be viewed fairly early in the process. Sometimes the total request is even required on a cover page. Objective information is considered factual, comparable, and statistical. **Objective analysis** attempts to compare apples to apples for competitive grants. For instance, the 2009 Fire Station Construction Grant (SCG) statistics showed that fire departments in the state of Texas made up for more than 5% of the total requests for funds, although Texas made up less than 2% of the states and territories that applied.[14] The statistics also put the average request for construction at $1.75 million. This information is important because if your department wanted to apply for a $2.5 million grant in that area, it could be a tough sell. Compare that with Nevada, which made less than 1% of the requests but averaged more than $3.5 million per request. Competition and the total amount requested does play into the bottom line for competitive grants.

Subjective Analysis. It's also important to ensure that your amount requested is reasonable for what others would expect to pay. Human evaluators bring personal opinions, bias, and experience to the table while evaluating your proposal. Evaluators' **subjective analysis** uses their interpretation of your application to rank and compare it to other requests. If your proposal was to purchase new fireground portable radios for $600 each with a 5% match, evaluators might wonder what kind of radio system you might plan on purchasing for so little money. Requesting $55,000 for a pickup truck would cause evaluators to question your responsibility as well.

AWARDING OF A GRANT

When you successfully achieve funding for your proposal and are awarded a grant, there are three additional aspects you should be aware of: the contract, deadlines, and reports.

The Contract

Usually you will receive a contract with the notification of your award. The contract is a legal document that needs to be signed and returned to the granting agency. It lays out the agreement for the grant, and usually includes such components as the time period, the amount awarded, in-kind contributions, the program goals, and reporting requirements. Although the grant is an agreement between the grantor and the grantee (agency or organization), it usually identifies a designee that is responsible for ensuring that the program or purchases follow the agreement. It's important to maintain a copy of the contract with other paperwork from the grant application.

Deadlines

Deadlines were important in the application process, and are just as vital throughout the life of the grant. Specific deadlines for certain aspects of a program are sometimes established, and at times ties funding to these benchmarks. For instance, a grant award for a smoke detector blitz program in your city might provide the first half of the funding up front, while the second payment is held until a midway report is received. This tiered funding allows the grantee to commence the project immediately, but ensures that funds are released as progress is made.

Reports

The last deadline involves submission of a final report. As stated earlier, other status reports might be required periodically throughout the grant period, depending on the size and length of the program. The final report is your chance to prove to the granting agency that they made a wise investment. It should directly address the goals and objectives listed in your grant application, as well as specific information they request. Because the final report doesn't generally follow a tight template like the application may have, it can be written with a more informal touch that includes stories and highlights. Pictures and graphs are useful if they convey a message, and copies of newspaper articles or thank-you letters sometimes help prove the value of the program. Remember that your final report will be used as a tool if you ever apply again and wish to be considered for future grants.

SUMMARY

One of the greatest opportunities for improved firefighter safety is the substantial increase in the number of grants available in recent years. Emergency services can still look to community organizations and businesses for support, but with the introduction of federal grants for equipment, personnel, and even fire stations, opportunities have proliferated. However, even with the influx of grant money available, grants are still competitive in nature. The need for finances will always outnumber the available funds, and organizations can easily be passed over if they don't understand how the grant process works and how to increase their chances of being successful.

Initiative 10 calls for the blending of grants with firefighter safety. Not only should grant requests that improve safety receive preference over other grants, but organizations should first demonstrate safe practices to be eligible. Maintaining a current list of fire and emergency service agencies that have a history of safe practices is something that needs to be developed. Once criteria have been established, agencies can verify their conformity and submit an application to be listed as an FLSI 16–compliant organization, ensuring that investments made would be primarily for the safety for our personnel—the most valuable of our assets.

KEY TERMS

AFG - Assistance to Firefighters Grant.

call for proposals (CFP) - A formal request by a grantor for projects to consider funding.

Central Contractor Registration (CCR) - A registry of organizations permitted to obtain grants, as well as to submit bids and proposals.

cost-sharing - A requirement of some grants that expect the grantee to invest a specific amount of money into the project.

Data Universal Numbering System (DUNS) - A unique numbering system for identifying organizations.

dry hydrant - A water appliance consisting of a pipe installed adjacent to a water source with a fire department connection to speed up the process of drafting.

electrocardiogram (EKG) - A machine capable of recording the graphical representation of a cardiac cycle. Also referred to as an ECG.

EMSC - Emergency Medical Services for Children.

Everyone Goes Home (EGH) - A prevention program created by the National Fallen Firefighters Foundation in an effort to reduce future line-of-duty deaths. One of the major accomplishments was the creation of the 16 initiatives, the basis of this text.

FEMA - Federal Emergency Management Agency.

Fire Prevention and Safety (FP&S) - An Assistance to Firefighters Grant funding fire prevention, fire and life safety, and research projects.

Fire Station Construction Grant (SCG) - An AFG grant with funding priorities for fire station renovation or construction.

foundation - A public or private organization with a stated purpose that often provides or receives funds to support its goals.

goal - A measurable and achievable ambition comprised of several objectives to ensure its success.

grantee - An individual, organization, or governmental unit receiving a grant.

grantor - An individual, organization, or governmental unit administering and awarding a grant.

in-kind contributions - Similar to cost sharing, grantees are sometimes required to provide such contributions as equipment, materials, labor, or other investments such as insurance.

International Association of Fire Chiefs (IAFC) - Organization of fire chiefs from the United States and Canada.

match - A type of cost sharing where matching funds (50/50) must be provided by a grantee. Some grants permit a match of in-kind contributions.

mission statement - An organizational proclamation stating the reason it is in existence.

National Fallen Firefighters Foundation (NFFF) - A nonprofit organization created to honor and assist families of firefighters who die in the line of duty, and create programs to prevent future events.

National Fire Fighter Near-Miss Reporting System - A voluntary, nonpunitive, fire service reporting system for near-misses.

National Volunteer Fire Council (NVFC) - A professional organization for volunteer firefighters.

NFPA - National Fire Protection Association.

NICHD - National Institute of Child Health and Human Development.

NIMS - National Incident Management System.

objective - A workable component of a goal.

objective analysis - Reviewing facts without bias.

qualitative research - Data or research utilizing feelings or opinions.

quantitative research - Date or research utilizing numbers or percentages.

Risk Watch - An all-hazard fire and life safety education program created by the NFPA for children.

SCBA - Self-contained breathing apparatus.

school resource officer (SRO) - An individual assigned to an educational institution as a liaison between students, faculty, and law enforcement.

SIDS - Sudden infant death syndrome.

Staffing for Adequate Fire and Emergency Response (SAFER) - An Assistance to Firefighters Grant funding staffing and volunteer recruitment and retention programs.

subjective analysis - Reviewing facts and personal opinions.

Sustainability Agriculture Research and Education (SARE) - A grant promoting research and education of agricultural procedures and techniques.

USFA - United States Fire Administration.

REVIEW QUESTIONS

1. What is the importance of safe practices as an eligibility requirement for obtaining a grant?
2. What are some of the companies that support the fire service on the national level?
3. How can a program seed grant benefit an organization?
4. What are some of the benefits of using a SAFER grant for volunteer recruitment?
5. What is the importance of compiling a list of FLSI 16 registered organizations?

STUDENT ACTIVITY

1. Pick a piece of new firefighting equipment you would like to purchase for your department, and write a narrative for an imaginary grant application. Be sure to include the components of an effective narrative described in this chapter, including:

    - Strategic level (significance or problem statement)
    - Tactical level (objectives)
    - Task (procedures)

FIREFIGHTER WEBSITE RESOURCES

http://www.fema.gov/firegrants/
http://www.riskwatch.org
http://www.lowes.com/cd_Workplace+Lowes+Heroes_149851308_
http://sites.target.com/site/en/company/page.jsp?contentId=WCMP04-031880
http://www.firemansfund.com/heritage/Pages/welcome.htm
http://www.fmglobal.com/page.aspx?id=01060200
http://www.statefarm.com/aboutus/community/grants/grants.asp

ADDITIONAL RESOURCES

http://www.firegrantsupport.com/
https://www.rkb.us/

NOTES

1. Wallace, M.J. 2007, October 1. First breath. *Fire Chief Magazine.* http://firechief.com/training/apparel/firefighting_first_breath/
2. NFPA standards for first responder personal protective equipment adopted by U.S. Department of Homeland Security. NFPA News Release, March 26, 2007.
3. http://www.grants.gov/aboutgrants/grants.jsp
4. http://www.mbwcinfo.com/id1.html
5. Hawkins, A.M. 1986. Keeper of the forest. History of the Black Forest community club. http://www.blackforest-co.com/bfcc/
6. The Heritage Program at Fireman's Fund Insurance Company, Grants and Donations Awarded to Date (report), June 30, 2010.
7. http://www.hfpg.org/GrantmakingPrograms/Overview/tabid/163/Default.aspx
8. Assistance to Firefighters Grants; Guidance and Application Kit. 2010, April. U.S. Department of Homeland Security, Federal Emergency Management Agency.
9. 2010 COPS Safe Schools Initiative Grant Owner's Manual. U.S. Department of Justice; Office of Community Oriented Policing Services. U.S. Department of Justice Office of Community Oriented Policing Services.
10. http://www.jacksonf.org/index.php/types-of-grants
11. Western SARE.
12. Minnesota Council on Foundations, Minneapolis, MN. http://www.mcf.org/mcf/whatis/common.htm
13. Paeno, L.M. 2008. *Professional grant writer; the definitive guide to grant writing success.* Kansas City, MO: Landmark Publishing.
14. 2009 Station Construction Grants Application Statistics Chart, AFG. http://www.firegrantsupport.com/docs/2009AFSCGAppStats.pdf

CHAPTER 11

Establishing Response Standards

LEARNING OBJECTIVES

- Define a standards developing organization (SDO) and how it can affect safety.
- Explain the importance of terminology in policies and procedures.
- Differentiate between standard operating procedures (SOPs) and standard operating guidelines (SOGs).
- List the seven components of an effective procedure or guideline.
- Explain the differences among the three series of procedures.
- Discuss the assembling, modeling, and adoption of procedures.
- Explain how safety would be improved with national standards.

CHAPTER 11
ESTABLISHING RESPONSE STANDARDS

 Initiative 11: National standards for emergency response policies and procedures should be developed and championed.

The first 10 chapters of this book have established a broad base of ideas using the 16 initiatives that, when utilized, can allow us to do our job effectively and with an increased level of safety. Chapter 3, "Applying Risk Management Techniques," described ways to assign risk to specific duties we could be expected to perform, while factoring in the benefits and likelihood of a successful outcome. Chapter 9, "Investigating Fatalities, Injuries, and Near-Misses," examined the ways that experience gained from near-misses and investigations could be shared with others in an effort to prevent similar occurrences. Through the implementation of Initiatives 3 and 9, we are led to the next logical conclusion, establishing standard policies and procedures that draw upon the experience we have gained in an effort to make the next incident safer. For instance, a concrete contractor that landed a bid for installing new sidewalks around the elementary school would learn pretty quickly what happens to unattended wet cement near a playground. It's unlikely that after the first incidence of bike tire tracks, handprints, and third-grade artistic renditions with a stick the contractor would follow the same **procedure** on the next day of the pour. Instead, a reasonably intelligent adult would change the timing of the work performed or protect the area with a barrier or a guard. It's also likely that the company would change the procedure in the future to avoid further losses.

Initiative 11 takes this concept one step further by encouraging similar organizations to adopt policies and procedures similar to their own. Many concrete finishing companies might already have procedures, but they may not be written out for all employees to understand. Other companies may never have run into the problem in the past. It could be that the practices they follow may not use the most efficient and cost-effective methods. It's understandable that not all contractors would need the entire **policy** if they only poured basement footers or floors on skyscrapers. However, some sections of the policy could be valuable to all contractors. The value of sharing this information is much more obvious when we are talking about firefighter deaths rather than mischievous sidewalk art. Industry, business, education, and government all use the "best practices" as a way to mutually benefit from knowledge and experience. Generic **risk management** situations and knowledge learned from investigations and near-miss situations can be strategically placed in policies and procedures that can be utilized by numerous emergency agencies, making our standard responses easier and safer.

It would be impossible to create standard policies and procedures within every emergency response organization for every conceivable situation. However, it's just as impractical to suggest that no policies and procedures could be adapted to organizations. Nationally accepted "base model" policies could be used as voluntary consensus templates to ensure that vital information is utilized industry-wide. In Chapter 2, "Enhancing Accountability," we looked at the possibility of creating our own safety standards similar to the ISO 9001 registry. A standard such as this would normally created by a **standards developing organization (SDO)**. For instance, the **International Organization for Standardization (ISO)** is an example of an international SDO. Leonard P. Connor, managing director of the **American Welding Society (AWS)**, put it best when he wrote that American business and industry is successful in part due to our "voluntary consensus standards."[1] Regardless of whether we are referring to the **American Society for Testing and Materials (ASTM)**, the AWS, or even the **NFPA**, SDOs create the standards, which in many ways influence our laws. For example, ASTM E1354-10a, Standard Test Method for Heat and Visible Smoke Release Rates for Materials and Products Using an Oxygen Consumption Calorimeter, directly affects the testing process of new materials proposed for acoustic wall covering materials in night clubs. How well the materials perform affects the requirements of building and fire codes. Before we get into the benefits and application of standards, it's important to explain the different terms we use and the intention of each.

TERMINOLOGY

There have been many recent discussions about the legal ramifications of maintaining **standard operating procedures (SOPs)** versus **standard operating guidelines (SOGs)**, and then deviating from them on scene. Although this book is not intended to provide legal opinions on one or the other, the common terminology and understanding of the choice of words is explained. For instance, it's understood that the word *shall* is a very strong word that gives little room for variance, whereas the word *should* gives more leniency and flexibility in decision making.

Specific words chosen are intended to have a specific meaning. Suppose you are watching an infomercial on television for dietary weight-loss supplements that appears to have fantastic results. The before and after pictures of real-life examples are impressive. The doctor giving a statement about the results of a trial looks the part, complete with a white lab coat, stethoscope, and certificates on the wall behind him. He explains, "The results of our study were simply astounding. Nearly 60% of the subjects who followed our program saw a significant weight loss of 10–15 pounds in the very first week of the study. Even more impressive is that up to a third of the weight that some lost was unhealthy fat!" Although it sounds impressive, they left themselves a lot of wiggle room when they wrote the script, which ultimately results in a lot of questions.

- How many people were in the study? If there were 12 people, only 7 would be nearly 60%.
- Who were the subjects? Were they first-time moms who enrolled to lose baby weight or possibly less enthusiastic people who were warned to lose weight by their doctors?
- What does it mean to "follow" the program? If they didn't abide by every rule, were they eliminated from the statistics? Were there 80 people in the study who took the pills but only 12 "followed" the program?
- If you ever "saw" a significant weight loss, did you ever "see" it come back just as quickly? How long did the subjects keep the weight off?
- What was their average weight before the loss? An 11-pound loss to a 170-pound person compared to a 315-pound person is quite different.
- Were the subjects encouraged to eat healthier and exercise more? How many did?
- Why did he choose the phrase "a third" when explaining about how much fat they lost? Is a third exactly 33.33% or just a loose calculation?
- How many is "some" that lost a third of the fat? Isn't that less than "most"?

The only thing we are pretty sure of is that some of the people lost between 10 and 15 pounds by following the program, some of which was fat. The terminology they chose left holes in their "research" and avoided many facts that could have made the statistics relevant. Although it sounds impressive, it would be difficult for someone who saw no results from buying and using the product to prove the company's claims were fraudulent and misleading. Similar potential problems in the terminology chosen for standard policies must be eliminated, while maintaining its original intent. Policies and procedures, along with SOPs and SOGs, are the most common ways of relaying this information.

Policies and Procedures

Many organizations have mission statements that describe what they do. Policies are specific clarifications and extensions of the mission statement that illustrate the beliefs

FIGURE 11-1
Policies proclaim organizational goals. It's not uncommon for several procedures to address a specific policy of safety.

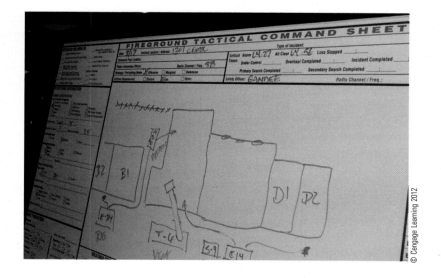

of the organization (Figure 11-1). When comparing policies and procedures, policies explain "why" something is done and procedures describe "how" it is to be done. For instance, a towing company might have a policy to provide 24-hour service (to benefit the customer) and another policy to impound the vehicle until payment is received (to protect the company). The procedures would address how the dispatching, scheduling, and access to drivers on call will be accomplished. Others would explain the company's payment collection process as well as the vehicle impound and release procedure.

Many times the word *policy* makes people think of an order that results in discipline when it's not followed. Although that can be true, let's set that definition aside for a moment and define the word as "a broad goal, or statement that guides actions." For example, let's say your **assistant chief (AC)** just put out a new policy requiring that a minimum of two ground ladders be placed for emergency egress at upstairs windows for all fires in two-story homes. Some members who read the policy may jump to the conclusion that the boss is just trying to look for reasons to write them up if they don't follow it. Some might feel that, although important, taking time to set ladders in every situation could detract from more important functions, such as ventilation or shutting off the utilities. If you were to sit down and discuss the policy informally with the AC, you might find that the primary goal or intent of the policy is to ensure that crews have a safe way out of the second floor if they need it. The policy could be the result of past fires, when the incident commander forgot to call for preemptive ladder placement and the crews prioritized the task too late to be effective if they had been needed. By creating a new policy, the department commits to ensuring that the task will be completed every time the situation arises.

Once the policy is in place for laddering buildings, the next step is to develop the procedure to be certain that it is adhered to consistently. Some departments achieve this step by adding the task to an incident checklist. Besides the obvious advantage of an increase in compliance, a policy and procedure for setting ladders in a prefabricated **incident action plan (IAP)** reduces the number of fireground decisions that have to be made by the **incident commander (IC)**. By spending less time considering ladder placement, which tends to be a decision that is static and invariable, the IC is able to devote more attention to important decisions involving risk management and tactics—both of which are dynamic and variable. Many departments develop a formal SOP or SOG to address the policy.

Standard Operating Procedures and Guidelines (SOPs and SOGs)

SOPs and SOGs are the accepted way to ensure consistency, effectiveness, and professionalism in an emergency response organization. Many times they dictate what equipment responds on what type of calls and what duties are required of the crews. Some departments also use them as a training tool to teach new members what is expected for given situations and to ensure that specific command procedures are followed. Many departments use them as a way to ensure that important safety measures are followed to prevent department losses or personal injuries. For instance, one SOP/SOG might try to prevent the catastrophic failure of equipment by ensuring that an aerial device is only set up on hard surfaces rather than on a gravel driveway or grass lawn. Others are intended to prevent injury or death by forbidding the aerial operator from extending or retracting the device with a firefighter on the ladder.

Both SOPs and SOGs use the common terms standard and operating. Many references cite the definition of standard as normal, accepted, average, or usual. Although considered a benchmark, the word does not mean "absolute", "definite", or "unequivocal". Therefore, using the word *standard* in SOPs or SOGs allows some deviation for unusual circumstances. The term *operating* is understood as a process or action of work that rarely is disputed. The issue of "procedure" versus "guideline" is much more controversial. In the past several years, many organizations have loosened the terminology of an SOP by changing them to SOGs, utilizing the softer term *guideline* rather than the more rigid term *procedure*. For example, your department may have an SOP for rooftop ventilation that states that among other equipment, a rope bag is to be carried to the roof with the crew. The intent of such a procedure could be to allow a way to raise additional tools, secure a piece of equipment such as a ladder on the roof, or to provide an emergency means of egress. Because the word *procedure* can be defined as "a course of action," it could be interpreted as mandatory. If so, a ventilation crew that did not have a rope bag would not be permitted to perform the job. It's possible that the intent was mandatory, and those members not taking the rope with them when ventilating would be subject to discipline. However, it could simply be a poorly worded procedure that was merely a suggestion to take some rope with them.

Some feel that guidelines are more easily deviated from than procedures, which could reduce the legal implications of not following the standard to the letter. An SOG that lists the rope bag as necessary equipment would be less enforceable, and act more as a suggestion. A crew that did not take the rope bag would still be permitted to complete the job. Others feel that by using the word guideline, it softens it too much in certain situations. The guideline that suggests a rope bag could be interpreted to only suggest an airpack and portable radio. Regardless of whether your department uses SOPs or SOGs, they have to be written in a way that addresses most possible scenarios. A good rule of thumb is to use strong words such as *procedure, shall*, and *must* when safety is involved, and use *guideline, should*, and *may* for recommended practices. NFPA references procedures in most of their standards, whereas **NIOSH** references both procedures and guidelines.

Interestingly, the word *standard* is also defined as an accepted point of reference. In other words, the very fact that we have SOGs and SOPs in organizations means that they should be standard or similar in content. If two organizations have standards that conflict, what makes one more acceptable than the others? Occasionally we see **strategic weapons attack teams (SWAT,** also sometimes known as special weapons and tactics) come under fire in the media for using excessive force when forcing entry into a home where a fugitive is suspected of hiding. Victims and the media use a "standard of care" approach to compare the policies and procedures of the SWAT team with other SWAT teams in the country. In Chapter 5, "Implementing Training and Certification Standards," we looked at the advantages of having similar training and standards of care regardless of your geographical location. Similarly, if we are going to use the word *standard* in our policies, procedures, and guidelines, we should make sure they are standard. This not only ensures that we are doing the right thing, but it also reduces our liability because we are following a standard.

Writing Style

SOPs and SOGs must be written using common terminology with a common writing style, or format. Many fire departments already follow a proven method from other industries with several components. Steven B. Page, author of *Establishing a System of Policies and Procedures*, urges writers to use seven specific headings in an effort to maintain consistency.[2] These include purpose, revision history, personnel affected, policy, definitions, responsibilities, and procedures.

Purpose

Let's say your fire department decides to draft an SOP regarding the securing of all equipment in the passenger compartment of their apparatus to prevent injuries. The first component of the policy should be the purpose of the document, which is clearly stated in one or two statements. The purpose should leave no questions about the intent of the SOP while giving the reader a hint about what is left to follow in the balance of the document:

> "The purpose of this document is to ensure the safety of personnel in the event of unexpected vehicle movement or a crash. It is vital that all portable equipment be properly stowed to prevent injury while the vehicle is moving."

Revision History

It's not uncommon for SOPs/SOGs to undergo changes over time, and tracking the history of revisions is an important component. Sometimes personnel are surprised to hear that the policy they had been following had been updated years ago. Noting the history can be accomplished through a typical revision/version decimal system (v2.4); however, many fire departments simply use a "supersedes" date. A complete index should be distributed with each revision that contains the latest revision number or date for each SOP. This allows personnel to know if the policy they are reading is the latest revision.

Personnel Affected

This section lists which specific personnel or ranks are involved in acting on the SOP. It could be intended for all personnel, all salaried employees, or all members of the trench rescue team.

Policy

The policy section explains the rationale behind the SOP or "why" it was written. Some fire departments use the word *purpose* rather than *policy* for this section. A policy statement for our SOP might read as follows:

> "It is the policy of the FLSI-16 Fire Department to protect its members while traveling in department vehicles. Emergency apparatus contain numerous types of portable equipment, map books, and various types of PPE, all of which can become a missile in the event of a crash or sudden vehicle movement if not properly secured. This policy will outline the procedures for protecting our members."

Definitions

The advantage to having a definition section in the SOP is to ensure consistency in application. For instance, does a fire prevention inspector's car fall under the policy? If so, staff vehicles must be listed under the definition for emergency apparatus or department vehicles. If no specific definitions for the document are needed, the heading is used but the subsequent section is left blank. This not only ensures consistency between documents but also allows the addition of definitions in future revisions. An example of a definition for our SOP is:

> *"**Positive Securing Device (PSD)**—A positive securing device is any retaining equipment used to hold a piece of portable equipment or PPE in place. It could consist*

of a commercially made bracket, or an approved in-house fabricated bracket, strap, or compartment used in normal operations. In certain situations, it can also consist of a seatbelt, backboard, or cot strap, or a rope, netting, or other homemade device used as a temporary measure to allow usage during patient transport, response, or return to the station. Personnel holding a piece of equipment is not considered a positive securing device as it could still slip or become dislodged in a crash."

Responsibilities

The responsibilities section is intended to explain who is responsible for acting on the SOP, and what specific action he or she should take. For instance, an SOP regarding a morning apparatus check might outline that the engineer would be responsible for the check and follow-up actions such as refueling. Our sample responsibilities for securing equipment that could become airborne might consist of:

- *"All members are responsible for the securing of equipment prior to vehicle movement."*
- *"Officers are responsible for ensuring that vehicles that need brackets or repairs are reported."*
- *"Officers are responsible to make changes to equipment storage as needed to keep the vehicle in a state of readiness while maintaining safety for the members."*

Procedure

The final component of our SOP should demonstrate the steps necessary to achieve the goal of the policy and expand on the responsibilities. You might say the policy is a vague statement of intent, whereas the responsibility section outlines who needs to do something about it (Figure 11-2). If the policy affects your job and the responsibilities point to you, the procedure section is something you must fully understand. Some procedures can be several pages in length. When procedures become extremely drawn out, consideration should be made to splitting it up into several procedures. A complete sample of this procedure in SOP form is found in the appendix of this book. However, a sample of the procedures would include:

1. *"Portable equipment in the cab or patient care compartment must be secured prior to vehicle movement. This can be accomplished in any one of four ways:*
   a. *Equipment will be secured in its normal compartment, bracket, or strap.*
   b. *If equipment is needed for patient care or is unable to be repacked on scene to fit in its normal location, it must be contained with a Positive Securing Device.*
   c. *If the normal location and a PSD are not options, the piece of equipment must be stowed in an outside compartment.*
   d. *If the equipment cannot be transported safely and securely, the officer in charge will call for another apparatus to complete the transport."*

FIGURE 11-2
Procedures outline how to achieve the goals of the policy.

LEVELS OF STANDARD POLICIES AND PROCEDURES

In order to provide standard policies for emergency response agencies, the first step would be to create a system based on the different levels of applicability. **Series One (S-1)** policies and procedures would be applicable to virtually all emergency response agencies with little or no alterations. **Series Two (S-2)** would act as a template for the majority of agencies, whereas **Series Three (S-3)** would be used for specialized planning and responses.

Series One (S-1: National)

All emergency response organizations have similar risks as a result of specific response duties that can apply to a volunteer member of a fire/police organization in Pennsylvania, a critical care transport team in Arkansas, or a ladder company in Los Angeles. Therefore, common industry-wide standards for procedures would be applicable with very limited local changes. Some examples of these commonalities include emergency response, apparatus positioning, utilizing the command system, and risk management.

Emergency Response

One of the core objectives of Initiative 11 is to develop standards for emergency response. Before standards can be developed, it's important to define what an emergency is. There are vast differences in response procedures for similar incidents in different areas of the country, but a national standard would be an effective way to improve safety.

Discussions about the classification of an emergency versus a nonemergency are usually passionate and opinionated. Most emergency responders would agree that the reason for the dilemma is due to liability. Many times, local procedures are based on the legal opinions of the fire chief, mayor, or law director. This could lead some to argue that there is no way to establish national standards because of different legal opinions and the belief that their way of doing business is the best. Others feel that the best way to protect your organization from legal threats is to comply with national response standards. National standards could improve safety based on the type of call, the apparatus, and the personnel responding.

Type of Call. The way an emergency call is reported and dispatched affects response. Calls such as building fires, heart attacks, and trench rescues are usually presumed to be emergencies based on the dispatcher matching the caller's information to a predetermined response procedure. Service calls for animal rescues, water leaks, and other investigations or complaints generally are regarded as nonemergency responses. Sometimes additional information categorizes it to an emergency or nonemergency response. For instance, a carbon monoxide alarm with no illness in many jurisdictions gets a nonemergency response to check for the poisonous gas, whereas a carbon monoxide alarm with people ill upgrades it to an emergency response with **ALS**. Other calls aren't quite as cut and dry when it comes to determining the severity of the emergency, but some specific types of calls could benefit from a national standard for response, including automatic alarms, odor or smoke investigation calls, and ambulance transports.

- Automatic Alarms
  - What is the standard response to an automatic fire alarm with no supporting information at a hotel? Some departments deploy multiple stations responding with lights and siren, whereas others wait for confirmation from hotel security before dispatching. It could be that your department responds as the first-due company with lights and siren (emergency), whereas secondary units continue nonemergency care until further information cancels them or has them upgrade the response. A risk-benefit model would have to take into account the number of false alarms, the potential for the loss of life and property with a delayed response, as well as the potential risk of an apparatus crash. Although motor vehicle crashes are a significant cause of death for firefighters, hotel fires have the potential to cause numerous deaths as well.
  - Many commercial and industrial buildings these days have automatic fire alarms, and sometimes automatic alarm response needs more planning. For example, a childcare center or school might require a heavier response during occupied hours than when it is closed. The chance of a real emergency at the city

outdoor pool pavilion in January that requires a heavy response for a fire alarm is unlikely. Sometimes the alarm is better at describing the problem. Thunderstorms are notorious for setting off false alarms due to power outages or surges. However, alarms can't be assumed false because lightning can be the cause of an actual fire. An automatic alarm that comes in as a water flow alarm at an unoccupied store in the middle of the night during a thunderstorm should be treated a little more seriously than a **pull station.** The reliability of an alarm that distinguishes the specific problem before we arrive depends on the design and performance of the system.

- Odor and Smoke Investigations
  - Sometimes smoke investigations turn out to be structural fires, but usually they don't. Many times the source is never specifically identified. An effective risk management system would consciously look at what a 911 caller reports and develop procedures for dispatchers to classify the call more accurately with regard to the number of units responding and whether they respond emergency or nonemergency.
- Ambulance Transport
  - One of the most unclear "emergency versus nonemergency" decisions is the transportation of patients to a medical facility. At times, state laws and organizational policies are vague in describing what constitutes an emergency, and hospital protocols occasionally conflict with textbooks. For example, many protocols suggest an emergency transport for a seizure or heart attack patient, whereas some **EMT** and paramedic textbooks suggest no lights or sirens to keep the patient calm. Some department policies dictate transporting all patients with lights and siren to reduce the legal implications of a patient's condition getting worse during a slow transport. Others reserve emergency transports for injuries or illness that directly affect the loss of life or limb. Many times the decision is based solely on the opinion of the ambulance crew at the time. The decision might be based on patient condition, traffic, call volume, or even the time of day. In many cases, the risk-benefit may not even be a factor at all. A standard policy could be constructed with a primary concern about patient treatment and a secondary emphasis on risk management.

Apparatus. Aside from the type of call, the apparatus responding to an incident affects the overall risk. This not only has to do with the type of apparatus, but also the condition of the vehicle. Rules vary from state to state regarding apparatus safety inspections, with many having no requirements at all. An 18-wheeler on an interstate might get randomly pulled over and inspected for brake adjustment, tire condition, or cracked springs. On the other hand, a pumper may not get a brake adjustment until the parking brake can't hold on a hill. Cracked springs on an aerial ladder might not be noticed until a ladder testing company performs its recertification test. National

FIGURE 11-3
Although a critical component of rural firefighting strategies, water tenders or tankers are one of the most dangerous pieces of equipment used.

policies and procedures that fire departments could adopt for apparatus inspection would be a step in the right direction. Two types of apparatus specifically are worth mentioning and should have their own procedures: Water tenders (or tankers) and personally owned vehicles (POVs) are involved in a significant number of firefighter fatalities every year.

- Water Tenders
  - NIOSH reports that one of the "most prevalent" vehicles for firefighter injuries and death is a tanker, or water tender.[3] Besides the fact that hauling a liquid in bulk proves to be an unstable load that can shift easily with a high center of gravity, tenders are probably the heaviest vehicle a firefighter will ever drive (Figure 11-3). Sometimes tenders are old military surplus vehicles, or converted milk or fuel tankers. Although Chapter 19 in NFPA 1901, Standard for Automotive Apparatus, requires water tanks to have baffles to prevent sloshing, food grade (milk) tankers are prohibited from having baffles due to cleaning regulations. A converted 3000-gallon gasoline tanker would carry an additional 6400 pounds in load due to the fact that water weighs almost 2.5 pounds more per gallon. If a water tender was converted from another use, it's vital that alterations be made to the vehicle to ensure that it is adequately safe. Regardless of its origin, the intent of a tender is to haul a large amount of water, which will have a large influence on the operational characteristics of the vehicle. Consider that even a small tanker with a full load of water would be easily carrying a load of 7 tons.
  - A policy should include rules about driving the vehicle full of water, or completely empty. A partially loaded tender can be an unstable vehicle that is unsafe to drive. SOPs/SOGs should also define when water tenders should

respond on an emergency basis and when response with the flow of traffic is a better decision. There has been a recent push in some areas of the country to take the drastic step of declaring water tenders nonemergency vehicles, equipping them with only yellow warning lights and no siren. Unfortunately, it's not quite that simple.

- Take a reported structural fire in a rural area with no pressurized water system, for instance. If the second unit out behind the attack engine is a 3000-gallon pumper/tender that provides water and manpower for the initial fire attack, it may need to be considered an integral component of the engine company in order to provide the resources needed to complete an interior attack that supports search and rescue. In this case, an emergency response might be prudent. NFPA 1720, Standard for the Organization and Deployment of Fire Suppression Operations, Emergency Medical Operations, and Special Operations to the Public by Volunteer Fire Departments, groups combinations of vehicles together into companies such as this.[4] On the other hand, a third-alarm tender box that provides an additional 10,000 gallons of water for a shuttle to surround and drown a stubborn fire in a barn that is full of hay and that is close to collapse might be a little lower on the risk-benefit model. Although most incidents are not as clear-cut as these two examples, a risk-benefit assessment can be made by categorizing the purpose of the water tender. In other words, is the reduction in response time going to affect the outcome of the preservation of life and property, and at what risk?

■ Personally Owned Vehicles (POVs)

- POVs are another consideration when establishing response priorities. State laws and departmental policies vary greatly on the response of personal vehicles. The risk versus benefit is always questioned when a young firefighter crashes in his or her own car responding to what turns out to be an incident that didn't require any more help. Most POVs are not equipped with the same caliber of emergency warning devices and visible identification as that of an emergency apparatus. Two-way radio communications are not as common, preventing important information, such as additional incident information or even a call cancellation. Standard procedures might call for an additional page or alert to relay such information. Add the fact that the condition of the POV's tires, brakes, steering, and suspension will not have the same minimum maintenance safety requirement standards as a department's fire apparatus leaves an extremely large potential for mishap and liability on the department if the POV is involved in an incident resulting from a response.

Personnel Responding. The final component of emergency response procedures relates to the personnel who are responding. The primary cause in most NIOSH reports of apparatus and POV crashes is operator error. Sometimes it's a high rate of speed that may have resulted in a rollover. Other times it's leaving the roadway and overcompensating. As discussed in Chapter 5, training and certifications vary greatly throughout the country, although operating an emergency vehicle has many of the same hazards wherever you respond. An S-1 SOP could address several of the components of operation that should be mandatory for all emergency vehicle operators, such as:

- Maintaining a valid state operator's license
- Completing an emergency vehicle operator course
- Completing periodic emergency vehicle refresher training
- Achieving departmental check-off for a specific vehicle
- Being free of alcohol, and illegal or impairing drugs
- Wearing a seatbelt (all occupants)
- Responding at a safe speed for the existing road, weather, and traffic conditions
- Stopping completely at all red lights, stop signs, and railroad grade crossings

Apparatus Positioning

Once an emergency vehicle has arrived on scene, proper positioning for safety is essential. Most organizations have specific procedures for apparatus placement to achieve tactical priorities. S-1 procedures for positioning would deal primarily with proximity dangers and scene protection.

Proximity Dangers. Most firefighters remember the standard rule for hazardous materials responses, which generically instruct us to place apparatus uphill and upwind from the incident. Standard policies could address other dangers that we don't necessarily have standard rules for, such as unsafe scenes (Initiative 12).

Scene Protection. With the introduction of high-visibility vests and more emphasis on using emergency vehicles as a protective barrier on roadway incidents, there has been a significant promotion of scene protection. Several textbooks suggest proper positioning for emergency vehicles at crash scenes. If an emergency vehicle is the first to approach a crash, it should position itself between the scene and approaching traffic. The second emergency vehicle should proceed past the crash and position itself to protect the other direction. In a standard engine and ambulance response, the engine would be better suited to block the lane and the ambulance better equipped to travel past the scene and work off the back for patient care. In addition to passerby traffic, emergency apparatus can be dangerous as well during water shuttles or mass casualty incidents.

Command System

Most organizations have some sort of policy and procedure for utilizing the incident command system based on scene requirements. It could be that command is initiated by the first-arriving units at an incident requiring multiple companies. Although the **National Incident Management System (NIMS)** was introduced in 2004, many discussions continue to ensue regarding what incidents require its use. According to the NIMS 700 course materials, NIMS provides "standardized organizational structures, as well as requirements for processes, procedures, and systems designed to improve interoperability."[5] A declaration of NIMS compliance is necessary to apply for most governmental grants, affirms that the organization's membership has been trained to specific levels, and affirms that the command system's use is "institutionalized." NIMS documents define institutionalized as utilizing the command system in both policy and practice.[6]

Two specific aspects of NIMS are important to note when discussing the concept of national standards for response policies and procedures. The first is that the command system and its associated terminology are proven to be effective on both strategic and tactical levels, as well as on conformance to safety standards. Any policies or procedures developed by an emergency service organization must follow this format. The second advantage of NIMS is that it provides the framework for expansion and interoperability. Any incident that uses a similar command system format early on, whether or not it is referred to officially as NIMS, can grow to meet the needs of the incident (Figure 11-4). This could include **EMS,** police, government, public utilities, hospitals, and additional county, state, or federal response teams or other agencies.

Some S-1 procedures could be developed that would meet the requirements of NIMS while considering the specific needs of first responders. Ideal SOPs would clarify the expansion process and the incident priorities as additional resources arrive. Different standards could address many of the most common emergencies. For example, a fire in an occupied elementary school or a mass casualty incident involving a tour bus could happen in almost any jurisdiction. It wouldn't take long for initial resources to become overwhelmed, forcing the incident to expand quickly. By guiding this growth

FIGURE 11-4
All incidents have the capability to expand into a significant event. The command system you use must be capable of expanding ahead of it.

process, standard procedures can assist in making sure it meets NIMS requirements, without changing our initial tactics.

Risk Management

As the incident grows, risk management must continually be reconsidered. An advantage of instituting S-1 procedures is that risk-benefit analysis for all levels—strategic, tactical, and task—becomes second nature. As this occurs, we begin to alter the culture of the fire service. For example, let's suppose it's winter and you respond to a reported chimney fire. You are the acting officer of the engine and arrive just after the captain in an SUV. You look at the house and see sparks and a dark red glow emitting from the chimney cap. With the family safe outside, the captain initiates command and orders your crew to extinguish the fire in the wood-burner and ladder the roof as your SOP states. You split your crew, sending two inside with floor runners and a pressurized water can while you and Thomas ladder the roof. You are familiar with the normal procedure, which involves taking off the chimney cap and dropping dry chemical "bombs" to extinguish the fire. The problem is that the chimney is a masonry "tower" that clears the ridge of the roof and proceeds 6' straight up with no support. To complete your task, you'll have to lean the extension ladder directly on the chimney, creating a lateral load on a questionable structure.

When policies and procedures do not build in risk-benefit components, we can set ourselves up for task-oriented actions (tunnel vision). Some firefighters would lean the ladder on the chimney, remark to the other "I hope it's strong enough" or "I'm not too sure about this," and start climbing. What's at issue is what we hope to achieve by even climbing the ladder. If things go perfectly, the cap is easily removed undamaged and the fire is extinguished quickly with one or two bombs. Worst case, the chimney cracks at a seam above the roof line and crashes through the roof with a firefighter close behind. Risk management makes us stop and think of other ways to accomplish the task of extinguishing the fire without any components of the worst-case scenario occurring. If your normal procedure was to extinguish the fire from inside homes equipped with freestanding chimneys, it takes the human factors out of play.

Series Two (S-2: Template)

The S-2 category of standard procedures would apply to a smaller audience of responders, and would act as a template for an organization to develop a proven but applicable procedure. Let's say you have been given the assignment of creating an SOP for your fire department for the creation of a rapid intervention team. Although your research uncovers hundreds of procedures and guidelines from other fire departments, many of them differ in content and some directly conflict with one another. You are now forced to make decisions based on your own opinion of which standard is standard. An S-2 procedure would begin with **OSHA**'s requirements, but would also contain the most up-to-date facts about rapid intervention based on research, near-misses, and lessons learned around the country. The procedure could then be tailored to your specific needs. ("Rescue 1 will be immediately dispatched for all structural fires to provide the necessary rapid intervention equipment.") Besides rapid intervention, other procedures that could easily be incorporated based on need include PPE, EMS, communications, incident safety officers, and emergency traffic.

PPE

Although **PPE** is global in its use, particular equipment for specific hazards makes it an S-2 procedure. Wildland fire turnout gear, structural firefighting ensembles, and airport proximity suits would each have their own template. The procedures could utilize information from the latest NFPA standards, which could be altered based on the applicability to the department. For instance, let's assume that there is a new release of NFPA 1971: Standard on Protective Ensembles for Structural Fire Fighting and Proximity Fire Fighting. You are a district chief in charge of the airport fire department in your city in the Southwest and are also responsible for the updating of SOPs. You discover that a new S-2 procedure for proximity suits is released, which is accompanied by a summary sheet of changes to the NFPA standard. In this case, you find that the radiant heat resistance tests have changed, and that both your proximity suits and helmet shells do not meet the new standards. After investigating costs, you determine that the helmets can be replaced now, but the proximity suits will have to stay in service for at least another year until funding is available. The S-2 procedure could be used as a template to include the updated reference to the new helmets. Proximity suits, however, would be referenced to the existing ensembles for the time being.

You may have just noticed another advantage to following this model. This type of a system would be effective at not only providing an SOP template for S-2 procedures, but also provides updated information on changes to the standard in a formalized and expedient manner. Email alerts could be set up to notify all interested firefighters and manufacturers of a new standards change or even a near-miss pertaining to proximity suits. Let's say airport firefighters from Miami-Dade Fire Rescue were inspecting their proximity suits and found significant damage to the seams in most of the suits, and reported it as a near-miss. A well-thought-out system could send out email alerts with a suggested change to the S-2 proximity suit inspection procedure. It could be that an accepted cleaning agent for proximity suits was actually breaking down chemicals in the thermal barrier seams, and that a change in cleaning and inspection was immediately warranted. We've suddenly taken great information from a near-miss, and possibly prevented a catastrophe somewhere else. Similar templates could be generated for hazardous materials response, medical or biological contact, and training for weapons of mass destruction.

EMS

Emergency medical services is another aspect of emergency response that would benefit from S-2 procedures. Most firefighters agree that the motto of the fire service is "the preservation of life and property." Regardless of whether those firefighters are EMTs or not, they are still involved in EMS. If they aren't responding with EMS on medical calls, they could be cutting victims out of cars or plucking them off the side of a cliff. Others might be limited to occasionally pulling a victim out of a fire or helping a fellow firefighter who gets injured at a fire scene. EMS is fairly well established in the fire service now, and standard policies should include it.

Sometimes law enforcement is dispatched or simply stumbles across a medical emergency in which they give patient care prior to the arrival of trained medical personnel. Procedures could be outlined on what their duties would or would not be. It could be that your fire department is dispatched first responder to all medical calls, but transport is handled by a third-party, hospital-based ambulance system. Whatever your situation, a choice of six to eight templates would cover the majority of the systems in place in the country, allowing your organization to ensure that interagency cooperation is maintained and patient care is optimal.

Communications

In Chapter 3, we looked at the rules of effective communication and compared emergency communications with other industries. Lessons learned from NIOSH reports related to communications should be incorporated into standards that can be used to make our job safer.

Incident Safety Officers

Many fire departments have a policy for appointing an **incident safety officer (ISO)** at all structural fires, but NIOSH reports continue to show that we can improve in this area. It might not be feasible to create an S-1 procedure for the appointment of an ISO, but an S-2 procedure would be beneficial and would provide vital information from OSHA and NFPA while allowing leeway to meet a local jurisdiction's needs and abilities. The base procedure would contain information from NFPA 1561, Standard on Emergency Services Incident Management System, including:

- When an ISO and **assistant safety officers (ASOs)** are appointed
- Authority of the ISO
- Responsibilities of the ISO

Emergency Traffic

Another example of a template that needs to be developed is a procedure for emergency traffic. Many departments refer to a firefighter who is lost, trapped, or otherwise in trouble as a mayday. As demonstrated earlier in this book, some sources use the term *emergency traffic*. The first step in creating a standard is to establish a standard definition. Once that is accomplished, an S-2 procedure can be developed that is adaptable to all applications and departments. The procedure would include:

- Criteria for declaring an emergency
- Steps to call and acknowledge an emergency
- Incident command's duties
- Rapid intervention's duties
- Trapped or lost firefighter's duties
- Communications or dispatcher duties

Series Three (S-3: Specialized)

There are certain specialized SOPs that your department might need, but most organizations don't. They could have to do with the terrain or severe weather native to the area you protect. It could be for high-security locations, such as the Louisiana State Penitentiary at Angola, or high-tourism locations, such as Liberty Island in New Jersey. It could be for a permanent trench rescue team or a one-time national event. A stock of specialized S-3 procedures would be invaluable for a department that needed one, and would preload safety at its inception. Some examples include special operations teams, officer promotions, and special events.

Specialized Operations Teams

Hazardous materials response teams, high-angle rope rescue, water rescue, and urban search and rescue (USAR) teams all have unique standard procedures. These and other similar teams operate in situations with an increased risk, and thus demand an increase in safety. Although each type of team is unique, similar teams throughout the country could see an increase in safety by sharing information. For instance, many high-risk businesses have industrial fire brigades. Fire brigades are in the unique position of strict adherence to OSHA and insurance company rules and regulations, compliance, and close cooperation with their local fire department, all while following their own corporate policies. S-3 procedures would allow fire brigades to research and share SOPs, allowing them to adopt new procedures and reconsider their existing ones (Figure 11-5).

Officer Promotions

Not all procedures have to be incident oriented. One of the most extensive arrays of procedures in the fire service is officer promotions. Some departments appoint or vote in their leaders, whereas others "try out" for the position each year. Others test based on written test scores and assessments or board reviews. It could be that a civil service commission makes the rules or that the fire chief is solely responsible for the requirements. Unfortunately, the application of safety compliance is seldom high on the list of any promotional procedures. Sample procedures could be developed for different rules

FIGURE 11-5 Industrial fire brigades would benefit from S-3 procedures, allowing them to blend occupational rules with the latest in firefighting standards.

and regulations, and could incorporate more safe practices into the promotional decision. For example, we've already pointed out how our fire service culture can stand to improve when it comes to safety (Initiative 1). An interview or testing policy should include questions about their cultural beliefs with safety questions, scenarios that involve risk-management decisions, and specific plans and goals to improve safety in the organization.

Special Events

Occasionally, special events catch a town by storm. Bethel, New York, wasn't ready for treating more than 5000 medical cases as a result of Woodstock in 1969.[7] A presidential visit, military funeral, or even the grand opening of a controversial business can bring colossal crowds, unprecedented media attention, and unattainable logistical needs to virtually any town. FEMA defines a special event as having the following criteria:[8]

- It is nonroutine.
- It places a strain on local resources.
- It may involve a large number of people.
- It requires planning and preparation.

S-3 procedures could be used in conjunction with NIMS forms, such as ICS-215, for operational planning. An effective SOP would clarify the completion of the forms and the process of implementation while addressing local needs rather than simply referencing the ICS forms.

IMPLEMENTING THE STANDARDS

There is no shortage of sample policies, SOPs, and SOGs to assemble. Besides the fact that many departments place their SOPs/SOGs on their website for all to see, numerous fire training, fireground safety, and even NFPA sites have web pages devoted to submitted samples. This excess of information is more cumbersome than picking out the straightest 2x4 in a dozen different lumberyards. The problem is we have no clearinghouse for SOPs/SOGs and, even if we did, we have no way to pick the best or the safest.

Policy Collaboration

Suppose the fire service decided that SOPs/SOGs should be standardized. We might first look at the way that other collaboration is succeeding in fire and emergency services. In Chapter 5, we discussed the need for college degrees (education) and standards certification (training) to be incorporated together. Two specific organizations have worked independently for their own cause, but are now working toward collaboration. Fire and Emergency Services Higher Education (FESHE) has allowed the coordinators of postsecondary educational institutions to create common outcomes for the standardization of college degrees. From a separate direction, **Training Resources and Data Exchange (TRADE)** has worked together to share training lessons between state programs and large metropolitan fire departments. Together, they have assembled their common goals, modeled them into a format that complements each other, and are currently adopting standards that incorporate each into a matrix of professional development. By following these three steps, standards could be developed.

Assembling

As stated earlier, many websites already have banks of policies, procedures, and guidelines available, and it's conceivable that they could be reassembled in one location with some effort. Most of the websites that have SOPs/SOGs promote firefighter safety, but placing them at one neutral location, such as NFPA, Everyone Goes Home, or the Firefighter Near-Miss Reporting System, would help with that goal. Once assembled, the SOPs/SOGs would be categorized into S-1, S-2, or S-3, and then grouped by subject.

Modeling

Once they were arranged, duplicates would be eliminated and they could be ranked or commented on in an effort to work toward a model standard. Another solution would be to create an SDO assigned to the task of sculpting model procedures, similar to those we looked at in the first few paragraphs of this chapter. This could be accomplished by a branch of an NFPA standards development committee or by a stand-alone group interested in establishing safe, effective, and legally defensible SOPs/SOGs. Regardless of how it's set up, definitions and a writing style would have to be established to ensure ease of reading. Applicable standards to reference, such as NFPA and OSHA, would be researched and incorporated.

Modeling should also look to the standards that are already established. For instance, NFPA 1500, Standard on Fire Department Occupational Safety and Health Program, provides information for many of the SOPs/SOGs that we have already pointed out. Additionally, NFPA has produced several standards that pertain to specific types of emergency response, including:

- NFPA 450: Guide for Emergency Medical Services and Systems
- NFPA 471: Recommended Practice for Responding to Hazardous Materials Incidents
- NFPA 1201: Standard for Providing Emergency Services to the Public
- NFPA 1407: Standard for Fire Service Rapid Intervention Crews
- NFPA 1710: Standard for the Organization and Deployment of Fire Suppression Operations, Emergency Medical Operations, and Special Operations to the Public by Career Fire Departments
- NFPA 1720: Standard for the Organization and Deployment of Fire Suppression Operations, Emergency Medical Operations, and Special Operations to the Public by Volunteer Fire Departments

Adoption

Although national standards for the majority of procedures would solve a lot of safety problems, there would be a heated debate about why everyone should have to follow them and by what authority they could be enforced. National standards would be most effective if compliance was mandatory, and that would take a national law. It's much more likely that individual states or **authorities having jurisdiction (AHJ)** would adopt them or voluntarily comply, much like they might comply with an NFPA standard. Compliance could also be an essential component of **FLSI-16** certification used to obtain grants discussed throughout this book.

SUMMARY

Fire and emergency services have been challenged on three different fronts in the past 20 years, all with similar results:

1. Terrorism and the threats of weapons of mass destruction have forced us to plan for large-scale events we never expected to perform.
2. Budgets have taken a plunge, forcing us to make do with less. As a result, many departments are forced to rely on mutual aid and automatic mutual responses like never before.
3. Finally, our desire to improve has created better regional response to natural disasters, such as earthquakes and hurricanes.

All three challenges have forced us to work together more to maintain firefighter safety and align the way we do our job. The time has come for creating standard policies and procedures for many of our functions.

To create common SOPs and SOGs, we have to realize that not all of them are interchangeable, and that certain policies are easier to apply. By breaking all policies into three different series, we can develop procedures for all (S-1), procedures for most (S-2), and procedures for some (S-3).

It's vital that all SOPs/SOGs follow the same format for ease of understanding, as well as using consistent terminology. It's also important to use existing standards from NFPA and OSHA to build the foundation of the procedures. SOPs/SOGs must also have safety and risk management built into each one. Collaboration by interested parties can assist in the assembly and modeling of national standards, allowing interested departments to adopt them. At that point, we can call them standard.

KEY TERMS

AC - Assistant chief.

AHJ - Authority having jurisdiction.

ALS - Advanced life support.

ASO - Assistant safety officer.

ASTM - American Society for Testing and Materials.

AWS - American Welding Society.

EMS - Emergency medical Services.

EMT - Emergency medical technician.

FLSI-16: Fire and Life Safety Initiatives 16 - Proposed rating for fire departments and EMS agencies based on their dedication to safety; could be used for grant eligibility.

IAP - Incident action plan.

IC - Incident commander.

International Organization for Standardization (ISO) - A global organization that provides registries of companies that comply with certain standards.

ISO - Incident safety officer.

NFPA - National Fire Protection Association.

NIMS - National Incident Management System.

NIOSH - National Institute for Occupational Safety and Health.

OSHA - Occupational Safety and Health Administration.

personal protective equipment (PPE) - A generic term used to describe the minimum apparel and gear needed to safely perform a specific duty.

policy - A broad goal or statement that guides actions, usually an extension of a mission statement that illustrates the beliefs of an organization.

positive securing device (PSD) - A bracket, strap, or other holding mechanism used to store portable equipment or PPE in a vehicle.

POV - Personally owned vehicle.

procedure - A course of action, sometimes with steps of how it is to be accomplished.

pull station - A device used by occupants to activate a fire alarm manually.

risk management - Identification and analysis of exposure to hazards, selection of appropriate risk management techniques to handle exposures, implementation of chosen techniques, and monitoring of results, with respect to the health and safety of members.

Series One (S-1) - A policy or procedure that could be applicable to virtually any fire and emergency service organization.

Series Three (S-3) - A policy or procedure used for specialized incidents that may be applicable to some fire and emergency service organizations.

Series Two (S-2) - A policy or procedure that could serve as a template for adoption by most fire and emergency service organizations.

SOG - Standard operating guideline.

SOP - Standard operating procedure.

standards developing organization (SDO) - An organization that creates rules for a specific industry.

SWAT - Strategic weapons attack teams (also called *strategic weapons and tactics*).

Training Resources and Data Exchange (TRADE) - An association of fire training professionals exchanging training information.

REVIEW QUESTIONS

1. Why does common terminology play such a significant role in policies and procedures?
2. What is the difference between standard operating procedures (SOPs) and standard operating guidelines (SOGs)?
3. When reviewing the seven specific headings included in a consistent standard operating procedure (SOP), what should be included in the policy section?
4. What factors contribute to the significant safety risk of water tenders?
5. In the proposed series standardization for SOPs, what application would an S-2 policy or procedure have?
6. Once assembled, what is the most likely way to accomplish the modeling of standard operating procedures (SOPs)?

FIREFIGHTING WEBSITE RESOURCES

http://everyonegoeshome.com
http://cdc.gov/niosh/fire
http://www.iso.org/iso/home.html
http://www.fema.gov/pdf/nims/nims_training_development.pdf

ADDITIONAL RESOURCES

http://www.myfoxdc.com/dpp/news/Md
Matt Pearce http://www.columbiamissourian.com/stories/2010/05/10/columbia-police-announce-further-changes-swat-policy/
http://www.limaohio.com/articles/swat-4367-team-commander.html
http://www3.whdh.com/news/articles/local/BO86502/
http://www.lacp.org/2008-Articles-Main/041608-SWATreport.htm
http://ag.ca.gov/publications/swat.pdf
http://www.astm.org/CDSTAGE/PAGES/E1354.htm

NOTES

1. http://www.aws.org/wj/2001/02/commentary/
2. Page, Steven B. *Establishing a System of Policies and Procedures*, Mansfield, OH 2002.
3. Fire Fighter Deaths from Tanker Truck Rollovers. National Institute for Occupational Safety and Health. NIOSH Publication No. 2002-111.
4. NFPA 1720; Standard for the Organization and Deployment of Fire Suppression Operations, Emergency Medical Operations, and Special Operations to the Public by Volunteer Fire Departments.
5. IS700 NIMS Course Summary; Summary of Course Content. National Incident Management System. 2006.
6. NIMS National Standard Curriculum Training Development Guidance. July 2005. http://www.fema.gov/pdf/nims/nims_training_development.pdf
7. Tiber, Elliot. 1994. "How Woodstock Happened." *The Times Herald-Record- Woodstock Commemorative Edition.* http://www.woodstockstory.com/how-woodstock-happened-7.html
8. FEMA Independent Study Program: IS-15 Special Events Contingency Planning for Public Safety Agencies. Emergency Management Institute.

CHAPTER 12

Examining Response to Violent Incidents

LEARNING OBJECTIVES

- Explain the differences between terrorism and other violent events such as riots.
- Describe the similarities between standard procedures developed by journalists in dangerous areas and procedures that can be developed for emergency services.
- Explain the importance of establishing better training to be prepared for violent incidents.
- List specific types of violent incidents that should have national standards.
- List the reasons fire and EMS uniforms should be professional yet unique.
- Explain the benefits of using Level II staging for violent events.
- Describe a situation in which "staging for a safe scene" is not an option.
- Describe how to best prepare for violent incidents.

Initiative 12: National protocols for response to violent incidents should be developed and championed.

In Chapter 11, "Establishing Response Standards," we outlined a plan to improve safety through the use of standard response policies. Safety has always been a concern for emergency responders working in violent situations. For example, the urban riots of the 1960s put firefighters in danger in numerous cities around the country. Fueled predominantly by political, economic, and racial issues, crowds could quickly turn on the police and anyone else associated with the government. Stores were looted, fires were set, and people were killed. Riots still can erupt today, but are usually foreseeable and loosely coordinated. They can be initiated by a police action, a court decision, or sometimes even a sporting event. In 1991, a videotape caught police beating Rodney King after a police chase. When the police officers were acquitted of local charges, rioters burned buildings, looted stores, and killed 55 people.[1] Emergency responders also became targets. However, riots are not the only danger to emergency responders. The rise in the number of school and workplace shootings in the past 30 years has also put fire and **EMS** at risk. But the most significant threat of violence to emergency responders is **terrorism**. In 2001, a single incident took 343 firefighters, more than three times the annual average of **LODDs**. As a result, it's imperative that a risk-benefit analysis be built into all violent incidents through standard policies and procedures.

TERRORISM

Terrorism has added a new dimension of danger for emergency responders because terrorist plots tend to be more elaborately planned. This is evident because many successful attacks are not only well funded, but are designed for a specific purpose. It's not uncommon to read a newspaper article about a terrorist attack somewhere in the world that was planned for quite some time before all the pieces were in place. Take the planning involved in detonating a bomb, for instance. Bombs are usually designed to have the most lethal effect on bystanders by containing shrapnel such as bolts, nails, or ball bearings. Placement of the bomb is strategically planned to have the greatest effect on a building, a crowd, emergency responders, or another target. Many times the terrorists will visit the scene well in advance of the attack to take photos or observe the actions of targets. Many times the site is visited over and over again. Concealment of the bomb in a vehicle or trashcan is a common technique used to blend it into the surroundings and avoid suspicion. Finally, the timing of the blast is often synchronized for the most devastation. The threat of this type of attack does not compare to that of a roving mob randomly beating people, overturning cars, and setting dumpsters on fire.

Weapons of Mass Destruction

One common term used when discussing terrorist attacks is **weapon of mass destruction (WMD)**. Intended to kill or injure as many people as possible with one strike, WMDs can come in many different forms. Fortunately, the dangerous agents needed for many WMDs are both difficult to obtain and disperse in a successful attack. The most probable form of agent to be used in a WMD has been categorized as **chemical, biological, radiological, nuclear, or explosive (CBRNE)**.[2]

Chemical

Some attacks in other countries have been successful with chemical agents. Chemicals used in a potential attack could be from military origins, but also could be simple industrial chemicals. Some specific types of chemicals are blister agents like mustard gas and Lewisite, pulmonary or choking agents such as chlorine and phosgene, and nerve agents such as VX, tabun, and sarin gas. Sarin was successfully used in the domestic terrorist attack in a Tokyo subway in 1995.[3]

Biological

Biological agents proved effective as a weapon as recently as 2001 in the United States, when anthrax was delivered in the mail. Although 22 confirmed cases were identified and five people were killed, 32,000 were treated for possible exposure.[4] The fear raised with possible exposure overtaxed responders and medical personnel. Other types of possible biological agents consist of smallpox, the plague, and ricin.

Radiological

Radiological dispersal devices (RDDs) are any mechanism for spreading radiological materials. Many times considered a "dirty bomb," an RDD uses an explosion or other force to distribute radiation over a larger area. Although successful attacks using RDDs have been rare, the potential is there for terrorists who can get the materials they need and devise an effective dispersal system.

Nuclear

The threat of nuclear weapons at the hand of terrorists became a reality in 1974, when John McPhee authored his book, *The Curve of Binding Energy*. In it, he posed the possibility that with the right ingredients, nuclear weapons were not out of the reach of any group who has "a common interest or a common enemy."⁵ It's unclear how many of the ingredients needed to construct a nuclear bomb are available; therefore, most efforts in the United States have concentrated on keeping them out of the country.

Explosive

Although some sources use only the first four classifications of WMDs, many now have added explosives to this list. Explosives are nothing new as far as weapons are concerned, but "high-yield" explosives have recently proven especially dangerous, and are now considered a WMD. For instance, the terrorist attacks of September 11, 2001, used jet fuel as a high-yield explosive weapon, and an explosion of fertilizer and diesel fuel was used in the domestic terrorist attack in Oklahoma City in 1995 (Figure 12-1).

Secondary Devices

Terrorism is also more dangerous because emergency responders are not only planned into the equation, but are easily targeted. When bombs were planted at an abortion clinic and a nightclub in Atlanta early in 1997, a **secondary device** was placed in a location apparently to disrupt rescue operations by injuring or killing emergency responders. In 1999, two teenage shooters at Columbine High School in Colorado placed bombs in two cars in the parking lot, timed to explode near emergency responders after they blew up the school.⁶ Those bombs failed to detonate.

FIGURE 12-1 Terrorism and WMD will continue to change the landscape of emergency response.

If terrorists can take the time to learn how to fly airliners, they certainly can spend some time observing where our first-due engine parks for emergencies at the high school or where we would likely put staging for a large incident at the airport. Not only was September 11 the deadliest terrorist attack in the United States, but it was also a significant hit to emergency workers, with roughly 10% of the victims being responding firefighters. Because of the direct way terrorism affects emergency responders, it's vital for us to develop standard procedures. To do that, we first must evaluate what it is that we do.

JOB DESCRIPTION

The first step to improve safety at violent incidents is to establish exactly what our job description is. Battalion Chief Mike Alder and Deputy Chief Mat Fratus explained it best in a class on assessing risk when they said, "It's not about limiting what we do, it's about defining what we do."[7] In other words, policies and procedures can be adapted to the needs of the department and community. Other industries have used a similar approach to violent incidents for exactly the same reason. For instance, according to the **Committee to Protect Journalists (CPJ),** Mexico is one of the most dangerous countries for visiting journalists who simply investigate and report the news.[8] Between drug cartels that are fighting for control and law enforcement that is less than effective, murders of journalists do occur and tend to be swept under the rug. As a result, many news organizations have taken steps to define what they do. Following their example, we might examine three specific ways to define what fire and EMS do by the way they address reporting, investigating, and extracting the news.

Reporting the News/Responding to Emergencies

Journalists have a desire and responsibility to report the news. One suggestion to reporters from the CPJ is to continue to cover stories of violence, especially when the violence is directed at the media. In other words, their duty is to report and that duty cannot be compromised. When reporters are threatened, stories still run. Fear of a violent backlash for simply doing their job cannot affect journalists' behavior. If it did, the validity of reporting would be questioned. It's been said that one intention of terrorism is to scare people into changing their behavior and abandoning their duty. This is as true for journalists as it is for emergency workers.

Emergency responders have a duty and desire to protect lives and property. We are also well aware of the secondary device, the term used to describe a bomb that is placed and timed in a way to target, injure, and disrupt emergency responders. An emergency call for shots fired at a middle school with people injured could entice responders to make a poor decision. Creating standard procedures for response to violent incidents allows us to respond intelligently, safely, and effectively.

Investigating the News/Assisting Law Enforcement

Some news agencies have decided that although their job is to report the news, investigations may not be necessary in these types of incidents. Some people consider investigative reporting crossing the journalistic line by trying to create news rather than simply reporting it. You've probably seen a reporter on the evening news stick a microphone in the face of a person suspected of a wrongdoing when the reporter suddenly becomes the target. Investigating individuals suspected of violent crimes for a drug cartel are not the type of people you want to back into a corner. By defining their duty as reporting rather than investigating, they put themselves in a safer and more defendable position. If reporters limit themselves to fair and unbiased reporting, they maintain a distinct line between their function and that of law enforcement. For example, it's not uncommon for criminals to talk to the press anonymously if they don't feel threatened by the reporter, but the same questions from a detective would likely yield no information.

Many fire departments and emergency medical service agencies have defined their role as *assisting* law enforcement, and allowing them to be the lead agency for violent incidents. Law enforcement has the tools, the training, and in most cases the legal authority to instill order. It's not uncommon for fire and EMS agencies to have a policy stating that responders will not immediately enter an unsafe scene, but instead stage nearby until the police clear it first. **Tactical EMS (TEMS)** is a designation of personnel trained to assist law enforcement with the medical needs of the team. Some TEMS teams may operate in unsafe areas and, as a result, should have their own procedures while being trained and equipped to defend themselves.

Extracting the News/Extricating Patients

Many journalists who report the news don't actually obtain the information from the scene. It used to be that after a crime was committed, a journalist would travel to a town in Mexico and ask witnesses what happened. Journalists were in constant danger as outsiders asking questions, while local people who provided information also became targets. It's not uncommon now for an elaborate network of reporters and **fixers,** local reporters who provide information, to spread the news. Fixers have historically provided open support to visiting journalists, but due to increasing dangers have predominantly gone underground. If journalists do need to make a trip to interview someone or obtain more information, they use fixers to plan their moves to get in and out quickly and safely. They basically have adopted the theory that if the drug cartels can secretly move drugs and cash across the border, there's no reason information can't be passed the same way on the same route.

Tactical EMS crews that are assigned to help **SWAT** or bomb teams work on the same principle. They know what they can do and when they can do it. For instance, the first steps for standard EMS on the scene of a male with a gunshot wound on a crowded street is to wait for a safe scene, then assess airway, breathing, and circulation while considering that the patient might have a cervical spine injury. If a potential neck

FIGURE 12-2
Procedures for violent incidents differ in many ways from everyday incidents.

or back injury exists, the patient is normally secured to a backboard before being moved. The same scenario during a riot utilizing TEMS might instead wait for a semi-safe scene with officers providing **cover fire** and protective gear to allow a crew to quickly access the patient. Extracting the victim to a safe location would be the first priority, after which airway, breathing, and circulation are then assessed.

This scenario is actually no different than what firefighters do when removing victims from a house fire. Little attention is given to minimizing injuries from patient movement. At the time, the unsafe atmosphere is the most pressing danger for the victim, so pushing, pulling, or even dropping the patient is less serious than being slow and careful. Hazmat scenes and violent incidents have similar priorities, and should form the backbone of standard procedures (Figure 12-2).

STANDARD PROCEDURES

In Chapter 11 we looked at ways to develop standard procedures based on three different "series" of applications. Several components of violent incidents could benefit from S-2 standards, where slight modifications are made based on a local need. They include the use of uniforms, protective equipment, and training.

Uniforms

Once we define what our job is, we have to ensure that we actually live by that definition. This includes the way we act as well as the way we look. If journalists declare that they want to report rather than investigate in violent situations, it wouldn't make much sense for them to dress like police officers. Similarly, a firefighter should look like a firefighter and the uniform should not be mistaken for anything else. Some leaders of fire departments and EMS agencies feel that a duty uniform should show the professionalism of the organization by maintaining a dress military or law enforcement look. However, even the military and law enforcement drop their badges when violence emerges. A soldier in desert fatigues and a SWAT member in a **battle dress uniform (BDU)** still look professional. Past incidents have shown that a firefighter or **EMT** who looks like a police officer can be mistaken for a police officer. The word *association* explains why.

Over the years, some researchers and safety advocates have suggested that the color yellow, lime green, or even pink is much more visible than red. Therefore, yellow, for instance, is a safer color for emergency vehicles such as fire trucks. The problem is that the general public doesn't associate yellow with emergencies, but with caution. Consider the following associations:

- A yellow school bus picking up children
- A yellow strobe light on a tow truck working on the side of the road
- A yellow traffic light that will change to red soon
- A yellow sign warning you where the edge of the bridge is located

As a result, the public might *see* the yellow fire truck first, but might not identify it as such. **NFPA** has included the use of more visible colors to apparatus, such as yellow lights and chevrons on the rear of emergency vehicles. This helps to catch other drivers' attention, and still allows for an organization's identifying color.

Similarly, uniforms must be distinguishable. Let's say you are the lieutenant on an engine company responding with EMS and police for a shooting in an apartment building. After the police declare the scene safe, you proceed to the hallway where the victim is injured. You have two firefighter EMTs with you assisting an EMT and paramedic from the private ambulance. Four police officers are in the hallway and one is inside the victim's apartment. Suddenly, a door from an adjacent apartment opens and a man with a gun emerges. He takes aim and begins to shoot. At a glance, he has several targets to choose from, but only half of them have weapons and are a direct threat to him. If he can't differentiate between them, he probably will shoot at those closest to him and will hit as many as possible.

This example is not to suggest that law enforcement is expendable and should be the ones to take the hit. It simply makes the point that law enforcement is responsible for investigating the facts, protecting people on the scene, and detaining suspects and criminals while we are only responsible for patient care. There's a good chance the five law enforcement officers in the building are wearing PPE consisting of ballistic vests to take a bullet, while we might be limited to blue latex-free gloves. If our uniforms appear the same, luck is the only thing keeping us from being shot. A victim or a suspect with a weapon should not have any trouble differentiating between us at a glance. On the outside, this means a unique but professional uniform. On the inside, it means body armor and other means of protection (Figure 12-3).

FIGURE 12-3
Emergency responders would benefit from designing a professional, yet readily identifiable uniform.

Protective Equipment

Most agencies do a sufficient job of providing the appropriate PPE for the incidents that they respond to. Experience has shown the importance of providing the equipment needed to keep personnel safe. However, because violent incidents are less frequent, many departments lack adequate PPE for these types of incidents. NFPA 1500 specifically states that at these types of incidents, "The incident commander shall ensure that appropriate protective equipment (e.g. body armor) is available and used before members are allowed to enter the hazard area."[9] Standard policies could define different types of body armor and the usage of different levels much like in a hazardous materials response. Level A would provide the highest level, including protective equipment such as ballistic helmets, whereas Level B might be limited to the use of a bulletproof vest.

It's interesting to note that law enforcement officers hide their ballistics vest under their uniform. The reason isn't necessarily because they are ugly or look unprofessional. The primary reason is because if they wore them on the outside it would remind a criminal to aim for the head. Firefighters and EMS workers who wear bulletproof vests should also hide them. Because we will only don them on specific emergency calls, it would be most effective for the vest to be concealed as a component of a high-visibility jacket with reflective striping and lettered with the agency's name worn *over* a uniform. They could be carried on vehicles for easy access, and not allow a criminal to mistake us for someone else. However, they would be in place to protect us when needed.

In some situations, protective equipment might include a weapon. Depending on the needs of the team, some members might be trained and equipped to protect themselves. For example, let's say the sheriff visits your county fire chief's meeting and expresses his desire for active medics to join the county SWAT team. TEMS is new to your area, and although you have interested members and think it would be a great way to work with law enforcement, you are concerned for the medics' safety. Although you feel they shouldn't be involved in storming a fugitive's hideout, you believe they should be able to defend themselves if the gunman escapes past TEMS. In this case, firearms training and range time might be warranted. Other departments might feel that nonlethal weapons would be sufficient. Available procedures for either choice would allow departments to follow national standards and choose what level would best suit their needs, provide protection for their personnel, and limit liability.

Training

If we really want to keep our personnel safe, we have to train for the unexpected. Every fire department and EMS agency has the potential to be involved in a violent situation. Besides training with law enforcement for violent incidents, general safety training should include size-up, positioning, self-defense, and evasive maneuvering.

Size-Up

Firefighters are trained to read the stability of a building during a rapid size-up. A good EMT can quickly determine if an overturned vehicle is stable enough to enter by simply walking around it. Unfortunately, we have less training in determining the intentions of a potentially violent person we encounter than other dangerous situations. Training systems can be developed with the assistance of psychology experts, law enforcement, or mental illness professionals to ensure that textbook examples of unstable or potentially unstable situations are easily identified. This could include the identification of situations that tend to induce violence toward others, warning signs of impending violence, or calming techniques to diffuse a tense situation.

For example, law enforcement does a phenomenal job of training police officers to approach vehicles safely during traffic stops. Many park their vehicle on an angle, with the front wheels turned to the right. If struck from behind, their car will tend to veer off to the right rather than striking them or the parked vehicle. If they approach a vehicle that was already stopped, some turn on their siren to alert the occupant of their presence. They turn on bright "take down" lights to blind the driver, and report the license plate to dispatch. They approach cautiously in the blind spot and put their finger-prints on the trunk lid when they walk by. They maintain an awareness of other passengers in the car, as well as traffic on the roadway. Can we learn anything from this account? As EMS approaches a vehicle for an apparent medical emergency or firefighters walk up to a car that is "smoking" from the radiator, we should adopt some of the practices of law enforcement.

Let's imagine you have been dispatched late one night to a rollover motor vehicle accident on a rural highway in which the sole occupant has climbed out prior to your arrival. Although he has a severe laceration to his cheek that will surely require stitches, he is a pretty big guy who is both belligerent and borderline confrontational. You and your partner attempt to calm him and diffuse the situation, but it seems that everything you say tends to make him angrier. You call on the radio for the state troopers to step it up, but he begins poking you in the chest as he rants about how he is tired of being harassed, and is declaring war on "the system" and everyone involved.

Positioning

In the event that diffusion was not effective, it's important for emergency responders to be in a position to delay an attack or at least make it more difficult for the perpetrator to connect with a blow. Some law enforcement officers suggest watching both hands of the person in question. If you don't see both hands, make him show both hands. Having one person devoted to watching the crowd could be necessary. In static situations such as transporting a patient in an ambulance, personnel should understand where the most strategic place to sit is that will offer the most protection while still facilitating patient care. Many ambulances carry soft restraints that can be used when the patient is a threat to self or others. Positioning in semi-open areas, such as a living room, would likely condition responders to use furniture and doorways to their advantage. Rolling a cot into his path or closing doors helps to slow an attacker. In a dynamic situation such as this example in which a perpetrator is moving toward you on an open roadway, maintaining an adequate distance and establishing defensive postures would be most important. Also realize that your escape routes are limited because of close access to the roadway. Escaping into traffic may not be the best option.

Self-Defense

Even with proper positioning, emergency responders must be able to defend themselves if an attack takes place. Utilizing available equipment, such as forcible entry tools, clipboards, and oxygen cylinders could be advantageous. A multipurpose dry chemical extinguisher not only puts out most types of fires, but like a carbon dioxide extinguisher is also far more effective at slowing down an assailant than a pressurized water extinguisher. By adapting the martial arts saying "I usually aim for the face," the extinguisher can be a great diversion by clouding the assailant's vision. Equipment not only can be used as a shield or weapon for a human, but can be effectively used to ward off an attacking animal. Most dogs can be corralled and locked into a spare bedroom with a decent sized trauma bag. In the case of our attacking motorist, equipment in hand may allow you to protect yourself. Some EMS agencies still use aluminum clipboards for that very reason. Back on the highway, if our assailant did grab hold of you, certain defensive moves could knock the victim off balance and allow you time to escape. Establishing standard procedures for declaring emergency radio traffic and the securing of apparatus from theft by an assailant should be included.

Evasive Maneuvering

Self-defense may only be effective for a short amount of time. In most cases, your best action is to distract the attacker and retreat. In his book, *SWAT Leadership and Tactical*

Planning, Tony Jones suggests moving in unexpected paths, such as a zigzag motion.[10] For example, let's assume that your confrontational car crash victim who grabbed you was knocked to the ground and you were able to get away. As you call for emergency traffic, he pulls out a handgun from an ankle holster. Still about 10' from your vehicle, you throw your jump bag at him as a diversion, and then run with your partner for cover. The shortest distance between two points is a straight line, but if you take the shortest route you increase the chances of being hit. Two people running away in a direct line make pretty easy targets. It's most advantageous to split up, one for the front of the vehicle and one for the rear and weave often. It's easier said than done, and should be occasionally practiced. Once behind the vehicle, either get in it and use it as an escape mechanism or continue past it to a building, the woods, or another place where you have more protection and concealment. The author also suggests not following anticipated routes such as sidewalks, driveways, and other paths.

TYPE OF INCIDENT

Most firefighters know that different types of building fires rely on different types of tools and tactics. The same exists for violent incidents. Having one vague policy for all types isn't effective or efficient. Several different types of violent incidents call for their own type of response. Some that could benefit from standard procedures include civil disturbances, a school or workplace shooting, a bomb scare, and an animal attack.

Civil Disturbance

In the past 50 years, riots have struck cities that were forced to deal with fire and EMS needs under unique circumstances. As a result, varying response policies have been adopted. Let's assume that you are a career firefighter currently assigned as a driver/operator of an engine company in a major southern city that has recently garnished national attention. A drug bust that went sour ended in a police officer being injured and two teens being killed. This morning, the court ruled that the police officer in question followed proper policies and would not face any punishment. Although the city had planned for a possible uprising, an unexpected riot broke out in a different area of the city where a similar instance occurred several years ago. Your station has just received word that a "state of emergency" has been declared from the mayor. It's likely that any of your responses to the area in question will dictate that special procedures will be used, including preparation, response, firefighting, and crowd control.

Preparation

There's a good chance your procedure will include preparations needed to reduce the impact of the disturbance on your safety. Some departments require full turnout gear throughout response to and return from an incident as an added level of safety and identification. In some locations, bulletproof vests are distributed in this instance for any type of call. Sometimes preparation calls for removing any portable equipment from outside the vehicle and stowing it inside a compartment or under the hosebed cover. In a riot situation, it's advisable to not provide the rioters with any more weapons in the form of pike poles or axes from the side of your engine.

Response

Many procedures call for multiple apparatus incident response. It's unlikely your engine will respond without other units together, following a police escort. It might be common for calls to be held until a police unit is available for escort. Sometimes cars and dumpsters are left to burn if they pose no exposure hazard. Some departments suggest using lights but no sirens for emergency responses through or in an area of civil unrest to avoid the announcement of our arrival. The procedure may call for a rapid response without stopping while passing through violent areas. If audible warning devices are necessary, it may be prudent to limit them to a windup siren and air horn, if equipped, rather than an electronic siren in an effort to make clear who we are.

Firefighting

Once you arrive at the scene of a building fire, safe procedures must continue. A **risk-benefit analysis** has to weigh a successful attack and extinguishment against the risk of the personnel being attacked. A common practice is to use all equipment and hose needed from only one engine. In a worse-case scenario, that vehicle could be abandoned if others aren't committed or blocked in. Sometimes procedures call for extra security at the hydrant to ensure it's not tampered with. Many times, the use of multiple hoselines is limited, instead using master streams to minimize the time on scene. Additionally, overhaul and investigations are extremely limited. The **incident action plan (IAP)** supports a "put it out and get out" strategy (Figure 12-4).

Crowd Control

Law enforcement usually prefers to use nonlethal weapons when their goal is dispersing unruly crowds. That could include rubber bullets, chemical irritants, or even water cannons. Many law enforcement agencies don't have water cannons and might suggest using a deck gun or fire hose. For example, let's assume that you are the **OIC** on a pumper that was dispatched to assist the police. On your arrival, the police sergeant explains to you that a mob has become violent and that they want you to use your fire engine and deluge gun to help disperse the crowd. It's not that uncommon for fire departments to assist law enforcement with equipment. Ladders, forcible entry tools, and thermal imagers are equipment that they can be quickly trained to use in support of their mission. It starts to get a little more questionable if they want to use an **SCBA** or a deck gun. Fire hydraulics formulas allow us to convert **gallons per minute (GPM)**

FIGURE 12-4 Limiting the time on scene reduces the exposure of emergency personnel at violent incidents.

flowing from a hoseline to **British thermal units (BTUs)** being produced by a fire and calculate the pump pressure needed to overcome friction loss and still provide 80 psi at the smooth bore tip. No fire formulas exist to calculate the nozzle pressure needed to disperse a crowd and not knock them unconscious. Although we want to support the police, using a fire truck to break up a crowd might not be something we want to get involved in.

A different situation might also lead us to the same conclusion. Let's assume you are the chief of a fire department and law enforcement asks for help in apprehending a known fugitive. They have the apartment building staked out and know that he is home with his girlfriend and her children. They come up with a plan to create a diversion using the fire department with a fake emergency response to the apartment next door. SWAT members will don fire gear, respond on a fire truck, and force an evacuation of the floor, which will force out the fugitive. Other SWAT members in the hallway will apprehend him before he can exit the building. It seems as though the plan is well thought out, and that providing a pumper, a firefighter to drive, and some extra gear and SCBAs is really not much to ask.

Similar to dispersing the crowd with a master stream, the problem is the public's perception. If this was a football game, we'd call it a gadget play. It's something so off the wall and unexpected that it just might work. If the police department executes the raid to look like a fire response, they may make the arrest and nobody gets hurt. Unfortunately, gadget plays don't always work and if they do there is a cost. In football, the other teams will prepare for it and if they try to execute it again, the runner will get hit with a loss. In this case, the next time the fire department responds to the apartment building, we may get hit with a loss. It's unlikely that residents of the complex will ever look at us in the same way again. The long-range trust we lost will last long after the police leave and the fugitive is in jail. Standard procedures must outline what we can do for law enforcement and what we can't. Although a pike pole or a hoseline might provide sufficient self-defense in some situations, the use of firefighting tools as offensive weapons must be strongly cautioned.

School or Public Shootings

Shootings create an unstable situation with a lot of unknowns. Many times shootings take place in large public buildings, which can be more difficult when the shooter is mobile. Information on the location and number of shooters is often vague and inaccurate. Although the Columbine High School attack in 1999 took place as live video was played on television, information reported from the scene was not very accurate.[11] Many times the assailant(s) has a specific motive or target in mind, but occasionally acts randomly. Although each shooting situation is unique in many aspects, standard procedures can assist responders to use available options to reduce risk and save lives. At face value, shootings are considered a police matter. A simple one-page policy would state that fire and EMS stage away from the scene until law enforcement clears the scene. Once that is accomplished, we support them with whatever they need. Unfortunately, a policy this cut and dry may not be as thorough as we need. The policy should clarify staging, protection, and triage and treatment.

Staging

The first consideration for a shooting event is that staging has to be distinctly identified. Many sources suggest that staging be **Level II,** away from and out of sight of the scene. By remaining uncommitted a block or two away, we place ourselves in an invisible position out of the line of fire, but available for quick response to the scene. For instance, let's imagine that a shooting has taken place at a shopping mall, and your department stages in the far corner of a parking lot. It's Saturday afternoon, and a masked gunman began shooting at people in the main corridor of the enclosed mall. A mass exodus is occurring, with people running from multiple exits. Parking in clear view of shoppers and employees would not be a good position to be in. It's very likely that evacuees would run to us to report the shooting and tell us which store the victims are in. Our expectation would be to wait for a safe scene, while their expectation would be that we belong inside. Level II staging also keeps us from getting caught up in the incident and self-deploying. Imagine if your captain on the engine had two teenagers working at the mall and watched as people fled in all directions. It's pretty easy to *say* that we will wait for a safe scene, but in reality it might turn into a risk-benefit decision where your captain decides to take the risk for an unknown benefit. The point is that by staging away from the scene, there is a better chance that we will remain in the best possible position to do our job.

Another reason staging should be addressed is dispatch information is not always accurate, with some past shooter incidents being first reported as fireworks, people injured, or even a fire alarm. A standard procedure must take into account that emergency responders may have already arrived on scene when a violent event was identified. Many schools are located very close to a fire station. What if the shooting occurred during fire prevention week, when the fire department was visiting classes at the middle school? A thorough procedure should address the duties of crews that find themselves in the epicenter of a violent event.

Protection

Consider that you are the officer on a ladder company that was returning from an emergency call when an automatic fire alarm is received from the high school. As you arrive on scene less than a minute later, you are greeted by a small number of students waving you down, but don't see the normal flow of people from all exits as you would expect from a fire alarm. The students that run to you frantically explain that somebody got shot just as dispatch advises you to stage for a possible shooting at the school. You only have two choices, stay or leave. Without standard procedures for this type of an incident, little forethought will be used to make a proper risk management decision.

In this case, you decide that the benefits of remaining on scene could be beneficial. The first unit on scene generally establishes command, so you establish unified command, pending arrival of the police. It is important to realize that, as a whole, fire and EMS are much more proficient at initiating command because we do it much more often than law enforcement. By inserting yourself in the unified command system from the onset, you set up the incident for sufficient management and organization. The ladder truck is fairly large and should provide some cover, so you decide to get out of the apparatus and concentrate on moving evacuees behind it as a temporary safe haven until they can be removed. You post firefighters in a protected location near the front and rear of the truck to keep watch and flag down escaping students, and to assemble the first aid and EMS equipment from the cab to prepare a temporary triage area. You question students and teachers who self-rescue about the number of shooters, the number of victims, and their last known locations. It's important to remember that sometimes assailants pretend to be victims and blend in with the crowd. Be alert for suspicious people or bystanders. As police officers arrive, they secure the area by the ladder truck and you share pre-fire plans with them in an attempt to track the shooter(s) and victims.

Offensive actions, such as entering the school to remove victims, should be discouraged. Besides the possibility of becoming a victim yourself, there's a good chance entering will make the scene more difficult for law enforcement. A situation like this might also contain booby traps. It's no different than if a police officer runs into a house that is on fire. The officer could become another victim for firefighters to drag out, or could vent the fire, allowing it to spread and grow. Defensive actions on the other hand should be encouraged. It could be that students are running from an exit at one end of the school while a shooter pumps off rounds from a second-floor window. Risk-benefit analysis might suggest moving the ladder truck between them to create a shield for escaping students (risk a lot to save a lot). What about students who are jumping from a second-floor window? It might be beneficial to pull an ambulance under the window to help them escape. It could be that reports of several downed students in one wing of the school lead to police clearing the wing and covering for you while your crew removes the victims. If a crew unknowingly ends up in a shooting event, a standard policy should encourage specific defensive duties. It follows that all emergency vehicles should be provided with adequate **PPE**.

Triage and Treatment

One problem with many **SOPs/SOGs** is that they don't transition well into other procedures or guidelines. For example, a shooting procedure such as this should blend seamlessly into a **mass casualty incident (MCI)** SOP, rather than simply cross-referencing it. This is another advantage of using standard procedures. In our last scenario, your ladder truck was inadvertently placed in front of a high school where an active shooter was creating chaos. You established command and a safe haven for evacuees and attempted to gather as much information for the police as possible. It's likely that you would soon begin receiving victims, branching your forward position into an initial triage and treatment sector. A normal MCI procedure establishes separate triage, treatment, and transport sectors in a safe location. Unfortunately, your crew of three may be receiving a fairly steady flow of patients prior to the scene being declared safe. Obtaining medical supplies and moving victims away from the scene may also be a problem. A well-thought-out procedure would not only outline what we should do, but how to smoothly evolve one SOP into one or multiple additional SOPs that address the given situations.

One other important thing to remember during triage or treatment for any incident is to actively look for a weapon. More states have concealed carry programs for handguns, not to mention the illegal carrying that can occur. Dropping a weapon off a backboard during patient transfer of an unconscious victim should never happen. During treatment, it is perfectly acceptable for an EMT to ask the victim if he has a weapon, and complete a thorough pat-down of the victim in the form of a hands-on trauma assessment. In the case of a shooting with multiple victims, we may be the first to touch the victim who may have a weapon. This is especially true in the case of an assailant who was injured.

Bomb Threats and Possible Bombs

Many response agencies have generic procedures for bomb threats and suspicious devices. If a fire or EMS apparatus is dispatched, it's usually to stand by, with Level II staging being preferred, possibly by a hydrant or other water source. Most procedures suggest maintaining radio silence, especially within 1000' of the suspected location. Portable radios, cell phones, and other electronic equipment can set off an unstable bomb. If a suspicious device has been located, emergency workers generally establish a larger hot zone and dispatch additional companies that may assist with evacuation. If a detonation occurs, fire and EMS take over with their respective roles of extinguishment and medical care, but responders must remain vigilant for a secondary device.

Although intentional bombings aren't as common in the United States as in other countries, when they do occur, they rarely are identified and dispatched as such. Many times the initial call received is for an explosion, a plane crash, or even a vehicle crash. Like our school shooting example, it isn't until after the first units are on the scene that the pieces come together to suggest a bomb. As a result, many times units are committed and actively involved on scene when they finally realize that the explosion may actually be a crime scene. Standard procedures would put an emphasis on clarifying caller information and the number of reported calls for more information, while factoring in the type of structure or location. For instance, an explosion at a propane storage facility or flour factory would tend to be less suspicious than an explosion in a bank parking lot or the atrium in the mall. Standard procedures that *assume* an explosion is a bomb until disproven with other caller or on-scene information would change the mindset of responders and force more caution in their response and approach.

Any procedures we have must also keep pace with the changing needs of society. With the risk of global terrorism growing, it is possible that car bombs and suicide bombers may become more common in the United States. Situations like these have an even greater demand for national response standards. In this case, tapping into the experience of fire and emergency services in other countries would give us a head start on ensuring the safety of our personnel.

Animals

A violent attack doesn't have to take place as a result of humans (Figure 12-5). Let's assume you respond to an elderly male who looks as if he is **dead on arrival (DOA)**. He appears to be deceased from the window, but access to the victim is prevented by a pit bull that may have mauled his owner and won't let anyone in the home. Standard procedures might have you wait for a law enforcement officer who puts the dog down so you can safely reach and assess the owner. Although the dog appeared violent, you find out minutes later from his family that "Tessie" did not bite and was very gentle. It then becomes evident that she was only protecting her owner who had died of natural causes. The family is upset about the death of two family members, and the newspaper reports some unflattering remarks about your department and the police.

As a result, you look into other options for dealing with dogs in the future. Your search ends when you discover that some fire departments are already prepared to handle these types of emergencies without waiting for law enforcement. Some phone calls result in an offer by the local humane society to train your department in safely controlling dangerous dogs with the proper equipment. For some departments, it could be poisonous snakes, Africanized honey bees, or even alligators. An S-3 standard that any department can "pull off the shelf" and put into action immediately saves time and includes all the safety aspects of a complete policy.

FIGURE 12-5 Violent incidents can include dealing with animals.

SUMMARY

When it comes to violent incidents, we need to have a solid plan in place. It's vital that we define our duties, plan our response, and train for the execution of our action plan. We must provide the needed logistics and tools for the activities that fall into our scope, and identify outside resources to provide what we can't. Violent incidents tend to pull the emotions of responders into poor risk-management decisions. A rioter may provoke a responder to return violence and a school shooting could tempt a rescuer into gambling away his or her safety on a losing bet. An intelligent gambler walks into a casino with a standard plan of how much money is available to wager. Likewise, we can eliminate some of the emotions of a violent incident if we have a standard plan of what we will do and what we won't.

Emergency incidents change over time, as do the dangers to emergency responders. Fires in oil-soaked heavy timber machine shops and balloon-frame construction residences are slowly being phased out of existence and someday will be rare. Most of us will never respond to a school boiler explosion or a man trapped in the gears of a grist mill. However, other emergencies, such as those involving terrorism, are likely to increase in the future and are a greater threat to our safety. History has proved that when a new threat was identified to a community, the best response was to assemble the troops and arm them with the proper weapons. In this case, terrorism is one of our biggest threats and the most applicable weapon is knowledge. We need to be able to dispense knowledge efficiently through the use of standard procedures to maintain a high level of success and safety.

KEY TERMS

BDU - Battle dress uniforms.

British thermal unit (BTU) - A unit of measure that can be used to describe the heat being produced by a fire.

CBRNE - Chemical, biological, radioactive, nuclear, or explosive.

cover fire - A tactic used by law enforcement that involves continual shooting in an effort to draw the attention away from an offensive move.

CPJ - Committee to Protect Journalists.

DOA - Dead on arrival.

EMS - Emergency medical services.

EMT - Emergency medical technician.

fixer - A person who assists journalists in obtaining information confidentially.

gallons per minute (GPM) - A common form of identifying a specific amount of water being pumped in a specific amount of time.

IAP - Incident action plan.

Level II staging - A term used to describe an area located a safe distance away from an incident, preferable out of view of the incident, where emergency vehicles park while awaiting orders.

line-of-duty death (LODD) - Fatalities that are directly attributed to the duties of a firefighter.

MCI - Mass casualty incident.

NFPA - National Fire Protection Association.

OIC - Officer in charge.

personal protective equipment (PPE) - A generic term used to describe the minimum apparel and gear needed to safely perform a specific duty.

RDD - Radiological dispersal device.

risk-benefit analysis - The weighing of the facts, determining the advantages and disadvantages of a certain activity.

SCBA - Self-contained breathing apparatus.

secondary device - The term used to describe a bomb or other hazard that is intended to target victims that are escaping or emergency responders that arrive to an incident.

SOG - Standard operating guideline.

SOP - Standard operating procedure.

SWAT - Strategic weapons and tactics.

TEMS - Tactical emergency medical services.

terrorism - A broad term used to describe the act of bringing on fear to a group of people, usually through violence, destruction, injury, or death.

weapon of mass destruction (WMD) - A device used to inflict significant damage to a large number of people.

REVIEW QUESTIONS

1. What are the two main reasons why terrorism is more dangerous for emergency responders than other violent events such as riots?
2. What makes secondary devices so dangerous?
3. What is an important safety component sometimes overlooked when choosing how a firefighter's uniform looks?

4. What are some of the procedures that might be instituted for emergency responses in an area of civil unrest and rioting?
5. What are some components that should be addressed in standard operating procedures (SOPs) and standard operating guidelines (SOGs) for violent incidents?

FIREFIGHTING WEBSITE RESOURCES

http://everyonegoeshome.com
http://cdc.gov/niosh/fire
http://cdp.dhs.gov/
http://www.everyonegoeshome.com/research/pages/Initiative_12__Violent_Incidents/index.html

ADDITIONAL RESOURCES

http://www.67riots.rutgers.edu/d_index.htm
http://www.religioustolerance.org/abo_viol.htm
http://www.disastercenter.com/birmingh.htm
http://whyy.org/cms/radiotimes/2010/12/20/endangered-imprisoned-journalists-around-the-world/
http://newsdesk.org/2010/05/mexico-the-most-dangerous-country-for-journalists/
http://www.cmemsc.org/news/wmd-overview-final.pdf
http://www.nationmaster.com/graph/mil_wmd_che-military-wmd-chemical
http://www.nationmaster.com/graph/mil_wmd_che-military-wmd-biological

NOTES

1. Kavanagh, J. 2011, March 3. "Rodney King, 20 years later" police car. CNN. http://articles.cnn.com/2011-03-03/us/rodney.king.20.years.later_1_laurence-powell-theodore-briseno-king-attorney-milton-grimes?_s=PM:US
2. Homeland Defense Planning, Resourcing, and Training Issues Challenge DOD's Response to Domestic Chemical, Biological, Radiological, Nuclear, and High-Yield Explosive Incidents. 2009, October. Report to Congressional Requesters, United States Government Accountability Office.
3. An arsenal of invisible weapons. 1998, March 11. CBS News. http://www.cbsnews.com/stories/1998/03/11/gulfwar/main4755.shtml
4. Bartlett, J.G., MD. 2002. Foreword. *PDR guide to biological and chemical warfare response.* New York: Thomson Healthcare.
5. McPhee, J. 1974. *The curve of binding energy.* New York: Farrar, Straus, and Giroux.
6. Toppo, G. 2009, April 14. 10 Years later, the real story behind Columbine. *USA Today.* http://www.usatoday.com/news/nation/2009-04-13-columbine-myths_N.htm
7. Battalion Chief Mike Alder, Deputy Chief Mat Fratus, FDIC Workshop "Assessing Dynamic Risk," April 20, 2009.
8. Silence or death in Mexico's press; Crime, violence, and corruption are destroying the country's journalism. A special report of the Committee to Protect Journalists. 2010, September. The Committee to Protect Journalists.
9. NFPA 1500 Chapter 8.2.6 NFPA 2002.
10. Jones, T.L. 1996. SWAT; Leadership and tactical planning. The SWAT operator's guide to combat law enforcement. Boulder, CO: Paladin Press.
11. An arsenal of invisible weapons. 1998, March 11. CBS News. http://www.cbsnews.com/stories/1998/03/11/gulfwar/main4755.shtml

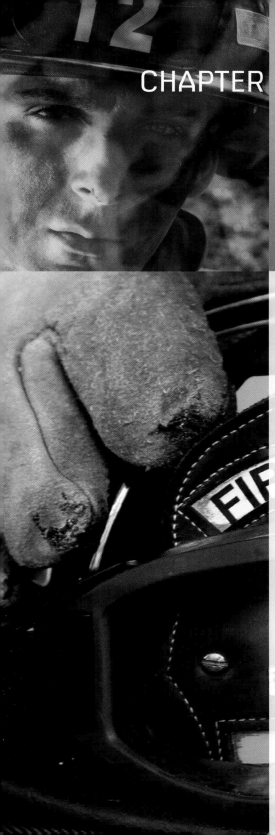

CHAPTER 13

Providing Emotional Support

LEARNING OBJECTIVES

- Compare the effects of physical and psychological stresses of a critical incident on the health and well-being of responders.
- Explain the myths of relying on alcohol or drugs to cure depression.
- List the differences between critical incident stress and post traumatic stress disorder.
- Compare the components of an Alcoholics Anonymous (AA) meeting and a critical incident debriefing.
- Define psychotherapy, and how it is normally applied to fire and emergency services.
- Explain the differences between demobilizing, defusing, and debriefing in critical incident stress management (CISM).
- List some of the ways individual therapy is offered to emergency responders.
- Explain the features of an effective chaplaincy program.
- Describe how technology influences confidentiality for psychological support.

Initiative 13: Firefighters and their families must have access to counseling and psychological support.

It's been said that an organization is defined by the people who make it up. The recent push for improving our health and wellness is evidence of this ideology. The Fire and Life Safety Initiatives also recognize its importance, and have appropriately devoted an initiative to address responders' physical health (Initiative 6) and a separate initiative to their emotional health (Initiative 13). Although each is equally important, it is interesting to note the ties between the physical and psychological sides of human beings. For example, two 54-year-old men can call 911 in two different cities at the same time. Both could be complaining of severe pressure of the chest with accompanying shortness of breath. They might each be overweight, out of shape, under stress, and have a history of hypertension. When emergency responders arrive on scene, both patients could be sweating profusely with pale skin and a feeling of impending doom. The difference is that one could be having a heart attack, and proceed into cardiac arrest before arriving at the hospital, while the other is diagnosed with an anxiety attack and is sent home. Many times it's difficult to ascertain the source of our problems.

STRESS: OVERVIEW

In Chapter 6, "Developing Medical and Fitness Standards," we looked at the way our bodies react to the emotional stress of being dispatched to an emergency call with people trapped. We explored how our bodies dump adrenaline into the bloodstream, thus increasing our heart rate and strength. We looked at how the physical activity created a need for more oxygen, which in turn raised our heart rate even more. Our response to stress is fed from both the physical and emotional strains of the incident. For instance, assume your fire department has responded to an explosion in a mixed occupancy hardware store on a Saturday morning in August. When you arrive on scene, it is evident that the explosion was significant and probably originated in the basement of the building. The storefront is literally "in the street," and the second-floor apartments are perched over an unstable pile of masonry, wood, and plaster. Surprisingly, there are only a couple of spot fires, which firefighters are already extinguishing with handlines. EMS is attending to three victims who were found in the street. You are assigned to begin a search of the rubble for the owner of the hardware store and any customers who may have become trapped. Crews upstairs successfully remove two people from the apartments. After some time of searching, you uncover the right arm of the owner, which appears to be reaching out of the rubble. You call for help and frantically dig down to expose the rest of his body. You hardly notice when you twist your left knee carrying the owner out of the rubble. Later, you assist in digging out the body of a customer. Although additional mutual aid companies assist, the incident isn't declared under control until early afternoon. At the completion of almost five hours of intense work, you are both physically and emotionally drained. Your crew has been released, but it's a slow walk back to the apparatus.

The constant flow of adrenaline allowed you to continue working for a long duration. There were highs and lows of emotions as victims were located. Both the physical and emotional stresses combined to exacerbate the effects of each other, taking a toll on your body. This stress has both short-term and long-term effects. Immediately, you feel exhausted and lethargic. You are slow to respond to questions, and find yourself resting for a few minutes on the side step of the engine after getting out of your gear. Your knee is a little sore, and you decide you should ice it down when you get back to the station. When you make it to the locker room for a much needed shower, you stand under the hot water motionless for quite some time. Muscles are sore, and visions of the store owner's arm sticking out of the debris are burned in your memory. But it was the owner's wristwatch you are likely never to forget. Similar to the type your dad always wore, it was originally what caught your attention. It was gold and shiny in a sea of gray masonry rubble and dusty plaster. The hands on the watch showed 12:17 when you first saw it and called for assistance. It was 12:25 when you helped strap his arm to a backboard as he was removed. How could the watch survive without a scratch on it and still keep time? How was it that the man's life is over but the watch lives on?

The long-term effects of the stress of the incident are largely unknown at the time. It could be that your knee never was right again after twisting it in the rubble. Thirty years later, you end up having it replaced, and still refer to it as your "hardware store injury." The emotional effects could continue to creep up on you, especially the wristwatch. It's amazing how many times a clock said 12:17 when you looked at it in the years following the incident. Sometimes when you see debris piles at emergency incidents, you remember that arm. It's not that you have nightmares about it, but, like your knee, it will be an occasional but constant reminder to you of that hot summer morning (Figure 13-1).

Coping with Stress

We must be able to accept the fact that we all are victims of both physical and psychological stresses, and we can set ourselves up for failure if we're not prepared. Not all responders will be affected in the same way. Some can walk through the aftermath of

FIGURE 13-1
Any day can consist of an emergency incident that has the potential to inflict a physical or emotional wound.

a grisly plane crash and eat a sausage sandwich. Others struggle through certain calls for the rest of their lives. However, it is safe to say that all emergency responders feel some sort of emotional pain as a result of the duties they perform. Couple this with the fact that responders are human and are forced to deal with other stresses, such as abuse, divorce, or the death of a family member. Although we are trained to accept the death and destruction we witness, we must also realize that we are only human and that it's impossible for us to not be affected at some time.

Boot camp, or basic training, for the United States Marine Corps is designed to initiate soldiers into the duties and culture of the organization, but also tends to "weed out" the recruits that couldn't handle the physical and emotional stress of combat. Although most fire and emergency services agencies don't have the same initial training regimen of Parris Island or San Diego, you may recall that some of your classmates never completed it. Others who did graduate may have lasted only a short while before deciding that the job was not for them. Still others may continue to respond to emergencies, but have deep emotional scars left from past incidents that they can't forget. **Burnout,** as it is referred to, many times is the direct result of the psychological stresses of responding to emergencies. As a result, many of us have developed ways to cope with the stress, with some being healthier than others. Two specific examples of unhealthy coping mechanisms are alcohol and drug abuse.

Alcohol Abuse

At one time, many believed that alcohol was the answer for both physical and emotional pain. It wasn't uncommon for surgeons in the Civil War to administer whiskey or ether to a soldier who was about to undergo emergency surgery to have his leg amputated. Similarly, it was fairly common for soldiers to use alcohol to numb the mental anguish of losing friends and fellow soldiers in battle. Like many military traditions, alcohol carried over to the fire service. Although better treatment and medications became available for physical pain, alcohol was still presumed to be beneficial for psychological pain. It wasn't uncommon for rescue workers to try to ward off despair with alcohol. Unfortunately, alcohol itself is classified as a depressant, and therefore proves ineffective in the fight against depression.

In more recent times, we have come to understand that there is a difference between alcohol use and alcohol abuse. Alcohol use, or social drinking, is accepted in many places by emergency responders, and is generally considered harmless. Unfortunately, many responders have seen a "harmless" social drinker drive under the influence of alcohol and suffer the consequences. Others seem to evolve from alcohol use to alcohol abuse. What seems to be less clear is where the line between the two is drawn. According to **Alcoholics Anonymous (AA),** some of the symptoms of alcohol abuse or alcoholism include:[1]

- A physical and emotional desire to drink beyond control
- A need to continue drinking after the first drink
- Drinking when there is every reason to remain sober

Additionally, alcohol abuse can lead to other destructive behaviors. Say your brother, for example, is going through a rough period in his personal life. Alcohol initially acts as a stimulant, which tends to improve his mood. Unfortunately, we know that he will build a tolerance to the alcohol, and it will take more drinks to get him to that high. Worse yet, the peak of the high is "pointed," with a sharp depressive slope behind it. He is likely to blow right past the peak, and actually become more depressed. You may notice that a person who drinks too much has a sudden mood change, from jovial to angry or depressed. In a desperate effort to climb back up the hill, some have a desire for more alcohol or to try to get there with drugs.

Drug Abuse

In the past 40 years, illegal drugs and controlled medications have become much more prominent as an influence on emergency responders. Sometimes it's because drugs are quicker or more effective at making people forget their pain, or because they feel drug use is less obvious to family, friends, and coworkers than alcohol is. Additionally, many emergency responders are positioned closely to controlled substances, such as prescription medications, as part of their job function, which makes them fairly easy to obtain. It's not uncommon to see a report of a health-care worker, such as a firefighter, accused of stealing medication from a patient.[2] The real question is how many times did he or she respond to emergencies under the influence of drugs before being caught? It's likely that anyone in that position puts citizens and other emergency responders at extreme risk.

Failure to Cope with Stress

When alcohol and drugs don't help an emergency responder cope with the stress, it's likely to bring on severe depression. Sometimes the story ends in suicide, but how often that happens is just an estimate. Take police officers, for instance. In 2008, there were 132 **LODDs**, yet a conservative estimate of suicides is over 300.[3] When suicides occur on duty, there is no secret and statistics are clearly recorded. But when a suicide occurs off duty, the statistics aren't tracked. The police officer suicide study showed that many times suicides are recorded as accidental or undetermined in an effort to preserve the honor of the family and department. At the very least, suicides are kept as quiet as possible, but some are beginning to take notice.

According to *USA Today*, emergency workers committing suicide are emerging as a "disturbing trend."[4] High-visibility rescues involving the trapped miners in Pennsylvania in 2002, the events of September 11, the Oklahoma City Bombing in 1995, and even the Jessica McClure rescue in 1987 all resulted in later suicides by emergency personnel. Many times, other events, such as family problems or the death of a loved one, contributed to the suicide, and it's unclear how much influence the rescue had. Regardless, many firefighters and EMTs can name fellow responders who committed suicide. As a result, it's imperative that we look at ways to support Initiative 13, and expose the causes, signs, and symptoms of stress.

POST-TRAUMATIC STRESS DISORDER

Although not officially labeled as such until the last 30 years, **post-traumatic stress disorder (PTSD)** has probably been around as long as humans were chased by wild animals and villages were overtaken by volcanoes. Much like carbon monoxide poisoning, it surely affected people who simply didn't know what it was. One of the earliest definitions of the affliction in the United States was in Webster's 1913 dictionary for "railway spine," which described unique physical and mental injuries sustained in railroad crashes.[5] Presumably from a delayed shock to the spinal cord, symptoms varied and occasionally wouldn't present themselves for quite some time. Similar labels were assigned to soldiers in battle, victims who were raped or tortured, and police officers who survived a shootout.

There are varying severities of critical incident stress, based on how an individual reacts and copes with the events.[6] For example, **acute stress disorder (ASD)** is the term used to describe a possible precursor to PTSD, with symptoms that generally last less than a week. Some people can proceed with no intervention at all. Others are affected, but return to normal in a short period of time with brief periods of treatment. Still others can develop PTSD if symptoms persist for a month. PTSD is generally considered the most severe form of acute stress, many times incapacitating the person from living a normal life. Individual treatment or medications may be required. Signs of PTSD could include avoidance, seclusion, flashbacks, as well as significant changes

TABLE 13-1 PTSD Signs and Symptoms

1. Re-experiencing symptoms:
   - Flashbacks
   - Bad dreams
   - Frightening thoughts
2. Avoidance symptoms:
   - Staying away from places, events, or objects that are reminders of the experience
   - Feeling emotionally numb
   - Feeling strong guilt, depression, or worry
   - Losing interest in activities that were enjoyable in the past
   - Having trouble remembering the dangerous event
3. Hyperarousal symptoms:
   - Being easily startled
   - Feeling tense or "on edge"
   - Having difficulty sleeping, and/or having angry outbursts

Source: National Institute of Mental Health, http://www.nimh.nih.gov/health/publications/post-traumatic-stress-disorder-ptsd/what-are-the-symptoms-of-ptsd.shtml

in eating or sleeping (Table 13-1). Another possible sign of PTSD is sudden and unexplained rage, as a result of deep-seated feelings being aroused by a memory, a sound, or the geographic location of an incident.

One such example of first responders reliving a traumatic experience is seeing a **roadside memorial.** Family members or friends of people killed in motor vehicle crashes often erect crosses, signs, or flower arrangements at the crash scene. On some occasions, these can appear at the scenes of fires or violent crimes, especially ones where a child has died. Although their intent is to ease the suffering for family or friends and provide closure, it can have the opposite results for emergency workers. For example, let's assume that your fire department responded to a motor vehicle crash on a state highway where a teenager was ejected and killed. You were among the first responders who found her in full cardiac arrest and began **CPR.** Although the paramedics did everything they could, she did not survive. Although you try to avoid it, a couple weeks later you drive by the scene with your friend and see that a small memorial has appeared in the area near the accident site. A small wire wreath with flowers and a cross stands between the skid marks that leave the highway. Immediately you are filled with rage and become visibly upset. It's not because of the memorial or because you are forced to relive the emergency call, but you find yourself angry that the memorial is at least 100' from where you found her. In fact, the skid marks weren't even made by her car. Your friend is surprised that you are so upset about such a seemingly meaningless detail, but it's clear that the detail is far from meaningless to you (Figure 13-2).

FIGURE 13-2
Roadside memorials can open up old wounds for emergency responders.

Whatever you want to call it—anxiety disorder, critical incident stress, or even railway spine—it makes it difficult for you to pass a scene you've been avoiding. When you are forced to face it, it may affect you more than the night of the crash. Like an infection that is ignored, it may heal on its own or worsen. However, it's likely to heal faster and more assuredly with the proper treatment. Recognizing the signs and symptoms is the first step to moving toward a cure, followed by treatment. The most common treatment for this type of stress can be broadly classified as **psychotherapy,** which encompasses varied forms of treatment and uses counseling as an avenue for healing. As with physical injuries or illness, the needs of the victim dictate the appropriate level of care. For instance, traditional psychotherapy would be at the upper end of the scale and is conducted by a trained professional for long-term rehabilitation. **Crisis intervention,** on the other hand, refers to brief communication therapy as a preventative or acute response to a stress. Although there are many different levels and specialties involved with the subject of psychotherapy, this text will use the term in the context of therapy for the treatment of critical incident stress. The two types of therapeutic communications most appropriate for emergency responders are group therapy and individual therapy.

Group Therapy

Although we might not realize it, psychotherapy in the form of brief group therapy is very common for emergency responders when it comes to postincident **debriefing.**

For quite some time, emergency responders have used this effective tool to talk or "open up" about an incident. We've learned that bottling it up and keeping it inside is not an effective way to deal with mental stress. Many have noted that discussing it with others who were in a similar situation and experienced it together encourages healing. Some responders have used it as a way of ensuring that others were able to deal with the stress. According to the *Handbook of Psychotherapy and Behavior Change,* brief psychotherapy in a group setting has proven successful in workplaces, as well as for critical incidents and PTSD.[7]

For instance, let's suppose your company responds to the suicide of a teenager. You are the senior firefighter **EMT,** and respond with four others. On your arrival with law enforcement, you find a 15-year-old boy who is **dead on arrival (DOA).** Apparently, he got home from school and hung himself in the garage with a tow strap. His mother found him when she pulled in the drive and opened the garage door. When you return to the station, you bring everyone to the kitchen for a postincident evaluation or "hot wash." Even though you discuss the operational components of the call, which were very limited in this case, you use the critique to encourage dialogue with the younger members. You discuss past incidents that affected you, how it's important to expect stress in these types of situations, and what to do about it. You also make sure they understand that if they have problems from calls like this, that they need to talk to someone. This form of group therapy is very successful, but not limited to fire and emergency services. In fact, one of the best examples of group therapy is Alcoholics Anonymous.

Alcoholics Anonymous (AA)

Alcoholics Anonymous (AA) is a successful example of organized group therapy. The program is primarily voluntary and uses a formalized social framework to allow victims to come to grips with their addiction. The social networking is very similar to a formal type of critical stress debriefing held by some emergency services after a significant event. This could be the death of a child or a coworker, a hazardous emergency scene, a mass casualty incident, or even the death of a victim after a prolonged effort to save her.

For instance, let's assume that your fire department has responded to a residential fire at night in which a family of five has perished. Although the victims were rescued promptly by firefighters with resuscitative measures performed immediately, none survived. The crews are visibly discouraged by their failure to save any of them. Some of the firefighters are handling it fairly well, but others are having obvious problems. Although no one knows it yet, one of the fire victims had poisoned the rest of the family before setting the fire and taking his own life, making any life-saving efforts by responders futile. By taking a look at how AA performs successful group therapy, we might find ways to improve our department's debriefing for this fire.

Meetings. AA meetings are held locally as a matter of convenience for as few as a half dozen to well over 50 people. Many meet regularly, such as weekly on Saturday mornings, and are scheduled with a predetermined starting and ending time. The meetings follow a basic protocol in which a team leader explains the ground rules and facilitates the meeting. For instance, first names are used exclusively (thus Alcoholics *Anonymous*), and everyone is expected to give their full attention to whomever is speaking. Attendees are encouraged to speak, but are not required to. Sometimes it's more of an open format in which anyone can address the group when the floor is available, whereas other groups move around the room and take turns. A person may simply "pass" if they prefer to only listen.

In an example situation, your fire department's members have experienced stress from a critical incident they were deeply involved in. Initial debriefing, or crisis intervention, is designed to allow the group to treat the initial injury of the stress and identify further needs. This process is also an avenue for a **facilitator** to normalize the feelings and symptoms that some members are likely experiencing. One of the most important aspects is to schedule it as soon as practical after the event occurs. If the fire occurred at 3 a.m., personnel first need to get some food, some rest, and see their families. If they are volunteers, they may have to go to work. For this example, your department schedules a debriefing for 7 p.m. that night and encourages all involved to attend. Senior firefighters and officers should make every effort to be there, thus demonstrating the importance of involvement to the newer members. It should be open to any outside emergency agencies that were involved, such as mutual aid companies, dispatchers, and possibly even the police officers who assisted. A comfortable but familiar location should be chosen to keep the attendees at ease, and some refreshments should be made available. It should be a closed meeting, with the only uninvolved attendees being debriefing team members, counselors, or clergy.

The facilitator should explain that the purpose of the meeting is to give everyone a chance to tell their story and share information. Ground rules should be established, along with a schedule and a quitting time. An open-ended meeting is more difficult to plan for, more exhausting for the members, and may make participants feel like they need to say something to be permitted to go home. Like AA, members should be encouraged to speak, yet not required to. As members discuss aspects of the incident, it tends to make others join in and add their thoughts, concerns, and observations. For members who are having a difficult time with the incident, it is especially helpful to see that others are similarly affected.

Single Share. Most AA meetings follow a rule that they refer to as **single share.** Although the definition formally means that each person can only speak once, its basic intent is to keep one person from controlling the floor or its discussions. By limiting the number of times an individual can speak, it keeps everyone equal and encourages others to participate. Every group has different dynamics based on who is present. Some people enjoy the stage and having the undivided attention of everyone else. It's important that these people have their needs met, yet don't discourage others from participating.

At the incident debriefing, it might be that Sarah, a member of a search crew, explains that she is upset with herself that she missed one of the children on the first pass, and that she wishes she could have located him sooner than she did. The group should take her comments seriously and absorb what she said without cutting her off or developing an immediate response to her opinions. Larry, a firefighter/paramedic, may describe that, from his vantage point, he felt that the victims were rescued very quickly. He, however, is discouraged that "nothing went right" while treating his victim on the way to the hospital. It seemed that every time he tried an intervention, the patient would not respond. He fears that the victim may have survived had he been more successful with his treatments.

Cross-Talk. As individuals discuss how they feel or what has happened to them, human tendency is for someone else to try to clarify someone's thoughts for them or to directly address an opinion or statement someone made. This can easily evolve into a discussion, or worse, an argument, in which either can be detrimental to progression of the group. This also diminishes the comfort level of individuals and tends to deter others from making comments. People should be permitted to have their own opinions and should be protected from being directly challenged. Therefore, **cross-talk,** or direct response to an individual, should be discouraged from the beginning.

For instance, Sarah made a truthful statement to the group about how she felt:

> "When I got to the top of the stairway, it was hot, with zero visibility but no signs of fire. I yelled to Tom that I was going to search the bedrooms on the left and handed him the hoseline. The room was cluttered, and I pushed some clothes out of the way. I finished searching the bedroom and was crawling out when my foot kicked something. I stopped and found the victim in the clothes I had shoved over. I just wish I would have noticed him the first time."

Sarah publicly stated that she wished she had found him earlier. A typical response by other firefighters would be to disagree with her, and tell her she did everything right. Someone might try to help her by saying something that actually makes her feel worse. It's best to refrain from any response. Sarah explained how she feels, and nobody can understand or argue her point. The best listener in this case is simply a listener.

The facilitator, on the other hand, may choose to address statements made by individual members to the group if he or she feels it would be beneficial. For instance, a mental health professional might take note that several members are blaming themselves, and explain to the group that it's normal for a person to self-blame in a situation such as this. She might point out that they didn't set the fire, and they likely performed as well as anyone would under similar circumstances. She could describe the conditions of the search as she understood them, and make an observation that firefighters went into an atmosphere that most people wouldn't. They crawled through complete darkness and high heat in full gear, feeling for a possible victim through thick gloves. She might question how anyone could locate a child in that situation, and explains that nobody "missed" the child, but rather succeeded in finding and removing every victim.

She also could explain that it's her understanding that people die in hospitals and surgery every day, and that any interventions by prehospital medical crews in a moving ambulance can't be guaranteed effective. By addressing individual statements to the group in this way, she's not discounting their feelings, but making broad statements about the skilled performance of the crews from an outsider's perspective.

Closing. AA meetings end on time and close with some thoughts from the facilitator. Many times that includes the facilitator's reiterating some of the basic principles of AA, and explaining that what they are feeling about their dependency is both normal and to be expected. However, it is manageable and they are capable of overcoming it. Alcohol will never be gone from their memory, but it will not dictate or control any part of their future. Additional resources are also available for those who feel it would be beneficial for her or his individual situation.

Many critical incident counseling sessions end exactly the same way. Attendees need to understand that they will not forget the fire, and certain aspects will likely be etched into their lives forever. The feelings that they have expressed or kept to themselves are normal. They need to understand that time will make it easier, and that additional help is available if needed. Information on the signs and symptoms they may experience are explained, along with warning signs to look for.

Critical Incident Stress Management

Critical incident stress management (CISM) is an organized form of crisis intervention used to manage the stresses of critical incidents.[8] Probably one of the most popular, comprehensive stress management programs in emergency services, CISM has been continually promoted by Dr. George S. Everly Jr. and Dr. Jeffrey T. Mitchell and consists of several "core components" that could be categorized as preincident, incident response, or postincident.

Preincident CISM. Firefighters and EMTs are well aware of the advantages of prevention in an emergency response system. Likewise, CISM puts an emphasis on educating fire and emergency services personnel about the dangers of critical incident stress and its potential warning signs. The training may take the form of stress reduction, preparing for stressful incidents, or learning healthy coping mechanisms. Additionally, planning for a critical incident is vital to ensure that key personnel are in a position to minimize the effects of an event and have a response team ready.

Incident Response. When an event does occur, intervention crews will ideally respond to the incident to gather information. If needed, they can provide individual assistance to members who need special attention. For example, the two members of a three-person search team that found their way out of a structural fire should receive immediate support on scene while crews look for the remaining member. For large incidents, the first phase of crisis intervention could be **demobilization.** This occurs

FIGURE 13-3
The first phase of crisis intervention, demobilization, is likely to occur just before release from an incident. This might take place at the rehabilitation (rehab) unit.

just before emergency crews are released from the scene. Everly and Mitchell suggest establishing an area just outside the incident where food and rest can be combined with information about critical incident stress. Although most existing fire department rehab units don't carry preemptive information on stress management, it might be something beneficial to have available on the rehab unit for special incidents (Figure 13-3).

Postincident CISM. At some incidents, demobilization is simply not feasible. It could be that workers had been released prior to the arrival of CISM workers, or that the incident is not large enough to warrant a substantial rehab area. CISM therefore doesn't appear until a scheduled meeting. Some meetings will have the goal of defusing, whereas others will be primarily for debriefing.

- Defusing
  - **Defusing** is scheduled immediately, or soon after the incident and follows a group discussion format. Defusing can introduce members to the process of CISM after the event, and allow team members to assess the needs of individuals. Usually the discussion format concentrates on what the workers observed and did at the incident, and how they felt and feel about it. Although it can occur, defusing is not necessarily designed to provide **closure**. Additionally, signs and symptoms are identified and attendees learn about what they mean and how to proceed.
- Debriefing
  - A debriefing is likely to be scheduled several days later, or up to a week after a traumatic incident. Much like the format of an AA meeting, the debriefing is

designed to get information out into the open so that members can start the process of healing. Similar to defusing, it tries to establish what the attendees did at the incident and how they feel about it. Additionally, symptoms are identified and time is spent educating the attendees about what they mean and how to proceed. The debriefing session is intended to provide closure for most, if not all, attendees.

One major difference between our critical debriefing and AA is the anonymity. Although AA prides itself in the secrecy of the individuals, emergency responders usually know each other fairly well. This isn't normally considered a disadvantage, as other types of group therapy include members who are well known to each other. For instance, family counseling can be very successful in helping individuals to overcome certain emotional problems. History has shown that the lack of anonymity for emergency responders participating in critical incident group therapy has not affected its ability to be useful. If anything, emergency responders need to have peers of their own to understand the situation, duties, and the protocols the responders were dealing with during the response. Outsiders may not be accepted.

Psychological First Aid

Recently, some researchers have questioned the ability of CISM to reduce the symptoms of PTSD, and have suggested that psychological first aid (PFA) may be more effective. The primary difference between CISM and PFA is that CISM uses debriefing, or a discussion about the events that caused the distress, and PFA concentrates on supporting the needs of the victim without any peer counseling. In order to understand the research, it's important to recognize that there are two classifications of psychological trauma victims, primary and secondary. Primary psychological victims are those who witness an event, such as a woman who watches helplessly as her husband is electrocuted. Secondary psychological victims are those who are later involved in the event, such as emergency responders who unsuccessfully try to revive him. Many organizations, including the World Health Organization (WHO) have discouraged the use of routine debriefing for primary victims of psychological trauma.[9] In other words, we should not ask the man's wife to describe what happened or what she did while she waited for emergency responders to arrive. It's understandable that this type of counseling could have detrimental effects.

On the other hand, other research has shown that debriefing of secondary trauma victims is much more effective. In an article written for the *International Journal of Emergency Mental Health*, the authors defend CISM and explain that seven of the eight studies conducted on emergency responders showed "a reduction in symptoms or problem behaviors following debriefing, and none of the studies found a negative impact."[10] As discussed in Chapter 7, more research is needed in this area, specifically in respect to secondary psychological trauma victims.

Individual Therapy

Individual therapy is another option, and unlike group therapy, is more private. Trained counselors in different professions can offer more thorough treatment that may be needed in certain situations. Some of the more popular sources for emergency responders include mental health support, **employee assistance programs (EAP)**, and chaplain programs.

Mental Health Support

Numerous professionals deal with the specific mental health needs of individuals. Some include social workers, psychiatrists, and psychologists. These mental health counselors may be effective at treating the effects of PTSD, but also can treat an underlying illness that could manifest itself after a critical incident. Rather than brief therapeutic communications, treatment usually consists of ongoing psychotherapy and may include medications or other rehabilitation. For example, PTSD might be treated with relaxation and exposure, in which the patient is taught to relax before being subjected to simulations of the critical incident that is causing the stress. Much the way a doctor treats allergies by continually exposing a patient to the allergen, the incident eventually becomes normalized through conditioning.

One of the advantages of utilizing a professional counselor is the fact that the visits can be discrete. If discussing feelings in a group counseling environment was a barrier, utilizing a mental health professional in a private setting is far more effective. Sometimes these professionals could come from a referral, or could be a component of an employee assistance program through an employer.

Employee Assistance Programs

An EAP is usually a component of a health benefit, or a resource that is provided by an employer for employee problems such as substance abuse, smoking cessation, and mental health issues. An employee with personal problems that affect his or her work performance can be referred to an EAP as a condition of returning to work, or EAP can be contacted voluntarily. Usually a toll-free number is provided for employees to contact in private. The operator puts the employee in contact with the specific resource needed. In many systems, the employer is not aware of who is using the program. This is a must for an EAP to be accepted and utilized by the employees.

The primary goal of an EAP is to fix problems before they grow out of control. In the case of smoking cessation or drug abuse, helping the employee break the addiction improves the employees' overall health and work attitude, as well as reduces future medical expenses. Sometimes emotional issues continue to worsen until treatment is initiated. For example, suppose a coworker is having marital problems. He has moved out of his house, is away from his children, and is living in his friend's apartment. You notice he is drinking more off duty, and he appears ill all the time. An effective EAP allows him to obtain the emotional help he needs on his own or at the suggestion of friends.

Suppose he doesn't seek any help. As more aspects of his personal life continue to collapse, he begins to call off work fairly regularly. When he does come in, he appears to be preoccupied with his family problems. He disappears in the fire station periodically throughout the day to have long, private conversations on his cell phone. One day he finally reaches his breaking point and punches a fellow firefighter in the face after an argument on an emergency scene. A secondary goal of an EAP is to be a component of a progressive disciplinary process. He may have been a great employee who let personal problems take over to a point where his job performance is directly affected. The employer might order him into an EAP until he is cleared for return to work. Without an EAP program built into the disciplinary process, his actions might be severe enough that he gets terminated. Besides losing a good employee, his life continues to spiral out of control. It could be that the only thing normal in his life to this point was his job.

Departments that don't have access to EAP, such as volunteer fire departments, should research the available options. Contracting with an outside agency is similar to paying premiums on an insurance policy. When selecting an EAP, it's important that research includes what experience the agency has in dealing with emergency responders, as well as their capabilities. In turn, the emergency service agency can share with its members exactly what is available to them.

Chaplains

Another way for emergency service organizations to provide emotional support for their members and families is through the adoption of a chaplaincy program. Chaplaincy programs exist in many hospital systems, correctional institutions, and the military. As opposed to EAPs, which are more clinical in nature, these programs can provide unique guidance and assistance on a personal basis to those who need and desire it. Some people feel more comfortable talking to someone with religious opinions rather than a mental health professional, especially in matters such as death. Some chaplains are active with a church, but share their time with the emergency organization. Others are employed exclusively by an emergency service such as a fire department, as a chaplain or as a lay firefighter who has an interest in serving as a chaplain. Chaplains may be male or female, and may come from any religious denomination. Although the primary goal of securing a chaplain is for counseling the members, some chaplains provide emotional support to family members as well. This is especially important in LODD situations. Other chaplain duties may include providing grief support for victims and the families of victims, as well as normal emergency response functions.

Suppose your fire department created a position for a chaplain, and advertised to the local churches for interested candidates. Interviews by the officers have narrowed it down to the best prospects. It's important to choose a chaplain with specific qualities—specifically someone who is available, approachable, and adaptable.

Availability. The main consideration for choosing a chaplain is availability. Obviously, it's important for him or her to be available to respond when needed. If the chaplain's other commitments take precedence to emergency response on short notice, the program is likely to be less effective. But even more importantly, a good chaplain is well known by the members. If the chaplain is not a regular fixture who knows the members by name, it's unlikely that anyone will confide in him or her. A successful chaplain not only responds to emergencies and drills, but also takes time to get to know the members before the time of need.

Approachability. A successful chaplain must also be someone who is approachable, whom people feel comfortable with. If a chaplain is to be utilized, she or he must be trusted by the members. One of the advantages of a good chaplain over a mental health professional through EAP is that, in many situations, members know the chaplain. Many chaplains in existing emergency response programs have become friends of members, even performing weddings and funerals for families. If members have no religious connections, then they may feel more comfortable using a chaplain they know for important ceremonies in their lives.

Adaptability. Not every member of the clergy will be cut out for the duties of a chaplain. It must be understood that the job description for a chaplain differs in some aspects from that of a minister, pastor, priest, rabbi, or other religious leader. Specifically, the chaplain should not be in a position to try to convert people to a specific religion. In fact, many chaplain organizations, including the Federation of Fire Chaplains, make it a point to explain that a chaplain is not there to **proselytize,** or convert members.[11] The larger the organization, the more vast are the religious beliefs of its members. It's important to remember that, when selecting a chaplain, he or she must be aware of the broad range of the beliefs of the members and be adaptable to their individual needs. A chaplain must agree to support all creeds, as well as those who have no religious beliefs, treating all members equal.

THE FUTURE

In Chapter 7, "Creating a Research Agenda," we looked at ways that technology might help us perform our job more safely in the future. Technology also shows promise when it comes to better serving the emotional needs of emergency responders and their families. Although our existing systems for dealing with critical incident stress are effective, they predominantly use group therapy. Many emergency responders learn better in groups and work better in groups, but some are more comfortable solving problems on their own. Therefore, it would follow that some would heal faster in a private setting at their own pace. Technology could provide solutions both for **relative anonymity** and for complete anonymity, as well as providing a support system at home.

Relative Anonymity

Everyone handles the stress from varied emergency calls differently. Some open up and want to talk about situations and others want to keep to themselves. It could be that the interpersonal relations are the most difficult for those who prefer not to talk about it. Say you and your partner Debbie respond to an emergency for a baby not breathing. According to the baby's mom, she tried to awaken him from his nap but couldn't. You arrive in the bedroom and find a 2-month-old boy who is not breathing and has no pulse. The two of you begin resuscitative measures and get one of the firefighters to drive to the hospital, where the infant is pronounced dead.

As you clean and restock the ambulance, you compliment Debbie on her professionalism and skill but her responses are short and to the point. She refuses to talk about the incident and it appears she's taking it hard. With two boys of her own and one still in diapers, it's likely that she had a hard time *not* thinking about her children as she tried to resuscitate the baby. After you return to quarters, you tell Debbie to call you if she wants to talk, and report the incident to your supervisor. You explain that Debbie may need some help in dealing with the call.

That night, you receive a text message from Debbie about the emergency incident that simply says "this is a day i wanna 4get!" That short message evolves into a two-hour texting conversation where she opens the gates of her heart to someone who was there and understands how she feels. As the evening progresses, it's clear to you that she will get through this experience, and be back to work next shift. There haven't been many research projects exploring this impersonal form of communication that is becoming so prevalent, especially when it comes to critical incident therapy. However, the advantages and possibilities are real and have been noticed recently by some emergency workers. Although your partner may not have the courage to look you in the eye and discuss what's bothering her, the "relative anonymity" of texting or emailing allows her the ability to participate in a form of psychotherapy while remaining in her comfort zone (Figure 13-4). This in no way means that texting or emailing is a cure-all for critical incident stress or is the best way to communicate. It does, however, identify another form of counseling that uses technology and that may be even more effective in the future. Society continues to accept informal communication. We would be remiss if we failed to see when it could be useful.

Complete Anonymity

One other potential use of technology involved a study that used the Internet as the platform for a specially designed **cognitive behavioral therapy (CBT)** for military personnel suffering from PTSD.[12] According to the National Association of Cognitive-Behavioral Therapists (NACBT), CBT is a form of psychotherapy that uses scientific methods to help patients "unlearn" the way they react to a specific stimulus.[13] Unlike forms of psychoanalysis in which a patient lies on a couch and tells a psychiatrist about

FIGURE 13-4
Texting and exchanging emails are examples of how technology can help the communication process, providing relative anonymity.

personal issues, CBT is less subjective. Internet-based CBT provides the added bonus of confidentiality. Although cheaper than traditional CBT, the study showed it was just as effective. It also suggested that mass counseling could be available for disaster situations, while providing a level of privacy that allowed users to feel secure.

Suppose your department has an incident that claims the life of a fellow firefighter. Members who were involved could be provided user names and passwords for access to an Internet-based website run by mental health counselors trained in PFA, CISM, and the treatment of PTSD. The first time emergency responders log in, they would be required to identify the agency they responded on behalf of, and create their own username and password. In some instances, this would be the only required action. The counseling agency could provide a report of who initially accessed the system, but no additional information would be provided about how many times they logged in or what treatment they received. This would ensure that all members met the requirement to view the website.

After logging in, there could be message boards from other LODDs to view, as well as information about critical incident stress. A new message board about their incident would be started for members to post to. Much like officers and senior members taking the lead in a group session, they can start the conversations anonymously and encourage discussions. If individuals want to read what others are saying, that's great. If they choose to contribute, that's even better.

By using unique screen names, responders could communicate on message boards to others who had similar experiences. Facilitators could participate in the discussions, and "sort" specific users to other threads for similar problems. Other discussion areas could be controlled by therapists trained to a higher level in CBT. They could participate in the discussions, but also "prescribe" online activities for submitters who display specific needs. One advantage of using CBT in this setting is that sessions or assignments follow a specific format or series of methods for treating a patient. As a result, patients can either be released from the program or referred for more individualized treatment.

Imagine a massive earthquake hitting your region, killing 1000 people and injuring thousands more. The response by emergency responders would be staggering, with some rescues and body recoveries continuing for weeks. The advantages of a system providing mass group therapy and individualized care to hundreds of firefighters and EMTs on their own schedule in complete privacy cannot be underestimated.

Support System

Another advantage of an Internet-based system is the possibility of adding family members. Many times families aren't included in the counseling or critical incident process until the situation has grown to damaging proportions. Spouses rarely know what to say to an emergency responder who is experiencing psychological stresses from an incident, and may not even recognize the signs and symptoms of severe distress. Children may have difficulty coping with the fact that a parent is away for long shifts at a time. They may fear for their parent's safety when hearing on the news of a scene injury or death. A web-based program that includes the needs of spouses, children, and parents would be beneficial in many ways to the safety and wellness of emergency responders.

SUMMARY

Experience has shown that the psychological stress of emergency scenes can be as severe as the physical stresses. It's not only important to prepare ourselves for critical incident stress, but to identify and treat it when it appears. Our first line of defense is each other. We can improve scene tactics for the next incident by discussing our successes and failures in an open format. Likewise, we can improve our emotional well-being in similar ways. CISM has evolved into a successful form of group therapy, allowing emergency responders to discuss events and begin to heal.

Another way to improve our performance at the next emergency is to outfit ourselves with the most effective tools available, and be trained in their efficient use. There are many devices out there for psychological support, but our organization may not possess them. We need to research, obtain, and implement instruments such as CISM, EAP, and chaplaincy programs. Additionally, utilizing technology, we need to continue to improve the tools we have, and invent the ones that don't yet exist.

KEY TERMS

acute stress disorder (ASD) - A stress disorder with signs and symptoms similar to PTSD that generally last less than a week; can be a precursor to PTSD.

Alcoholics Anonymous (AA) - An organization that uses group therapy to treat alcoholism.

burnout - A term used to describe the effect of emotional stresses on emergency responders that are often career ending.

closure - A term used to describe the acceptance of an event, allowing an individual to move on and heal.

cognitive behavioral therapy (CBT) - A form of psychotherapy that helps participants unlearn the way they react to specific stimulus.

CPR - Cardiopulmonary resuscitation.

crisis intervention - A form of brief communication psychotherapy used as preventative or acute response to stress.

critical incident stress management (CISM) - An organized form of crisis intervention promoted by Dr. George S. Everly Jr. and Dr. Jeffrey T. Mitchell.

cross-talk - A term used by AA to ensure that participants in group therapy don't directly address comments made by another participant that might lead to an argument.

debriefing - The third phase of crisis intervention that occurs several days to a week after a traumatic incident with a goal of closure.

defusing - The second phase of crisis intervention that occurs soon after a critical event. Although closure can occur during this phase, it's not a goal.

demobilization - The first phase of crisis intervention that occurs just prior to release from the scene.

DOA - Dead on arrival.

employee assistance program (EAP) - A program to aid an employee with mental health, substance abuse, and other personal problems that may affect job performance.

EMT - Emergency medical technician.

facilitator - A person assigned to lead a meeting, discussion, or group therapy session. In most cases a facilitator doesn't add opinions or views to a discussion, but instead encourages others to continue to communicate and ensures ground rules are followed.

line-of-duty death (LODD) - Fatalities that are directly attributed to the duties of a firefighter.

post-traumatic stress disorder (PTSD) - The most severe form of stress disorder with signs and symptoms that persist for more than a month. Many times individuals are incapacitated from living a normal life.

proselytize - The word used to describe the action of attempting to convert an individual to a specific religion. Most chaplain organizations forbid the practice in emergency services.

psychotherapy - A form of psychological treatment utilizing counseling as a means of healing.

relative anonymity - A term used to describe a form of communication that minimizes personal interaction. Technology has allowed more avenues, such as texting and sending emails.

roadside memorial - A makeshift memorial placed at the scene of a incident by friends and family members of a victim.

single share - A term used by AA to ensure that some participants in group therapy don't take too much time away from the rest of the group, and that everyone gets a chance to speak.

REVIEW QUESTIONS

1. Describe why alcohol and drugs are ineffective at curing depression.
2. Explain the signs and symptoms of post-traumatic stress disorder.
3. Describe the similarities between physical and emotional injuries from incidents.
4. How can a meeting of Alcoholics Anonymous serve as an example for critical incident debriefing?
5. What are some potential benefits of a chaplaincy program?

ADDITIONAL RESOURCES

IAFF Guide to Behavioral Health; Achieve Total Wellness. 2005. Washington, DC: International Association of Firefighters.

http://www.alcohol-recovery-info.com/Alcoholics-Anonymous-Meeting-Protocol.html

http://www.nimh.nih.gov/science-news/2007/internet-based-ptsd-therapy-may-help-overcome-barriers-to-care.shtml

http://www.anxietyinsights.info/internetbased_therapy_for_ptsd_shows_potential.htm

Mannion, Lawrence P. 2004. *Employee Assistance Programs; What Works and What Doesn't.* Westport, CT: Praeger Publishers.

http://www.wltx.com/news/story.aspx?storyid=77195&catid=291

NOTES

1. This is A.A. An introduction to the A.A. recovery program. 1984. New York: Alcoholics Anonymous World Services, Inc.
2. http://www.8newsnow.com/story/7196806/i-team-firefighter-paramedic-drug-abuse-raises-questions?ClientType=Printable&redirected=true
3. Calvert, P. Police officer suicide: Where are we now? Institute for Criminal Justice Education, Inc. http://www.icje.org/police_suicide.htm
4. Hopkins, J. and C. Jones. 2003, September 22. Disturbing legacy of rescues—suicide. *USA Today.* http://www.usatoday.com/news/nation/2003-09-22-legacy-usat_x.htm
5. http://www.webster-dictionary.org/definition/Railway%20spine
6. Williams, T. 2004, March. Post-traumatic stress disorder among military veterans. Disabled American Veterans National Service Program Brochure. Washington, DC.
7. Bergin, A.E. and L. Sol. 1994. *Handbook of psychotherapy and behavior change,* 4th ed. New York: John Wiley & Sons, 676.
8. Everly, G.S. and J. Mitchell. 1999. *Critical incident stress management, CISM; A new era and standard of care in crisis intervention,* 2nd ed. Ellicott City, MD: Chevron Publishing.
9. IASC Reference Group for Mental Health and Psychosocial Support in Emergency Settings. (2010). Mental Health and Psychosocial Support in Humanitarian Emergencies: What Should Humanitarian Health Actors Know? Geneva. Pg. 11.
10. Jacobs, Julie, Horne-Moyer, H. Lynn, and Jones, Rebecca A. 2004, Winter. Volume 6, Number 1. The Effectiveness of Critical Incident Stress Debriefing with Primary and Secondary Trauma Victims. *International Journal of Emergency Mental Health.* Pg. 10. Chevron Publishing Corporation.
11. http://firechaplains.org/?page_id=18
12. Litz, B.T., C.C. Engel, R. Bryant, and A. Papa. 2007, November. A randomized controlled proof-of-concept trial of an internet-based, therapist-assisted self-management treatment for posttraumatic stress disorder. *American Journal of Psychiatry* 164(11): 1676–84.
13. http://www.nacbt.org/whatiscbt.aspx

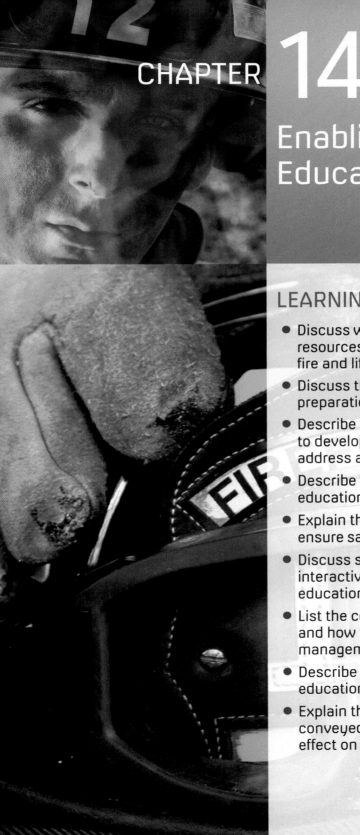

CHAPTER 14

Enabling Public Education

LEARNING OBJECTIVES

- Discuss why public education must receive more resources in order to be championed as a critical fire and life safety program.
- Discuss the difference between prevention and preparation.
- Describe the advantage of using local statistics to develop a fire and life safety program to address a need.
- Describe what resources are available for public education in relation to fire and life safety.
- Explain the purpose of using an action plan to ensure safety during equipment demonstrations.
- Discuss some of the advantages of using interactive learning in fire and life safety education programs.
- List the components of NFPA's Champion Model and how they can be applied to project management while building major programs.
- Describe the link between fire and life safety education and firefighter safety.
- Explain the three subjects that must be conveyed to the public that have a direct effect on improving firefighter safety.

Initiative 14: Public education must receive more resources and be championed as a critical fire and life safety program.

September 11, 2001, had such a profound effect on so many things relating to the fire service that our job will never be the same. Probably one of the most interesting of all is the federal dilemma of "prevention or preparation." Should we go after the terrorists where they congregate and train, or should we beef up security and plan our response? In 2005, our homeland again suffered devastation in several cities when hurricanes ravaged our southern coast. The worst of Hurricane Katrina wasn't the wind or the rain itself, but the failure of the levy system in New Orleans. Suddenly the government was again facing a federal dilemma of "prevention or preparation." Should we identify the risks to our communities and "strengthen the levies," or should we "stockpile bottled water and pumps" and plan our response?

PREVENTION VERSUS PREPARATION

Most emergency responders agree that the answer is to try to prevent something from happening *in addition to* preparing for the big one. Fire departments have used this philosophy since buckets were front-line. Most fire chiefs see that we are a significant piece to the solution of any manmade or natural disaster. We can assist in identifying risks to assist in prevention, as well as plan our response. However, we also know that not all disasters are weighted the same. We know that no terrorist attack or natural disaster in the history of the United States has come close to the fire deaths we have seen. As a result, we are as prepared as ever for the number-one killer, residential fires, with equipment only dreamed of 35 years ago. The hard question is, have our emergency calls, equipment, training, and facilities (the preparation aspect) improved since the 1970s? If so, have our fire prevention and fire and life safety education programs (the prevention aspect) improved equally?

One of the oldest traditions within the fire service is still one of the most important, yet sometimes the most overlooked (Figure 14-1). Since the first volunteer fire companies were formed in America, fire safety education has been an integral part. As buildings burned and people died, primitive statistics were kept by the departments and were used to warn the public of the dangers of unsafe acts. Fire codes were developed to help prevent future catastrophic fires. It's probably safe to say that most, if not all,

FIGURE 14-1 Many fire departments have a history of being involved in fire safety education, which can directly affect firefighter safety.

present fire codes and fire safety education presentations were a direct result of a tragedy. The fire service has the unique advantage of attracting a captive audience anywhere we go. Children look up to firefighters, and parents are proud to point us out and explain to their children what the fire department does. Many adults can't help but spend a little time looking at a fire truck or asking a firefighter questions whenever they get the chance. This desire by the public to listen is a "dream come true" to a person in marketing, and the goal of every advertiser. Even the media wants to put a microphone in front of us. As a result, we have a golden opportunity to spread our message, which historically has concentrated on fire safety.

So do prevention efforts really work? The problem is that calculating how many times something *didn't* happen is a difficult thing to do. Although we can't determine the actual number, there are several ways to get a close estimate. The first is with statistics (Initiative 7). By keeping better records and tracking changes, we can see prevention in action. The second way is by recording close calls, or events that were *almost* disastrous (Initiative 9). In previous chapters, we looked at how a **root cause** and **contributing factors** can align to allow a catastrophic event. All it takes is one component to fall out of place, and it becomes a near-miss. First responders have numerous stories of emergency calls in which a smoke detector, an exit light, a seatbelt, a belay line, or even a chin strap saved a life. There is sufficient proof that prevention does work. Initiative 14 calls for more resources to be devoted to life safety education; however, it is written in the context of determining how we can use life safety education as a tool to improve firefighter safety. In simple terms, prevent the people from being trapped in a fire and we won't need to search for them. Before we can do that, we must first look at what life safety education is and what opportunities exist.

PROGRESSIVE MESSAGE

In the past several years, many fire departments began to refer to their public education programs as fire and life safety education. Although fire safety education was still the focus of their attention, many began getting more involved in injury prevention. The reasons firefighters have been willing to expand the prevention message beyond fire safety include a diversification of duties, a history of problem solving, and a proven success rate with safety education.

Diversification of Duties

These days, most fire departments are at least partially involved in **EMS**. Some serve as first responders, treating and stabilizing victims prior to an ambulance arriving. Others provide the complete package with a system of **BLS** and **ALS** transports. As a result, firefighters are directly involved in responding to emergencies and treating patients at emergencies that don't even involve a fire. For example, your engine company might get dispatched to a bicycle versus car crash and find an 8-year-old girl unconscious. After assessing her, you find a significant wound to the back of her head and a broken left wrist. She is breathing, so you hold her neck still while another firefighter bandages the wound on her head. When the ambulance arrives, you help immobilize her and prepare her for transport to the trauma center. As the ambulance pulls away, the driver of the car asks you if the girl is going to be okay. You answer that you aren't sure, and silently wonder how severe her head injury might actually be. At that point you realize that if only she had been wearing a bike helmet, the driver never would have had to ask you that question.

Firefighting has evolved into providing whatever our citizens need when they need it. Now more than ever, we are directly involved in children who choke, middle-aged people who have heart attacks, and elderly patients who fall. We are called on to respond as professionals, and tend to follow the same approach we always have had for firefighting, one of problem solving.

History of Problem Solving

Firefighters have always been problem solvers, and many times can identify the cause of an event others might call an **accident.** For example, suppose your department responds to a structural fire in a one-story residential home. The sole occupant, an elderly female, was able to escape the home prior to your arrival. Your crew finds a working fire in a spare bedroom and extinguishes it fairly quickly. Overhaul is a little more difficult due to the large amount of newspapers and magazines stored in the room. The fire investigators find the probable cause of the fire, an overloaded and damaged extension cord under a table. Firefighters might blame the fire on the misuse of an extension cord and unsafe storage practices, but the homeowner and the insurance company might simply blame it on *electrical causes*.

This ability to look for and identify the root cause and contributing factors of an incident is a skill we have developed over hundreds of years of experience. This skill applies to other types of emergencies to which we respond as well. If the same woman fell and broke her hip, she may attribute it to clumsiness, but we might identify the cause as a loose rug. Contributing factors could include slippers with no tread and a medication that makes her dizzy if she doesn't eat first. In other words, we tend to look past the obvious, and determine how an event could have been prevented. This natural inclination to solve problems allows us to take it to the next step, attempting to prevent it from happening again.

Proven Success Rate with Safety Education

When it comes to fire safety education, we are very good at what we do. Consider for a moment the probability of how many 6-year-olds who live in your town will actually have their clothes catch on fire this year. Now estimate the percentage of the 6-year-olds who know how to Stop, Drop, and Roll if it happens. This simple but effective message can reduce the severity of burns to children as a result of safety education, and can also be effective with older adults. When we set our minds to fix a problem, we can use our popularity to promote a solution, and encourage a behavioral modification. Firefighters consistently top the polls in surveys as the most trusted profession. They trust both our character and our knowledge, which makes us *appear* to be qualified to teach safety lessons. It is true that all firefighters should learn to talk to the public in this regard at some level, but personnel in some ranks and positions should obtain the necessary credentials to ensure that the programs we develop are successful.

NFPA 1035, Standard for Professional Qualifications, for the **fire and life safety educator (FLSE), public information officer (PIO),** and **juvenile firesetter intervention specialist,** outlines three different levels of certification for the fire and life safety educator.[1] It also defines the certification for a public information officer and the two levels of certification for a juvenile firesetter intervention specialist. **NFPA** has also acknowledged the change in our progressive message and has provided the resources necessary for fire departments to take on an all-risk educational program. They already had years of proven success following the development of the **Learn Not to Burn**® educational program when they introduced injury prevention to the fire service with the **Risk Watch**® curriculum. Focusing on the top eight preventable injuries, Risk Watch® provides lesson plans and activities for fire departments to introduce in elementary schools[2] (Figure 14-2). Risk Watch® also provides educational materials for natural disasters.

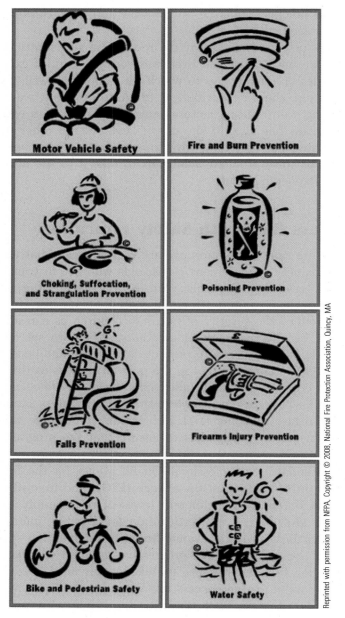

FIGURE 14-2
Risk Watch® encompasses the eight top preventable injuries for children.

DEVELOPING A PROGRAM

Many fire and life safety education reference books are available to the fire service to assist in building programs as a result of community risk analysis. When developing a program, it's important to remember that although the merits of such research are obvious when it comes to the message (what), it's important to also take into consideration the audience (who) and the atmosphere (where).

The Message

Determining what the subject of the lesson will be is very important. The amount of time we have with the public is extremely limited, so the last thing we want to do is squander the time on a subject that is not important to what we are trying to achieve. The primary goal of a fire and life safety education presentation is to convince them to change a behavior in an effort to make them safer, or teach them how to react once an emergency occurs. We can use statistics to determine the message, but "hot topics" must also be considered.

Statistics

In Chapter 7, "Creating a Research Agenda," we looked at how the utilization of research and statistics could guide us in the way we do our job. Statistics can be used in a similar way when deciding what fire and life safety education programs should be instituted. This information can be gained from a broad area, such as national or state statistics, but can also be obtained from local statistics, from the fire department, law enforcement, a hospital, or the health department.

National and State Statistics. On a national level, NFPA issues annual data on fire and life losses in *NFPA Journal®*. They report the top causes of fires and the fires that kill, with the statistics to back them up. For example, data indicate that in recent history, unattended cooking is the primary cause of residential fires.[3] Other statistics prove that smoking materials are the most common cause when fatalities occur.[4] Unattended cooking and the improper disposal of smoking materials are both subjects that are caused primarily by a human decision, and therefore can be successfully addressed with an effective educational program. Engineers might design a new stove that shuts off when a certain temperature is reached or provides its own suppression system, and safer cigarettes might self-extinguish when unattended, but the fires are still caused primarily by human behavior. As a result, your department might decide to put together an educational prevention program based on these two statistics.

Many states have developed task forces or safety coalitions that collect data from other sources, such as motor vehicle crashes or occupational injuries, which also could be helpful. Your state might be heavily involved in tracking **sudden infant death syndrome (SIDS)** statistics or infant car seat injuries. It could be that the program you are considering based on a local need already is being addressed at the state level. They may have funding and materials available, but nobody at the local level to deliver the program. Researching what is out there at this stage is vital.

Local Statistics. With the computerization of fire and EMS records, local statistics are becoming increasingly more available. Hospitals often generate local statistics in an effort to prevent injuries. Many health departments are active in gathering data and creating educational programs. Your own fire department statistics might expose potential problems. Let's assume you respond to a **carbon monoxide (CO)** alarm in a home that is only about two years old. Because it's a summer afternoon and the home has new appliances, you might expect that the reason for the alarm is a cheap detector that is reaching the end of its short lifespan. However, when you arrive on scene, your meter does detect elevated levels of CO in the basement. Although you check the appliances and other potential sources, you can't pinpoint the cause. It appears to be more concentrated at the sump pump pit, and you surmise that a gas was migrating in through a drain due to the pit's low water level. You run some water into the pit and ventilate the house. The readings drop to zero, and you clear the scene.

Over the next several months, you note that your department continues to respond to the same subdivision (and sometimes the same homes) for seemingly the same problem. Finally, one officer identifies the source in the basement as fumes emitting from the automotive battery used as a battery backup for the sump pump. When the lieutenant removed the caps on the battery, he found the battery cells dry, burning up from the trickle charger running constantly. A little chemical research showed that some CO detectors, including the fire department's, could not differentiate between CO and the sulfur dioxide emitting from an overcharged battery. Apparently the builder had not installed maintenance-free batteries in any of the homes, and any battery that was not checked and topped off with water periodically was having the same problem.

As a result, your department creates an informational flyer and distributes it to the entire subdivision. Besides offering manufacturer recommendations for using a maintenance-free battery when replaced, it also explains how to check and maintain the existing battery. It contains nonemergency contact numbers for the fire department in the event that homeowners would prefer to have someone check it for them. As quickly as they started, the CO alarms in the neighborhood subside. Tracking these statistics also proves the program efforts worked and are worth the investment of time and money.

Hot Topics

Finally, hot topics provide hints as to what information our citizens need. The media is a great source for identifying hot topics, especially after a news story exposes a potential risk. A great example of a hot topic from several years ago was **supplemental restraint systems (SRS)** in vehicles, more commonly known as airbags. When they became common on most vehicles in the late 1990s, some people were under the false impression that they took the place of seatbelts. Public education programs were initiated to explain that the first "S" in SRS stood for *supplemental*, meaning *in addition to* seatbelts. Other people didn't realize the dangers of placing a rearward-facing infant carseat behind an airbag. Many public safety organizations found themselves teaching injury prevention by addressing the hot topic. Some other hot topics included carbon monoxide detectors, bicycle helmets, and 10-year lithium batteries for smoke detectors. Departments that remain vigilant in the ever-changing needs of the public can provide the information needed to put them at ease.

The Audience

Once a subject is selected, the audience must be considered. Some factors that should be taken into consideration are the age, number, culture, and educational level of the audience. Ideally, the lesson being delivered will build on what they already know. Awareness of their present knowledge is much easier with children in school than it is with adults. This is especially true when a department visits the school each year and uses knowledge gained in the fourth-grade program next year in the fifth-grade classes. Another important note is to consider **secondary audiences.** Suppose your department is asked to teach a group of preschoolers. Remember that there will likely be parents, teachers, or camp counselors in attendance who would likely benefit from messages directed to them. Teachers and counselors should understand the basic messages and their meaning to reinforce the message later on. If parents or other caregivers are in the audience, it's important to tie in the importance of testing the smoke alarms at home and practicing a fire drill with their children.

When it comes to adults, it's helpful to consider the general background of the audience. For instance, suppose you have been asked to talk to two different groups of adults in one week, and have decided to discuss fire alarms and smoke detectors, as well as to remind them of the special hazards of winter and the upcoming holiday season. The first group is approximately 15 business owners at a chamber of commerce meeting. You could assume that they have a good idea of what businesses are in for with the holiday season, and that they would be familiar with fire alarms. You could tie fire alarms back to having a smoke detector at home, but discussions would likely include keeping store aisles and exits clear. The second group might be for approximately 60 senior citizens in a *senior living* apartment. You could assume that very little of them know what to do when the fire alarm in the apartment building sounds, and that most of the discussions would include how to keep their own apartment safe. Discussing portable heaters, cooking techniques, and the proper use of extension cords might top the list. Regardless of the audience, we must remember two important points when preparing our message to a specific audience. The message must be in their own language and address their own concerns.

Their Own Language

Besides the obvious benefit of not speaking in a foreign tongue, it's important that they comprehend what you are saying. Speaking to young children is much different than talking to senior citizens. Children respond best to someone speaking at their level. Vocal changes and facial expressions can make the difference between your audience getting or missing the message. It may be necessary to keep your distance with very small children in order to not frighten them. This is where a firefighter in protective gear can either help or hinder your program. Older children and adults usually enjoy a little humor in the presentation. It keeps their attention and can tend to drive important points home. Avoid jargon unless the audience understands it. Seniors sometimes prefer a more formal presentation with a slower pace. Subtle humor is also effective, but always be on the alert to prevent talking down to a group.

Their Own Concerns

Equally important, the discussion must be related to their own concerns. Take the time to break down the lessons you want to teach to those of the same age. For example, with a stock subject such as smoke detectors, we can lay out a plan to address the concerns of each age group. Preschoolers need to know the noise that smoke detectors make, and what to do if they hear it. First-graders should understand that detectors have batteries like their toys, and that they too lose charge and they need to be changed. Third-graders should know the value of an escape plan and the procedure for testing the alarms. Fifth-graders should know the value of not disarming a detector for any reason, the danger of false alarms, and taking all alarms seriously in public places. Adults must also hear of the dangers of removing a battery, and possible solutions for common false alarms, such as relocating the detector. They also need to be reminded to take the time to test and change the batteries, and that hard-wired detectors sometimes have batteries as a backup. Seniors must know safe ways to test and change their batteries without falling. It is also helpful to have the phone number of an agency that can help with battery changes if they are unable to do it themselves.

The Atmosphere

A common mistake made in delivering fire and life safety education programs is the belief that "one presentation fits all atmospheres." For instance, a table set up with smoke detectors and a **CPR** mannequin might be effective at a health fair in a church community room, but not a wise application of resources at a community festival or county fair. Trying to demonstrate CPR over the screams from the tilt-a-whirl or the music of a band is fruitless. The audience is there for fun, and a table with a fire extinguisher holding down some pamphlets on carbon monoxide cannot compete with carnival rides, games, and elephant ears. However, altering your efforts to fit the atmosphere can make a significant difference. To choose an appropriate program, first decide what category of atmosphere you will be presenting (or performing) it in. The most common include educational, workplace, community group, recreational, and media events.

Educational Facilities

Educational presentations can be delivered at preschools or childcare centers, elementary, secondary, or high schools, and sometimes even college. The class is usually in a controlled location, with attentive students, and calls for a straight forward presentation with limited visual aids. Visual aids are a vital part of all presentations, but you must be careful not to overwhelm or overshadow your message in an educational situation. Videos can be successful in the classroom, but if they don't teach what you want in the limited time you have, consider eliminating them. Large classes generally restrict hands-on activities. The lecture or demonstration should be outlined in advance and rehearsed to fit perfectly in the allotted time. When a schoolteacher tells you that you have 18 minutes, she means 18 minutes. Sometimes prepared games are effective as a recap or posttest to drive home important lessons. These games can be purchased commercially or can be homemade with a little creativity. Finally, remember the techniques of your favorite schoolteacher, who probably showed enthusiasm, interest, humor, and concern.

Workplace Locations

One of the most important components of a successful public education program is to teach adults in the workplace. Sometimes this is the only location you can ever speak directly to them. These can be manufacturing, office, retail, medical, or even food service establishments. Because adults are not typically interested in hearing a lecture about fire safety, the workplace is sometimes the only place where they will listen. Be prepared for some interested attendees, as well as some disgruntled workers. A good speaker can reach some of the most irritable employees when they are obviously there because they have to be. Usually the company will invite you for a specific topic, but take the opportunity to bring up some of the subjects that affect them as a result of your risk analysis. Say, for instance, that you are called to speak about the dangers of radon in the home. Rather than decline based on the relative insignificance of radon, instead research radon and maybe even buy a cheap home test kit. Explain that radon is not much of a threat to them, and devote some time talking about things that do make a difference, such as smoke alarms and escape plans. Take time to check out their evacuation and emergency plans in advance and reiterate them to the employees. Note the specific hazards you notice in their building and on their grounds, offering possible solutions or suggestions in the event of an emergency.

NFPA Journal® magazine has a section with news briefs on recent fires broken down by occupancy. Cut out articles that are similar to some of your buildings and file them in the fire prevention file for the local business. If you get called to teach fire safety to employees at Walmart®, it's a great idea to hold up an article about a fire at a store in another state. If the employees came into the class thinking they could never have a fire in the store, they left with a different opinion. Remember that adults are generally interested in fires and current events. If you can, relate a story about an emergency or a fire in a similar building and talk about how it could or couldn't happen there. It will help tie in the practicality of the issues being discussed, which, in turn, keeps their attention longer and helps them retain the lessons.

Community Groups

Churches, scouts, rotaries, business associations, senior citizen associations, and coaches are all examples of community groups that make up common audiences. It is essential to tie in your topic with how it applies to them. Coaches should learn first aid and CPR, along with the possible pitfalls of cellular 911. Business associations should hear what your fire department has to offer, and senior citizens should understand the accidents and injuries that they are most at risk for, and how to prevent them. Keep a record of leaders or contacts in the groups. You may need to contact them later for donations, or for something as simple as providing victims for a mock crash or mass casualty drill. Find out as much as possible about the group before your meeting. If you are invited to address the local Shriners® chapter, it is helpful to know in advance that they are one of the nation's leading burn prevention and care organizations for burned children. What a way to open a safety talk. First, thank them for their own work in the community, and then tie in your desire to eliminate burn victims as well. You will have their undivided attention and respect.

Recreational Locations

It's important to realize that a very creative approach in planning is necessary to ensure recreational opportunities will be a success. County fairs, community picnics, mall events, and even day camps need to have a unique program. The size and location of the event will help you establish what will work and what won't. Some recreational events involve small groups indoors and others are large groups held outdoors.

Small Groups Indoors. The month of October is the official month of fire prevention, but in some parts of the country it can already be too cold for outdoor recreational activities. Additionally, we have to share the month with Halloween. Rather than compete with it, a little imagination might allow you to team up with the holiday. NFPA announces a theme each year for **Fire Prevention Week (FPW).** Let's suppose they come up with "Hunt for Holiday Fire Hazards in Your Home." You decide a scavenger hunt of sorts would be a great way to tie the message in with a fun activity and Halloween.

You know that the historical society of your town owns an old mansion that it sometimes uses for community events, and you partner with them to create your own *haunted house* filled with scary fire hazards. It would be a night of fun for families in which, as a group, they use a **thermal imaging camera (TIC)** to move through the dark house and find hidden heat sources. They travel room to room with a firefighter escort and find hot extension cords, fireplaces, and space heaters. They also discover simulated hot electrical outlets, cigarettes, lighters, and matches. They learn that dark images are cold, and are taught to locate cold things that can help, such as smoke detectors, fire extinguishers, and exits. Many objects can be simulated using ordinary items in conjunction with dry ice and heat tape. For example, you figure out that if you attach a small plastic bowl filled with ice to the ceiling, it looks just like a cold smoke detector on the TIC. After their search, the families enjoy cider and doughnuts while they watch a short safety video about what they learned.

Large Groups Outdoors. The most common type of recreational setting is a large outdoor festival, community picnic, or fair. Some of the limitations include ensuring personnel and citizen safety, providing adequate electrical needs, and providing an effective sound system. Whenever possible, plan ahead and get your events on the entertainment schedule with the civic committee. This will bring you a larger audience, and prevent you from competing directly with other events. When planning the event, remember that the most important component is the lesson, not the entertainment (Figure 14-3). Once these items are addressed, a little ingenuity can provide the resources for equipment demonstrations, clowns, and even interactive learning.

- Fire Equipment Demonstrations
    - An obvious way to catch the attention of the public is by conducting equipment demonstrations. Your department owns some unique and fascinating tools, but safety must always be the first priority. Many citizens know what the "Jaws of Life" are, but how many actually have watched them cut a car apart in less

FIGURE 14-3
Planning for large events must be based on the educational objectives.

than 15 minutes? How many have seen your new aerial truck being washed in front of the fire station or respond to alarms, but never saw it rise 75 feet in the air and spray 1500 gallons a minute?
- Demonstrations must be planned out well in advance. Say your department decides to set up the tower ladder and spray water at a target in an open field for the crowd. Planning and conducting the demonstration should follow some of the same safety components utilized at emergency incidents. These would all be components of an action plan that should be prepared. The plan would describe what events will occur in a specific time period, and would likely include the goals and objectives of the demonstration.
- Build a lesson plan.
  - Develop a lesson plan to ensure that key messages are relayed.
  - Identify goals that follow the lesson plan.
  - Create objectives to work towards the goals.
- Obtain permission and cooperation.
  - Have the police department assist with crowd control.
  - Have fire dispatch monitor the radio frequency.
  - Have the water department allow the use of the hydrant.
- Maintain command and control.
  - Have an incident commander oversee the entire operation.
  - Utilize a safety officer who will be able to stop the demonstration at any point.
  - Secure a tactical or special radio channel for all operations.
- Ensure a safe setup.
  - Maintain continuous scene control.
  - Stabilize the aerial truck in a safe location 30 minutes prior to the demonstration.
  - Practice the demonstration several times, days in advance.
  - Monitor weather conditions, including wind and temperature.
- Provide an announcer to ensure education of the crowd.
  - Greet the audience and encourage more people to come watch.
  - Discuss the training that emergency vehicle operators receive and what to do as an emergency vehicle approaches.
  - Explain the different types of apparatus that respond, and the specific purpose of each.
  - Describe and demonstrate the functions of an engine company and a truck company, as well as their duties.

- Explain that fire spreads so quickly that a working smoke alarm and an automatic sprinkler system are the best friends you can have if you have a fire at home.
- List the extinguishing capabilities of fire department equipment compared to that of a pan of water or a garden hose. Explain how safety must be their first priority.
- Describe the safety equipment worn by the firefighters and how it compares to ordinary clothing. Explain the limitations of our PPE.
■ Conduct an exciting demonstration.
  - Provide a quick air horn blast from the engine company at a safe distance from the crowd.
  - Raise and extend the ladder to 70' at 70 degrees toward a target with the lights on.
  - Have the engine company forward lay from the hydrant with lights but no siren.
  - Have firefighters flush the hydrant, connect supply line, and charge it. Flow water to the aerial master stream where it is directed at the target.
  - Have the target hit with a straight stream and get knocked off its stand.
■ Give closing remarks.
  - Thank the public for their support of the fire department.
  - Explain that we hope this is the only time they ever see us flow water.
  - Welcome them to talk to the firefighters and see the equipment once the hose has been reloaded and the ladder has been stowed or maybe even visit a fire station.

■ Clowns
  - Some fire departments have turned to clowns to spread their message in an entertaining show. Magic and jokes are often involved, and are scripted around specific fire safety messages. A successful fire clown will understand that the safety message is the primary goal, and that the antics should complement, not detract from, the message.

■ Interactive learning
  - Some of the most promising forms of life safety education come in the form of interactive learning. If you've ever been to any of the Disney® theme parks, you probably walked away from it in awe. They combine imagination and engineering into what they refer to as Imagineering.[5] By mixing sights, sounds, and movement, they create an atmosphere that erases the line between real and imagination. They can make you fly, make you swim, and take you to a place that doesn't exist, all while tricking you into thinking that it's really happening.

The fire service is a full 30 years behind. It's hard to compete with the technology our children are growing up with. We simply cannot hold up a smoke detector or show a doll house with a pretend hot stove to a group of fifth-graders and impress them. We cannot scare them into practicing a fire drill at home by telling them a story. This generation is used to instant gratification, realism, and multimillion-dollar video productions. A stand-alone puppet show simply cannot compete.

- The answer is interactive learning. We need to transport them into a world where they want to be and from which they can still learn. Fire safety trailers and fire sprinkler trailers are a good start, but even those can be improved. This need might best be addressed through the use of **animatronics.** Sometimes referred to as *robots*, the term comes from mixing animation and electronics to create a realistic creature or experience. Animatronics have been introduced to teach small children in the form of fire trucks and smoke alarms, but we need to expand their use to adolescents. For instance, animatronics could be used to create simulators incorporated into fire safety classes for babysitters. This takes a serious change to the way that we think and the programs that we run.

- Take, for example, the fire extinguisher simulators that are now becoming more common. Much like an interactive video game, an extinguisher is aimed at the video screen of a simulated fire and extinguishes (or spreads) the fire based on the technique of the student. Some educators argue that teenagers should not be taught how to use a fire extinguisher, but instead should be taught to exit. The problem is that almost any 13-year-old boy who causes a fire in the kitchen while cooking macaroni and cheese after school will try to extinguish it. It's far better to teach him when to fight and when to escape than to tell him nothing. We need to mix entertainment and learning into one package that keeps their attention and helps change their behavior.

- Animatronics are now affordable and are becoming more common. A simple robot movement is nothing more than the result of a servo, or small motor. A servo can be controlled by radio control or by robotic software on a computer (Figure 14-4). Radio control allows flexibility in the program, but requires someone to run it. Software allows it to run on preset scripts, reducing a need for extra personnel. This opens up a world of possibilities. Suddenly a retired fire truck pump panel can come to life, allowing children to learn how to run a pump. Instead of brochures on a table, computers with touch screens can show pictures of medications and candy, allowing children to test their ability to tell the difference. You *can* compete with the festival and entertainment, if you level the playing field.

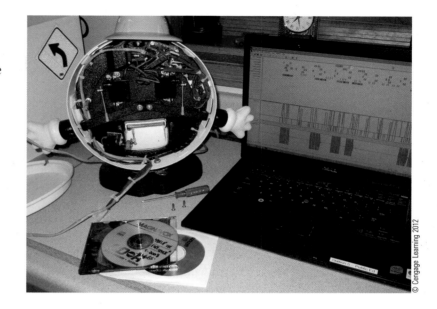

FIGURE 14-4
Animatronics can be controlled by readily available and inexpensive software.

Media

The media is an audience that needs special considerations. It's no secret that they love hard news. They want the specifics on the fire last night, but don't necessarily care about what you're doing for Fire Prevention Week. Train your department's officers to choose their quotes wisely after an incident. House fires must report detector function and outcome, and car crashes should always report the difference that seatbelts made or didn't make. Another approach is to follow up hard news with another press release a day or two later. For example, let's say a fatal fire occurred in a house that was lacking a smoke alarm. Several days after the fire, meet with the investigator to discuss new information about the cause or the factors leading up to the fire. Call a press conference or send out a press release, giving new facts and public safety information weaved throughout. If the safety tips are brief and included in the press release, they'll run it. If it gives new information about the fire, people will read it.

Press releases for new equipment are not very exciting on their own, but give it a twist and you'll get more press than you bargained for. Imagine that several years from now your department is one of the first to receive a new type of firefighting equipment that allows the tracking of firefighters in 3D from outside the structure. Rather than simply submit a press release and some pictures of a firefighter holding the unit in the station, *really* educate them, which, in turn, will educate the public. Suppose you devise a plan to use an abandoned home that the training division is in the process of readying for fire training. After inspecting the house and ensuring it is safe, you put in some furniture to make it realistic and charge it with some theatrical smoke. You invite the media at staggered times to come use the equipment themselves. After demonstrating the equipment, you provide them with whatever firefighting gear or equipment they need to allow them to tell the story to the public how they want. They can wear the tracking equipment themselves, and search a room with firefighters while the camera compares smoke conditions with a monitor in the command vehicle. Ensure that throughout their taping, they add important safety information for the public.

PARTNERSHIPS

After you walk out of the planning stage with a blueprint of a great program, you may be stuck with one big question: Now how am I going to do this? The answer is, with help. There are three specific ways that partnerships can be utilized to ensure your project will be a success, including project management, funding, and personnel.

Project Management

Managing a fire and life safety project of any magnitude takes good management skills. A great example is NFPA's Risk Watch® curriculum, but the amount of time and resources needed to provide the education is daunting for one or two firefighters. Suppose your fire department wanted to focus on water safety for children after some adolescents drowned without wearing lifejackets. NFPA suggests the **Champion Model** as a way of project management. It consists of a coalition, a champion, careful planning, a compelling case, credentials, collaboration, continuity, creativity, camaraderie, and commitment.[6]

Coalition

The organization of a coalition is a great way to link resources from other groups for an effective program. If your department has a swift-water rescue team, they would be a great place to start for building a coalition. You also find that your state has a watercraft division that teaches water safety. Besides having experience, they also have grants available for educational projects. Having a schoolteacher involved also ensures that the educational components will be complete. As word spreads, other interested people join your coalition.

Champion

Every project needs to have one champion, or leader, who is responsible for overseeing the coalition. Some other responsibilities might include calling the meetings, leading the planning discussions, obtaining funding, and evaluating the progress. For our water safety project, you are the most likely candidate for the position.

Careful Planning

As champion, you have the responsibility for carrying the project through completion. A strategic plan should be developed that establishes the goals. Remember that goals are measurable and attainable, and utilize objectives and a timeline to complete. At your first coalition meeting, you establish two important goals to get started.

1. Create a plan for an interactive learning environment to teach water safety.
   - The lesson must be realistic and make the students think they are involved.
   - The lesson must be portable to be used at different schools.
   - The lesson must teach several components of water safety.
   - The lesson must include the use of lifejackets.
2. Provide the agenda for a follow-up meeting.
   - The meeting will be held in 30 days.
   - Research what other programs are out there.
   - Each member is required to return with five ideas, even if they aren't complete.
   - Brainstorming will take place, and follow-up meetings, goals, and objectives will be established.

Compelling Case

You had originally based your decision to promote water safety due to an incident that occurred in your town. It's important to also look at statistics and other data to strengthen your cause. You may find that another county had a similar occurrence several years ago, and that they may be interested in joining your coalition. It could be that members of the yacht club where the children drowned are interested in helping with the project. Motivated individuals can assist you in pushing the project through and helping with any hurdles that appear in your way.

When you approach them, it turns out the yacht club had been considering some way to help and are very excited about the project. You share your goals with them and they toss out the idea of building a boating simulator that can be towed around the county. They suggest that they could help with some funding, including the donation of an old boat. Additionally, the club owns a truck that could tow it to schools or festivals. You are excited about their enthusiasm and take the idea back to the coalition's next meeting.

Credentials

It's vital that your program use educational materials that have been proven. You review the Risk Watch® module for water safety with the members at the next meeting and explain the idea of the boating simulator with the coalition. Your group decides that any interactive lessons must match up to the classroom component in the water safety module. A boating simulator is nothing without the educational component to back it up.

Collaboration

After the meeting, the school teacher arranges a meeting among you, other elementary teachers in the school system, and the administrator in charge of the curriculum. You discuss the project and its benefits with the educators and they agree to participate in the program. The success of Risk Watch® requires close collaboration between the advocates of your coalition and classroom teachers.

Continuity

Continuity is important in several different aspects of safety education. Children aren't necessarily going to learn a subject and change their behavior after one lesson. Repetition is needed over several years to ensure that the message is ingrained in their thought process. In other words, just because they put a lifejacket on once in the third grade does not mean they will use one several years later when the opportunity presents itself. It must become so natural that they couldn't imagine going on a boat without one.

Another aspect of continuity is that the lesson must be the same for all students. Your coalition recognizes this and applies it to the boat simulator idea. Say the students come out to the school parking lot and climb aboard an open boat with a lifejacket on. First of all, we are depending on good weather and no distractions in the parking lot. Busses pulling in or recess occurring on the playground is sure to detract from the lesson. Secondly, we are relying on an instructor to lead the discussions, which surely will vary from class to class and instructor to instructor.

Creativity

Whatever the program, creativity must be part of the process. The yacht club made a novel suggestion, but continuity and some logistical problems have made it problematic. Brainstorming allows your coalition to solve some of the problems associated with the plan. One of your members suggests an enclosed trailer, much like a fire safety trailer. The trailer could look and feel like a boat on the inside, but would keep students out of the weather. Another suggests a movie theater inside, which depicts safe and unsafe actions near and on the water. Suddenly a good idea turns into a great one, and just keeps getting better. Your group decides to look at the yacht club donation boat, and come up with a plan to cut the hull and put it in an enclosed trailer. One member offers to research some small hydraulic cylinders to make the boat rock with the waves. Another puts together a plan to shoot video from a boat to play in the theater. Creativity is the difference between something kids forget and something they remember.

Camaraderie

The team you have assembled is vital to the success of your project. Delegation not only gets specific jobs accomplished in a timely fashion, but also keeps all team members involved. As funding becomes available and the design of your water safety education unit starts to fall into place, it's important to keep it fun and interesting for everyone involved. As Marsha Giesler states in her book, *Fire and Life Safety Educator*, a group such as this will "allow members to be part of something bigger than themselves."[7] An efficient team realizes this and remains committed to the project.

FIGURE 14-5 By following the Champion Model, projects that seem out of reach become possible.

Commitment

When the project first begins, everyone is excited, with their eyes on the goal. As construction begins and problems arise, it's easy for the team to get discouraged or even start to turn on each other. Anticipate struggles, but keep the momentum up and the finished project in mind. When the project is complete, the water safety lesson your group created is a four-year educational program in the elementary schools that teaches the Risk Watch curriculum. Starting in kindergarten, each year adds progressive lessons that reiterate what they learned in the previous year and adds new material. The students finally "graduate" from the program in the third grade, and celebrate with a "boat ride" in the school parking lot (Figure 14-5).

Funding

In Chapter 10, "Blending Grants and Safety," grants were exposed as a source of funding. Some grants give preference to educational programs that show promise in prevention-themed activities. It's important to note that most funding agencies will not fund a stand-alone "prop" without an associated lesson. In other words, funding agencies would not consider buying you a boat and an enclosed trailer unless there was a comprehensive educational component. The investment must support the project objectives and a lesson plan. By convincing them that the prop is simply a platform to teach the lesson, it is much more likely to be funded. In the case of our water safety education unit, one of your members obtains a grant for the enclosed trailer and a generator to provide power. The yacht club agrees to pick up the expenses of outfitting the boat inside the trailer and making it look real. They also agree to tow the trailer, and store it inside a building when it is not being used. The local electrician's union agrees to complete all the electrical work at no charge, and a body shop offers to paint the trailer graphics. One of your coalition members devises a plan to raise money by allowing individuals to sponsor the lifejacket on the boat for $100 each. Creativity with any project involves fresh ideas and approaches to teaching the lessons, constructing props, and even raising money.

Personnel

Projects of this magnitude also call for time. Besides the time commitment to plan and build the water safety trailer, teaching water safety lessons will continue to tax the volunteers. Besides members of your coalition and their agencies, additional personnel can sometimes be gained from Explorer or cadet programs, or community emergency response teams.

Explorer or Cadet Programs

Many fire departments have a cadet program that allows teenagers with an interest in emergency services to learn about firefighting and EMS. The National Exploring Division of Learning for Life®, or **Explorers,** is a national association of "posts" around the country with a goal of teaching young men and women about careers in emergency services. Some departments use the programs as a type of *farm league* to recruit and train future members. Another benefit is the help they can provide at fire department functions or educational programs. Having teenagers help teach younger children can have added benefits. If third-graders see 15-year-olds wearing lifejackets during the program, they will have little resistance to wearing them as well.

Community or Citizen Emergency Response Teams

Another possibility for personnel is the use of a **community emergency response team (CERT).** The idea is simple. Train and utilize interested and talented members of your community to assist the fire department in large-scale emergencies. In the event of a community-wide emergency, these people can assist emergency crews in their duties. As you know, there are far more small emergencies in our towns than large ones. Although the public is being trained for the big one, their training will be used at the soccer field, the Cub Scout camping trip, and the workplace. Many CERTs are used for natural disasters, especially in areas prone to specific disasters such as earthquakes and hurricanes, but they are effective in any area and can be catered to local needs.

FEMA lists more than 300 CERT programs on its website.[8] Most volunteers are trained to the awareness level in recognizing and identifying potential problems, removing potential or actual victims, making the appropriate notifications, and securing the scene. Averaging approximately 20 hours, the students are taught other skills such as first aid, fire extinguisher use, and an introduction to the **National Incident Management System (NIMS)**. More advanced classes could teach triage, evacuation plans, logistics dispersal, and shelter management. Labor, professional, and other citizen groups could be trained to plug even larger "holes" in a major disaster response. It doesn't take much of an imagination to realize how beneficial it would be to contact the local electricians union, clergy association, or ham radio group if needed, *and* have them trained in NIMS. As a supplemental support staff for fire and life safety education, CERT members can make your program a reality.

LIFE SAFETY EDUCATION FOR FIREFIGHTER SAFETY

The first step in utilizing fire and life safety education as a benefit for firefighter safety is to realize that the principle of prevention by education works. If we believe that an educational program is effective in reducing the number of injuries and deaths of civilians along with the number and severity of fires, it can be assumed that the reduction of our injuries and deaths will also be reduced. Therefore, every official fire and life safety education program and impromptu public interaction must be devoted to dispersing information that is vital to reducing our injuries and deaths. Firefighters need to learn how to seize the opportunity and teach the public what we need them to know to keep themselves safe, which, in turn, keeps us safe.

Let's assume you are invited to teach a fire extinguisher class at a machine shop. Although most of your students are over the age of 40 and are generally unhappy to be there, you do your best to get them to listen and learn something. You finish the classroom portion of the class and decide to take a couple minutes to teach them some safety lessons for home that will improve their safety, as well as that of the responding firefighters. They need to know how to prevent fires and injuries, how and when to call for help, along with what specific actions to take and not to take prior to our arrival.

Preventing Emergencies

Teaching people how to prevent fires will reduce the number of times we will respond to the most deadly of fires. For example, we understand how important smoke detectors are to the early recognition of a fire in a home. Although they are prevalent in homes, we still see fatalities because a detector was disabled or had a dead battery. Suppose one of the workers at the machine shop asked you whether he should jump from a second-story window or wait to be rescued. The best answer might be that the early detection of a fire will likely not trap you upstairs at all. Having working smoke alarms and practicing an escape plan allows you to exit *before* being trapped. If your department is utilizing a proper risk management plan, a house fire with all occupants accounted for in the driveway will call for a much safer response than one with occupants trapped upstairs.

Prevention must also include information about how to prevent injuries. Take fall prevention for senior citizens as an example. If an elderly couple learns how to remove some of the trip hazards from their home, they may never fall and therefore won't need us to pick them up. How many firefighters and EMS workers have received a back, shoulder, knee, elbow, or ankle injury carrying a patient? Have any rescuers ever drowned because a driver decided to drive across a water-covered roadway and got swept away? It's a pretty simple concept; educating children about the dangers of going on the ice prevents us from having to go get them when they get stranded or fall through.

For your machine shop audience, you take a couple minutes to talk about some of the fire hazards in their own garage. You assume that because they use tools all day, they probably own quite a few themselves and spend a significant portion of their free time working in the garage. Using fire inspections as an example, you draw some parallels from what the fire inspectors are looking for in the machine shop to similar hazards, such as extension cords and flammable liquids, in their garage. By tying in a fire prevention message to things that are important to them, the message is better received.

Reporting Emergencies

When people witness an emergency or are a part of one, they must be trained to recognize it as an emergency and contact us as soon as possible. Earlier alerting of an emergency allows us to respond quicker and catch the incident at a more manageable size without putting ourselves in as much danger. The special considerations of cellular phones should be discussed, as well as the emergency procedures when calling 911 from a company phone system.

Although the class at the machine shop went well, you end up meeting one of the men, Larry, again the next year. Larry was at home attempting to melt an ice dam on his roof with a small torch. Although he realized there was a possibility of catching his house on fire, he also knew that his roof was leaking pretty badly and that if he couldn't clear the ice, the entire ceiling might have collapsed. He was proceeding carefully and was satisfied with his progress, but eventually smelled smoke. He looked to his left and saw small wisps of white smoke coming from behind the gutter. If he called 911 immediately, you might have arrived in time to stop the fire at its incipient stage, with nobody getting hurt. However, he instead tried to extinguish it himself. By the time he returned from the garage with a pry bar to pull off the fascia, he found black smoke and a flicker of flames coming from under the first course of shingles. He returned to the garage to grab the garden hose, but found it frozen. He unhooked the hose and filled a 5-gallon bucket with water. As he returned to the ladder, he found that the flames were now several feet high and brown smoke was emitting from the ridge vent at the top of the roof.

Generally speaking, the public is not aware of how fast fire spreads or how manageable scenes can quickly deteriorate. The only way for them to understand is for us to convince them. Their experience with fire is in a controlled situation, such as a fireplace, a candle, or a campfire where the fire cannot spread. When a small fire is noticed in a three-dimensional void space, like the area behind a gutter, most people have no idea what can happen. We know that when the fire hits that open attic space between the rafters, it's following it all the way up to the peak. In fire extinguisher classes, we teach the acronym **RACE: Rescue, Alarm, Confine, Extinguish.** The first steps are always to get people out of harm's way and call for help. If they fail at extinguishing, at least backup is on the way, the building is being evacuated, and the fire may be confined.

Educational programs must be able to convey the speed at which fire travels and convince the audience that they need to call for help early. Video footage or computer-animated simulations that depict real time and fire conditions are effective ways to demonstrate it. We may never prevent Larry from making the poor choice of melting ice off his roof with a blowtorch, but we may convince him to call 911 if it does catch on fire. By convincing him that he probably won't receive a bill for our response or a ticket for his indiscretion, we can have him concentrate instead on what he should do *after* he calls us for help.

First Responder Actions

Many times the actions of the occupants prior to our arrival are what really put us in the most danger. If Larry decides to pull down the stairway and enter the attic to extinguish the fire, we find ourselves in a far more precarious position. He could easily become trapped in the attic or be overcome by smoke. A rapidly spreading attic fire is nowhere for a homeowner to be incapacitated. Educational programs should outline the actions that could be completed prior to our arrival. Simply telling Larry to stay put after calling 911 will probably not work. He's a person who is used to fixing things and he's not likely to stand at the road and watch his house burn down. An effective training will teach him some of the actions that might be helpful and some that should be discouraged.

For example, say you are talking to a group of children about evacuating a house fire and getting everyone in the family to a meeting place. One of them asks what to do if the dog is still in the house. It's easy to tell them that they should do nothing and wait for the firefighters to arrive. Would you do nothing and wait for help? It's much better to give them something safer to do that could help their dog rather than have them decide to reenter the structure. History has shown that when citizens reentered a building on fire, many had to rely on firefighters to bring them out. Instead, they can stand in front of the house by an open door or window and call for their dog. They can clap or whistle or make noises that the dog understands. You can explain that their dog is probably scared, and may be hiding so the more noise the better. It's important for them to tell the firefighters that the dog is still inside, as well as where the dog likes to hide. All these things allow the child to be active in the dog's rescue, but to significantly reduce the chance of the child (or the rescuer) being injured.

Other information citizens should know while waiting for emergency responders could be to maintain a safe scene, provide basic first aid or CPR training, and help emergency vehicles find the address or the incident. Remember that most people want to help with an emergency, so it's far better to assign them tasks that will help the situation rather than make it worse.

SUMMARY

Sometimes it's difficult to prove that prevention works. It's hard to say how many children *didn't* play with a lighter because of a fire safety lesson. There are, however, numerous success stories out there. By tracking these success stories, we can get a better idea of how effective fire and life safety education is. Whether your department has a designated fire and life safety educator, the fire chief makes an occasional press release, or you are asked a fire safety question by your next-door neighbor, it's vital that all emergency responders take the opportunity to spread accurate knowledge. Many of us took an oath to protect life and property when we joined the fire department, but nowhere does it say that it has to be done exclusively with turnout gear and a hoseline. We can also meet our obligation with education.

But there's more to it than that. In a sort of self-serving way, the next fire we prevent or reduce in severity might actually save our own life. There is a direct relationship among fires, how the public reacts, and our safety. By teaching the citizens how to be fire safe, how to react to emergencies, and how to call us, we actually protect ourselves. Fire and injury prevention will never put us out of business. Society continues to amaze us with new and interesting ways to destroy their property and injure themselves. Rest assured we will be there to answer the call and try to prevent the next one.

KEY TERMS

accident - An event that is not predictable, preventable, or avoidable.

ALS - Advanced life support.

animatronics - A term used to describe the combination of animation and electronics to create a robot or other lifelike movement; a form of interactive learning that is extremely effective.

BLS - Basic life support.

CERT - Community or citizen emergency response team.

Champion Model - Developed by the NFPA, this guideline for project management assists in the planning, development, and implementation of a fire and life safety program.

CO - Carbon monoxide.

contributing factor - A tertiary component of an event that may have encouraged an event to occur or worsened the outcome.

CPR - Cardiopulmonary resuscitation.

EMS - Emergency medical services.

Explorers - A national cadet program that allows teenagers to explore a career in emergency services.

FEMA - Federal Emergency Management Agency.

FLSE - Fire and life safety educator.

FPW - Fire Prevention Week.

juvenile firesetter intervention specialist - An individual trained to interact with juveniles who have a history of setting fires.

Learn Not to Burn® - A fire safety education program developed by the NFPA in 1979, and is still used today in many areas.

NFPA - National Fire Protection Association.

NIMS - National Incident Management System.

public information officer (PIO) - An individual designated to be a liaison with the media, usually at emergency incidents.

RACE: Rescue, Alarm, Confine, Extinguish - An acronym taught during fire extinguisher classes which provides the student with four vital actions required when discovering a fire.

Risk Watch - An all-hazard fire and life safety education program created by the NFPA for children.

root cause - The primary cause of an event. Without it, the event would likely have not occurred.

secondary audience - The term used to describe a group of people who, although not a primary target of a fire and life safety education program, are likely to be in attendance. School teachers would be an example while addressing students.

SIDS - Sudden infant death syndrome.

SRS - Supplemental restraint system.

TIC - Thermal imaging camera.

REVIEW QUESTIONS

1. Using a house fire as an example, how can both prevention and preparation save lives?
2. How can local statistics uncover a need for a specific fire and life safety program?
3. Why is it so important to develop an action plan for an equipment demonstration?
4. What is the benefit of using interactive learning in fire and life safety education programs?
5. What are the three subjects that must be conveyed to the public that have a direct effect on improving firefighter safety?

NOTES

1. NFPA 1035: Standard for Professional Qualifications for Fire and Life Safety Educator, Public Information Officer, and Juvenile Firesetter Intervention Specialist, 2010 Edition. National Fire Protection Association.
2. http://www.nfpa.org/categoryList.asp?categoryID=1050&;URL=Safety%20Information/For%20public%20educators/Education%20programs/Risk%20Watch®
3. An Overview of the U.S. Fire Problem. National Fire Protection Association. http://www.nfpa.org/assets/files/PDF/Research/Fire_overview_2009.pdf
4. Selected Published Smoking-Material Fire Incidents. One-Stop Data Shop; Fire Analysis and Research Division, National Fire Protection Association, October 2008.
5. Disney Imagineering. http://corporate.disney.go.com/careers/who_imagineering.html
6. Risk Watch Presentation Script, NFPA.
7. Geisler, M. 2011. *Fire and life safety educator*. New York: Delmar Cengage.
8. http://www.fema.gov/news/newsrelease.fema?id=3131

CHAPTER 15

Advocating Residential Fire Sprinklers

LEARNING OBJECTIVES

- Discuss the importance of code enforcement and home fire sprinklers.
- Cite some examples of how NFPA 13D makes it easy to comply with.
- Discuss some of the arguments against residential sprinklers.
- Explain how most arguments against residential sprinklers are flawed or misleading.
- List some engineering obstacles that must be addressed with a fire protection system installed in a home without access to municipal water.
- Develop logical arguments for residential sprinklers using statistics. Describe the code enforcement duties that directly affect firefighter safety.
- Explain how fire and life safety education can promote the message and importance of residential fire sprinklers.
- Discuss the advantages of partial systems in addressing immediate safety concerns.

Initiative 15: Advocacy must be strengthened for the enforcement of codes and the installation of home fire sprinklers.

It's no secret that a change for safety takes time. Aircraft used seatbelts as early as 1930. The first seatbelt in a vehicle was offered in a 1956 Ford, but they weren't standard until 1968.[1] Similarly, the first airbag for a passenger car was installed in 1971, but they weren't required for another 27 years. So the big question is why was the first sprinkler code adopted in 1896 by what is now the **NFPA** and we still are wrestling with its installation in homes today? In his book *Alpha to Omega, The Evolution in Residential Fire Protection*, Chief Ronnie Coleman explains that a combination of factors has been responsible throughout history, including:

- A fire service that wasn't convinced of the need for sprinklers
- A building community that wasn't ready to embrace sprinkler systems
- An insurance industry that wasn't sure it would reduce risk
- Sprinkler manufacturers that didn't aggressively pursue a residential market

Although the list could be used to describe the attitudes toward residential fire sprinklers in 1939, they are still prevalent today. We are much closer to making them standard in all homes, but resistance is still the general rule.

Initiative 15 addresses both fire prevention code enforcement and promoting residential fire sprinklers. Recent changes approved to the national model building code are sure to bring the adoption of more state residential sprinkler mandates, but it does little for existing homes. Many states are involved in disputes to try to reverse the decision. Therefore, we should work diligently to find ways to get fire suppression systems in all buildings, whether they are required or not.

RESIDENTIAL SPRINKLERS

At face value, the use of home fire sprinklers would be an accepted practice and would be installed in every home today. Statistics provided by the NFPA and fire sprinkler advocates claim that smoke detectors and sprinklers increase the chance of surviving a home fire by more than 80%. Design improvements have allowed installation in remote and rural homes with no "city water" and even those with no electricity by utilizing solar power. The cost is relatively economical for most installations, many times less than the cost of adding a hot tub or spa. Most fire departments today recommend the installation of residential sprinkler systems, and some governing authorities require it for new construction. Still, residential sprinklers are estimated by NFPA to only be approximately 2% of new construction, and virtually nonexistent as a retrofit in existing homes. It might be easy to understand why many realtors, builders, and financial brokers would be leery of adding sprinklers, but it's a little more difficult to explain why appraisers, insurance agents, and even firefighters don't necessarily embrace them.

Maybe it's just bad luck. In an article in *NFPA Journal* explaining how the sprinkler code was assigned the number 13, Russell P. Fleming explained that the numbering system was probably intended to be 1-3, but later had the dash removed.[2] Although the origins of the standard date back to 1896, the pamphlet wasn't officially assigned the number "13" until 1936. It's doubtful that the number is solely responsible for the difficulties in acceptance, even though superstitions prevent many buildings from having a thirteenth floor and some cities from having 13th Street. More likely, the case is probably that residential sprinklers are not fully understood by most people. Many new homebuyers don't consider looking for the added protection, and many builders simply don't offer fire suppression systems. In an effort to provide a standard that would be both effective at saving lives and reasonably priced, NFPA developed a specific standard for residential structures.

NFPA 13D: Standard for the Installation of Sprinkler Systems in One- and Two-Family Dwellings and Manufactured Homes

Many jurisdictions that have mandatory residential sprinkler requirements refer to NFPA 13D for design and installation criteria. Unlike most NFPA standards, 13D takes components of NFPA 13, Standard for the Installation of Sprinkler Systems, and simplifies it for residential use. For example, NFPA 96, Standard for Ventilation Control and Fire Protection of Commercial Cooking Operations, has not been rewritten and applied for residential use in the kitchen. Additionally, it's unlikely that any NFPA standard has ever been described as **affable,** the user-friendly residential fire sprinkler standard comes as close as one might be. There are several examples of how easy it is to comply with the standard, including intent, design factors, and cost savings.

Intent

The intent of NFPA 13D is to save lives and property, but also consider the financial impact of such systems on homeowners. 13D systems have a design based on the assumption that a fire that originates from a single ignition source can be extinguished, or

FIGURE 15-1
Residential systems are designed to allow occupants time to escape and prevent flashover rather than extinguish the fire. As a result, sprinkler piping doesn't resemble commercial systems.

contained for enough time to allow the occupants to safely escape. The standard also takes into account the furnishings of a home, and thus provides information to reduce the chance of an accidental discharge. For instance, it cites two choices for sprinkler heads: one that is ordinary temperature-rated for areas of the home with an ambient temperature of less than 100°F (38°C), and one that is intermediate temperature-rated for areas such as an attic that could normally be more than that temperature. The ordinary temperature-rated head won't discharge until it rises to a minimum temperature of 135°F (57°C), and an intermediate temperature-rated head will open at no less than 175°F (79°C). The standard also lists normal household appliances and the minimum distance a fire sprinkler head should be mounted away from it for both ordinary and intermediate heads.

Many commercial and industrial systems require the use of a fire pump to boost pressures higher than normal operating pressure (Figure 15-1). NFPA 13D requires no fire department connection, and no booster pump over normal operating pressure. In other words, if the normal pressure of the water system in the home is adequate for the hydraulic needs of a second-story shower and a sprinkler head in the adjoining master bedroom, no additional equipment is needed. Additionally, the sprinkler system is hydrostatically tested at normal operating pressure.

Design Factors

Several design factors of residential suppression systems are borrowed from commercial and industrial systems that can be utilized if desired by the builder or homeowner, yet some are specifically permitted in residential systems. The standard is also worded in a way that encourages installers to get creative when it comes to design improvements.

For instance, when addressing the type of pipe, it cites the performance standards for pipe:

> *Nonmetallic pipe used in multipurpose piping systems not equipped with a fire department connection shall be designed to withstand a working pressure of not less than 130 psi at 120 degrees F. (5.2.1.3)*

By citing a performance standard, it opens up opportunities for new materials that might be offered in the future. Choices are also the norm when it comes to the type of system, allowing the installation of wet, dry, or even preaction systems.

Wet System. **Wet systems** In many situations, a wet system is the most economical. **Wet systems** are filled with pressurized water to each head. The tactical advantage of a wet system is that there is no delay. When the fusible link on the head melts, water immediately is discharged on the fire. The wet system can be a stand-alone or can be combined with the domestic water system as long as the sprinkler requirements are calculated into pipe sizing. In areas that are prone to freezing, antifreeze is an acceptable option for stand-alone systems. Some may choose instead to install a dry or preaction system.

Dry System. **Dry systems** charge the sprinkler lines with pressurized air, and are effective in unheated areas that are susceptible to freezing, such as attics, porches, and garages. Generally speaking, dry systems are more expensive due to the requirement of having a clapper valve that holds back the water with air pressure, along with the addition of an air compressor to keep the lines pressurized with air. When a sprinkler head activates, it discharges air until the valve opens and water can fill the lines, resulting in a short delay.

Preaction System. **Preaction systems** are hybrid systems that are basically dry, but use other technology to charge the system sooner than a dry system. NFPA 13D permits the use of three different preaction systems: single interlock, noninterlock, and double interlock. All three are essentially dry systems that are charged automatically when a combination of sprinkler head or detection device senses a fire. For example, if a single interlock system is installed in your home, the pipes would contain no water and no pressurized air. A sensitive detection device would notice a fire, and would open an automatic valve on the water line, charging the system. No water would flow through the head until a fusible link on a sprinkler head melted.

Cost Savings

Several components of NFPA 13D reduce the costs when compared to a normal fire suppression system. For example, it is possible that a family that is building their own home could complete the work themselves.

> *The layout, calculation, and installation of sprinkler systems installed in accordance with this standard shall only be performed by people knowledgeable and trained in such systems. (4.8)*

With engineering help from the architect and some research about installation requirements, the couple could theoretically install their own system with no "certification." Another cost-saving feature is that the standard does not require protection in small bathrooms, garages, carports, porches, crawlspaces, or attics. Although fires can and do occur in these areas and some jurisdictions require the protection, NFPA has concentrated on installations for life safety.

Finally, the requirements for an automatic fire alarm are less stringent than other automatic suppression systems. The standard states:

> *Local waterflow alarms shall be provided on all sprinkler systems in homes not equipped with smoke alarms or smoke detectors in accordance with NFPA 72, National Fire Alarm and Signaling Code. (7.6)*

Although the standard is meant to encourage the installation of fire sprinkler systems by reducing costs, some people completely oppose them and concentrate their efforts on stopping any movement for acceptance. When faced with opposition, it's a good idea to first examine what the resistance is based on.

THE OPPOSITION

In 2008, the **International Code Council (ICC)** voted to mandate residential sprinklers in all new one- and two-family dwellings. Although it became effective in January 2011, states were individually required to adopt the amendments to the **International Residential Code (IRC)**. The code references NFPA 13D, although some jurisdictions have more stringent requirements. Opponents of the change launched numerous attacks in an effort to try to stop the change from taking place in different areas of the country.

The political and legal challenges are likely to continue for some time, but regardless of their outcome, the fire service needs to work together and educate everyone involved with facts rather than emotions. History and statistics already have proved that residential fire suppression systems are effective at minimizing damage from a fire, improving the likelihood of survival for the occupants, and are cost-effective when standardized. Firefighters need to first become convinced that residential sprinklers are needed, and then advocate their use globally. Some of the general arguments against mandatory installation in all single- and double-family homes include the costs, the availability of water, the fear of water damage, a decreased resale value, freedom of choice, and, ultimately, the need.

Cost

It's no secret that the building industry has taken a financial hit in the past several years. As a result, it makes sense that one of the biggest forces of resistance to the addition of mandatory residential fire sprinklers is from the building industry. From their side, it seems counterproductive to increase the cost of a new home by another couple percent at the same time they are trying to sell more. In 2008, the **Fire Protection Research Foundation (FPRF)** completed a thorough cost analysis for residential sprinklers.[3] Using 30 different construction projects around the country, they compared the complete cost of installation for options such as:

- The utilization of both municipal water systems and wells
- Square footage that ranged in size from 1900 to 6500 square feet of sprinklered area
- The utilization of varying materials, including copper, CPVC, and PEX
- The combination of domestic water and fire suppression along with stand-alone systems
- Installation in both cold and warm climates

This list also provides the individual factors that affected the installation price. For example, the use of well water, copper pipe, and antifreeze all contribute to a more expensive system. Some of the communities required an additional investment, such as relatively high inspection fees (nearly $800 in one instance), and other requirements that are not included in NFPA 13D, such as protecting unheated garages, attics, and porches. The actual costs ranged from $0.38 per square foot to $3.66 per square foot, with an average of $1.61. For the homes studied, the final cost of a residential fire sprinkler system ranged from $2386 to $16,061. As a result, many builders look at those numbers and are against the idea from the start. On the other hand, other groups are more likely to weigh the facts, taking into consideration the value of a fire sprinkler system and what is in their best interest.

Value

On the surface, adding 1% to 3% additional costs as an unfunded mandate tends to upset consumers, but we all know that value goes much deeper than the simple cost of a product. In times of financial crisis, it's common for buyers to cut back on things they don't necessarily need. If you get laid off from your job, you are far less likely to go out and buy a new car, even if you had planned on buying one. However, as soon as the economy turns around and you get a new job, it's likely that you will go out and buy what you want. Suppose that when you do return to the car dealer, you find that the new vehicle you had your eye on had a 2% increase in price due to newly mandated safety systems such as side curtain airbags or a more environmentally friendly emissions system. It's ridiculous to suggest that you would walk away and decide to keep your old car based on the unanticipated expense. If consumers simply refused to invest in safer and better designs, they would still be using chain saws without antikick components and letting their children throw lawn darts in the backyard.

Many people believe that consumers approve of investing more into safety, and many different statistics suggest the same. In a poll of 14,000 drivers, statistics showed that they were willing to pay more for fuel economy as well as additional safety advances such as rollover protection.[4] A study in Rhode Island showed that seafood consumers were willing to pay as much as 10% more for safety assurances such as requiring a "catch date" on all seafood.[5] The Business Travel Coalition conducted a survey of corporate travel managers and learned that most of the companies are willing to pay more in airfare for higher safety standards in regard to regional airlines.[6] There's no evidence to suggest that informed consumers would feel any different about residential fire sprinklers.

Some builders have pointed out the number of existing new homes that have not sold, and allude to the fact that a sprinkler law would be detrimental to those sales as well. However, an interesting fact about products that undergo a change for new technology or safety is that sales of the discontinued products actually increase as the new ones are being released. For example, many firefighters can remember when significant emissions standards took effect on diesel engines in the United States in January 2010. This change added a significant increase to the cost of a new truck chassis, including larger fire apparatus. Fire truck manufacturers immediately bought up every engine they could in anticipation of the changes. The demand for the old style of engines jumped when the changes were announced, and fire departments hurried to get their order in under the wire. Other fire equipment companies are very familiar with the fact that when NFPA comes out with a new standard for **PPE,** such as a structural firefighting boot, the old ones sell fairly quickly. Whether we agree with the actions of consumers purchasing old stock to avoid new safety requirements and get a good deal or not, it regularly occurs. There is no reason to assume that homes built without sprinklers will be any different (Figure 15-2).

Personal Interest

Although home builders may be interested in protecting their profits by leaving residential sprinklers on the shelf, not all trades would benefit from fighting the addition of suppression systems. Numerous architects, plumbers, and sprinkler contractors along with their associated suppliers would benefit from a change in fire suppression requirements. For example, the **Plumbing-Heating-Cooling Contractors Association (PHCC)** has supplied information to its members about how to properly position themselves to take advantage of the growing opportunity.[7] Consider what happened to the highway construction industry when President Obama approved the American Recovery and Reinvestment Act of 2009 as a way to stimulate the economy. A significant portion was allocated for highway projects. As a result, road materials were produced, trucks delivered asphalt, and laid-off workers went back to work. As more residential structures need fire sprinkler systems, more fire protection engineers will be needed to design them, more pipes will be ordered, and more installers will be put to work. To argue that the construction industry would suffer as a whole with the addition of a mandatory sprinkler law is short-sighted and misleading.

FIGURE 15-2
A residential sprinkler law may actually increase the sales of new vacant homes.

As more contractors are able to submit quotes for jobs, the average price will come down. In the FPRF study, the least expensive suppression systems were in California, in a city that has required residential fire sprinklers for more than 20 years. The builder who participated in the study attributed the low cost to significant competition among sprinkler contractors in the area. The law of supply and demand will surely decrease the cost of residential fire suppression systems as they become common, and more qualified installers are able to competitively bid on projects.

Water Availability

The availability of water is a valid concern for some people. New homes with access to municipal water systems may have to increase the size of the service line supplying the home from the main water line at the street. A larger water meter or automatic bypass line could be required to provide the necessary water in the event of a fire. Aside from that, any water line on the street that can supply fire hydrants will have no problem serving a residential sprinkler system. However, homes that have no access to a public water system will need additional support.

Suppose you live in a home that has no access to public water, and is therefore supplied by an underground well and pump. Wells have a specific amount of water available to draw from, along with a recovery time to refill after the main supply of water is pumped out. The underground aquifers that supply water to the well dictate the amount of additional gallons available each minute. For instance, your well may have 200 gallons available, and your pump may be capable of displacing 20 gpm. If you began filling a swimming pool, the pump and water supply would be sufficient for the first 10 minutes. If your recovery time only allowed 10 gpm, your water supply simply couldn't keep up with your pump.

NFPA 13D requires that a single sprinkler head be capable of discharging 18 gpm, or a pair of sprinkler heads operating would each provide 13 gpm. Therefore, the well and the pump would both have to be capable of supplying 26 gpm. The standard also requires that the minimum flow be sustainable for 10 minutes. If the well and pump were unable to meet the requirements for 260 gallons in 10 minutes, an additional storage tank and pump may need to be installed. Obviously, this is a significant portion of the costs associated with installing a system in a home without access to municipal water supply and a weak water supply. Many underground wells can supply the required 26 gpm for 10 minutes.

As stated earlier, NFPA 13D has relatively lenient requirements, including water supply, when compared to the much more stringent NFPA 13. This is due to the fact that NFPA 13 is designed to extinguish a high percentage of fires with a significant safety factor, whereas the residential standard is designed to hold a fire in check and prevent flashover, which can allow occupants time to escape. In other words, NFPA 13 is designed for a high *fire* risk and NFPA 13D is intended to address a high *life* risk. This results in relatively reasonable requirements. For example, someone might think the requirement for storing 260 gallons of water in their basement is excessive, and that it would take up too much space. However, many homes over the years in colder climates stored 275 gallons of fuel oil in their basement with no problem. Similarly, a fire booster pump that is required to pump 26 gpm will surely add cost, but when compared to many lawn sprinkler pumps that put out double that with an inexpensive 1.5 horsepower motor, the requirement is very reasonable.

Water Damage

One of the standard arguments against residential sprinklers is that the deluge-type systems that are commonly portrayed on television and in movies will cause significant water damage. Although some humorous scenes play out when smoke sets off an entire sprinkler system in a building, it's about as realistic as an actor successfully defibrillating someone's heart with a lamp cord plugged into a wall outlet. Firefighters realize that most systems are designed so that only the sprinkler head that is exposed to a specific amount of heat, not smoke, from a fire will discharge a predetermined amount of water on the fire to extinguish it or confine it. Regardless, it is plausible to argue that a fire suppression system has the ability to create water damage as a result of an operating sprinkler head, or a leak in the system.

Water Damage from an Operating Head

It is true that the longer water flows from an open sprinkler after the fire has been extinguished, the more water damage will occur. NFPA statistics have continually shown that many fires are contained with only one sprinkler head flowing. As stated earlier, a single head must be capable of flowing 18 gpm. Compare that with the fact that many attack lines used today for interior firefighting are capable of flowing 180 gpm. Simple math shows that a single sprinkler head flowing for 10 minutes while we respond to the fire puts out the same amount of water as one handline does in 60 seconds. It's also reasonable to assume that the broken stream flowing at 18 gpm on an incipient fire is much more efficient at cooling and extinguishing a fire than a blast of 180 gpm at a room and contents fire that is "rocking." As a result, water damage is likely less prominent with a working residential sprinkler than if we attack it in a normal fashion.

Another consideration when comparing the cost of water damage with fire damage is the construction techniques and materials used today. Many fire departments have noted in the past 20 years that a home that they thought they *saved* from a fire was actually a total fire loss and was instead torn down. Compare that with how many homes that receive water damage from a clean source of water, such as a bathtub overflowing or a water line freezing and bursting, are destroyed beyond repair. Although storm flooding is a significant concern for many areas, very few occupied homes are total losses as a result of *clean* water damage. In fact, water damage from a sprinkler system is much less of a threat to a home than a fire that is left uncontrolled until you arrive.

Water Damage from a System Leak

If you've owned a home before, you might recall all the times you've had water damage. It could have been caused by storm flooding, a roof leak, a window left open, or an ice dam under the shingles. You may have had a bathtub or a washing machine overflow. When it comes to drain lines and traps under sinks and tubs, they sometimes fail or can be knocked loose. Dishwashers, washing machine hoses, and toilets sometimes leak, and occasionally sump pumps fail. Water damage can come from many different sources in the home. Now think back, how many times have you had a pressurized water line burst? If you're like most of us, you can remember several water leaks, but have to admit that a copper or plastic water line rarely fails. Permanently installed pressurized water lines are incredibly trustworthy when compared to all the other water problems we can have in the home. It is true that doubling the amount of piping in a home doubles your chances of a leak, but if the water lines are installed correctly, the chance of a leak is still extremely low.

Sprinkler heads, on the other hand, are more likely to leak. If you have industrial or commercial buildings in your response area that have fire suppression systems, you may have responded to leaks from a damaged head as the result of a forklift striking it or something bumping into it. Although accidental damage can occur to a residential head, remember that there are far fewer forklifts and ladders in your kitchen than at the local trucking company's dock. If a child were to throw a ball or a shoe at a normal commercial head, it could break and start flowing water. As a result, most residential heads take this into consideration and are more similar to the "pop-up" design of lawn sprinklers. This hidden design not only reduces the chance of a mishap, but provides for better aesthetics by concealing the sprinkler head.

If we still aren't convinced of what the risk of water damage from a sprinkler system is compared to the risk of having a fire without the added protection, all we need to do is ask a person who calculates risk every day. Check with an insurance agent. If an insurance company believes their financial risk for fire and water damage is reduced by having a residential sprinkler system in one of the homes they cover, it will reduce the rates if you have one. If the insurance company believes that the risk of water damage from a sprinkler outweighs the risk of not having one at all, the rates will rise. Estimates from around the country show that most insurance companies give a discount up to 15% based on how much of the home is protected. For example, the Office of State Fire Marshal in the state of Maine lists 13 different insurance companies that give discounts in the state ranging from 5% to 12%.[8] Therefore, it's evident that many professionals in the insurance industry don't worry about water damage from fire sprinklers as much as they worry about the fire.

Resale Value

Realtors have the duty to sell homes. It is therefore a safe assumption that most realtors know what sells. For instance, let's say you decide to sell your home and invite a realtor to come see it and tell you what she thinks it is worth. After a look around, she'll give you plenty of suggestions to improve its value immediately, and make it more marketable. She's likely to mention curb appeal, open spaces, convenience, and storage. But what does she think of the fire sprinklers? She may be very good at anticipating that a new front door will improve the chances of the home selling, but may not be educated enough about fire sprinklers to make an informed decision. If she believes that lawn sprinklers are a feature buyers are looking for but fire sprinklers are a liability, she probably won't be as passionate about the sale. If realtors aren't educated about the advantages of fire sprinklers, they are likely to divert potential buyers to the spacious master bathroom rather than what could be the best feature of the home.

Freedom of Choice

One of the arguments against mandatory residential fire sprinkler systems is the word *mandatory* is in the same sentence as *residential*. Many people have a political belief that governmental requirements imposed on our lifestyle at home are an infringement of our rights. They argue that the installation of sprinklers is a matter of personal choice. The argument is valid only to the degree to which their decision doesn't infringe on the rights of others. For instance, a person staying in a hotel should not be in danger because the people who are staying below her are careless. Many times, these same basic rules apply to those who can't make decisions for themselves. For example, children who travel in vehicles must be secured in child safety seats. Sometimes decisions are

FIGURE 15-3
Many requirements may be considered by some as a loss of personal freedom but usually are based on a level of safety that the public would expect.

made based on the overall well-being of the public. This occurred when the ICC voted that residential fire sprinklers must be installed in all homes for public safety. By making that decision, they essentially put active fire protection on the same plane as staircase railings and minimum window opening dimensions for emergency egress from a bedroom (Figure 15-3).

Need

A final argument posed by the opposition is whether residential sprinklers are even necessary. The NFPA released comments in regard to concerns about the IRC adoption of a residential sprinkler mandate and in dispute of a claim by the **National Association of Home Builders (NAHB)** that residential smoke detectors by themselves are sufficient.[9] The NAHB had been using NFPA statistics in an effort to convince the public that they were *safe enough* without the added expense of sprinkler systems. NFPA estimates that working smoke detectors raise your chance of surviving a fire to 99.45%. Although it sounds like an acceptable percentage in itself, the actual number is still roughly 3000 civilian fire deaths per year. If you take a look at NFPA's statistics prior to their big push for smoke detectors 40 years ago, the survival rate was 99.20% with 6000 deaths per year without smoke alarms. Using the NAHB's reasoning, we should have never bothered with smoke detectors because the public is only a quarter of a percent more likely to survive a fire due to the smoke alarm effort. It's probably safe to say that the estimated 3000 people saved every year as a result of the standardization of smoke detectors disagree with the opposition's opinion.

Proving a need for residential sprinklers comes down to facts. Data from municipalities that have had residential sprinkler requirements for more than 20 years prove how successful they really are. Although other locations may have had sprinkler laws for more time, Scottsdale, Arizona, has been a model that sheds light on the effectiveness of residential sprinklers. Their single-family home law was enacted in January 1986, which equates to over 45,000 homes being protected.[10] Prince George's County, Maryland, has more than 15 years of data from their mandatory sprinkler law.[11] Some of the most important facts uncovered in Scottsdale, Prince George's County, and other similar studies throughout the country include life safety, actual fire damage, and actual water damage.

Life Safety

The primary goal of the installation of residential sprinkler systems is life safety. NFPA has no record of any fire in any fully operational sprinklered building in which more than one person has died. In Scottsdale, about half of the homes are protected by fire suppression systems. From 1986 to 2001, data show 13 people died in nonsprinklered dwellings and none died in sprinklered homes. Prince George's County's results are similar, with no deaths in protected homes and townhouses, as opposed to more than 100 in unprotected residences.

There's no question that smoke alarms have done a great job of saving lives by alerting occupants of a fire. However, a smoke alarm is a passive device that can only warn people of a smoke danger so they can attempt to escape. Fire sprinklers take protection one step farther as an active device that can usually stop a fire. This is especially beneficial if the occupant is unable to escape. Smoke alarms have the advantage when it comes to incipient fires people may not yet know about. Sprinkler systems, on the other hand, take advantage of the heat emitted from a fire and attack it by cooling it. Fire sprinklers won't ever replace smoke alarms, much like airbags will not replace seatbelts. They instead complement each other to improve an occupant's chance of survival.

Actual Fire Damage

As you would expect, the cost of fire damage in a sprinklered home is much less than the cost in a home that is not protected. Average fire loss in Arizona for unprotected homes was a little more than $45,000 per fire, but the houses protected by a sprinkler system had an average loss of less than $2500. An interesting statistic gained from the Maryland study was that the average fire loss from a nonsprinklered home was $10,000, but rose to five times that when a fatality occurred. Conversely, homes that were protected averaged less than $5000 per fire. In other words, the incidents with the most significant fire loss also were the causes of life loss.

Unfortunately, even with sprinkler ordinances in certain areas of the country, very few new homes have them installed. In fact, annual estimates from the NFPA, USFA, and the Home Fire Sprinkler Coalition place them at less than 3% of new homes.

Because fire loss data from the homes that are protected shows such a drastic reduction in fire loss, it's likely that more insurance companies will notice it as more homes are equipped with sprinkler systems. As they see the reduction in their costs, it follows that they will encourage more homeowners to choose fire sprinklers in the form of reduced rates.

Actual Water Damage

Some of the most interesting statistics from the Arizona study are in regard to water damage. Although it ranks high on the misconceptions of sprinkler systems, the actual numbers are staggering. Normal firefighting operations with hoselines discharged an average of almost 3000 gallons of water at each fire, but sprinklered fires used little more than 10% of that. Additionally, 92% of the fires were controlled by one or two sprinklers. Not only is water damage minimized with fire sprinklers, but less water is wasted. In these ecology-minded times, it's important to stress the green side of residential sprinklers. The quicker a fire is extinguished, the less water is wasted, and the amount of toxic smoke and carbon dioxide emitted into the atmosphere is reduced.

DIFFERENT PATHS

When trying to institute a change of this magnitude, it's important to tackle the problem from many different angles. We need to support the professionals who have dedicated great time and resources to making residential fire sprinklers standard in other ways. For example, new construction and retrofitting existing homes will make a significant difference in the ways we fight fires 100 years from now. However, there's still work to do now in the buildings we will fight a fire in next year. We need to look at ways to save civilians' and firefighters' lives now. Some ways we might accomplish this include code enforcement, the adoption of sprinkler mandates, sprinkler advocacy, and future planning.

Code Enforcement

Most of this chapter has concentrated on residential fire sprinklers, but Initiative 15 also calls for the enforcement of existing codes. Concentrating on new laws doesn't make much sense without enforcing the ones we have already established. Fire department leaders need to constantly reevaluate how the fire prevention and inspection side of their department is performing. The job description of a fire prevention bureau or division covers a broad range of duties. Some fire departments rely on company officers to complete the necessary functions while others have a division specifically charged with the responsibilities. In regard to code enforcement, fire prevention obligations generally include plan review, fire inspections, fire investigation, and fire and life safety education.

Plan Review

In some cases, fire departments are involved in building plan reviews, but many are completed by building officials. It's important for all fire departments to be active in building relationships with representatives of zoning or the building department. Building officials must be viewed as an integral component of firefighter safety. Plans can be analyzed for buildings or even for complete subdivisions. The primary goal of plan review is to **objectively** ensure compliance with existing codes, but it's also important to look at them **subjectively.** By looking at them subjectively, you can try to envision problems encountered by occupants trying to escape, as well as difficulties in firefighting. For instance, a plan review for a new auditorium for the high school would have to meet exit number and size requirements based on the fire code. Suppose that the plan showed several of the exits in close proximity to the **fire department connection (FDC)** for the fire suppression system. A subjective look at the plans might lead to suggestions for better locations of the exits or the FDC based on fire department duties when they arrive.

Plan reviews must consider firefighter safety. Imagine that you are asked to review the plans for planned improvements to a mobile home park. The project is limited to new sewers and roads, but there are questions regarding lane widths and turnarounds for fire apparatus. Subjective review might also compile a list of existing firefighter safety concerns. The list could cite zoning issues such as clearance between homes, the locations and sizing of propane tanks, and unsafe electrical service lines. By bringing the concerns to the owner or developer, some may be addressed or compromises can be established.

Fire Inspections

Conducting fire inspections is not only a way to reduce the risks of fire and injury through code enforcement, but also gives the fire department a chance to become familiar with their buildings. Whether designated fire inspectors prepare a newsletter for the line firefighters or companies complete inspections on their own, information must be passed on. Hazards to the safety of firefighters can be identified, even if the unsafe condition meets the code requirements.

Fire Investigations

Although many times fire investigations are conducted by different personnel than fire inspections, the two jobs are closely tied. For example, let's assume a fire occurs at an automobile body shop. The origin of the fire appeared to be outside the structure in a vehicle parked against the building, but flames spread to the building before the fire was reported. An investigator may have questions about the prefire conditions of the building, including vehicle storage. Many times it's helpful to talk to the person who conducted the last fire inspection prior to questioning the business owner. This is especially valuable when contributing factors to the fire are identified as past concerns or violations brought to the attention of the building owner yet are not addressed.

Fire investigations should be as much a part of preventing the next fire as they are finding the cause of the last one. Identifying the cause and contributing factors allows fire inspectors to spot potential problems before they can injure a firefighter. Sharing the concerns with line firefighters also helps reduce the chance of future injury.

Fire and Life Safety Education

Many times a fire inspector is also assigned the duties of fire and life safety educator. If we really believe that Initiative 14 is vital to firefighter safety, then all positions in the fire department should include the duties of a fire and life safety educator. Personnel assigned to code enforcement should be even more involved in safety education because of a unique opportunity they have. Educators refer to a **teachable moment** as a point in time when students are especially likely to receive a message and understand it. If there ever is a teachable moment with a business owner in your jurisdiction, it's when he or she is learning of a fire safety violation that needs immediate attention.

Suppose you conduct an annual fire inspection at an insurance agency. Although it appears the agent is less than excited to see you, she assigns one of her clerks to walk you through the office. Upon completion, the only violation is portable fire extinguishers, which have expired inspection tags. You hand the clerk a violation notice, and inform him that you'll be back in 30 days to check on the extinguishers. At that point, the insurance agent bolts from her office and questions the importance of the extinguisher tags. She is upset about being forced to pay "some extinguisher guy to put a new tag on it" every year. Your response could be that you didn't write the code, you just enforce it and be on your way. Hopefully you recognize this as a teachable moment rather than the opening bell of a three-minute bout in a ring. You admit that you hear the complaint often, and that you, too, would be upset to be forced to buy a tag every year. However, you explain that she isn't buying a tag, but is recertifying the extinguisher. The "extinguisher guy" is actually certified to inspect the extinguisher, and has other duties than just changing the tag. You explain how the technician might check the pressure in the tank and ensure that the gauge is reading accurately. You might point out that the extinguisher is strategically placed by the exit so that occupants are in the process of exiting when they consider using the extinguisher. The extinguisher technician is supposed to periodically clean the outside of the hose and nozzle and ensure that it doesn't conduct electricity. The inside of the hose must be checked for bee nests and other obstructions. The technician may shake the extinguisher to ensure that there is indeed dry chemical in the extinguisher. You explain that sometimes extinguishers don't work as planned, and that your back should be to an exit. In fact, the fire department conducts free classes for businesses such as hers. Finally, the technician does change the tag and stamps it with his name and certification number. You point out that this information is especially useful if the extinguisher fails to discharge when there actually is a fire. You suggest that as an insurance industry professional, you're sure she understands the concept of liability in a situation such as that.

The Adoption of Sprinkler Mandates

Fire service leaders have been successful in getting a residential sprinkler law passed at the national level. It's now up to every fire service organization and firefighter to get it adopted at each state and local level. The sooner homes are built with the added protection, the sooner we'll start to see a reduction in civilian deaths. It also follows that more residential sprinklers will reduce the severity of fires with the added protection, and will allow occupants to escape prior to our arrival. If fires are less severe and civilians are already out of the home, it reduces our risk and we can expect to see a reduction in firefighter deaths as well (Figure 15-4).

One past problem associated with sprinkler mandates on a local level is the possible detriment to a municipality that is the first to enact a sprinkler law. For example, suppose your county had no mandatory residential sprinkler laws, but your town had been considering it. It's likely that purchasers of a new home would have built in a neighboring town in an effort to save the money on the added protection that wasn't required there. As a result, municipalities wouldn't have been able to compete fairly when it came to the cost of a new home. The most successful mandates are widespread, as the IRC adoption is at state level.

Sprinkler Advocacy

Besides being involved in the adoption of sprinkler laws, every fire service organization should adopt the policy of advocating residential sprinkler systems. This could take place in the form of public promotions and trade-offs.

FIGURE 15-4
All firefighters must realize that the installation of residential sprinklers has a direct effect on their safety.

Courtesy of Lt. Rob Gandee

Public Promotions

Fire and life safety education programs should include information about residential fire sprinklers. Promotion could include public service announcements on local cable television channels, information, and links to sites such as the Home Fire Sprinkler Coalition® on department websites, and fire sprinkler slogans on apparatus. It's not that smoke alarms don't need our support anymore, but home fire sprinklers need our attention right now. Ensure that your department's mission statement or list of organizational values includes the promotion of residential sprinklers. Consider that any post-fire press releases include information about how beneficial a fire suppression system would have been.

One of the most interesting possibilities for the promotion of residential fire protection is with the realty industry. Fire departments should look at ways to reach out to realtors to answer questions they might have and promote residential sprinklers with facts. Offering to visit a company's staff meeting for the purpose of educating them about existing or future sprinkler requirements can provide them with the information they need. You can also take the opportunity to discuss different types of smoke alarms and carbon monoxide detectors. Imagine the benefits of a realtor who was in your class explaining the benefits of a home to a potential buyer. Not only would he sound like an expert when discussing the safety features of the home, but would likely promote your fire department as well.

Trade-Offs

Trade-offs can occur during the preliminary or planning phases of construction, and are an effective way to increase fire protection. In many instances, builders or developers meet the minimum letter of the law with a literal interpretation of the code. For instance, suppose a commercial building code requires fire sprinklers in all new buildings of more than 50,000 square feet. An architect could essentially design the store at 48,000 square feet to avoid the rule and added expense of sprinklers. At a plan review, the fire official might ask the builder to add in the protection, which, in this case, is likely to be refused. A trade-off could allow the builder to save money in other areas of construction by permitting a variance of the code. In return, the builder would add fire sprinklers. The idea is far from new. In 1979, Chief Dave Hilton of Cobb County, Georgia, cut a deal with a builder in an effort to add sprinklers to a multifamily development when they were not required.[12] He created a **consortium,** a group of professionals working toward a solution that had not been accomplished before. They met, debated, and completed test burns. In the end, the fire department eliminated costly requirements for compartmentalization of the buildings and instead invested the money in fire protection.

Sometimes the trade-off is not in the form of building requirements. Zoning and other developmental requirements might also be negotiable. If a fire department is successful in obtaining code changes for larger fire lanes, a cul-de-sac turning radius, or fire hydrant spacing, they may be able to use them to improve fire protection in the buildings. The theory might be that if all homes in the subdivision are fully sprinklered, fewer hydrants will be needed in the event of a fire. Trade-offs can also take place between other organizations or departments. For instance, many times water **purveyors** (private water companies or municipal operators) charge significantly more for larger meters and the ability to provide fire suppression. Negotiating with the water department to reduce or eliminate the costs for residential fire protection can go a long way in making protection financially possible for builders and homeowners. Because many water purveyors don't charge for water used in firefighting, demonstrating the reduction of water usage with sprinkler systems will help them in reducing water loss and reducing fire flow demand.

Future Approaches

In an effort to keep the tide moving toward fire sprinklers being the accepted norm in all buildings, we must look to the future for other ideas to promote integration. Some ideas might follow the next logical step to acceptance, but others could offer suggestions about how we can encourage installation in situations that aren't mandated. These include new home design, rural auxiliary use systems, and the ease of retrofitting.

Home Design

One way to promote residential sprinkler systems is to encourage affiliated industries to expand their market. For instance, many people who are considering building a new home review magazines or online selections of home plans. These selections offer virtually every size, room layout, and architectural design available. Full sets of blueprints can be ordered, including plumbing and electrical drawings. Encouraging these types of blueprint companies to offer residential sprinkler drawings would increase the awareness of buyers and builders, while making it easier for them to obtain the information they need.

This opens up an entirely new opportunity, to design homes with fire protection and economics in mind. For instance, suppose a certain residential wall-mounted sprinkler head is capable of discharging in an 11' radius. An architect who designs fire protection in with the house plans could limit a normal bedroom width to 11' rather than 12'. Wall-mounted sprinkler heads in a bedroom or stairway could be designed against a shower wall in the master bathroom to minimize the cost of sprinkler coverage. Water line size could be calculated to provide the residential water needs as well as the fire protection needs. Rather than design sprinklers for a home, design a home and sprinklers together to reduce the overall cost. This not only can improve aesthetics, but also efficiency.

Rural Sprinkler Components

Some fire protection equipment manufacturers provide a complete fire protection system to supplement an existing well water system. Usually these consist of a plastic water tank and pump on a skid unit complete with electrical controls. Although effective, some might argue that it takes up a significant portion of space and may be a deterrent to would-be homebuyers. A new design would not only take up less space, but also would be more attractive. The most efficient use of space is by using a tank with square sides. A 300-gallon tank could be constructed that is 4' tall by 5' wide and 2' deep, not much larger than a dresser lying on its side. A tank, pump, and water control could be easily hidden in a closet in the basement behind a wall. It also could be installed by a floor drain, which would be helpful in the event of a leak.

Stagnation of the water might also be a concern for these units. New designs might allow the pumps to have an "irrigation cycle," which would allow the homeowner to change a couple of valves and pump the tank to a garden hose for watering plants or the lawn once a month. An automatic fill valve could refill the tank when the system is switched back to fire protection mode, and possibly even use a rainwater collection system in a parallel tank for water conservation. This type of an application would not only allow the pump to operate and water to cycle, but would offer the homeowner an added benefit besides fire protection. It could be designed in a way to eliminate the chance of accidental disarmament. Valves could reset automatically when the pump kicked off or an alarm could sound if a homeowner forgot to switch it back.

Ease of Retrofitting

When considering home fire sprinklers, we have to admit that lives can be saved by making small improvements. An incident in a mobile home in 2009 had a script written for tragedy.[13] A 30-year-old woman was asleep in the back bedroom when a fire started in the front of the home. The superheated gases quickly spread down the hallway, where they pushed past the beeping smoke detector. The resident reportedly had medical problems, which contributed to her sleeping through the alarm. As heat from the fire started melting the plastic trim at her bedroom door, water suddenly burst from the furnace closet in the hallway, stopping the heat and fire in its tracks. One sprinkler head, installed 30 years prior on a single, dead-end piece of half-inch copper pipe held the fire in check. The resident awoke to find thick black smoke in the hallway and exited a bedroom window. When firefighters arrived, they had no rescue to perform, which allowed them to simply extinguish the fire.

It's not clear who made the decision to install the sprinkler head and why. Firefighters occasionally come across a lone "soldier" protecting a furnace room, but they aren't very common. It's probably safe to say that the sprinkler head and associated piping cost well under $100 in materials and labor to install when the mobile home was being manufactured. There's a good chance it wasn't ever flow tested, and wasn't even fed by an adequately sized branch line. It wasn't engineered for the closet and surely had problems with the way it was mounted in regard to clearance and obstructions. It was probably a poor design with no maintenance but it still saved someone's life. Stories like this are actually not that uncommon. Reports of fires have been published in which washing machine supply hoses have burned through and extinguished the fire. Plastic water lines in the basement have melted and held a fire in check. These extinguishing systems were neither engineered nor designed, but still were effective.

In no way should we downplay the importance of a properly engineered, installed, maintained, and tested residential fire suppression system. We should continue to fight for sprinkler adoptions for new and existing construction. The adoption by all states of a national sprinkler law for new construction will save numerous lives over the next century. Our grandchildren's grandchildren will benefit from our hard work, but unfortunately it will have little effect on making our parents any safer. As a result, we need to look at how to make changes now. Many fire service professionals are addressing this problem by working on a plan to retrofit existing structures. However, retrofitting existing homes is estimated to cost twice the amount as new construction and likely will meet even more opposition. Even if a law was passed requiring retrofitting, it would likely only be enforceable when the house is being sold, referred to as **point of sale.** Some houses don't sell for 20 or 30 years, which doesn't help us at all in our lifetime. Realistically, if you live in a single-family home, it's likely that it will *never* have a residential sprinkler system in it as long as you live there. We need to realize that any change we make now has the potential to save lives and make our job safer tomorrow. A comprehensive plan to get sprinklers into every home should encourage the voluntary installation of sprinklers in regard to partial systems and retrofitting equipment.

Partial Protection

A homeowner who is willing to install a couple of sprinkler heads in his basement before he finishes the ceiling should be encouraged to do so. He should be able to easily purchase the materials at a home improvement store in a kit that he can install himself (Figure 15-5). You can already purchase small tap kits and fittings to tie your refrigerator water line into existing pipe without even turning off the water. They are both inexpensive and simple to use. There is no reason why a fire sprinkler head and tap kit can't be manufactured on a saddle clamp that attaches to an existing ½-inch or ¾-inch copper line (hot or cold). A homeowner should be able to buy the kit for less than $35, and visit your fire station or your website and obtain recommendations for the best installation practices. He should be able to call your nonemergency phone line and have an inspector or other trained firefighter stop by and give him installation suggestions. Rather than treat residential sprinklers as a high-tech gadget that is taboo to touch or talk about, we should be totally involved in voluntary residential sprinklers the same way we are for smoke detectors, carbon monoxide detectors, and fire extinguishers.

Remodeling projects is another area we should concentrate on. According to the NFPA, a review of statistics showed that more than 75% of fire deaths occur in homes in which the room of the fire origin is the kitchen, living room (family room or den), or a bedroom.[14] These rooms should be specifically targeted to have the greatest effect on life safety. If a family is remodeling their kitchen, they should be encouraged to install a

FIGURE 15-5
The fire service should promote sprinklers in many different ways, including getting materials in stores and encouraging homeowners to purchase and install them.

sprinkler. The cost of extending a water line to provide fire protection while the cupboards and drywall are removed would be a wise investment. According to do-it-yourself magazines and websites, a kitchen remodel can cost between $5000 and $25,000, depending on the features chosen. Twenty feet of copper pipe and a couple of sprinkler heads is a drop in the bucket. Factoring in that kitchen fires are the number-one cause of home fires, it makes you wonder why we haven't been pushing them in remodeling projects all along.

  Of course, the question of liability always comes up in discussions such as these. Are we legally liable for supplying a sprinkler head and distributing it to a homeowner? The answer is probably no more than supplying a smoke detector that fails to awaken a family. Many departments distribute free batteries. What if they go dead a month later? Some firefighters are permitted to install smoke detectors for residents who can't do it. What if they run a screw into a wire that eventually leads to an attic fire? How many firefighters tape off a melted wire from a small fire, and then turn off the power and tell the occupant to have it checked before using it again? Are firefighters liable if they mistakenly turned off the wrong breaker? How many times have members of your department told someone it's safe to go back into a building? What's safe? Are you liable if you teach employees how to use a fire extinguisher and one of them accidently discharges it at someone or drops it on his or her toe? What about a company drill in which a hoseline is stretched to a hydrant in front of the fire station and a jogger trips over it? The fact is that we work knee-deep in liability every day. We won't be charging occupants a fee to design and install a suppression system in their basement, and we aren't running a business by rewiring their main electrical panel. We are simply providing residents with the equipment and information they need to better protect themselves in any opportunity that presents itself.

Finally, there's another self-serving reason to get involved in partial protection. Consider that firefighter close calls, injuries, and deaths often occur when a firefighter falls through the floor of a building on fire. Usually it's a result of a fire below them, usually in unprotected joists, and usually in the basement. If the home is relatively new, it may very well be the result of a truss floor failing that didn't hold up to fire very long. Firefighters enter the front door and begin a search when gravity takes over. If a homeowner is encouraged to spend $200 and one afternoon installing *unengineered* random sprinkler heads in existing copper lines around *engineered* floors, why would we try to prevent it?

Retrofitting Equipment

Another way to make it easier for homeowners and contractors to retrofit is by encouraging the introduction of more building supplies to assist them (Figure 15-6). You may have noticed that many homes in the past used hot water pipes and radiators around the baseboards to heat the home. Small bedrooms may have only used one or two walls, and large rooms used the entire perimeter. It doesn't take much imagination to realize that this setup is very close to what we need for a residential sprinkler system. Sprinkler heads take up less space, but need to be mounted high on the walls, making them less than desirable. As a result, we have a genuine need for ways to hide the suppression system. Crown mold should be developed that allows plastic water lines to be concealed behind it. Decorative accent features of the crown mold could conceal adapted sprinkler

FIGURE 15-6 Retrofitting in some structures is fairly simple by enlarging ceiling beams and columns to hide the pipe.

heads. Techniques and materials to conceal vertical water lines are also needed. Suppose you need to install a vertical ½-inch copper line in your dining room, and prefer not to tear down all the drywall. Decorative "columns" could be designed to snap into the crown mold in the corners of the room, concealing the riser. If you desired to hide the line in the middle of the wall, a snap-in vertical chase might be the answer. It might be installed by cutting a ¾-inch trench cut from floor to ceiling, and then snapping in a flush-mounted plastic "gutter" that pushes insulation and wiring back. The gutter could have ½-inch round plastic clips in it that secure the water line, and a flush mount cover that snaps in place to conceal it. The cover could easily be painted or wallpapered over, making it virtually invisible.

The problem is that we haven't looked for a solution; it's not that one doesn't exist. The fire service has many connections to the building industry, and it's time for us to start explaining to them what we need. The opportunities exist for creative individuals to develop ways to mask residential systems in a retrofit situation that is both cost effective and aesthetically pleasing.

SUMMARY

Residential fire protection systems have been in the background for quite some time. It's probably safe to say that firefighters in general aren't against them, but we certainly haven't done an adequate job of supporting them, either. The facts are clear. They not only are cost effective, but in most cases allow the occupants to escape. They minimize fire loss and make our job safer. There really is no plausible reason for firefighters to not back them.

We now have a national building code to assist us in getting residential fire sprinklers installed in all-new construction. It's time to get our own states and local building officials to adopt them. We need to convince the public that residential sprinkler systems are exactly what we need in times like these, and increase our support of retrofitting sprinkler systems for existing buildings. We need to look at new ideas, such as partial and multiuse systems, in an effort to protect civilian and firefighter lives immediately. The sprinkler advocates have been out there long enough, and it's time for us all to jump in and help push.

KEY TERMS

affable - User friendly or easy to comply with.

consortium - A group of professionals working toward a solution that had not been accomplished before.

dry system - A fire protection system that is charged with a gas (usually air) to the sprinkler heads. When a fusible link on a head reaches a predetermined temperature, the air dispels, allowing water to enter the piping and be dispensed directly on the fire.

FDC - Fire department connection.

FPRF - Fire Protection Research Foundation.

ICC - International Code Council.

IRC - International Residential Code.

NAHB - National Association of Home Builders.

NFPA - National Fire Protection Association.

objective - (analysis) Reviewing facts without bias.

PHCC - Plumbing-Heating-Cooling Contractors Association.

point of sale - A term used to describe the time when a home sells, usually requiring an inspection or upgrade.

PPE - Personal protective equipment.

preaction system - A fire protection system that is charged with a gas (usually air) to the sprinkler heads. When a fusible link on a head reaches a predetermined temperature, the air dispels, allowing water to enter the piping and be dispensed directly on the fire. Although similar to a dry system, a preaction system utilizes a detection device such as a smoke or heat detector to prefill the piping with extinguishing agent.

purveyor - A term used to describe the managing agent of a specific utility, whether publicly or privately operated such as a water purveyor.

subjective - (analysis) Reviewing facts and personal opinions.

teachable moment - A point in time when students are especially likely to receive a message and understand it.

trade-offs - A practice in which an authority having jurisdiction (AHJ) permits a variance from a certain requirement in return for an option that achieves similar goals.

wet system - A fire protection system that is charged completely with a fluid (usually water) to the sprinkler heads. When a fusible link on a head reaches a predetermined temperature, the extinguishing agent is dispensed directly on the fire.

REVIEW QUESTIONS

1. Explain some of the ways that a residential fire protection system differs from a commercial or industrial system.
2. Although cost is a significant concern when considering the installation of a residential fire sprinkler system to a new home, explain some of the reasons why they can be affordable.
3. Explain the extra components needed to provide a residential fire protection system in an area without a municipal water supply.
4. Describe the importance of using fire and life safety education programs to promote residential fire sprinklers.
5. Explain the benefits of a residential fire suppression system to firefighter safety.

ADDITIONAL RESOURCES

http://www.nfpa.org/publicColumn.asp?categoryID=2058&itemID=47749&src=NFPAJournal&cookie_test=1

http://www.usfa.dhs.gov/citizens/focus/residentialsprinklers.shtm

http://www.homefiresprinkler.org/releases/Scottsdale15.htm

NOTES

1. http://www.edinformatics.com/inventions_inventors/seat_belt.htm
2. Fleming, R.. 2010, July/August. Why 13? A primer on how NFPA's sprinkler standard hooked up with an 'unlucky' number. NFPA Journal.
3. Home Fire Sprinkler Cost Assessment; Final Report.
4. Consumers willing to pay for fuel economy, safety, survey finds. consumeraffairs.com. December 8, 2004.
5. Roheim Wessells, C., and J.G. Anderson. Consumer willingness to pay for seafood safety assurances. http://www.thefreelibrary.com/Consumer+willingness+to+pay+for+seafood+safety+assurances.-a017036118
6. Hughes, J. 2010, October 26. Airline passengers fail to realize some trips flown by partners. http://www.bloomberg.com/news/2010-10-26/-confusing-regional-airline-role-needs-review-ntsb-says.html?cmpid=yhoo
7. Leavitt, R. Residential fire sprinklers coming soon! Part 1 of 5. http://phccweb.org/files/PDFs/FireSprinklerRequirements1.pdf
8. http://www.state.me.us/dps/fmo/sprinklers/residential_systems/insurance_discounts.html
9. NFPA Comments on IRC Proposals RB53-09/10, RB54-09/10, RB56-09/10, RB57-09/10. National Fire Protection Association.
10. http://www.usfa.dhs.gov/citizens/focus/residentialsprinklers.shtm. Statistics as of January 2001.
11. Weatherby, S. Benefits of residential fire sprinklers: Prince George's County 15-year history with its single-family residential dwelling fire sprinkler ordinance.
12. Coleman, R.J. 1985. *Alpha to omega; the evolution in residential fire protection.* City, CA: Phenix Publications.
13. http://www.woodtv.com/dpp/news/local/kent_county/Sprinkler_saved_woman_in_home_fire
14. NFPA 13D, Installation of Sprinkler Systems in One and Two Family Dwellings and Manufactured Homes. Table A.1.2(b) Fires and Associated Deaths and Injuries in Dwellings, Duplexes, and Manufactured Homes by Area of Origin: Annual Average of 1986-1990 Structure Fires Reported to U.S. Fire Departments.

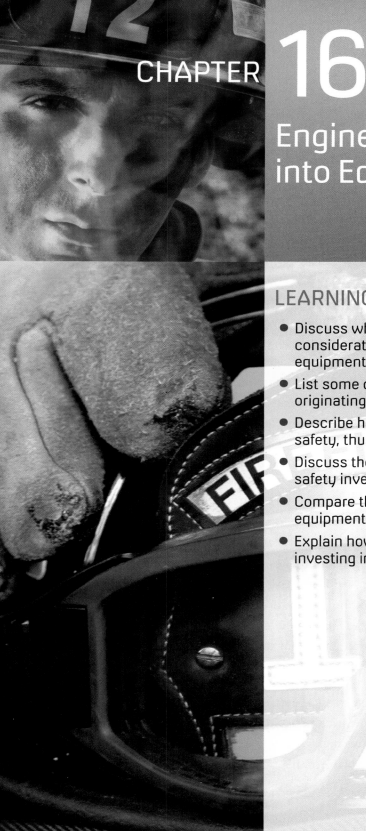

CHAPTER 16

Engineering Safety into Equipment

LEARNING OBJECTIVES

- Discuss why safety should be a primary consideration in the design of apparatus and equipment.
- List some of the groups responsible for originating safety.
- Describe how manufacturers can improve safety, thus better serving the fire service.
- Discuss the advantages of creating a fire-based safety invention clearinghouse.
- Compare the level of safety when designed into equipment versus safety that is retrofitted into it.
- Explain how a cost-benefit analysis can justify investing in safety techniques.

Initiative 16: Safety must be a primary consideration in the design of apparatus and equipment.

Safety is an evolution. Did you ever notice how some equipment seems to continue to become safer with each version? Consider guardrails on the side of a highway, for example. They are primarily designed to protect vehicles from striking immovable roadside hazards, such as bridge abutments or light poles, and to keep vehicles from leaving the roadway and ending up in a river or oncoming traffic. A tertiary design factor is to cushion the collision, and thus absorb some of the energy that could otherwise be transferred to the occupants of the vehicle. Over the past 50 years, they've been tested with wood poles, steel poles, steel cables, and box channel. Beginning in the 1960s, tests began to prove that the horizontal "W" shape was the best at containing vehicles and absorbing impact.[1] However, safety design has most evolved in regard to the end treatments, where the guardrail begins. This area has proven to be especially dangerous because of the angle of impact.

Guardrails are most effective at cushioning a blow at a low angle of deflection, such as a vehicle traveling parallel with it. All guardrails must start somewhere, meaning the end treatment tends to take a direct hit. A simple idea was to simply flare the end, or start at a safe distance off the highway and slowly bring it closer to the direction of travel on an angle. Unfortunately, longer flaring of the guardrail takes up more space and is expensive, and short and sharp-angled flaring proved ineffective at containing a vehicle. Other end-treatment designs took the leading end of the guardrail and rotated it down to the ground, where it was anchored to a footing. Although effective at diverting a careening vehicle from a direct hit, the diversion was usually in an upward direction, at times launching the automobile into the hazard the guardrail was intended to protect.

Most end treatments used today are designed to absorb energy from the crash, rather than divert an errant vehicle. Some are now even capable of sustaining a head-on collision at 70 miles per hour.[2] Other improvements include strategically placed failure points, such as weak poles and truss-like structures. It's now common to see a guardrail

CHAPTER 16
ENGINEERING SAFETY INTO EQUIPMENT
425

FIGURE 16-1 Although vehicles are designed to absorb energy from a crash, modern highway components such as water-filled crash attenuators also contribute to safety.

system installed with holes drilled in the first several wooden posts at ground level, which allow them to sheer off while the guardrail crumples above. Significant research, testing, and design changes have allowed guardrails to improve safety on the roadway (Figure 16-1).

It's probably safe to say that guardrails and end treatments received so many design changes since their inception based on their poor performance. Once engineers came up with a design that solved one problem, another presented itself. Essentially, the evolution of safety was driven by failure. The advantage of this change model is that trial and error eventually can improve safety, but the downside is that once a reasonable amount of success is attained, the design changes tend to cease. In many instances, this is why safety designs sometimes stall for firefighting equipment. Once there is a perception of "safe enough," attention tends to be diverted to another problem.

THE SAFETY CURVE

Electric motors, hydroelectric dam turbines, and even fire pumps use efficiency and power charts to demonstrate what output can be predicted based on input. The standard representation is an x,y graph, with a horizontal and a vertical axis that each represents different data. The horizontal plane (x) can represent effort or power (input), and the vertical plane (y) represents gain in efficiency (output). By placing specific data obtained during testing on the graph, a curve can be drawn to demonstrate what other data could be expected. In most instances, the chart starts with a significant gain in output relative to little improvements in effort. Eventually, the line levels off at a point where further increases in energy produce little improvement. We can use a similar chart to graph different types of improvements, including safety or even tee-ball.

For example, suppose you agree to coach your daughter's tee-ball team. The first practice of 5- to 6-year-olds is quite an eye opener for you. Some children can throw the ball and some can catch, but others have trouble figuring out which hand to put the mitt on. It's clear you have your work cut out for you, and you won't be spending too much time this year teaching them "steal" signs or how to execute a flawless run-down. Instead, you sit down and create a list of priorities to ensure that the kids will progress in their training so they can actually play a game. Similar to an efficiency chart, you rank the skills in an order that will give you the most chance for success and put them in the first phase of improvement.

Improvement Phase 1

The first phase of progression involves the basic skills of throwing and catching, as well as learning the infield positions. Before the next practice, you chalk out circles on the dirt where the infielders are supposed to stand. You draw a big "1" by first base, and do the same for second and third. You also create a drawing similar to what you might find in a coloring book of the baseball field with all the positions on it for them to take home, hang on the refrigerator, and study. At the next several practices, you decide to concentrate on having them wear their mitts but not use them. They instead take turns in the fielding circles, run up to a ball on the ground, pick it up, and throw it to a parent at first base. After they understand the mechanics of picking a ball up from a position and throwing it to first base, they are taught to use their mitt to stop or catch the ball before throwing it. After a couple weeks, they begin to look like they understand how the game will progress.

These improvement phases can also be applied to safety. If you want to increase the level of safety at your training academy, you might rank the factors involved in a similar way to the baseball field. The first of the progressive phases with the best chance for improvement might be identified as wearing sufficient **PPE** for the task involved and maintaining an attentive behavior. Your first day of class would concentrate on the dangers of firefighting and the limitations of the protective equipment. Rather than teaching what the PPE *can* do, you emphasize what it *can't* do. That lecture is expanded even more by concentrating on what it can't do if it's not used properly. By creating a presentation of scene and training-ground pictures, you demonstrate different levels of protection that are required for specific tasks, and what is expected. You also take some time to discuss the appropriate behavior of the students that is required during evolutions. You concentrate on the need for vigilance in regard to attention while using equipment, personal discipline, and the reasons horseplay will not be tolerated.

Improvement Phase 2

At your daughter's tee-ball practice, you now devote your attention to batting, running, and getting the players to listen to base coaches. Your best chance of success is to continue building off what worked in improvement phase 1. For instance, you continue to use your fielding circles, but now attempt to wean them off by scratching the circle into the dirt with a stick rather than using white lime. They practice running the bases, following the orders of the base coaches while other players field a ball and throw to first base. They begin to understand the progress of the game and some of the basic strategies involved. By the time their first game comes along, their progress is apparent as you compare them to the other team.

The fire academy is also ready for another level of safety training, so you decide to use the teamwork approach to teach solidarity and empowerment. You break the class into companies with specific goals and rewards for success. The class learns that on emergency scenes, there's no "franchise player," and that success is gauged on the group's accomplishments. They learn to check each other's gear before entering a dangerous area and watch each other's backs when they are in it. They practice using communication and empowerment with the challenge and response model of **crew resource management (CRM),** and are constantly forced to adapt and overcome obstacles. Failures are evaluated and successes are rewarded.

Although we may not see drastic improvements in safety in the second step, the skills they are learning are setting them up for success in the future. The safety curve has made its arc and is starting to level out. This is also the point in which safety designs might be considered good enough. The problem is that nobody maintains that level without constant attention. The firefighters in the academy will tend to forget what they learned over time, especially if they aren't forced to continue to learn. Building construction, chemical hazards, new technology, and public expectations are all dynamic influences on our safety. If we remain static, our skills will slip and the competitors will overtake us. This is where phase 3 comes in.

Improvement Phase 3

It's obvious that your daughter's tee-ball team will be required to practice as much as possible after the games start if they want to improve. Their skills have leveled off, and they can make it through five innings without any major catastrophes. Unfortunately, some players have forgotten some of what they learned, some have picked up bad habits, and others just simply got lazy. In a similar fashion, once a firefighter completes basic training, the learning cannot stop. In Chapter 5, we looked at how many certification organizations are using continuing education and recertification to improve the skills and knowledge of practicing firefighters. It's important that continuing education is treated as more than just *time*, but actually improves the skills and knowledge of those who attend.

Initiative 16 calls for continued improvements to safe design, and that safety is a primary consideration when purchasing equipment. Just as there are phases of progressive improvement in safety at the fire academy, we can apply the same safety curve to equipment and apparatus. Rules, regulations, standards, and liability all work to ensure that safety improves until it levels out. This is where vigilance on our part can step in and continue to push to make it even safer. The first step is to look at who needs to be encouraged to keep improving.

THE ORIGINATION OF SAFETY

When it comes to safe equipment, procedures, and apparatus, the push for a safer design can and must come from several different directions. The primary reason is because each can be successful in its own way. For example, Initiative 6 looked at ways to improve the health of firefighters by developing health and fitness standards. We examined ways that individuals, departments, doctors, researchers, and technology could all influence positive changes. Designing safety into new and existing equipment also needs to be a team sport.

Working together can also increase the awareness of the improvements. For instance, Chapter 9 examined how near-miss reporting can benefit the fire service. It also showed how part of a comprehensive plan must include ways to disseminate the information. Statistics continue to show that injuries and **LODDs** continue to occur long after similar situations and threats had been identified, yet not communicated throughout the fire service effectively. Initiative 16 could be better supported with the institution of a safe design award program. Providing rewards to groups that bring safe design into use would not only encourage those who strive to make safer equipment, but would also serve as a platform to communicate the efforts and improvements to safer design. Categories of awards could include those who influence safe design, including NFPA committees, research institutions, government agencies, manufacturers, and individual firefighters.

NFPA Committees

The **NFPA** has led the way for standards that are used by fire departments everywhere. According to the NFPA, technical committees are the "consensus body responsible for the development and revision of an NFPA document."[3] They are comprised of manufacturers, end users, and other interested parties. They review existing standards and support revisions every five years. The three major categories of NFPA standards directly affecting operations of the fire service include equipment standards, apparatus standards, and personnel standards.

Equipment Standards

One of the most familiar accomplishments for NFPA when it comes to increasing safety is through the use of equipment standards. Common portable equipment such as **SCBAs** and thermal imagers have been researched and tested in an effort to ensure the safety of anyone who uses them. Sometimes it's not just the equipment, but how it's used in conjunction with other similar equipment. For example, PPE components are commonly tested individually. According to NFPA 1971, Standard on Protective Ensembles for Structural Fire Fighting and Proximity Fire Fighting, 2007, a structural firefighting boot must meet specific criteria to be considered safe. The minimum height for the boot is 10".[4] Structural firefighting pants must also meet certain requirements, such as the trousers inseams being available in 24–36" lengths.[5] However, no test or requirements ensure that hazardous chemicals cannot enter the ensemble between them. As stated in a newsletter released by the Fire Service Section of the NFPA, "PPE items must be worn together and therefore evaluated together."[6] As a result, NIOSH initiated a research project to review ways of conducting "leakage tests" by combining various components of PPE that are normally worn together.

Apparatus Standards

In addition to things such as portable equipment and PPE, technical committees also make recommendations for mobile equipment and apparatus. NFPA 1901, Standard for Automotive Fire Apparatus, covers everything from the weight of personnel on the apparatus to foam tank drains. Although revisions to the standard are made every five years, **tentative interim amendments (TIAs)** are passed as needed between revision dates for important changes. For example, a TIA was issued in August 2009 regarding removable monitors on ladder waterways that could inadvertently fall or be ejected.[7] It called for a redundant securing method that did not have to be performed by an operator. The TIA was likely influenced by the recommendation of a NIOSH LODD investigation in early 2009.[8] The report suggested that manufacturers should "provide aerial ladder trucks with secondary stops or other mechanical means of preventing inadvertent waterway separation or launch." The investigation had also uncovered 10 separate incidents involving four different manufacturers in which a removable monitor had been displaced or launched.

You might notice that the TIA doesn't specify how a removable monitor should be attached or how it might be accomplished without the operator being involved (such as inserting a secondary pin). NFPA standards many times place that responsibility back on the manufacturer. This allows manufacturers and engineers to be part of the solution, without excluding specific companies from competing. Apparatus manufacturers might start by compiling a list of ways to provide the protection, and then weigh them based on the capabilities and operational requirements of their equipment, the cost, and the ease of retrofitting. It could be, in this example, that the design changes involve the manufacturer of the monitor. It's possible that the monitor itself could be designed in a way to assist apparatus manufacturers in complying with the new standard more economically.

Personnel Standards

In Chapter 5, we looked at the professional qualification standards for personnel. Like equipment and apparatus standards, personnel standards change as well. For example, in 2009, a new standard was developed to assist in the application of **NIMS**. NFPA 1026, Standard for Incident Management Personnel Professional Qualifications, describes the **job performance requirements (JPRs)** of each position in the command system. Suppose you are assigned as the **staging area manager (SAM)** at a large incident. Section 9.3.3 of the standard lists some of your initial duties, which include the following:[9]

- Assemble strike teams, task forces, and single resources
- Assemble documentation supplies and communication equipment
- Assemble priorities and coordinate the work within staging

Without standards for personnel, it's likely that some duties might be overlooked at a complex incident, or different positions would overlap. For major incidents such as these, it's imperative that everyone is on the same page.

Research Institutions

Research institutions that are the most involved with firefighting are government-based and educational institutions. As referenced in Chapter 7, the **National Institute of Standards and Technology (NIST)** and **Worcester Polytechnic Institute (WPI)** are two such institutions. Many times, the fire service makes decisions based on what we've been taught or what feels right. Research institutions, on the other hand, prove or disprove what we thought we knew (Figure 16-2). Although they operate from outside the fire service, their studies contribute greatly to firefighter safety.

Governmental Regulations and Standards

Many times, safety designs are adopted from automobiles and are configured for emergency apparatus. Over the years, the **anti-lock braking system (ABS), supplemental restraint system (SRS),** and rollover protection have all contributed to safer apparatus. Additionally, the U.S. **Department of Transportation (DOT)** will periodically make safety requirements to trucks that carry over to fire apparatus or other aspects of our job.

For instance, in 2009, the **Federal Highway Administration (FHA)** made changes to the **Manual on Uniform Traffic Control Devices (MUTCD)** in regard to roadway signs. The new rule required the phasing in of fluorescent yellow-green for school area signs, and permitted its use for bicycle and pedestrian signage.[10] It's clear that the intent of using the new color is for identifying people in traffic, especially when observing its prominent use on emergency responder traffic vests. Unfortunately, many companies have begun to produce traffic cones, election signs, and advertising

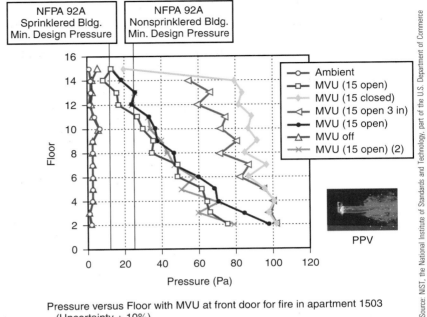

FIGURE 16-2 Research institutions use scientific methodologies to prove or disprove what we thought we knew about things we use every day, like positive pressure ventilation.

Pressure versus Floor with MVU at front door for fire in apartment 1503 (Uncertainty ± 10%)

signs with the same color. It may be only a matter of time before drivers become accustomed to the color much like we did orange. A push to restrict its use for *people* near a highway may be a significant benefit for safety.

Manufacturers

Industry has also been a significant factor in bringing safe design to the consumers. Outside the fire service, Cirrus Aircraft® was the first manufacturer to provide an airplane parachute standard on all models. The **Cirrus Airframe Parachute System**™ **(CAPS**™**)** is integrated into the construction of the plane. The parachute and deployment method were created by Boris Popov, owner of Ballistic Recovery Systems®, after he survived a hang gliding accident.[11] Although he had a *personal* interest in seeing parachutes become standard, all manufacturers have a *corporate* interest in sales. It's likely that the safety record of a specific airplane will play into the decision of a customer. Although many safety standards are required through regulatory agencies, sometimes manufacturers such as Cirrus Aircraft® are responsible for significant design changes for safety.

Some of our equipment has rounded out at the top of the safety curve and has been ignored for quite some time. We need to realize that there is still plenty of room for improvement in equipment we take for granted. Take an aluminum ground ladder, for example. Many firefighters can recall a raised ground ladder sliding, kicking out, or possibly even tipping over at an incident. Although it should be secured at the bottom and tied off at the top, sometimes anchors aren't readily available and sufficient personnel aren't on hand to assign to the duty. The fire service has predominantly moved toward aluminum ladders due to their strength-to-weight ratio as well as their cost. One downside to the ladder that is not commonly considered is the lack of friction when resting against a wall. An aluminum extension ladder that leans on an aluminum-wrapped parapet or an aluminum gutter has little surface area of contact and virtually no friction to help hold it in position. Add rain or ice into the equation and you have a situation in which an unsecured ladder can slide laterally with the slightest of wind. Besides damage to the ladder, it could strike someone below or trap firefighters on a roof or upstairs.

Aluminum ground ladders haven't changed much since they were mounted on a 1942 Mack fire engine, and they could use a safety upgrade. Some sort of friction device stamped into the edges of beams or adhesive strips similar to the tape used on stair treads might increase the holding power. Framing hammers have used a "waffle" head for years to improve the nail-driving ability of a less than perfect swing by increasing lateral friction. Pompier ladders were a single-beamed ladder designed to hook into a window on a floor above, allowing a firefighter to climb up. The hook included a series of teeth similar to a handsaw. If the top 5' of the last fly section had beams consisting of similar teeth or another friction device, the ladder might be more stable than the design we've been using for 70 years. It could be that a total change in design could allow better ways to raise, extend, and secure a ladder against a building.

One way that manufacturers could assist fire departments in improving safety is to identify how they involve safety in their design process. Departments that purchase equipment could be provided with information from the manufacturer about common injuries associated with similar equipment, and what they have done to reduce the risk of injury with theirs. Many already supply training information on a CD or DVD, and could further address the reduction of injuries with their equipment with greater detail. As discussed in Chapter 4, the safety messages would be interwoven throughout the manual rather than being tucked in the front. The benefit to the firefighters is an improved cognizance to safety during design and production, and the advantage for the manufacturer would be to reduce injuries with their model. Besides limiting their liability for injuries, a company that can claim they have the safest roof ventilation system would likely sell more than the others. This would be especially true if fire departments had Initiative 16 in mind when purchasing (Figure 16-3).

Firefighting Design

Not all equipment needs to follow the industry's or NFPA's lead for safety standards. A large portion of equipment and techniques used on the fireground today were adopted and adapted, or created and perfected by the people using them. Both portable and mobile equipment have been invented or improved to make a job easier or safer.

FIGURE 16-3
Manufacturers are in a position to build safety components into equipment, such as a cable cutter on a medical helicopter. This final layer of safety may be the last chance to prevent a crash in the event of the chopper striking a cable in flight.

Portable Equipment

Although countless portable tools have been invented by firefighters, there isn't a much more appropriate choice for an example than the Halligan tool. Designed by Hugh Halligan of the FDNY, it soon became the industry standard for forcible entry.[12] Many companies now sell variations of the Halligan, but even 70 years later, the fire service has been unable to perfect the tool much better than its original design. In a similar fashion, other firefighters have introduced new nozzles, hoselays, fire retardant gel, and even home extinguishing agents over the years.

The fire service could benefit from a clearinghouse for inventions or suggestions to improve safety. If private invention submitting companies can help people get their inventions to market, why can't a nonprofit extension of the fire service accomplish the same thing? The main advantage would be that safety could be evaluated and designed into it. If properly trained and interested individuals were assembled into a committee, they would be able to review the equipment or technique, compile specific concerns, and even submit it for testing. If a firefighter invented an ingenious piece of equipment that actually had financial incentives, it could be easily submitted to fire manufacturers involved with the committee.

Apparatus

Along with the many inventions created by firefighters in the line of hand tools, occasionally members of fire departments join forces to invent their next apparatus. Although it's not uncommon for the membership to give suggestions to the fire chief or an apparatus committee, it is possible to take the model one step further and design an apparatus that never existed before. Suppose your volunteer fire department protects a rural area with no pressurized hydrants. As a result, you respond to structural fires with an engine and two water tenders. You also have automatic aid dispatched, sending more water tenders for reported structural fires. The second apparatus out after the attack engine is a water tender with a front-mounted pump. Its primary duty is to back into a driveway or park on the roadway, and supply the attack engine that completed a forward lay in a driveway. The front-mounted pump has proven effective at pumping water from a portable tank in front of it, at the same time keeping the roadway open to accommodate the water shuttle.

Although your fellow members have become proficient at streamlining water shuttles, the primary complaint is the safety of the personnel. When the truck was built at the factory, no safety override was installed on the front-mounted pump. Essentially, an operator could set the parking brake and get out of the cab while the automatic transmission was still in drive (pump and roll). The operator could walk to the front of the truck, activate the manual pump shift, and begin to pump water. As the throttle was increased, the truck could eventually overpower the parking brake, allowing the apparatus to lunge forward and crush the operator. As a result, your department added a warning light and buzzer to the pump panel to warn if the vehicle was not in park. Wheel chocks were also provided in an effort to keep the truck from moving. However, it still had the potential for injury.

As far as safety is concerned, another problem is that the portable water tank stored on the side of the vehicle is difficult to unload with two people. The aftermarket bracket swings down to near-ground level, but the 150-pound tank still has to be lifted out of its cradle. It then has to be dragged around the truck to the front where the pump is. Worse yet, history has proven that for most of the responses in which the portable tank was needed, only one person was on the vehicle. As a result, one person unloaded and deployed the tank. Although no injuries have been reported, they are sure to occur if the practice continues.

Your research of modern water tender design doesn't help much. When you contact dealers and manufacturers, you find that there are no companies that offer a way for one person to safely unload a portable tank. One manufacturer offers a way to set a 2000-pound cribbing box on the ground, and another builds a rescue truck that can deploy a full stairway off the back. You find one company that can unload ground ladders at the touch of a button to the ground, and even one that can lower the entire hosebed for ease of loading. However, portable tank brackets are still the same as when you bought yours.

Discussions at the fire house lead to sketches on napkins and plastic models of a better design. A small working model of a portable tank bracket is constructed that unloads easier off the back of the truck. The firefighters agree that a rear-mounted pump might be safer than one that is front mounted:

- It puts the pump right where the portable tank will unload.
- It allows the driver to pull into the driveway, keeping the operator off the roadway.

CHAPTER 16
ENGINEERING SAFETY INTO EQUIPMENT
435

FIGURE 16-4
Firefighters have always excelled at creating better ways to do their job. The time has come to use that gift to make the job safer.

- It still keeps one lane of traffic open for the water shuttle.
- It eliminates the chance of running over the operator.

You then pick an apparatus manufacturer that will accommodate your design. Although none agree to build the bracket for you, you pick a local fabricating shop that builds it for you. When the truck is completed and the bracket system is installed, training ensures that one person can unload the portable tank and supply water from a draft in less than three minutes[13] (Figure 16-4).

The downside to the plan is that it was never officially tested or certified. You believe that significant thought and testing was completed and that the new product is much safer than the old, but you still don't know. Our existing fire service has no mechanism for introducing it to the fire service for testing, improvements, and acceptance. In your case, you weren't interested in a patent or a check. You just wanted to be able to do the job better and safer. If more departments can utilize the idea and prevent injuries, you are even more satisfied.

DESIGNING SAFETY

One of the most effective ways to incorporate safety is during the initial design process. This book has referenced the comparison of bolt-on protection versus built-in protection numerous times. Like the other examples, designing safety is no different. Other organizations have understood the benefits of built-in protection, including **NIOSH** when they established a program called **prevention through design (PTD).** By reviewing injury data, all new equipment and techniques can be engineered to prevent future injuries. The process involves four functional areas of stakeholder input, including research, education, practice, and policy.[14]

Research

PTD sometimes works with research organizations in an effort to establish the injury prevention needs of industry, and design equipment better. When researching injuries, they can be categorized as either specific (acute) injuries or repetitive motion (chronic) injuries.

Specific Injuries

Acute injuries are produced when a specific event causes an immediate injury, such as a burn, a fractured leg, or a laceration. The cause and effect of specific injuries are relatively easy to investigate, but care must be taken to ensure all contributing factors are exposed. Suppose your fire department was dispatched to an unattached garage fire. As a part-time firefighter, you and Bobby are on duty. You don your gear, verify the address, and climb aboard in the officer's seat. You calm him down and ensure both of you are seat-belted. Although he's been on the department for a year and a half and is checked off to drive the engine, he doesn't have much fire experience and you can tell that he is pumped up. You call dispatch and advise your response, and the dispatcher replies that there are several calls on it and that a propane grill and a car inside are involved. You ensure that automatic aid departments have been dispatched and return your attention back to Bobby's driving. Your calming him has done an effective job of making sure he drives safely.

As you turn the corner into the subdivision, you can see the black smoke rising by the cul-de-sac and review what the plan is. Bobby is to stop at the hydrant before the driveway and you will exit the engine to wrap it with the supply line. On your signal, he'll forward lay up to within about 200' of the garage and set the pump. After he has the pump engaged, you tell him to hook up the supply line to his intake and you'll pull a 2½" handline off the crosslay with the smooth bore nozzle to knock down the fire. The second unit on scene will charge the supply line. Everything goes according to plan, and as you make it up the driveway to the engine, you find that neighbors are helping Bobby pull your attack line off the driver's side. You ask them to stand back while you don your SCBA mask and Bobby charges the line. You are surprised to find the line still uncharged by the time you are ready to advance, and look up to see him standing on the side step by the pump panel, yanking the discharge valve with both hands. Suddenly, the valve opens and the hoseline jumps, striking Bobby in the cheek. He appears okay, so you open the line and begin to knock down the fire from the driveway. You note that the pressure is much too high, and motion for him to lower the pressure. It's then that you see the red streak running down the front of his turnout gear.

After the other companies arrive and the fire is extinguished, you make it over to check on Bobby. He's got a bloody paper towel covering his cheek, and shows you a pretty deep laceration high on the jaw that you determine is going to need some stitches. You get someone to take his spot, and send him off to the hospital for treatment. It's pretty obvious that the cause of the injury was a hoseline that suddenly jumped when it was being pressurized, but Initiative 9 calls for the thorough investigations of all near-misses, injuries, and fatalities. As a result, you and Bobby both write statements and you assist the fire chief in conducting a full investigation. Your standard investigation worksheet addresses several possible contributing factors that are addressed, including proper equipment, positioning, technique, and design considerations.

Proper Equipment. Safety equipment or PPE is designed to be the last line of defense. Although the results of your investigation showed that Bobby was wearing proper PPE, including his helmet with the chin strap attached, it did little to prevent the injury in front of the ear where it occurred. The investigators agree that the helmet could have prevented the hose from causing an injury in other areas of the head, but existing safety equipment could not have prevented the injury.

Proper Positioning. Bobby was definitely in a risky position while charging the line. Although he wasn't straddling the hoseline, his head was dangerously close to the hoseline as it became charged. Statements by Bobby explained that he had tried several times to pull the valve handle from the ground, but was unable to open it against the pressure. As a result, he climbed onto the step and leaned out to use his weight to open the valve. When the valve did move, it opened quickly and the hose reacted violently to the water hammer. The hoseline jumped upward, striking Bobby in the face. Therefore, the positioning was correct to complete the task the way he intended, but the technique was a contributing factor in the injury.

Proper Technique. The investigation discovered that Bobby's technique to open the valve was wrong, primarily because he didn't reduce the pressure on the valve before attempting to open it. Further investigation showed that many other firefighters on the department were trained the same way to use the electronic pressure control on the pump panel, specifically, the preset button. In other words, the pump operator would engage the pump and open the tank discharge valve, and then allow the pressure governor to increase the pressure to a preset amount against closed valves. The preset on this particular engine was determined to be 180 psi, which had been based on a friction loss calculation using a 200' preconnected, $1^3/_4$" handline flowing 160 gpm with an automatic nozzle. The hoseline chosen for this particular fire, however, was larger in diameter and was equipped with a smooth-bore nozzle and a $1^1/_4$" tip. You calculate the required pump pressure at only 91 psi, but the preset was double that. This not only explains why the valve was difficult to open, but why you had to have him reduce pressure before fighting the fire.

The investigation identified the root cause of the injury as poor training. When you first received the new engine several years ago, the sales representative conducted some extensive drills on the new electronic throttle, but most of the members who attended that session don't even pull shifts anymore. In fact, it appears the newest firefighters were only trained on what sequence to push the buttons to get water, rather than on specific instances when the preset button might be advantageous and when it should be avoided.

Proper Design. The final consideration of investigating the incident relates to the design considerations that contributed to the injury. The first and most obvious is misuse of the electronic throttle. Although the intention of a preset button is to slowly raise the pump pressure to a predetermined pressure, valves aren't designed to be opened and closed easily against higher pressures. The valves themselves haven't undergone any significant changes in the past 40 years. The ball in the valve might be plastic now rather than brass, but little has changed in the way a quarter-turn $2\frac{1}{2}$" valve operates under pressure. NFPA requires speed controls on larger valves, but this ball valve would still be acceptable. It is possible that a better design for all valves is achievable that would allow charging lines slower at higher pressures without the dangers of water hammer.

Another design consideration is the location of the hoselines. Crosslays are intended to provide quick and easy deployment of preconnected attack lines in a convenient spot close to the pump. Unfortunately, they put pressurized lines in close proximity to the pump operator. In Chapter 8, we looked at how technology might make the area around pump panels safer to operate (Figure 16-5). Changing the way a pump panel is designed could reduce the chance of injury from a hoseline. For example,

FIGURE 16-5
One of the areas with the most potential for injury is around operating fire apparatus.

the area where speed lays are deployed could be more isolated from the pump panel, or a solid face could be developed that hinged downward to the ground as the hose was deployed. This could house an oversized duct that the hose was loaded through, providing some burst protection close to the vehicle. It's also possible a similar idea could be used by loading hoses through some sort of flexible protective sheath that also reduced whipping of the lines.

Finally, existing helmets don't necessarily protect operators from hoselines that can suddenly jump or burst. Very similar to the guardrail example in the opening paragraphs of this chapter, we have to ask how safe is safe enough? Structural firefighting gear is designed to protect firefighters from the hazards of firefighting. The design also assumes that other equipment will be worn. The mask is designed to cover the face, the helmet strap is designed to not interfere with the mask, the helmet shield swings down over the mask, and the hood is designed to make a tight fit around the mask. Bobby only needed his helmet in this case, which may not be an effective design for stand-alone use. A thorough review of the nonfirefighting needs of head protection might expose a fairly significant design flaw. When discussing the needs of a pump operator's head protection, we can identify the following:

- Blunt force protection to the face and head
- Redundant eye protection
- Hearing protection and radio communications
- Climate control based on environment

Compare this list with the abilities of existing equipment. You'll probably find that some pump operators don't use hearing protection or that they take off their helmet when it's too hot outside. Most don't have shields down or goggles on, so the eyes are seldom protected. How effective is the brimmed helmet from preventing injuries caused by slips on the ice or trips over hose? Pump operators have no protection that could have prevented Bobby's injury. It may very well be that pump operators need a significant makeover when it comes to helmet design. As discussed earlier, the European models offer much more protection, but have little chance of acceptance without a documented need. Research may show that pump operators have such a need.

Repetitive Motion Injuries

Repetitive motion injuries, on the other hand, are cumulative in nature. **North Carolina State University (NCSU)** found that numerous repetition injuries were occurring in the construction industry as a result of continuous nail gun use.[15] For example, construction workers who were building a condominium would build the floor structure, and then sheet the floor with plywood or particle board. To complete the task, they were forced to bend over and pick up the nail gun to fasten the decking in place numerous times. As they secured the sheet to each floor joist, they were forced to walk in a hunched-over position, leaning forward to place the nails. With each fastener, the kick of the nail gun would create symptoms that could lead to carpal tunnel syndrome. In the short term, workers notoriously complained of back pain and took frequent breaks. In the long run, injuries and even disabilities were possible.

As a result, researchers at NCSU developed a prototype by installing a long, curved handle and a pair of rear wheels to the nail gun. The front of the tool that deployed the nails would rest on the floor. The workers found that they could walk upright, and operate the tool from a trigger on the handle. Not only was bending over minimized, but the design of the handle absorbed most of the recoil. Workers who tested the new device approved of its ease of use, although the main intent of the improvements was for safety.

Exposure to sirens, patient lifting, and pulling a ceiling during overhaul can all lead to a chronic injury. Suppose a research organization noticed that firefighter injury data show an above-normal chronic shoulder injury rate for overhaul operations when firefighters are using hand tools to pull ceiling. According to Debbie Swartz, a registered nurse and ergonomics educator at Palo Alto Medical Foundation, there are several risk factors to an activity that can lead to a repetitive motion injury, including forceful exertion, awkward position, compression, vibration, and poor physical health.[16]

Forceful Exertion. Pulling ceiling to expose hidden fire with a plaster hook or pike pole takes forceful exertion. The first motion is to thrust the tool upward in an effort to pierce the ceiling material, and then rotate the tool and pull down to remove the ceiling. If you've completed the task before, you'll know that different building materials make the job easier or more difficult. Piercing a lath and plaster ceiling is difficult enough, but pulling it can be equally as troublesome. The job can be one of the most exertive duties on the fireground and therefore is a component of many physical agility tests. Unfortunately, it also takes place later in structural fires when fatigue has already set in and injuries are more likely.

Awkward Positioning. Another problem with pulling ceilings is that completing physical tasks in a bad position puts unnecessary strains on the body. Leaning forward to reach over a stairwell or reaching for a vaulted ceiling puts you in an unstable position. Many areas, such as closets or hallways, don't allow a stable stance while working. Pitching coaches constantly evaluate the mechanics involved with baseball pitchers to ensure that they maximize the physics of their bodies and minimize injuries. Firefighting is no different.

Duration. The amount of time involved in completing a task also contributes to the likelihood of an injury. Limited numbers of personnel on the scene force existing firefighters to complete more tasks and work longer. Modern architectural designs put steep roofs and large attics in new two-story homes. As a result, many fire departments are finding themselves attacking a fire in a bedroom, and then immediately pulling as much ceiling as possible to open up the attic. The only thing keeping a room and contents fire from extending to the attic and burning the top off the house is ½" of drywall.

Compression. Reducing blood flow to muscles by compression also can lead to repetitive injuries. PPE such as firefighting turnout gear and SCBAs can cause compression. SCBA straps take the weight of the cylinder and pack and distribute the weight on the shoulders and waist. By lifting your arms over your head, you move most of the weight to the shoulders. As a result, blood flow can be reduced to the muscles that need it the most. Compression also has to be taken into account when realizing that much of the time the neck is in a hyperextended position to look at ceiling. The weight of the helmet, hood, and mask contribute to more pressure being exerted on the cervical spine. The research study in Indianapolis referenced in Chapter 6 noted that compression was much more prominent on the fireground than expected, especially when completing search and rescue.

Vibration. One of the most common forms of repetitive motion injuries outside the fire service is carpal tunnel syndrome. Many times, vibrations are a contributing factor. Jackhammers, chainsaws, and other machines that vibrate the hands are often to blame. Although most overhaul activities involve hand tools such as a pike pole, the constant force exerted (especially when the hook inadvertently strikes a ceiling joist) can also take a toll on hands in a similar matter. Sometimes overhaul includes the use of reciprocating saws and other gasoline-powered saws that produce similar vibrations.

Poor Physical Health. The final risk factor identified to effect repetitive motion injuries includes a medical history that can exacerbate the effects of everything mentioned above. Conditions such as diabetes, obesity, hypertension, heart disease, and lack of exercise all contribute to the body's inability to prevent injury and heal itself.

Education

The next step in PTD is education. Take the Alaskan fishing industry, for example. In an effort to reduce fatalities in one of the most dangerous professions in the country, the United States Coast Guard first reviewed data on fatalities. In a report issued in 2010 regarding fishing industry risk factors and recommendations, NIOSH identified many ways to reduce deaths, as well as the success of implementing the recommendations. Besides more inspections of the vessels, increased education of the crews was initiated. Those recommendations led to a 60% reduction in the fatality rate to crab fisheries in the Bering Sea.[17] Mandatory drills included "man overboard" and "abandon ship" to ensure that everyone knew exactly what to do in an emergency. The standard use of personal flotation devices and maintaining a watertight integrity to the ship were also encouraged.

Reviewing firefighter injury data shows that a significant cause of injuries is slips and falls. Similarly, the navy noticed a significant portion of lost time injuries were in their firefighting divisions. As a result, the **Firefighter Injury Prevention Training (FIPT)** project was initiated.[18] It included a two-year study that found that roughly 40% of the injuries were slips and falls, and 40% of those were attributed directly to lifting patients or getting on and off fire apparatus. Although these three duties can be made safer with design, the most drastic improvements in the safety curve can be gained by education. The FIPT concentrated its efforts on education, creating videos that display proper techniques. It is true that better footwear, more hand holds, lower steps, and a smarter location for equipment can all make apparatus safer, but it still comes down to how you climb up and down.

Practice

Once research has shown a way to reduce injuries and workers are educated, the next step is to put it into practice. Some forward-thinkers in the construction industry have implemented interesting changes that take designing safety to the next level. Safer Design® is a nonprofit organization with a goal of "designing out" hazards through stages of construction and the entire life cycle of a building.[19] Through research at the University of the West of England, the **safety targeted assessment via gateway evaluation (STAGE)** was created. Two specific examples include worksite layout and future planning.

Worksite Layout

Planning the worksite ahead of time can increase safety and productivity for large commercial and industrial buildings. Many of the features used by a safety conscious construction company can be identified and considered by the fire service while completing fire inspections, prefire plans, and company walk-throughs.

Hazard Identification. The first step for a safe worksite is to identify the existing hazards. Underground and overhead utilities must be identified and marked. Significant hazards, such as overhead high-voltage lines, might be fenced off the entire length between poles with orange construction fencing rather than just have a sign pointing up. The organization suggests that the entire site can be secured with chain link fencing, and security cameras and lights can be installed. Areas prone to flooding are identified and can be provided with extra drainage or pumps.

Plot Plan. The next goal is to produce complete drawings of the worksite. The drawings include a footprint of the new building, but concentrate on the site around it. Roadways are designed to bring workers into one area, where they can park and have easy access to restrooms and a weatherproof break room. A "delivery loop" roadway, designated for delivery trucks and emergency vehicles, should encompass the building. The loop eliminates the need for trucks to back up. Lay-down areas, where materials will be stored, should be located adjacent to, but on the outside of, the loop and access points and lift areas for cranes should be located on the inside of the loop in close proximity to the building. Fire and emergency access is designated and marked ahead of time to ensure that access is always available.

The difference between this type of plot plan and normal drawings is the amount of strategic planning that is involved. The same amount of effort should be placed into prefire plans used by fire departments. Suppose a four-story office building was being constructed in your city. If the construction company used a well thought-out plot plan such as this, it would be a great reference for responding companies during construction. It also would serve as a great template for a thorough prefire plan once construction is completed. Apparatus placement, incident command, and rehab could all be considered. Mass casualty incidents could also be planned, including the location for effective triage, treatment, and transport.

Site Preparation. Site preparation would take the components of the plot plan and build an area that reduces the chance for injuries and mishaps. Water lines and underground utilities would be installed first, ensuring that hydrants were available for fire protection. Compacted materials or concrete would be used for all roadways, crane lift points, and material lay-down areas. No cranes would set up in the mud, but would instead use the delivery loop to travel to the next access point that was strategically placed to reach specific areas of the building. Even pedestrian traffic would be planned, preventing personnel from being in an area where they could get injured. Crosswalks should be clearly marked across the delivery loop, and on the perimeter of the material lay-down area. Essentially, very few workers would have muddy boots, which in itself reduces slip hazards in the building.

Although we are unable to paint crosswalks at incidents, we could stand to improve the crowd control at an incident or training. If you've been to a live burn where an acquired structure was used for training and then allowed to burn down, you've probably seen firefighters in dangerous areas without sufficient protection taking pictures or looking in the windows (Figure 16-6). They may climb back into a crushed car after an extrication to retrieve a piece of equipment without their gear. It could be that they simply meander around a dangerous scene too long after their work is completed. We have to remember that just because the incident is over doesn't mean we are not in any more danger.

Building Practices. Safety continues throughout the building process. The biggest contributor is the prefabrication of building components on the ground whenever possible. Falls from a height have been a concern for the construction industry for quite some time. Their answer is really a pretty obvious one: to remove the height from the work. Any components that can be assembled on the ground and lifted can reduce the chance of a fall. For instance, a steel stairway could be fabricated on the ground,

FIGURE 16-6
The safe design of an incident scene or training site means planning for apparatus and emergency workers.

including handrails, landings, and even support columns. The structure can then be set in place as one unit by a crane.

Another change in the planning process is to build safety first. Stairways are now installed early in the building process (sometimes before the next floor) in order to reduce the need for ground ladders and to provide an anchor point for safety harnesses. Floors and roofs are completed sooner, which provide solid surfaces to work from and reduce the chance of falls. Where floors cannot be provided early, edge protection and safety netting can be used. Most components in the structure are designed to bolt together, reducing the need for welders and torches in the structure.

We might also adopt some of these practices. Whenever work can be done from the ground, we should do it. Ventilation saws must be started on the ground and warmed up before being shut down while climbing. Many times, videos will show crews on the roof trying to start a cold power saw that may be out of fuel before giving up on it and climbing back down. We need to put safety systems in place as early as possible as well. Windows should be laddered, exits cleared, and hazards identified as soon as possible. Some hazards should be removed, and where they can't be abated, they need to be clearly marked or blocked off.

Future Planning

The safe design of the building also takes into consideration maintenance and use of the building and grounds long after the cranes pull out. Effective drainage is designed to ensure no standing water that could freeze or breed insects remains. Planning for maintenance and service vehicle access is important and can be achieved by keeping components of the delivery loop or crane lift points. The service access could be restricted through the use of gates or removable barriers. Building components adjacent to the service access road would have crash protection on columns and other supports as a precaution.

Even window washing and other normal maintenance duties can be planned for to eliminate falls. Parking lot lighting poles can be designed to lower the lights, and wall lights in stairwells can be installed on landing surfaces, reducing the need to set up a ladder on an incline. Well-designed furniture details can be used to ensure that ceiling lights shine on walkways and aisles, and to prevent maintenance over filing cabinets and desks. Outside stairways can be protected from the weather and can encourage natural ventilation in order to reduce moss buildup. Parking lots can provide pedestrian safety, such as sidewalks with positive barriers from vehicles. Manholes should be located just off the pavement, eliminating the need for workers to work in close proximity to vehicles. Roof access points should be well-marked with reflective paint, lighting, and nonslip surfaces. Rooftop piping and conduits can have bridges over them to eliminate trip hazards, and all roofs can be designed with edge protection.

Policy

The last step is to make it policy. Many organizations have a safety policy, but an effective policy includes all levels of employees, and uses safety as a primary consideration with all decisions. In Chapter 11, we discussed having standard policies and procedures to ensure a consistent level of safety. One of the policies that should be standard is purchasing and equipment apparatus alteration guidelines. Obviously, cost and the ability to perform the desired functions may eventually lead the decision, but a checklist could provide guidance in ensuring that safety is a primary consideration and is thoroughly evaluated at the time of purchase.

RETROFITTING SAFETY

Initiative 16 urges emergency response organizations to incorporate safe design, even when it's not required. History is full of examples of tragedies that could have been avoided by providing safety equipment that was suggested, but not necessarily required. After two **Canadian National (CN) Railway** freight trains collided head-on in Anding, Mississippi, in July 2005, recommendations from the **National Transportation Safety Board (NTSB)** to the **Federal Railroad Administration (FRA)** included mandatory installation of specific safety equipment.[20] Although an exact cause for the crash was never identified, the investigation ruled out locomotive, freight car, track, and signaling as a cause, leaving only operator inattention. The crash occurred a minute after a northbound train being controlled by an engineer and conductor traveled past a stop signal at 4:14 a.m. The southbound train was slowing to pull off a siding to allow the other to pass when the collision took place, killing all four workers. The train that caused the crash was traveling at 45 miles per hour, with no indication of attempting to slow.

The report also shows that **positive train controls (PTCs)** were strongly recommended by the NTSB since 1990, and that the FRA did not have a system in place to control runaway trains when the operator failed to respond to expected actions. Although a PTC is a fairly elaborate system that would have to be phased in, a cheaper fix probably would have prevented the crash. **Alerters** have been available for retrofits on locomotives, which monitor signal controls as the train passes them. If the operator fails to react to a signal, the audible alarm sounds. In the event that an engineer is distracted or asleep, the alerter would gain his or her attention. If no action was taken, most systems would automatically apply the train's brakes. At the time of the crash, 43% of CN's locomotives had alerters, but were only being added on new equipment rather than retrofitted. Canada required the installation of alerters on all lead locomotives in the country, but the United States had no requirement for freight trains. It was assumed that most of the CN trains with the safety equipment were used in Canada, where they were required. If it was as simple as just ordering up another 300 for all their locomotives and installing them on a Saturday afternoon, they probably would have done it. Therefore, effective research needs to examine why they weren't installed.

In Chapter 15, we looked at ways to encourage easier and less expensive ways to retrofit residential fire sprinklers. In the same way, we need to investigate how we can retrofit safety into all our existing equipment. A review of the CN railway's decision not to retrofit their fleet voluntarily, or even an emergency response organization's decision not to supply all personnel with a new piece of safety equipment, could be a result of cost, availability, and need.

Cost

The primary reason for not complying with any voluntary safety standards is many times directly related to the cost. When mandates require safety equipment, it's considered an expense. Anytime voluntary safety costs money, it's weighed as a "potential return on investment." If it's not worth the cost, it's likely to get set aside. It can be assumed that the cost of retrofitting locomotives with alerters is a significant expense, especially if the electronics of older trains weren't compatible with the new equipment. Adding in the cost of installation and the locomotive being out of service for an extended period of time makes voluntary compliance even more unlikely.

Weighing costs against risks happens in the fire service as well. For example, suppose that you read about several failures of fire stream appliances (e.g., gated wyes, siamese fittings, and manifolds) around the country with injuries reported. Upon further review, you find that your fire department uses similar models to the ones that have recently been shown to fail. You are considering the replacement of all your questionable fittings and purchasing another style made of a different material. The cost to replace everything at one time is fairly significant, and would not be easy to accomplish.

Some ways you can help determine your options related to cost include research, phasing, and grants.

Research

The first step to evaluating your concerns versus the cost is to complete thorough research. In this case, you'll need to decide if the fittings you have put your personnel at risk, and how much. In Chapter 3, we used the **SPE model** borrowed from the United States Coast Guard to determine risk based on severity, probability, and exposure. Severity and probability were ranked on a scale of 1–5, exposure 1–4, and were then multiplied together to come up with a risk potential. In this case, you estimate the risk mathematically.

- Severity—Failure could kill a firefighter who is close to it when it exploded. (5)
- Probability—The number that has failed is becoming significant. (2)
- Exposure—A firefighter could be operating the valve when it failed. (4)

By multiplying the three numbers together, we come up with a risk potential of 40, or a slight-to-moderate risk. If we decided to keep the same equipment, we could reduce the risk simply by lowering any of the numbers above. Severity is the highest, but it's unlikely we can reduce it. PPE will only give so much protection from flying shrapnel. If the appliance had some sort of cover placed over it to reduce the severity of a failure, it might be reduced to 3 or 4, but something like that would have to be constructed, carried on the apparatus, and deployed. It also would have to be removed whenever a valve underneath it needed to be operated. Likewise, exposure is difficult to reduce, especially if the appliance in question is a wye or manifold that requires someone to operate the valves.

Probability is already a low number, but could easily be reduced even further. Cutting the probability in half will cut the risk in half. Regular and postuse inspection of the appliances could show cracks or other damage that could lead to failure. Proper setting of relief valves and vigilance while operating the pump can also contribute to a safer operation. The mathematical formula also proves that if we reduce one of the factors to zero, there is no risk at all. Therefore, eliminating the use of appliances whenever possible makes it a nonissue.

A new approach to safe design by manufacturers might start to include pressure relief valves standard on *all* appliances. Small relief valves similar to those found on hot water tanks and air compressors could safely vent some of the pressure in the event of overpressurization, reducing the chance of damage to the unit or firefighters (Figure 16-7).

Phasing-In

When soldiers started becoming regular victims of roadside bombs in Iraq and Afghanistan, the military worked on permanent solutions to the problem.[21] Unfortunately,

FIGURE 16-7 In an effort to prevent catastrophic failure, the use of relief valves similar to those on SCBA cylinders might be designed to work on water pressure appliances.

as new equipment was being designed and fabricated, more **improvised explosive devices (IEDs)** continued to detonate and kill unprotected troops. Brigades began improvising by welding steel plates and other armor to the bottom and sides of Humvees. Although IEDs continue to trouble the military, safer designs continue to be introduced, replacing the older equipment.

A similar strategy should be taken on a local level. For every unsafe piece of equipment that is replaced or improved, the mathematical chance of an injury or death is reduced. It's also important to phase in what you can when it comes to safety. It could be that the most active companies get the new equipment, or that the equipment is added when another component is replaced, such as fire hose.

Grants

Initiative 10 puts an emphasis on the awarding of grants based on their intent to improve safety. A grant program that endorses safety fits well with this example. If there is a genuine need to reduce risk by replacing equipment, many grants available to the fire service can be considered. It's important when specifying the new appliances that research is conducted to ensure that the new equipment is substantially better than the pieces being replaced.

Availability

Another consideration for not complying voluntarily is not having the appropriate equipment readily available. It could be that some locomotives have braking systems that are not compatible with the alerter that is trying to override it. It's also probably safe to say that there aren't a lot of places that manufacture alerters, and that the majority of their business is for new equipment. Voluntary compliance is much more attainable when it is easy.

When it comes to retrofitting fire apparatus, there are several examples of equipment that might not be very easy to purchase. For instance, much of the new apparatus being built has seatbelt warning lights to tell the operator if someone is not buckled in. The system is fairly simple on new vehicles, because the technology already is available to sense when a seatbelt is secured and when someone is sitting in the seat. However, if you wanted to retrofit your 1986 pumper, you would need parts that might not be readily available. It could be that the seats and seatbelts all need to be replaced, in addition to doing extensive electrical work. It just might be easier to look back and see if everyone is belted and order the feature on the next one.

Need

The final reason voluntary compliance can suffer is if the added level of protection is not viewed as a genuine need. We don't have to travel far to find a great example from the previous chapter. If citizens and builders were convinced of a need for residential sprinkler systems, it's likely many would be equipped. As demonstrated, builders continue to argue that there is no need for home fire sprinklers, and that buyers will not pay for them. The more that consumers hear that specific message from builders, the more likely they are to agree with them. The fire service needs to show consumers the facts, and prove that the protection is warranted and a smart investment.

Unfortunately, it's not always a black-and-white matter. Sometimes convincing people of a need is a matter of choosing one safety component over another. For example, some fire stations have a flashing yellow warning light in front to assist emergency vehicles in getting out into traffic. Upon receipt of an alarm, the light turns red. The problem is that if the light flashes yellow every day that commuters drive by on their way to work, it's likely they won't notice the change the one day that it's flashing red. A better option is to have a full traffic light that stays green until it announces your approach, when it turns yellow, and then red like regular traffic lights. In this case, the people most likely to notice the safety concern are those who respond from the station. If the problem is not conveyed, safer equipment will never prevail.

SUMMARY

Initiative 16 encourages using safety as a primary concern when ordering apparatus and new equipment. Although it sounds obvious, most of us put more weight on cost, aesthetics, and even opinions. New improvements to safety can come from several different sources, and each must be encouraged to participate in the evolution. The fire service needs to find better ways to streamline the process. Safety could benefit from encouraging manufacturers becoming more active in firefighter safety, and making it easier for firefighters and their departments to get new ideas tested and developed. The key is ensuring that we never give up on a better design. Even the NFPA utilizes a five-year rotation on the review of standards to ensure that they are as complete as they can be.

Although safe design is better achieved in the planning stages of new equipment and apparatus, a plan to retrofit safety is just as much a part of the process. It's hard to say that engineers and manufacturers need to step up to the plate if we refuse to invest in their improvements for our own safety. It is true that not all safety equipment is feasibly retrofitted. There are times when research will show that our time and investment would be better spent on something else. However, ignoring the fact that safety can be improved with our existing equipment and apparatus would be a mistake.

KEY TERMS

ABS - Antilock braking system.

alerter - A passive safety device installed on a locomotive that warns the engineer when a signal has been passed that appears to have been ignored.

CAPS - Cirrus Airframe Parachute System.

CN - Canadian National (Railway).

crew resource management (CRM) - A procedure adopted from the aviation and maritime industries that recognizes the value of all personnel.

DOT - Department of Transportation.

FHA - Federal Highway Administration.

FIPT - Firefighter Injury Prevention Training.

FRA - Federal Railroad Administration.

IED - Improvised explosive device.

JPR - Job performance requirements.

line-of-duty death (LODD) - Fatalities that are directly attributed to the duties of a firefighter.

MUTCD - Manual on Uniform Traffic Control Devices.

National Institute of Standards and Technology (NIST) - A government laboratory with a fire research division committed to the behavior and control of fire, and providing valuable information to the fire service.

NCSU - North Carolina State University.

NFPA - National Fire Protection Association.

NIMS - National Incident Management System.

NIOSH - National Institute for Occupational Safety and Health.

NTSB - National Transportation Safety Board.

personal protective equipment (PPE) - A generic term used to describe the minimum apparel and gear needed to safely perform a specific duty.

positive train control (PTC) - An elaborate system of active railroad safety devices that take control of a locomotive that fails to respond to a signal that appears to have been ignored.

PTD - Prevention through design.

SCBA - Self-contained breathing apparatus.

SPE (severity, probability, exposure) model - The USCG's risk management tool for assigning a numeric value to the three components of a risky task.

SRS - Supplemental restraint system.

STAGE - Safety targeted assessment via gateway evaluation.

staging area manager (SAM) - A position in NIMS responsible for duties associated with managing apparatus and personnel in staging.

tentative interim amendment (TIA) - A revision made to NFPA standards between revision dates.

WPI - Worcester Polytechnic Institute.

REVIEW QUESTIONS

1. Who are the groups responsible for building safety into our equipment and systems?
2. Why are manufacturers in a position to better serve the fire service by improving the safety of their equipment?
3. What would be the benefits of developing a fire-based invention clearinghouse?
4. What are the four functional areas of the prevention through design (PTD) process introduced by NIOSH?
5. What are some of the reasons why an organization may decide against retrofitting safety components to current apparatus or equipment?

STUDENT ACTIVITY

1. NFPA standards are updated every five years, with TIAs introduced as needed between updates. Pick a standard for equipment, such as NFPA 1901, Standard for Automotive Fire Apparatus, and research older versions to identify some of the ways it has evolved throughout the years. Pay special attention to features that are identified and then later required as technology is introduced.

ADDITIONAL RESOURCES

http://news.bbc.co.uk/2/hi/technology/8422942.stm
http://www.dbp.org.uk/welcome.htm
http://www.railroadforums.com/forum/showthread.php?t=8626

NOTES

1. Sicking, D.L., R.P. Bligh, and R.E. Hayes Jr. 1988, November. Optimization of strong post W-beam guardrail. Research Report 1147-1F. Texas Transportation Institute.
2. Trinity Highway Products Catalog . FASTRACC™. Revised January 2008.
3. http://www.nfpa.org/categoryList.asp?categoryID=819&;URL=Codes%20&%20Standards/Code%20development%20process/Technical%20Committees

4. NFPA 1971; Standard on Protective Ensembles for Structural Fire Fighting and Proximity Fire Fighting, Section 6.10.3, 2007.
5. NFPA 1971; Standard on Protective Ensembles for Structural Fire Fighting and Proximity Fire Fighting, Table 6.1.11, 2007.
6. Williams, Dr. J., and A. Shepherd. 2011, Spring. NIOSH projects. Inside leakage tests. Fire Service Section Newsletter. NFPA.
7. NFPA 1901, Standard for Automotive Apparatus, 2009 edition. Reference 19.6.4.6; TIA 09-2. Issued August 6, 2009.
8. http://www.cdc.gov/niosh/fire/reports/face200812.html
9. NFPA 1026, Standard for Incident Management Personnel Professional Qualifications, 9.3.3., 2009.
10. Manual on uniform traffic control devices for streets and highways. 2009. U.S. Department of Transportation; Federal Highway Administration.
11. http://www.brsparachutes.com/brs_history.aspx
12. Hashagen, P. 2002. *Fire Department City of New York; The bravest. An illustrated history 1865-2002.* Nashville, TN: Turner Publishing.
13. http://www.youtube.com/watch?v=6-VHjdTSYLs
14. www.cdc.gov/niosh/topics/ptd/
15. http://www.ise.ncsu.edu/ergolab/construction.html
16. http://www.pamf.org/healtheducation/media/toyourhealth/repetitivemotion.html
17. Fatal Occupational Injuries in the U.S. Commercial Fishing Industry: Risk Factors and Recommendations Alaska Region.
18. Quigley, S.L. 2009, November 25. Joint program aims to reduce firefighter injuries. http://www.af.mil/news/story.asp?id=123179541
19. http://saferdesign.org/default.aspx
20. Collision of Two CN Freight Trains; Anding, Mississippi, July 10, 2005. Accident Report NTSB/RAR-07/01. PB2007-916301. National Transportation Safety Board.
21. Lenz, R. 2006, January 13. Soldiers do it themselves, improve Humvees. USA Today.com. http://www.usatoday.com/news/world/iraq/2006-01-13-humvees-troops_x.htm?csp=34

APPENDIX A

SOP-1 No-Fault Management

1. Purpose:
    1.1. The purpose of this SOP is to define No-Fault Management (NFM), describe the indications and contraindications for its use, and qualify its adoption by the FLSI-16FD.
2. Revision History
    2.1. New SOP January 2012
3. Personnel Affected
    3.1. All staff and line emergency responders
4. Policy
    4.1. It is the policy of the FLSI-16 FD to invest time, training, education, and other resources into fire and emergency services personnel and expect professional attitudes at all times. It's also understood that many decisions made by emergency personnel are made with limited information, under significant stress, and in compressed time. As a result, it can be expected that decisions made while carrying out official duties may not always be optimal for a given situation. Therefore, the FLSI-16 FD adopts NFM as a core principle, as outlined in Section 7 of this procedure.
5. Definitions
    5.1. No Fault Management (NFM)—A theory that assumes that a failure of any member of a team is a failure of management, and that mistakes can be expected. NFM assures that mistakes that meet certain criteria will not be punished in an effort to make safety and customer service flourish.
    5.2. Event—Any occurrence that had the potential, or did cause injury (to a member of the department, other emergency agency, or the general public), or damage to equipment, vehicles, or station (public or private).

5.3. System Failure—An event that occurs due to a root cause and contributing factors.

6. Responsibilities

    6.1. It is the responsibility for all officers to understand this policy and its intent, as well as its application to foreseeable and unforeseeable events.

    6.2. It is the responsibility of all members to understand the intent and procedures included in Section 7.

    6.3. All members are responsible for safety, and to communicate any observable, imminent hazards.

7. Procedures.

    7.1. Intent

        7.1.1. The intent of this procedure is to produce a work environment that encourages personnel to make the best possible decisions under uncertain circumstances without fear of disciplinary actions.

        7.1.2. Officers must recognize that mistakes or questionable situations can lead to further mistakes and worse situations unless acted on immediately.

        7.1.3. Members must recognize that ignoring warning signs or attempting to hide undesirable effects, results, or outcomes of a decision has the potential to create safety hazards and impeded customer service.

        7.1.4. Properly used NFM will concentrate on investigating the event in an effort to identify the root cause and all contributing factors, and initiate changes to reduce the chance of similar events occurring in the future.

    7.2. Indications

        7.2.1. Any of the following or similar conditions are acceptable indications of applying NFM, provided no contraindications listed in 7.3 are involved.

            7.2.1.1. Any loss to equipment such as vehicle damage, station damage, or portable equipment damage.

    7.3. Contraindications

        7.3.1. NFM will not be used when any of the following apply:

            7.3.1.1. When mistakes or actions are covered up, or an attempt is made to hide the event or hinder the investigation.

            7.3.1.2. When the event is the result of dereliction of duty.

            7.3.1.3. When the individual involved is under the influence of drugs or alcohol.

            7.3.1.4. When the individual knowingly violated safe practices or chose to not use safety equipment.

            7.3.1.5. When there is a pattern of unsafe acts, infractions or events.

7.4. Fact-Finding Phase
  7.4.1. The following steps shall be taken when an event occurs:
    7.4.1.1. Immediately notify your supervisor, who will notify the department safety officer (DSO), the safety committee, and fire chief within 24 hours.
    7.4.1.2. If the event will cause a delay in response, immediately notify dispatch and ensure response coverage is available.
    7.4.1.3. If the event involves personal injury, ensure that medical needs are addressed per SOP-123.
    7.4.1.4. Remove the equipment involved in the event from service if there is any question about its safety, mark it, and secure it.
    7.4.1.5. Ensure that everyone involved completes a NFM-Event Statement Form as soon as possible after the event. Forms will be assembled and forwarded to the fire chief for review by the safety committee.
    7.4.1.6. System failures will be identified by the fire chief and safety committee, and used in the solution-developing phase.
7.5. Solution-Developing Phase
  7.5.1. Training Issues—The training officer will review the results of the investigation and make changes for implementation into an educational or training program.
  7.5.2. Procedure Issues—The fire chief will review the results of the investigation and make changes to the SOPs if it is determined that changes are needed.
  7.5.3. Equipment Issues—The equipment committee will review the results of the investigation and make suggestions for changes as needed.
7.6. Completion
  7.6.1. The fire chief and safety committee will ensure that all solutions have been implemented and are evaluated on an annual basis.
  7.6.2. The members involved in the event will not be disciplined unless conditions in 7.3 were noted, but a copy of the final report will be kept in their personnel files if the action is a pattern, as described in Section 7.3.1.5.

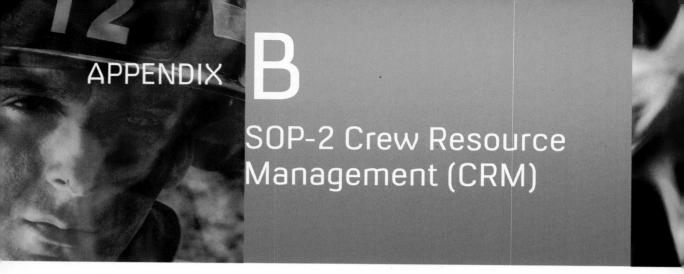

APPENDIX B

SOP-2 Crew Resource Management (CRM)

1. Purpose:
    1.1. The purpose of this SOP is to define Crew Resource Management (CRM), describe the procedure for its use, and qualify its adoption by the FLSI-16FD.
2. Revision History
    2.1. New SOP January 2012
3. Personnel Affected
    3.1. All staff and line emergency responders
4. Policy
    4.1. It is the policy of the FLSI-16 FD to invest time, training, education, and other resources in fire and emergency services personnel and respect each as a vital component of a safe operation. The FLSI-16 FD adopts CRM as a core principle as outlined in Section 7 of this procedure.
5. Definitions
    5.1. Crew Resource Management—A theory adopted from the airline and maritime industries in which all personnel have the responsibility to monitor changing conditions and the authority to present a specific observation to a supervisor or others.
    5.2. Challenge/Response—A method of communication in which a subordinate challenges a supervisor with an observation or concern in a respectful format, allowing the supervisor to make an informed decision and a respectful response.
6. Responsibilities
    6.1. It is the responsibility for all officers to understand this policy and its intent, as well as its application to foreseeable and unforeseeable events.

6.2. It is the responsibility of all members to understand the intent and procedures included in Section 7.

7. Procedures
    7.1. Intent
        7.1.1. The intent of this procedure is to produce a work environment that encourages personnel to maintain vigilance in observing dangerous conditions and forecasting dangers.
        7.1.2. Officers must realize that observations and concerns brought to them by firefighters under respectful conditions do not challenge their authority, but rather provide them with more data to make better decisions.
        7.1.3. Members must recognize that observations and concerns must be brought to their supervisor in a respectful manner, to provide as much information as possible.
        7.1.4. Properly used CRM will utilize all members to observe conditions, develop a prediction, and provide the platform for a challenge/response model.
    7.2. Challenge/Response
        7.2.1. The challenge and response model must be brief but respectful. CRM does not undermine the authority of a person in charge.
        7.2.2. When a member notices a danger that may not be apparent to his or her supervisor, the information should be presented in a challenge/response model as soon as possible. The model will usually include:
            7.2.2.1. A statement of understanding (example: "Captain, I know we need to get in there and search the building.")
            7.2.2.2. A statement of concern ("But look at the smoke coming out of the ground by the bushes.")
            7.2.2.3. A statement of explanation ("Do you think it's a basement fire?")
            7.2.2.4. A statement of alternative action ("What do you think about searching the second floor from a ladder until we find the fire?")
        7.2.3. The supervisor who is presented with a challenge, must consider the information and develop an appropriate response.
            7.2.3.1. The response agrees with the suggestion ("Good idea, we'll hit the second floor bedrooms first and report the observation to the incident commander.")
            7.2.3.2. The response disputes the suggestion ("The basement windows showed no sign of fire, just the kitchen in the back. We have to complete the primary as soon as possible, but we'll take extra time sounding the floor.")

7.3. Evaluation

    7.3.1. CRM is something that must be practiced to be fully successful.

    7.3.2. Post Incident Evaluation (PIE) worksheets include a section for logging all CRM communications as an evaluation tool.

    7.3.3. Suggestions for improving CRM in regards to the training program or SOP should be forwarded to the Assistant Chief of Operations.

APPENDIX C

SOP-3 Residential Sprinkler Retrofits (RSVR)

1. Purpose:
    1.1. The purpose of this SOP is to define Residential Sprinkler Voluntary Retrofits (RSVR), describe the procedure for assisting homeowners with adding fire protection, and supporting the addition of partial protection when sprinklers are not required. Residential fire sprinklers are a benefit to the both the FLSI-16 FD and the citizens we are sworn to protect.
2. Revision History
    2.1. New SOP January 2012
3. Personnel Affected
    3.1. All staff and line emergency responders
4. Policy
    4.1. It is the policy of the FLSI-16 FD to enforce existing fire codes when fire protection is required, as well as to encourage the voluntary retrofitting of homes with residential fire sprinklers when not required. The FLSI-16 FD will therefore assist homeowners and landlords in adding the protection as per Section 7 of this procedure.
5. Definitions
    5.1. Partial Protection—Fire sprinklers in some areas of a building.
    5.2. Site Visit (SV)—The process of promoting sprinklers through a visit by any fire company.
    5.3. Voluntary Retrofit—The addition of fire sprinklers when not required by code.
6. Responsibilities
    6.1. It is the responsibility of all officers to understand this policy and its intent, as well as its application to foreseeable and unforeseeable events.

6.2. It is the responsibility of all members to understand the intent and procedures included in Section 7.

7. Procedures
    7.1. Intent
        7.1.1. The intent of this procedure is to encourage homeowners, contractors, and business owners to recognize the value of fire sprinklers.
        7.1.2. FLSI-16 FD members must enforce existing codes, yet encourage homeowners, contractors, and business owners to add partial protection whenever possible.
    7.2. Training and Marketing
        7.2.1. All members will receive training on residential sprinklers annually by the training division.
        7.2.2. The fire prevention bureau (FPB) and fire and life safety education (FLSE) divisions will actively market and promote the idea of partial protection.
        7.2.3. Post-incident press release worksheets will include fire sprinkler activation and their effect on the incident.
        7.2.4. All stations will be provided with marketing materials and demonstration equipment.
        7.2.5. The public will be encouraged to contact the FPB and schedule a preliminary site visit (SV).
    7.3. Site Visit (SV)
        7.3.1. Interested parties will contact the fire department requesting information about partial protection.
        7.3.2. The first step is for the fire department to make a SV to the location.
        7.3.3. The SV can be completed by any trained fire or rescue company, staff member, inspector, or educator.
        7.3.4. The SV will include the following:
            7.3.4.1. A discussion about the benefits of fire sprinklers. (marketing flyer)
            7.3.4.2. Suggestions regarding sprinkler head locations. (marketing flyer)
            7.3.4.3. Explanation of materials supplied by the FLSI-16 FD. (current list from FPB)
            7.3.4.4. Explanation of materials and costs to be supplied by the customer. (current list from FPB)
            7.3.4.5. Information on requesting insurance reductions. (insurance flyer)
            7.3.4.6. Provide a list of local contractors. (current list from FPB)

7.4. Installation
- 7.4.1. Members are encouraged to assist homeowners and landlords in providing information and suggestions for installation.
- 7.4.2. If a homeowner or landlord is unable to complete the installation, they should be encouraged to contact a contractor on the list. (current list from FPB)
- 7.4.3. Members should not engage in any installation of voluntary retrofits.
- 7.4.4. Any questions should be forwarded to FPB for clarification.

APPENDIX D

SOP-4 Purchasing Policy

1. Purpose:
    1.1. The purpose of this SOP is to ensure that all equipment purchases are made with the consideration of safety, and qualify its adoption by the FLSI-16FD.
2. Revision History
    2.1. New SOP January 2012
3. Personnel Affected
    3.1. All staff and line emergency responders
4. Policy
    4.1. It is the policy of the FLSI-16 FD to provide its members with the best equipment available whenever possible. A significant component of the "best" equipment is the safest. Therefore, purchases will be made after considering safety when comparing new equipment.
5. Definitions
    5.1. PRW—Purchase Requisition Worksheet
    5.2. Significant Purchase—For the purpose of this policy, obtaining a piece of equipment, apparatus, or making a station alteration worth more than 100 dollars. It also includes all pieces of PPE, regardless of cost:
        5.2.1. Uniforms
            5.2.1.1. Class A and Class B Uniforms
            5.2.1.2. Duty Shoes and Boots
        5.2.2. Structural Firefighting Ensemble
            5.2.2.1. Turnout Pants and Coat
            5.2.2.2. Firefighting Helmet, Hood, Gloves, and Boots

5.2.3. Rescue Gear
- 5.2.3.1. Rescue Coat, Extrication Gloves
- 5.2.3.2. Rope Rescue Gear
- 5.2.3.3. Swift Water Gear

5.2.4. EMS Safety Equipment
- 5.2.4.1. EMS Coat, Traffic Vest
- 5.2.4.2. EMS Gloves, Goggles, etc.

6. Responsibilities
   6.1. It is the responsibility for all officers to understand this policy and its intent, as well as its application to all purchases.
   6.2. It is the responsibility of all members to understand the intent and procedures included in Section 7.

7. Procedures.
   7.1. Intent
      7.1.1. The intent of this procedure is to ensure that safety is considered when making all significant purchases.
      7.1.2. Performance, quality, and cost are also components that must be looked at.
   7.2. Process
      7.2.1. All significant purchases will begin with the completion of a Purchase Requisition Worksheet (PRW). The PRW will include six sections.
         7.2.1.1. Requesting Personnel—Included is the apparatus or station where it will be located, personnel completing the worksheet.
         7.2.1.2. Account Information—Explanation or account number where funds are available, including updated budget.
         7.2.1.3. Needs Statement—A brief description of the equipment, a summary of why the purchase is requested, what its intended use is, and if it replaces another piece of equipment. If a needs assessment was completed, attach to the PRW.
         7.2.1.4. Comparative List—A list of equipment considered, including ranking of capabilities, company, safety considerations, and cost.
         7.2.1.5. Suggested Purchase—An explanation of equipment suggested for purchase, including reasons why it was chosen. Safety should be weighted heavily.
         7.2.1.6. Suggested Vendor—An explanation of vendor suggested.

APPENDIX E

SOP-5 Secured Equipment

1. Purpose:
    1.1. The purpose of this SOP is to ensure the safety of personnel in the event of unexpected vehicle movement or a crash. It is vital that all portable equipment be properly stowed to prevent injury while the vehicle is moving.
2. Revision History
    2.1. New SOP January 2012
3. Personnel Affected
    3.1. All staff and line emergency responders
4. Policy
    4.1. It is the policy of the FLSI-16 FD to protect its members while traveling in department vehicles. Emergency apparatus contain numerous types of portable equipment, map books, and various types of PPE, all of which can become a missile in the event of a crash or sudden vehicle movement if not properly secured.
5. Definitions
    5.1. Equipment—For the purposes of this procedure, any firefighting or EMS equipment, PPE, map books, or supplies.
    5.2. Positive Securing Device (PSD)—Any type of retaining equipment used to hold a piece of portable equipment or PPE in place. It could consist of a commercially made bracket, or an approved in-house fabricated bracket, strap, or compartment used in normal operations. In certain situations, it could consist of a seatbelt, backboard, or cot strap, or a rope, netting, or other homemade device used as a temporary measure to allow usage during patient transport, response, or return to the station. Personnel holding a piece of equipment is not considered a positive securing device as it could still slip or become dislodged in a crash.

6. Responsibilities
    6.1. It is the responsibility of all officers to ensure that vehicles that need brackets or repairs are reported.
    6.2. It is the responsibility of all officers to keep the vehicle in a state of readiness while maintaining safety for the members.
    6.3. It is the responsibility of all members to ensure that *all* equipment is properly prior to vehicle movement as outlined in Section 7.
7. Procedures.
    7.1. Intent
        7.1.1. The intent of this procedure is to ensure that no equipment can be dislodged during normal operation of a vehicle, or in the event of a crash.
        7.1.2. Management must recognize that some equipment may need to be secured in devices that must be purchased, and some equipment will need to be stored in a location that could be a detriment to job efficiency.
        7.1.3. Members must recognize that they have a responsibility to secure and properly stow equipment in all situations.
        7.1.4. Properly used, this procedure will reduce the chance of our own equipment injuring us while in transit.
    7.2. Securing of equipment
        7.2.1. Equipment will normally be secured in its designated compartment, bracket, or strap.
        7.2.2. If equipment is needed for patient care (monitor/defibrillator) or is unable to be repacked on scene to fit in its normal location (rolled, wet hose), it must be contained with a PSD.
        7.2.3. If the normal location and a PSD are not options, the piece of equipment must be stowed in an outside compartment.
        7.2.4. If the equipment cannot be transported safely and securely, the officer in charge will call for another apparatus to complete the transport.
        7.2.5. Semi-annual inspection of PSDs shall be performed to ensure that no fraying of straps or other damage has occurred.
    7.3. Requests for equipment
        7.3.1. All vehicles should carry ratchet straps to be used as makeshift PSDs when needed.
        7.3.2. Requests and suggestions for permanent PSDs should be completed on the appropriate "Equipment Request" form.

APPENDIX F

Firefighter Injuries and Fatalities Report

APPENDIX F
FIREFIGHTER INJURIES AND FATALITIES REPORT

Year	Fire Scene Injuries	Other Scene Injuries	Total Scene Injuries	Other Injuries	Total Injuries	Total LODDs	Total Calls	Fires	Medical Aid	False Alarms	Mutual Aid	Haz Mat	Hazardous Cond	Other
1977	**	**	**	**	**	157	**	**	**	**	**	**	**	**
1978	**	**	**	**	**	171	**	**	**	**	**	**	**	**
1979	**	**	**	**	**	126	**	**	**	**	**	**	**	**
1980	**	**	**	**	**	140	**	**	**	**	**	**	**	**
1981	67,510	9,600	77,110	**	**	135	**	**	**	**	**	**	**	**
1982	61,370	9,385	70,755	**	**	126	**	**	**	**	**	**	**	**
1983	61,740	11,105	72,845	**	**	113	**	**	**	**	**	**	**	**
1984	62,700	10,630	73,330	**	**	119	**	**	**	**	**	**	**	**
1985	61,255	12,500	73,755	**	**	126	**	**	**	**	**	**	**	**
1986	55,990	12,545	68,535	**	**	121	11,890,000	2,271,500	6,437,500	992,500	441,000	171,500	318,000	1,258,000
1987	57,755	13,940	71,695	**	**	131	12,237,500	2,330,000	6,405,000	1,238,500	428,000	193,000	315,000	1,328,000
1988	61,790	12,325	74,115	**	**	136	13,308,000	2,436,500	7,169,500	1,404,500	490,500	204,000	333,000	1,270,000
1989	58,250	12,580	70,830	**	**	119	13,409,500	2,115,000	7,337,000	1,467,000	500,000	207,000	381,500	1,402,000
1990	57,100	14,200	71,300	**	**	108	13,707,500	2,019,000	7,650,000	1,476,000	486,500	210,000	423,000	1,443,000
1991	55,830	15,065	70,895	**	**	108	14,556,500	2,041,500	8,176,000	1,578,500	494,000	221,000	428,500	1,617,000
1992	52,290	14,645	66,935	**	**	77	14,684,500	1,964,500	8,263,000	1,598,000	514,000	220,500	400,000	1,724,500
1993	52,885	16,675	69,560	**	**	81	15,318,500	1,952,500	8,743,500	1,646,500	542,000	245,000	432,500	1,756,500
1994	52,875	11,810	64,685	**	**	106	16,127,000	2,054,500	9,189,000	1,666,000	586,500	250,000	432,500	1,948,500
1995	50,640	13,500	64,140	**	**	103	16,391,500	1,965,500	9,381,000	1,672,500	615,500	254,500	469,500	2,033,000
1996	45,725	12,630	58,355	**	**	99	17,503,000	1,975,000	9,841,500	1,816,500	688,000	285,000	536,500	2,360,500
1997	40,920	14,880	55,800	**	**	100	17,957,500	1,795,000	10,483,000	1,814,500	705,500	271,500	498,500	2,389,500
1998	43,080	13,960	57,040	**	**	93	18,753,000	1,755,500	10,936,000	1,956,000	707,500	301,000	559,000	2,538,000
1999	45,550	13,565	59,115	**	**	114	19,667,000	1,823,000	11,484,000	2,039,000	824,000	297,500	560,000	2,639,500
2000	43,065	13,660	56,725	27,825	84,550	105	20,520,000	1,708,000	12,251,000	2,126,500	864,000	319,000	543,500	2,708,000
2001	41,395	14,140	55,535	26,715	82,250	450	20,965,500	1,734,500	12,331,000	2,157,500	838,500	381,500	605,000	2,917,500
2002	37,860	15,095	52,955	27,845	80,800	101	21,303,500	1,687,500	12,903,000	2,116,000	888,500	361,000	603,500	2,744,000
2003	38,045	14,550	52,595	26,155	78,750	113	22,406,000	1,584,500	13,631,500	2,189,500	987,000	349,500	660,500	3,003,500
2004	36,880	13,150	50,030	25,810	75,840	119	22,616,500	1,550,500	14,100,000	2,106,000	984,000	354,000	671,000	2,851,000
2005	41,950	12,250	54,200	25,900	80,100	115	23,251,500	1,602,000	14,373,500	2,134,000	1,091,000	375,000	667,000	3,009,000

(continues)

** Accurate numbers unavailable at time of compilation

APPENDIX F
FIREFIGHTER INJURIES AND FATALITIES REPORT

Year	Fire Scene Injuries	Other Scene Injuries	Total Scene Injuries	Other Injuries	Total Injuries	Total LODDs	Total Calls	Fires	Medical Aid	False Alarms	Mutual Aid	Haz Mat	Hazardous Cond	Other
2006	44,210	13,090	57,300	26,100	83,400	107	24,470,000	1,642,500	15,062,500	2,119,500	1,159,500	388,500	659,000	3,438,500
2007	38,340	15,435	53,775	26,325	80,100	118	25,334,500	1,557,500	15,784,000	2,208,500	1,109,500	395,500	686,500	3,593,000
2008	36,595	15,745	52,340	27,360	79,700	118	25,252,500	1,451,500	15,767,500	2,241,500	1,214,500	394,500	697,500	3,485,500
2009	32,205	15,455	47,660	30,490	78,150	90	26,534,500	1,348,500	17,104,000	2,177,000	1,296,000	397,000	625,500	3,586,500
2010	**	**	**	**	**	72	**	**	**	**	**	**	**	**

** Accurate numbers unavailable at time of compilation

Sources: Numbers of calls and firefighter injuries are estimates from NFPA's National Fire Experience Survey. Numbers of firefighter deaths are taken from the annual NFPA tracking all of firefighter deaths. Statistics vary as additional reports are filed.
NFPA's "U.S. Firefighter Injuries - 2009," Michael J. Karter, Jr. and Joseph L. Molis, October 2010,
Firefighter Fatalities in the United States 2010, Rita F. Fahy, Paul R. LeBlanc and Joseph L. Molis, June 2011, and previous reports in the series.
USFA Reports and Statistics

Sources Updated: 8/2011
http://www.usfa.dhs.gov/fireservice/fatalities/statistics/index.shtm
http://www.nfpa.org/itemDetail.asp?categoryID=955&itemID=23605&URL=Research/Fire%20statistics/The%20U.S.%20fire%20service
http://www.nfpa.org/itemDetail.asp?categoryID=413&itemID=18238&URL=Research/Fire%20reports/Overall%20fire%20statistics

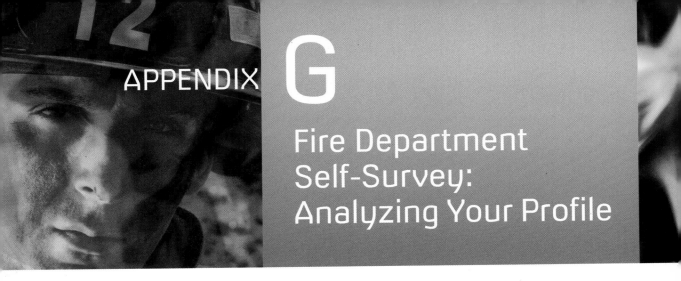

APPENDIX G

Fire Department Self-Survey: Analyzing Your Profile

One of the first steps to implementing the 16 Firefighter Life Safety Initiatives is to analyze your profile. Use the text to gain more information on individual statements. Rank each statement with a number from 0 to 4.

- 0- Never or Strongly Disagree
- 1- Seldom or Disagree
- 2- Sometimes or Neutral
- 3- Usually or Agree
- 4- Always or Strongly Agree

Add the score to see where your profile is (0–320). Identify where changes can be made to increase your profile, and rescore as your organization improves.

Chapter	Statement	Rank
1		
	My organization is receptive to change.	
	My organization is proactive when it comes to changes for safety.	
	My organization has a culture that promotes safe practices.	
	My organization reads and learns from near-miss and NIOSH reports.	
	My organization will likely adopt a safety culture.	
2		
	I value personal accountability.	
	My agency values organizational accountability.	

APPENDIX G
FIRE DEPARTMENT SELF-SURVEY: ANALYZING YOUR PROFILE

Chapter	Statement	Rank
	My organization encourages members to be involved.	
	My organization reviews incidents and continually improves.	
	My organization has reviewed NFPA 1500 and made changes for safety.	
3		
	My organization understands what risk management is.	
	My organization practices risk management on a daily basis.	
	My organization complies with OSHA's two-in, two-out rule.	
	My organization has SOPs or SOGs for rescuing a trapped firefighter and practices them regularly.	
	My organization makes effective communication a priority.	
4		
	My organization encourages empowerment.	
	My organization trains new responders how to address safety concerns.	
	My organization practices crew resource management (CRM).	
	My organization believes in "built-in" safety versus "bolt-on."	
	My organization uses checklists to ensure consistent safe practices.	
5		
	My organization has procedures in place to ensure proper training prior to performing specific duties.	
	My organization requires state and/or national certifications.	
	My organization encourages both higher education and outside training.	
	My organization continues to evolve its in-service training to meet local needs.	
	My organization values accreditation.	
6		
	My organization provides annual medical physicals.	
	My organization provides annual physical fitness evaluations.	
	My organization provides equipment and time for physical fitness.	
	My organization encourages a healthy lifestyle.	
	My organization supports national health organizations and their goals.	

APPENDIX G
FIRE DEPARTMENT SELF-SURVEY: ANALYZING YOUR PROFILE

Chapter	Statement	Rank
7		
	My organization requires accurate data reporting, and ensures its quality.	
	My organization continues to learn from other agencies in regard to new procedures and techniques.	
	My organization carefully examines all the facts before making changes.	
	My organization consistently reevaluates why and how we do our job.	
	My organization continues to improve its level of firefighter safety.	
8		
	My organization sees the benefits of technology for safety.	
	My organization does not rely exclusively on technology for safety.	
	My organization investigates new technology that can increase safety when purchasing equipment and apparatus.	
	My organization encourages input from manufacturers when purchasing equipment and apparatus.	
	My organization weighs the value of new technology with the risk of technology failure.	
9		
	My organization has a well-defined procedure for investigating injuries and property damage events.	
	My organization encourages the submission of all near-misses.	
	All members of my organization know how to submit a near-miss report.	
	My organization has a well-defined procedure for making changes based on the results of investigations and near-misses.	
	My organization regularly shares lessons learned with other similar organizations.	
10		
	My organization has personnel trained in successfully writing and obtaining grants.	
	My organization regularly applies for grants.	
	My organization uses unsuccessful grant requests as a learning tool for later success.	
	My organization puts an emphasis on grants used for the safety of personnel.	
	My organization commits time and resources to obtaining grants.	

APPENDIX G
FIRE DEPARTMENT SELF-SURVEY: ANALYZING YOUR PROFILE

Chapter	Statement	Rank
11		
	My organization uses SOPs and SOGs.	
	My organization ensures consistent use of SOPs and SOGs.	
	My organization differentiates between SOPs and SOGs.	
	My organization reviews national models when writing SOPs and SOGs.	
	My organization constantly reevaluates existing SOPs and SOGs and makes changes as needed.	
12		
	My organization has SOPs or SOGs for violent incidents.	
	My organization has SOPs or SOGs for members who find themselves in a violent incident.	
	My organization has SOPs or SOGs regarding what assistance will be provided to law enforcement and what will not.	
	My organization has uniforms that are not easily confused with law enforcement.	
	My organization constantly reevaluates existing SOPs and SOGs for violent incidents and makes changes as needed.	
13		
	My organization has programs in place for personnel suffering from substance abuse or personal problems.	
	My organization has SOPs or SOGs regarding how incident debriefing will occur.	
	My organization provides training in recognizing the signs and symptoms of acute stress disorder (ASD) and post-traumatic stress disorder (PTSD).	
	My organization can identify the need for, and refer personnel to, a mental health professional.	
	My organization can provide family members the needed resources for emotional support.	
14		
	My organization recognizes the need for fire and life safety education.	
	My organization trains its personnel in how to educate the public.	
	My organization puts a priority on safety at all public demonstrations.	
	My organization believes there is a direct correlation between fire and life safety education and firefighter safety.	

Chapter	Statement	Rank
	My organization continues to look at new ways to bring safety messages to the public.	
15		
	My organization is actively involved in fire prevention by conducting new building plan reviews and annual fire safety inspections.	
	My organization publicly supports residential fire sprinklers.	
	My organization promotes a residential fire sprinkler mandate for new construction.	
	My organization provides training to its personnel on residential fire sprinklers and their benefit.	
	My organization looks for new and innovative ways to promote residential fire sprinklers.	
16		
	My organization has SOPs or SOGs for the purchase of new equipment or apparatus.	
	My organization thoroughly reviews the safety record of new equipment or apparatus being considered.	
	My organization puts a high priority on purchasing safe equipment or apparatus.	
	My organization continues to look for safer ways to do our job.	
	My organization uses a cost benefit analysis to justify safer equipment or apparatus.	

GLOSSARY

A

ABS Antilock braking system.

AC Assistant chief.

accident An event that is not predictable, preventable, or avoidable.

accountability A process of tracking the location of firefighters.

accountable Similar to the term *responsible*, it describes a person who is held liable for completing a specific duty.

active fire protection A form of fire protection that extinguishes fire.

acute stress disorder (ASD) A stress disorder with signs and symptoms similar to PTSD that generally last less than a week; can be a precursor to PTSD.

adenosine triphosphate (ATP) A predominant source of energy for cells.

AED Automated external defibrillator.

aerobic Term used to describe metabolism in the presence of adequate oxygen; also used to describe sustained exercise requiring sufficient stamina.

affable User-friendly or easy to comply with.

affective domain A learning area defined by Benjamin Bloom involving feelings and values.

AFFF Aqueous film-forming foam.

AFG Assistance to Firefighters Grant.

AHJ Authority having jurisdiction.

AICPA American Institute of Certified Public Accountants.

AIDS Acquired immune deficiency syndrome.

air management The process of maintaining an awareness of the cylinder air level in a SCBA while using it, and ensuring that it will allow time to escape.

Alcoholics Anonymous (AA) An organization that uses group therapy to treat alcoholism.

alerter A passive safety device installed on a locomotive that warns the engineer when a signal has been passed that appears to have been ignored.

ALS Advanced life support.

amorphous silicon (ASi) A form of technology used in TICs.

anaerobic Term used to describe metabolism without the presence of adequate oxygen; also used to describe short bursts of exercise requiring sufficient muscle exertion.

animatronics A term used to describe the combination of animation and electronics to create a robot or other lifelike movement; a form of interactive learning that is extremely effective.

ANSI American National Standards Institute.

ARFF Aircraft rescue and firefighting.

A (Alpha) side The front of a structure.

ASO Assistant safety officer.

ASTM American Society for Testing and Materials.

ATC Air traffic control.

ATF Bureau of Alcohol, Tobacco, Firearms and Explosives.

atrophy The loss of muscle, generally from lack of use.

automatic distress signal unit (ADSU) A piece of equipment that sends a signal when a predetermined criterion is met. A PASS alarm is a form of ADSU.

Aviation Safety Reporting System (ASRS) A voluntary near-miss reporting system for airline incidents.

AWS American Welding Society.

B

baby boomers A term used to describe a generation of people born between 1946 and 1964.

barium strontium titanate (BST) A form of technology used in TICs.

barrier Any form of protection in place to reduce the chance of injury if an event occurs.

BDU Battle dress uniforms.

BLS Basic life support.

British thermal unit (BTU) A unit of measure that can be used to describe the heat being produced by a fire.

BSI Body substance isolation.

B (Beta) side The left side of a structure.

built in The process of building safety into a product or process. It's generally more effective than a bolt-on feature, and ensures that safety is a primary function and consideration.

burdened A perceived disadvantage of a person which, under certain circumstances, could encourage the person to make poor risk management decisions in an effort to try to prove themselves.

burnout A term used to describe the effect of emotional stresses on emergency responders that are often career ending.

C

CAF Compressed air foam (system).

call for proposals (CFP) A formal request by a grantor for projects to consider funding.

candidate physical abilities test (CPAT) A type of criterion task test involving several firefighting activities with a focus on agility testing.

CAPS Cirrus Airframe Parachute System.

cardiovascular collapse Overexertion, or the sudden inability of the heart and vascular system to meet the demands of the muscles.

CBRNE Chemical, biological, radioactive, nuclear, or explosive.

CDC Center for Disease Control.

CE Continuing education.

Central Contractor Registration (CCR) A registry of organizations permitted to obtain grants, as well as to submit bids and proposals.

CERT Community or citizen emergency response team.

chain of survival A series of steps identified by the American Heart Association that is needed to successfully resuscitate someone in cardiac arrest.

Champion Model Developed by the NFPA, this guideline for project management assists in the planning, development, and implementation of a fire and life safety program.

chevrons A method of diagonal striping consisting of opposing colors utilized on the back of vehicles intended to provide more visibility.

chronic obstructive pulmonary disease (COPD) A disease affecting the lungs and the ability to utilize oxygen. Some examples include emphysema and chronic bronchitis.

citizens band (CB) A public mobile radio system.

closure A term used to describe the acceptance of an event, allowing an individual to move on and heal.

CN Canadian National (Railway).

CO Carbon monoxide.

COA Committee on Accreditation.

cognitive behavioral therapy (CBT) A form of psychotherapy that helps participants unlearn the way they react to specific stimulus.

cognitive domain A learning area defined by Benjamin Bloom involving memorization.

cognitive radio A radio that chooses its own frequency based on importance, quality, and availability.

combat challenge A type of criterion task test involving several firefighting activities, many times in head-to-head format as a race.

Confidential Incident Reporting and Analysis System (CIRUS) A voluntary near-miss reporting system for the rail industry in many areas of the United Kingdom.

congestive heart failure (CHF) Medical condition in which the muscle of the heart is unable to meet normal demands of the body, often resulting in fluid buildup in the ankles (edema) and possibly in the lungs (pulmonary edema).

consortium A group of professionals working toward a solution that had not been accomplished before.

contributing factor A tertiary component of an event that may have encouraged an event to occur or worsened the outcome.

cortisol A hormone found in human saliva which may be an accurate indicator of emotional stress.

cost sharing A requirement of some grants that expects the grantee to invest a specific amount of money into the project.

cover fire A tactic used by law enforcement that involves continual shooting in an effort to draw the attention away from an offensive move.

CPJ Committee to Protect Journalists.

CPR Cardiopulmonary resuscitation.

creatine phosphate A molecule that is stored in muscles and can readily be converted to ATP to provide energy.

credentialing An official way of designating the qualifications of an individual by a governing body.

crew resource management (CRM) A procedure adopted from the aviation and maritime industries that recognizes the value of all personnel.

crisis intervention A form of brief communication psychotherapy used as preventative or acute response to stress.

criterion task test (CTT) A type of test used to determine the physical capabilities of a person based on specific job functions. A firefighting CTT would likely include tasks such as wearing PPE while dragging fire hose, rescuing a simulated victim, and carrying equipment.

critical incident stress management (CISM) An organized form of crisis intervention promoted by Dr. George S. Everly Jr. and Dr. Jeffrey T. Mitchell.

cross-talk A term used by AA to ensure that participants in group therapy don't directly address comments made by another participant that might lead to an argument.

C (Charlie) side The rear of a structure.

culture The values, customs, and traditions of a group of people or an organization.

curvilinear A term used to describe curved lines on a chart, especially helpful in estimating results. In this text, it refers to estimating the results of an action based on other influences which could affect the outcome in an unanticipated way.

D

data Information, usually in the form of numbers or statistics.

Data Universal Numbering System (DUNS) A unique numbering system for identifying organizations.

debriefing The third phase of crisis intervention that occurs several days to a week after a traumatic incident with a goal of closure.

defusing The second phase of crisis intervention that occurs soon after a critical event. Although closure can occur during this phase, it's not a goal.

demobilization The first phase of crisis intervention that occurs just prior to release from the scene.

diastole The resting stage of the heart, also the lower number (denominator) of a blood pressure.

DOA Dead on arrival.

DOT Department of Transportation.

dry hydrant A water appliance consisting of a pipe installed adjacent to a water source with a fire department connection to speed up the process of drafting.

dry system A fire protection system that is charged with a gas (usually air) to the sprinkler heads. When a fusible link on a head reaches a predetermined temperature, the air dispels, allowing water to enter the piping and be dispensed directly on the fire.

D (Delta) side The right-hand side of a structure.

due regard A legal term used to describe the professional and safe driving attitude of emergency responders as a standard of care.

DUI Driving under the influence of drugs or alcohol.

dynamic Changing or moving.

E

eddy A dangerous area of a fast-moving body of water.

education The process of learning through higher education.

electrocardiogram (EKG) A machine capable of recording the graphical representation of a cardiac cycle. Also referred to as an ECG.

emergency traffic The term used to describe a firefighter emergency.

employee assistance program (EAP) A program to aid an employee with mental health, substance abuse, and other personal problems that may affect job performance.

empowerment Granting permission to subordinates to exceed their normal authority in an effort to better achieve organizational goals.

EMS Emergency medical services.

EMSC Emergency Medical Services for Children.

EMT Emergency medical technician.

ETA Estimated time of arrival.

event data recorder (EDR) A "black box" used to record information prior to and during a crash to assist in the investigation.

Everyone Goes Home (EGH) A prevention program created by the National Fallen Firefighters Foundation in an effort to reduce future line-of-duty deaths. One of the major accomplishments was the creation of the 16 initiatives, the basis of this text.

Explorers A national cadet program that allows teenagers to explore a career in emergency services.

extrinsic Affected by an outside or controllable source, such as smoking's effect on a disease.

F

facilitator A person assigned to lead a meeting, discussion, or group therapy session. In most cases a facilitator doesn't add opinions or views to a discussion, but instead encourages others to continue to communicate and ensures ground rules are followed.

FAST Firefighter assist and search team.

fast twitch (Type II) muscles A type of muscle that predominantly operates quickly using little oxygen.

FDC Fire department connection.

Federal Aviation Administration (FAA) A division of the United States Department of Transportation responsible for civilian aviation oversight and safety.

FEMA Federal Emergency Management Agency.

FESHE Fire and Emergency Services Higher Education.

FHA Federal Highway Administration.

FIPT Firefighter Injury Prevention Training.

Firefighter Close Calls A website devoted to ensuring that near-misses and other important information are shared in an effort to prevent future events.

Fire Prevention and Safety (FP&S) An Assistance to Firefighters Grant funding fire prevention, fire and life safety, and research projects.

FIRESCOPE Firefighting Resources of Southern California Organized for Potential Emergencies; one of the early versions of NIMS.

Fire Station Construction Grant (SCG) An AFG grant with funding priorities for fire station renovation or construction.

Fire Suppression Rating System (FSRS) A system of grading used by the fire insurance industry to establish rates.

First responder authentication credentials (FRACs) A proposed national system of credentialing emergency responders.

fit for duty A term used to describe a rule or policy that requires firefighters to accomplish the duties that they could be expected to perform in a specific amount of time. This could be accomplished through periodic medical physicals or the use of a CTT.

fixer A person who assists journalists in obtaining information confidentially.

FLSE Fire and life safety educator.

FLSI 16 (Fire and Life Safety Initiatives 16) Proposed rating for fire departments and EMS agencies based on their dedication to safety; could be used for grant eligibility.

foundation A public or private organization with a stated purpose that often provides or receives funds to support its goals.

FPRF Fire Protection Research Foundation.

FPW Fire Prevention Week.

FRA Federal Railroad Administration.

friction loss The conversion of useful energy into nonuseful energy due to friction.

functional capacity evaluations (FCE) A series of tests commonly used to monitor the rehabilitation of workers to facilitate return to work.

G

gallons per minute (gpm) A common form of identifying a specific amount of water being pumped in a specific amount of time.

Generation Xer A term used to describe an individual from the generation of people born between 1965 and 1985.

geographic information system (GIS) A computer system such as ArcInfo® that uses aerial photographs and maps in conjunction to form layers of information.

goal A measurable and achievable ambition comprised of several objectives to ensure its success.

GPS Global positioning satellite.

grantee An individual, organization, or governmental unit receiving a grant.

grantor An individual, organization, or governmental unit administering and awarding a grant.

H

halon A chemical extinguishing agent that is no longer being produced due to its adverse effects on the environment.

hazmat Hazardous materials.

head and neck (HANS) device A device used by race car drivers.

heads-up display (HUD) A visual indicator of cylinder air level in the mask of a SCBA.

health and fitness coordinator (HFC) The person who, under the supervision of the fire department physician, has been designated by the department to coordinate and be responsible for the health and fitness programs of the department.

Health Related Fitness Program (HRFP) A comprehensive program designed to promote the member's ability to perform occupational activities with vigor, and to assist the member in the attainment and maintenance of the premature traits or capacities normally associated with premature development of injury, morbidity, and mortality.

hindsight bias A psychological reaction to an event in which a person falsely believes he or she has predicted or forecasted the event, such as "I knew that was going to happen."

hypothesis A scientific deduction used to predict or explain an outcome.

I

IAP Incident action plan.

IC Incident commander.

ICC International Code Council.

ICS Incident command system.

IDLH (immediately dangerous to life and health) An acronym used to describe an atmosphere that requires special precautions.

IED Improvised explosive device.

IFSAC International Fire Service Accreditation Congress.

impact attenuators Safety devices attached to immovable roadside structures or construction vehicles designed to disperse energy in the event of a crash.

incident with potential (IWP) Another name for a near-miss.

in-kind contributions Similar to cost sharing, grantees are sometimes required to provide such contributions as equipment, materials, labor, or other investments such as insurance.

Institute for Safe Medication Practices (ISMP) A nonprofit organization providing information to patients and the health-care community in regard to safe medication administration.

Insurance Services Office (ISO) A fire industry organization.

integrated rescue device (IRD) A rescue strap built into structural firefighting gear used to drag a firefighter in case of emergency.

International Association of Fire Chiefs (IAFC) Organization of fire chiefs from the United States and Canada.

International Association of Fire Fighters (IAFF) Labor organization that represents the majority of organized firefighters in the United States and Canada.

International Organization for Standardization (ISO) A global organization that provides registries of companies that comply with certain standards.

investigation A review of an event in which fact finding provides insight as to the root cause and contributing factors with the intent of preventing future events.

IRC International Residential Code.

IRS Internal Revenue Service.

ISO Incident safety officer.

IV Indicates an intravenous line.

J

JPR Job performance requirements.

juvenile firesetter intervention specialist An individual trained to interact with juveniles who have a history of setting fires.

K

kinetic energy Energy at work, as in moving, or expanding.

L

LDH Large-diameter hose.

Learn Not to Burn® A fire safety education program developed by the NFPA in 1979, and is still used today in many areas.

LED Light-emitting diode.

Level II staging A term used to describe an area located a safe distance away from an incident, preferable out of view of the incident, where emergency vehicles park while awaiting orders.

line-of-duty death (LODD) Fatalities that are directly attributed to the duties of a firefighter.

M

Maslow's hierarchy of needs A graphical depiction of the order of human needs.

match A type of cost sharing where matching funds (50/50) must be provided by a grantee. Some grants permit a match of in-kind contributions.

maximum heart rate (MHR) A theoretical number that an individual's heart can rise to before uncertain health risks can occur. It can be calculated by subtracting a person's age from 220.

mayday A term used by the aviation and maritime industries to declare an emergency.

MCI Mass casualty incident.

mentoring The process of an experienced person counseling someone else who is new to an organization or career.

metabolic equivalent (MET) A measurement of work, assuming 1 MET is equal to a resting metabolic rate.

mission statement An organizational proclamation stating the reason it is in existence.

MSHA Mine Safety and Health Administration.

MUTCD Manual on Uniform Traffic Control Devices.

myocardial infarction An event where heart muscle dies; a heart attack.

N

N-95 filter mask A dust mask that is effective for filtering some sizes of particles, but is ineffective at filtering dangerous gasses.

NAHB National Association of Home Builders.

NASA National Aeronautics and Space Administration.

National Fallen Firefighters Foundation (NFFF) A nonprofit organization created to honor and assist families of firefighters who die in the line of duty, and create programs to prevent future events.

National Firefighter Near-Miss Reporting System A voluntary, nonpunitive, fire service reporting system for near-misses.

National Institute of Standards and Technology (NIST) A government laboratory with a fire research division committed to the behavior and control of fire, and providing valuable information to the fire service.

National Registry of Emergency Medical Technicians An organization that establishes standards for the training and certification of EMS providers.

National Research Agenda An official plan that would organize and prioritize what research is needed to save lives in the fire service.

National Volunteer Fire Council (NVFC) A professional organization for volunteer firefighters.

NBFSPQ National Board on Fire Service Professional Qualifications, or ProBoard.

NCSU North Carolina State University.

near-miss An event that had the potential for serious consequences but avoided catastrophe. If the event is studied and used as a learning tool, future events could be prevented. Also called an incident with potential (IWP).

NFA National Fire Academy.

NFIRS National Fire Incident Reporting System.

NFM No-fault management.

NFPA National Fire Protection Association.

NHTSA National Highway Transportation Safety Administration.

NICHD National Institute of Child Health and Human Development.

NIMS National Incident Management System.

NIOSH National Institute for Occupational Safety and Health.

NIOSH Alerts Safety bulletins periodically released by NIOSH to the fire service with a focus on similar events that are causing injuries and deaths, and providing prevention suggestions.

NTSB National Transportation Safety Board.

O

objective A workable component of a goal.

objective analysis Reviewing facts without bias.

OIC Officer in charge.

OJT On-the-job training.

Operational Risk Management (ORM) A risk management program utilized by the USCG.

organizational accountability A term used to describe the ability of an association to be held to certain standards.

OSHA Occupational Safety and Health Administration.

overexertion Cardiovascular collapse, or the sudden inability of the heart and vascular system to meet the demands of the muscles.

oxidizer A material that emits oxygen as it burns.

oxygen uptake The ability of an individual to obtain and utilize oxygen during periods of high demand.

P

panoramic detection system (PDS) A system of cameras used to identify traffic congestion and adjust patterns accordingly.

parasympathetic A component of the autonomic nervous system in which the body rests and restores.

PASS Personal alert safety system.

passive fire protection A form of fire protection that alerts occupants or a monitoring station of a fire.

PDA Personal digital assistant.

peer fitness trainer (PFT) An individual certified to help develop a fitness program and assist the members in achieving their goals. Similar to a health and fitness coordinator defined by NFPA.

peripheral vascular resistance (PVR) A term used to describe the amount of pressure the heart must overcome to pump blood throughout the body. Results are commonly referred to as blood pressure.

personal accountability A term used to describe the ability of a person to be held to certain standards.

personal protective equipment (PPE) A generic term used to describe the minimum apparel and gear needed to safely perform a specific duty.

personnel accountability report (PAR) A verbal or visual report to incident command or to the accountability officer regarding the status of operating crews; should occur at specific time intervals or after certain tasks have been completed.

personal identification verification (PIV) Credentialing system used by some federal employees and contractors, usually in the form of a smart card containing an electronically embedded chip.

PHCC Plumbing-Heating-Cooling Contractors Association.

PIN Personal identification number.

PM Preventative maintenance.

point of sale A term used to describe the time when a home sells, usually requiring an inspection or upgrade.

policy A broad goal or statement that guides actions, usually an extension of a mission statement that illustrates the beliefs of an organization.

positive-pressure ventilation (PPV) A technique of forcing pressurized air into a structure or enclosed space in an effort to clear the area of smoke or gasses. It can also be used in conjunction with a fire attack in certain situations.

positive securing device (PSD) A bracket, strap, or other holding mechanism used to store portable equipment or PPE in a vehicle.

positive train control (PTC) An elaborate system of active railroad safety devices that take control of a locomotive that fails to respond to a signal that appears to have been ignored.

postincident critique An evaluation of an incident after it occurs, specifically examining successes and areas for improvement.

post-traumatic stress disorder (PTSD) The most severe form of stress disorder with signs and symptoms that persist for more than a month. Many times individuals are incapacitated from living a normal life.

potential energy Energy at rest.

POV Personally owned vehicle.

preaction system A fire protection system that is charged with a gas (usually air) to the sprinkler heads. When a fusible link on a head reaches a predetermined temperature, the air dispels, allowing water to enter the piping and be dispensed directly on the fire. Although similar to a dry system, a preaction system utilizes a detection device such as a smoke or heat detector to prefill the piping with extinguishing agent.

proactive A method of making changes to avoid an event before one can occur.

ProBoard National Board on Fire Service Professional Qualifications (NBFSPQ).

problem filtering During an investigation, the process of eliminating or not identifying contributing factors. It can occur inadvertently or in an effort to prevent fault.

procedure A course of action, sometimes with steps of how it is to be accomplished.

professional qualifications A series of NFPA standards identifying the requirements needed to attain specific certification levels.

proselytize The word used to describe the action of attempting to convert an individual to a specific religion. Most chaplain organizations forbid the practice in emergency services.

psi Pounds per square inch.

psychomotor domain A learning area defined by Benjamin Bloom involving mastery of skills.

psychotherapy A form of psychological treatment utilizing counseling as a means of healing.

PTD Prevention through design.

PTT Push to talk.

public information officer (PIO) An individual designated to be a liaison with the media, usually at emergency incidents.

Public Protection Classification (PPC) A number assigned by the insurance industry to fire departments and districts based on several components including equipment, staffing, and water supply.

pull station A device used by occupants to activate a fire alarm manually.

pulmonary edema A medical emergency where fluid accumulates in the lungs, often a result of congestive heart failure.

purveyor A term used to describe the managing agent of a specific utility, whether publicly or privately operated such as a water purveyor.

Q

qualitative data Data or research utilizing feelings or opinions.

qualitative research Data or research utilizing feelings or opinions.

quantitative data Data or research utilizing numbers or percentages.

quantitative research Date or research utilizing numbers or percentages.

Queens College Step Test A test that is capable of measuring oxygen uptake.

R

RACE: Rescue, Alarm, Confine, Extinguish An acronym taught during fire extinguisher classes which provides the student with four vital actions required when discovering a fire.

RAVE II An advanced military simulator used for combat training.

RDD Radiological dispersal device.

reactive A method of making changes to avoid future events after one has occurred.

recognition-primed decision making (RPD) A process of using experience to guide decisions at a later date.

rehabilitation (rehab) The designation of an area where emergency responders can rest and recover.

relative anonymity A term used to describe a form of communication that minimizes personal interaction. Technology has allowed more avenues, such as texting and sending emails.

remote patient monitoring (RPM) A method of using technology to monitor the vital signs or other physical findings of a patient outside a medical facility.

responsible Similar to the term **accountable**, it describes a person who has been given the authority to carry out a specific duty.

RIC Rapid intervention crew.

RIG Rapid intervention group.

risk-benefit analysis The weighing of the facts to determine the advantages and disadvantages of a certain activity.

risk management Identification and analysis of exposure to hazards, selection of appropriate risk management techniques to handle exposures, implementation of chosen techniques, and monitoring of results, with respect to the health and safety of members.

Risk Watch An all-hazard fire and life safety education program created by the NFPA for children.

RIT Rapid intervention team.

roadside memorial A makeshift memorial placed at the scene of a incident by friends and family members of a victim.

roentgen A measurement of radiation.

root cause The primary cause of an event. Without it, the event would likely have not occurred.

S

safety culture A philosophy that prioritizes safety as a paramount value and relies on it to guide many of an organization's decisions.

SCA Sudden cardiac arrest.

scalar A scale of ranking, such as 1–10.

SCBA Self-contained breathing apparatus.

school resource officer (SRO) An individual assigned to an educational institution as a liaison between students, faculty, and law enforcement.

SCUBA Self-contained underwater breathing apparatus.

secondary audience The term used to describe a group of people who, although not a primary target of a fire and life safety education program, are likely to be in attendance. School teachers would be an example while addressing students.

secondary device The term used to describe a bomb or other hazard that is intended to target victims that are escaping or emergency responders that arrive to an incident.

Series One (S-1) A policy or procedure that could be applicable to virtually any fire and emergency service organization.

Series Three (S-3) A policy or procedure used for specialized incidents that may be applicable to some fire and emergency service organizations.

Series Two (S-2) A policy or procedure that could serve as a template for adoption by most fire and emergency service organizations.

SIDS Sudden infant death syndrome.

single share A term used by AA to ensure that some participants in group therapy don't take too much time away from the rest of the group, and that everyone gets a chance to speak.

situational awareness A term used to describe the recognition of an individual's location, the surrounding atmosphere, the equipment being utilized, and the evolution of an incident.

slow twitch (Type I) muscle A type of muscle that predominantly operates slowly using oxygen.

software defined radio (SDR) A radio that uses software to change radio frequencies rather than being restricted to the hardware installed, allowing the user more choices for communication.

SOG Standard operating guideline.

SOP Standard operating procedure.

SOTER (supervisor of tracking employee resources) A person used on scene as technology improves who would be responsible for tracking the location, time on air, vital signs, and stress level of emergency responders.

span of control The ideal number of personnel that a person can effectively manage.

SPE (severity, probability, exposure) model The USCG's risk management tool for assigning a numeric value to the three components of a risky task.

specific gravity The weight of a liquid as it is compared to water.

spirometry A pulmonary function test that can be used to estimate oxygen uptake.

SRS Supplemental restraint system.

Staffing for Adequate Fire and Emergency Response (SAFER) An Assistance to Firefighters Grant funding staffing and volunteer recruitment and retention programs.

STAGE Safety targeted assessment via gateway evaluation.

staging area manager (SAM) A position in NIMS responsible for duties associated with managing apparatus and personnel in staging.

standard of care The care a person would expect to receive in a similar medical crisis in a similar area.

standards developing organization (SDO) An organization that creates rules for a specific industry.

static At rest, or unchanging.

STEMI S-T elevated myocardial infarction.

subjective analysis Reviewing facts and personal opinions.

Sustainability Agriculture Research and Education (SARE) A grant promoting research and education of agricultural procedures and techniques.

SUV Sport utility vehicle.

SWAT Strategic weapons and tactics.

sympathetic A component of the autonomic nervous system in which the body uses adrenaline to react to a stressful situation.

systole The pumping stage of the heart, also the higher number (numerator) of a blood pressure.

T

taratogens Chemicals that can create future birth defects.

teachable moment A point in time when students are especially likely to receive a message and understand it.

TEMS Tactical emergency medical services.

tentative interim amendment (TIA) A revision made to NFPA standards between revision dates.

terrorism A broad term used to describe the act of bringing on fear to a group of people, usually through violence, destruction, injury, or death.

thermodynamics The release of heat that occurs as energy is converted from one form to another.

TIC Thermal imaging camera.

trade-offs A practice in which an authority having jurisdiction (AHJ) permits a variance from a certain requirement in return for an option that achieves similar goals.

traffic preemption system A system that overrides traffic light timing devices to give emergency vehicles the right-of-way. Most use a strobe light or siren to trigger them, but some now use GPS.

training The process of learning a skill by practicing.

training-based decision making A process of using training to guide decisions at a later date.

Training Resources and Data Exchange (TRADE) An association of fire training professionals exchanging training information.

trending The identification of patterns in an effort to predict future occurrences.

TSA Transportation Security Administration.

U

unsafe act An action by an individual that is performed in a way that could result in an injury. Many times it involves an individual not using proper PPE, not following safe procedures, or rushing a task.

USCG United States Coast Guard.

USFA United States Fire Administration.

V

vanadium oxide (VOx) A form of technology used in TICs.

vapor density The weight of a gas as it is compared to air.

vasoconstriction A body's ability to shrink the blood vessels to increase peripheral vascular resistance and blood pressure.

VES Vent, enter, search.

victim syndrome The desire for an individual to be portrayed as a victim, usually for esteem benefits.

viscosity The thickness of a liquid.

VO_2 max Oxygen uptake.

W

weapon of mass destruction (WMD) A device used to inflict significant damage to a large number of people.

wellness A term used to describe the encouragement and promotion of a healthy diet, exercise, and life.

Wellness-Fitness Initiative (WFI) A fire service management initiative.

wet system A fire protection system that is charged completely with a fluid (usually water) to the sprinkler heads. When a fusible link on a head reaches a predetermined temperature, the extinguishing agent is dispensed directly on the fire.

Wildland Fire Lessons Learned Center An educational website devoted to the continued improvement of wildland firefighting, using near-misses and lessons learned.

WPI Worcester Polytechnic Institute.

WTC World Trade Center.

INDEX

A
"A" (Alpha) side, 79, 94, 124, 131
AA. *See* Alcoholics Anonymous (AA)
ABS. *See* Anti-lock braking system (ABS)
AC. *See* Assistant chief (AC)
Accident
 defined, 103, 131, 394
 energy conversion to injuries, 103–108
 kinetic energy, 103
 potential energy, 103
 problem solving and, 372
 safety equipment, 104–106
  HANS device, 105
  operational redundancies, 107–108
  PPE limitations, 106
  PPE perceptions, 105–106
Accountability
 blame, 35–38
 defined, 56, 94, 131
 health and safety, 49–56
 Initiative 2, 199–200
 organizational. *See* Organizational accountability
 PAR. *See* Personnel accountability report (PAR)
 personal accountability. *See* Personal accountability
 responsibility vs., 33–35
 scene accountability, 70–72
 task level risk management, 83
 worksheets or logs, 123
Accreditation
 Committee on Accreditation, 149
 IFSAC, 139, 148–149
 NBFSPQ or ProBoard, 139, 149
Acquired immune deficiency syndrome (AIDS), 105
Active fire protection, 215, 220
Acute stress disorder (ASD), 352, 367
Adenosine triphosphate (ATP), 172–174, 187
Adoption of response standards, 322
ADSU. *See* Automatic distress signal unit (ADSU)
Advanced life support (ALS), 13, 310, 372
AEDs. *See* Automated external defibrillators (AEDs)

Aerobic metabolism, 174–176, 187
Aerobics Program for Total Well Being, 174
Affable, 398, 420
Affective domain, 140, 156
AFFF. *See* Aqueous film-forming foam (AFFF)
AFG. *See* Assistance to Firefighters Grant (AFG); Equipment grants (AFG)
AHJ. *See* Authority having jurisdiction (AHJ)
AICPA. *See* American Institute of Certified Public Accountants (AICPA)
AIDS, 105
Air management, 86, 94
Air traffic control (ATC), 91
Aircraft rescue and firefighting (ARFF), 17
Alcohol abuse, 350–351
Alcoholics Anonymous (AA), 355–358
 closing, 358
 cross-talk, 357–358
 functions, 367
 meetings, 355–356
 overview, 355
 single share, 356–357
 symptoms of alcoholism, 351
Alerters, 446, 450
Alpha side, 79, 94, 124, 131
Alpha to Omega, The Evolution in Residential Fire Protection, 397
ALS. *See* Advanced life support (ALS)
Alveoli, 171, 187
American Heart Association chain of survival, 137
American Institute of Certified Public Accountants (AICPA), 150
American National Standards Institute (ANSI), 7
American Society for Testing and Materials (ASTM), 302
American Welding Society (AWS), 302
Amorphous silicon (ASi), 227, 245
Anaerobic metabolism, 174–176, 187
Analyzing learning process, 140
Animal control, 343
Animatronics, 384, 394
Annual medical physicals, 181–182
Annual pump testing, 184

ANSI. *See* American National Standards Institute (ANSI)
Anti-lock braking system (ABS), 18, 430
Apparatus
 firefighting design, 433–435
 NFPA standards, 429
 response standards, 311–313
  personally owned vehicles, 313
  positioning, 314
  water tenders, 312–313
 safety of, 21
Apparel, EMS worker safety, 7
Applying learning process, 140
Aqueous film-forming foam (AFFF), 146, 211
ARFF. *See* Aircraft rescue and firefighting (ARFF)
ASD. *See* Acute stress disorder (ASD)
ASi. *See* Amorphous silicon (ASi)
ASO. *See* Assistant safety officer (ASO)
ASRS. *See* Aviation Safety Reporting System (ASRS)
Assembling policies, 321
Assistance to Firefighters Grant (AFG), 275
Assistant chief (AC), 304
Assistant safety officer (ASO), 70, 318
ASTM. *See* American Society for Testing and Materials (ASTM)
ASTM E1354-10a, Standard Test Method for Heat and Visible Smoke Release Rates for Materials and Products Using an Oxygen Consumption Calorimeter. *See* American Society for Testing and Materials (ASTM)
ATC. *See* Air traffic control (ATC)
ATF. *See* Bureau of Alcohol, Tobacco, Firearms and Explosives (ATF)
ATP. *See* Adenosine triphosphate (ATP)
Atrophy, 174, 187
Authority having jurisdiction (AHJ), 212, 322
Automated external defibrillators (AEDs), 9, 45, 137
Automatic alarms, 310–311

487

INDEX

Automatic distress signal unit (ADSU), 227, 245
Auxiliary cooler (sweating), 169–170
Aviation Safety Reporting System (ASRS), 255, 269
Awareness level
　of certification, 28
　of training, 154
AWS. *See* American Welding Society (AWS)

B

Baby boomers, 204, 220
Bad habits, 10–11
Barcode scanning software, 71–72
Barium strontium titanate (BST), 227, 245
Barrier, 25–27, 269
Basic life support (BLS), 372
Battle dress uniform (BDU), 333
Biological weapons, 328
Blame, 35–38
Blink: The Power of Thinking Without Thinking, 68
Blood pressure, 169
BLS. *See* Basic life support (BLS)
Body composition, 186
Body metabolism, 164–176
　aerobic vs. anaerobic, 174–176, 187
　fuel, 172–174
　metabolic equivalent, 172
　overview, 164–165
　oxygen, 165–172
　　oxygen uptake, 171–172
　　the pipes, 170–171
　　the pump, 165–170
Body substance isolation (BSI), 112, 144
Bomb threats, 342–343
Briefings, unsafe act communications, 127–128
British thermal unit (BTU), 339, 345
Brotherhood, 21
Brunacini, Alan, 62
BSI. *See* Body substance isolation (BSI)
BST. *See* Barium strontium titanate (BST)
BTU. *See* British thermal unit (BTU)
Budgets for grants, 294–295
　direct expenses, 294
　in-kind contributions, 294
　total amount requested, 295
Building aspects, task level risk management, 83
Built in GPS, 232, 245
Burdened, 111, 131
Bureau of Alcohol, Tobacco, Firearms and Explosives (ATF), 261
Burnout, 350, 367
Business associations, 150
Businesses as grantors, 277–279

C

"C" (Charlie) side, 69, 94
Cadet programs, 390
CAF. *See* Compressed air foam (CAF)
Call for proposal (CFP), 280–282, 297
Canadian National (CN) Railway, 445
Cancer, 207
Candidate Physical Ability Test (CPAT)
　ceiling pull, 179
　criteria, 150
　defined, 156, 187, 220
　equipment carry, 178
　forcible entry, 178
　hose drag, 178
　Initiative 6 (medical and fitness standards), 208

ladder raise and extension, 178
　overview, 177–178
　rescue drag, 178
　search, 178
　stair climb, 178
CAPS. *See* Cirrus Airframe Parachute System™ (CAPS)
Carbohydrates for metabolism, 173
Carbon monoxide (CO) alarm, 376
Cardiac arrest
　chain of survival, 137
　sudden cardiac arrest, 164
Cardiopulmonary resuscitation (CPR), 353, 378
Cardiovascular collapse, 166, 187
Cardiovascular disease, 207
Career firefighter basic training, 143
CB. *See* Citizen band (CB) radio
CBRNE, 328–329
CBT. *See* Cognitive behavioral therapy (CBT)
CCR. *See* Central Contractor Registration (CCR)
CDC. *See* Center for Disease Control (CDC)
CE. *See* Continuing education (CE)
Ceiling pull test, 179
Center for Disease Control (CDC), 106
Central Contractor Registration (CCR), 285, 297
CERT. *See* Community emergency response team (CERT)
Certification
　Initiative 5. *See* Initiative 5 (training and certification standards)
　level of safety, 27–28
　　awareness level, 28
　　operations level, 28
　　technician level, 28
　NIMS, 138
　recertification requirements, 151–152
　recognition by other states, 135
　requirements, 143–145
　　career firefighter basic training, 143
　　volunteer firefighter basic training, 143–145
　transfers to other locations, 135–137
　used to obtain grants, 322
CFP. *See* Call for proposal (CFP)
Chain of survival, 137, 156
Champion Model, 386, 394
Change
　cultural compliance, 29
　emergency services, 2–3
　leadership during, 7–15
　　economic impacts, 8–9
　　managing change, 8–9
　　opportunities for change, 13–15
　　overview, 7–8
　　political impacts, 9
　　resistance to change, 9–13
　　social impacts, 9
　　technological impacts, 9
　proactive, 4–5
　reactive, 4–5
　reasons for, 7
　for safety, 5–6
　safety culture, 15–28
　　certification level, 27–28
　　components, 22–27
　　creation of, 17–19
　　cultural changes for safety, 19–22
　　evolution to, 16–17
　　overview, 15–16
　significant changes of the U.S., 2
Chaplains, 362–363

Charlie side, 69, 94
Checklists, 123
Chemical, biological, radiological, nuclear, or explosive (CBRNE), 328–329
Chemical weapons, 328
Cherry picker, 274
Chevrons, 5, 30
CHF. *See* Congestive heart failure (CHF)
Chronic obstructive pulmonary disease (COPD), 171, 187
CIRAS. *See* Confidential Incident Reporting and Analysis System (CIRAS)
Cirrus Airframe Parachute System™ (CAPS), 431
CISM. *See* Critical incident stress management (CISM)
Citizen band (CB) radio, 212, 220
Civil disturbance, 337–340
Closure, 359, 367
CN. *See* Canadian National (CN) Railway
CO. *See* Carbon monoxide (CO) alarm
COA. *See* Committee on Accreditation (COA)
Cognitive behavioral therapy (CBT), 364–366, 367
Cognitive domain, 140, 156
Cognitive radio, 244, 245
Coleman, Ronnie, 397
Collapse, firefighter sensors for, 242–243
Collapse prediction, 201
Combat challenge, 176–177
　defined, 187
　forcible entry, 177
　high-rise pack carry, 177
　hose advance, 177
　hose hoist, 177
　overview, 176
　victim rescue, 177
Committee on Accreditation (COA), 149
Committee to Protect Journalists (CPJ), 330
Common skills training, 146
Communication
　operability and interoperability, 212–213
　radio. *See* Radio communications
　response standards, 318
　of safety concerns, 18–19
　technology, 244
　unsafe acts, 125–129
　　briefings, 127–128
　　challenge and response, 128–129
　　overcommunication, 126–127
　　terminology, 125–126
Community emergency response team (CERT), 390
Community organizations as grantors, 276–277
Compressed air foam (CAF), 211
Computer-based training and education, 206
Confidential Incident Reporting and Analysis System (CIRAS), 256, 269
Congestive heart failure (CHF), 165, 188
Consortium, 414, 420
Continuing education (CE), 150
Contract for grants, 295–296
Contributing factor
　barrier for, 264
　defined, 25, 30, 269, 394
　investigations of, 250

INDEX

prevention vs. preparation, 371
problem filtering, 25
protection against, 263
Cooper, Kenneth H., 174–175
COPD. *See* Chronic obstructive pulmonary disease (COPD)
Cortisol, 229, 245
Cost-sharing, 279, 297
Counseling, 181
Cover fire, 332, 345
CPAT. *See* Candidate Physical Ability Test (CPAT)
CPJ. *See* Committee to Protect Journalists (CPJ)
CPR. *See* Cardiopulmonary resuscitation (CPR)
Creatine phosphate, 174, 188
Creating an "Open Book" Organization, 101
Creating learning process, 141
Credentialing, 154–156
Crew resource management (CRM), 121–130
    communication, 125–129
    defined, 30, 220, 450
    empowerment for safety, 121
    equipment safety, 427
    evolution to safety, 16
    policies and procedures, 123–124
        authority, 123
        checklists, 123
        training, 123–124
    problem solving, 129–130
    situational awareness, 124–125
    unsafe acts, eliminating, 203
Crisis intervention, 354, 367
Criterion task test (CTT), 176, 188
Critical incident stress management (CISM), 358–360, 367
CRM. *See* Crew resource management (CRM)
Cross-talk, 357–358, 367
Crowd control, 338–340
CTT. *See* Criterion task test (CTT)
Culture
    cultural compliance, 29
    defined, 2, 30
    Initiative 1, 198–199
    safety culture. *See* Safety culture
The Curve of Binding Energy, 328–329
Curvilinear, 185, 188

D

"D" (Delta) side
    defined, 56, 94, 131
    personal accountability situation, 38
    recognition-primed decision making example, 68
    situational awareness, 124
Daily operations, 54–55
Data
    analyzing, 217–218
    collection, 215–217
        data collection point, 217
        form of data, 216–217
        during investigations, 261–262
        plan for, 216–217
        type of data, 217
    defined, 4, 30
    hypothesis, creation of, 216
    related to exposure, 209
    research agenda, 208–209
    sharing and access, 209–210

utilizing, 218–219
    long-term goals, 219
    medium-term goals, 219
    short-term goals, 218–219
Data collection from HRFP, 181
Data Universal Numbering System (DUNS), 285, 297
Dead on arrival (DOA), 343, 355
Deadlines for grants, 296
Debriefing, 354–355, 367. *See also* Postincident critique
Decision making, risk management, 64–69
    recognition-primed, 67–69, 95
    training-based, 65–67, 95
Defusing, 359, 367
Delta side. *See* "D" (Delta) side
Demobilization, 358–359, 367
Denver Drill, 266
Department of Transportation (DOT), 4, 430
Diastole, 166, 188
Disguised unsafe acts, 109–110
Disseminated impacts of grants, 282
Distraction events, 111–112
DOA. *See* Dead on arrival (DOA)
Dodson, Dave, 49, 69–70
Donor-advised grants, 283
DOT. *See* Department of Transportation (DOT)
Driving under the influence (DUI), 110, 131
Drug abuse, 351
Dry hydrants, 281, 297
Dry system, 400, 420
Due regard, 195, 220
DUI. *See* Driving Under the Influence (DUI)
DUNS. *See* Data Universal Numbering System (DUNS)
Duties, diversification of, 372
Dynamic, 6, 30

E

EAP. *See* Employee assistance program (EAP)
Economic impacts of change, 8–9
Eddy, 88, 94
EDRs. *See* Event data recorders (EDRs)
Education. *See also* Training
    affective domain, 140
    computer-based, 206
    defined, 156
    empowerment and, 101–102
    grants, 280–282
    higher-educational requirements, 149–150
    Initiative 14. *See* Initiative 14 (public education)
    medical and fitness, 181
    prevention through design, 441–442
    psychomotor domains, 140
    public. *See* Public education
    teaching processes, 139–141
        analyzing, 140
        applying, 140
        creating, 141
        evaluating, 141
        remembering, 140
        understanding, 140
    teaching standards, 141–143
        job performance requirements, 142
        knowledge, 142
        open consensus-based method, 142
        professional qualifications, 142
        resources, 142–143
        skills, 142
        training and certification standards, 135

Efficiency factors of firefighting, 184–187
    extrinsic factors, 185–187
    movement speed, 185
    work rate, 185
EGH. *See* Everyone Goes Home (EGH)
Electrocardiogram (EKG), 294, 297
Emergency medical services (EMS)
    apparel for safety, 7
    diversification of duties, 372
    PPE, 105
    response standards, 315, 317–318
    situational awareness, 85
    training and certification standards, 135
    uniforms, 333
    use of technology, 226–227
    violent incidents, 327
Emergency Medical Services for Children (EMSC), 275
Emergency medical technician (EMT)
    no-fault management, 23
    personal accountability, 44
    postincident stress management, 355
    response standards, 311
    standard of care, 136
Emergency response plan, 264
Emergency scenes, 55–56
Emergency services
    changes, 2–3
    EMS. *See* Emergency medical services (EMS)
    EMSC, 275
    safety culture, 2
Emergency traffic
    defined, 94
    incident management, 72–73
    response standards, 318–319
    stressful communications, 91–92
Emotional support
    future technology, 363–366
        complete anonymity, 364–366
        relative anonymity, 363–364
        support system, 366
    for postincident stress management, 214
    post-traumatic stress disorder, 352–363
        group therapy, 354–360
        individual therapy, 361–363
        overview, 352–354
    stress, 348–352
        alcohol abuse for coping, 350–351
        coping with, 349–351
        drug abuse for coping, 351
        failure to cope, 352
        overview, 348–349
    for substance abuse, 214
Employee assistance program (EAP), 361–362, 367
Employee responsibilities, 27
Empowerment
    defined, 100, 131
    education, 101–102
    enablement, 102
    firefighter, 100–103
    personnel, 102–103
EMS. *See* Emergency medical services (EMS)
EMSC. *See* Emergency Medical Services for Children (EMSC)
EMT. *See* Emergency medical technician (EMT)
Enablement, 102

Energy conversion to injuries, 103–108
Enforcement (Initiatives 9, 11), 115
Engineering (Initiative 8), 113–114
Environment (Initiative 1), 114–115
Equipment
  design for portable equipment, 433
  designing safety, 435–445
    education, 441–442
    policy, 445
    practice, 442–445
    research, 436–441
  equipment carry test, 178
  equipment grants (AFG), 284, 286–287
  NFPA standards, 429
  origination of safety, 428–435
    firefighting design, 432–435
    governmental regulations and standards, 430–431
    manufacturers, 431–432
    NFPA committees, 428–430
    research institutions, 430
  personal accountability to, 46
  PPE. *See* Personal protective equipment (PPE)
  retrofitting safety, 445–449
    availability, 448–449
    cost, 446–448
    need, 449
  safety curve, 425–428
    Improvement Phase 1, 426–427
    Improvement Phase 2, 427
    Improvement Phase 3, 427–428
  safety of, 424–425
Essentials of Exercise Physiology, 184–185
Establishing a System of Policies and Procedures, 306
Estimated time of arrival (ETA), 125
Evaluating learning process, 141
Evasive maneuvering, 336–337
Event data recorders (EDRs), 233, 245
Everyone Goes Home (EGH)
  defined, 30, 57, 220, 269, 297
  Heritage Program funding, 278
  investigations, 211
  opportunities for change, 14–15
  personal accountability, 43
  website, 267
Exercise training program, 180
Explorers, 390, 394
Explosive weapons, 329
Extrinsic factors, 185–187, 188

F

FAA. *See* Federal Aviation Administration (FAA)
Facilitator, 356, 367
Falls, 233–236
  from apparatus, 233–234
  from buildings, 235–236
  from ladders and aerial devices, 234–235
Fass, Brian, 184
FAST. *See* Firefighter assist and search team (FAST)
Fast twitch (Type II) muscles, 186, 188
Fat for metabolism, 173
Fatality
  injury patterns, 252–253
  investigations of, 251–253
  MSHA investigations, 251
  root cause, 252
FCEs. *See* Functional capacity evaluations (FCEs)

FDC. *See* Fire department connection (FDC)
Federal Aviation Administration (FAA)
  defined, 30, 269
  near-miss reporting, 255
  safety culture, 15
Federal Emergency Management Agency (FEMA)
  AFG program report, 276
  CERT programs, 390
  near-miss reporting, 257
Federal Highway Administration (FHA), 430
Federal Information Processing Standards Publication 201, 155
Federal Railroad Administration (FRA), 445
FEMA. *See* Federal Emergency Management Agency (FEMA)
FESHE. *See* Fire and Emergency Services Higher Education (FESHE)
FHA. *See* Federal Highway Administration (FHA)
Fire academies, 54
Fire and Emergency Services Higher Education (FESHE)
  FESHE model, 152–153
  purpose of, 139
Fire and life safety education (FLSE), 412
Fire and Life Safety Educator, 388
Fire and life safety educator (FLSE), 373
Fire department connection (FDC), 411
Fire Department Incident Safety Officer, 69–70
Fire department physician, 181
Fire detection and extinguishing agents, 211
The Fire Fighter Conditioning Workout Plan, 163
Fire inspections, 411
Fire Prevention and Safety (FP&S), 283, 297
Fire Prevention Week (FPW), 380
Fire Protection Research Foundation (FPRF), 402
Fire Station Construction Grants (SCG), 284, 297
Fire station, safety of, 21
Fire Suppression Rating System (FSRS), 52, 57
Firefighter assist and search team (FAST), 74
Firefighter Close Calls, 268, 269
Firefighter orientation, 54
The Firefighter's Workout Book, 175
Firefighting Resources of Southern California Organized for Potential Emergencies (FIRESCOPE), 69, 94
Fireground safety, 93
Fireman's Fund Heritage Program, 278
FIRESCOPE, 69, 94
First responder authentication credentials (FRACs), 155–156, 157
Fit for duty, 179–180, 188
The Fit Responder, 184
Fitness assessment, 180
Fitness coordinator, 180
Fitness level, 186
Fixers, 331, 345
Flashover
  injuries by, 239–242
  simulators, 239–242
    realistic environment, 241
    realistic gear, 240–241
    realistic movement, 241
    realistic response, 241–242
  warning devices on gear, 239
Flashover prediction, 201
Fleming, Russell P., 398

FLSE. *See* Fire and life safety educator (FLSE)
FLSI-16 (Fire and Life Safety Initiatives 16). *See also specific Initiative*
  certification used to obtain grants, 322
  defined, 57, 323
  safety rating, 53–54
FM Global® fire prevention grant, 279
Forcible entry test, 177, 178
Foundations as grantors, 279, 297
FP&S. *See* Fire Prevention and Safety (FP&S)
FPRF. *See* Fire Protection Research Foundation (FPRF)
FPW. *See* Fire Prevention Week (FPW)
FRA. *See* Federal Railroad Administration (FRA)
FRACs. *See* First responder authentication credentials (FRACs)
Friction loss, 171, 188
FSRS. *See* Fire Suppression Rating System (FSRS)
Fuel for metabolism, 172–174
  adenosine triphosphate, 172–174
  carbohydrates, 173
  creatine phosphate, 174
  fat, 173
  proteins, 173–174
Functional capacity evaluations (FCEs), 208, 220
Fundamental Issues in Strategy, A Research Agenda, 197

G

Gallons per minute (GPM), 169, 188, 338–339, 345
Generation Xers, 204, 220
Geographic information system (GIS), 209, 220
Get in the Game: 8 Elements of Perseverance that Make the Difference, 40–41
Giesler, Marsha, 388
GIS. *See* Geographic information system (GIS)
Gladwell, Malcolm, 68
Global positioning satellite (GPS), 230, 232
Goal, 275, 297
Goldfeder, Billy, 167
Governmental regulations and standards for equipment, 430–431
GPM. *See* Gallons per minute (GPM)
GPS. *See* Global positioning satellite (GPS)
Grantee, 275, 297
Grantor, 275, 297
Grants and safety
  adaptations to situations, 274–275
  Assistance to Firefighters Grant, 275
  awarding of grants, 295–296
    contract, 295–296
    deadlines, 296
    reports, 296
    safety improvements, 448
  blending of, 212
  certification used to obtain grants, 322
  choosing a grant, 285–287
    additional equipment, 286
    eligibility of recipients, 285–286
    grant priorities, 286–287
    new equipment, 287

INDEX

replacement equipment, 286
sustainability, 287
classifications of grants, 279–285
    donor-advised grants, 283
    equipment grants, 284
    mini or quick-response grants, 282–283
    operating grants, 280
    program seed grants, 283–284
    research and education grants, 281–282
    staffing grants, 284–285
components of a grant, 287–295
    budget, 294–295
    narrative, 289–294
    objectives (tactical), 289–290
    procedures (task), 290–294
    research and evaluation, 288
    significance (strategic level), 289
education funding, 389–390
Fire Station Construction Grant, 284, 297
FM Global® fire prevention grant, 279
grant writer, 285
grantors, 276–279
    businesses, 277–279
    community organizations, 276–277
    foundations, 279
Initiative 10, 212, 274
State Farm® insurance safety grants, 279
Group therapy, 354–360
    Alcoholics Anonymous, 355–358
    critical incident stress management, 358–360
    debriefing, 354–355
    psychological first aid, 360
Guarantee of contractors, 293–294

H

Hallinan, Joseph T., 60
Halon, 211, 220
HANS device, 105, 131
Hazardous materials (hazmat), apparel for safety, 7
Head and neck (HANS) device, 105, 131
Heads-up display (HUD), 86, 94
Health and fitness coordinator (HFC), 180, 188
Health and safety accountability, 49–56
    implementation, 54–56
        daily operations, 54–55
        emergency scenes, 55–56
        fire academies, 54
        firefighter orientation, 54
        promotional exams, 55
        recruit schools, 54
        training, 55
    NFPA 1500, 50–51
    rating systems, 52–54
        FLSI 16, 53–54
        Insurance Services Office, 52
        International Organization for Standardization, 53
Health maintenance, 207–208
Health-Related Fitness Program (HRFP), 181, 188
Heart
    auxiliary cooler (sweating), 169–170
    PPE and, 169
        thermodynamics, 169
    cardiac arrest
        chain of survival, 137
        sudden cardiac arrest, 164

cardiovascular collapse, 166, 187
cardiovascular disease, 207
congestive heart failure, 165, 188
discharge pressure (blood pressure), 169
maximum heart rate, 175
myocardial infarction, 166, 188
oxygen uptake, 171–172
the pipes, 170–171
pulmonary edema, 169, 189
throttle (heart rate), 166–169
    cardiovascular collapse, 166
    diastole, 166
    myocardial infarction, 166
    parasympathetic nervous system, 166
    sympathetic nervous system, 166
    systole, 166
transfer valve (volume or pressure), 169
viscosity of blood, 171, 189
Heart-Healthy Firefighter Program, 163, 278
Heinrich, H.W., 252
Heritage Program, 278
Hero, 21–22
HFC. *See* Health and fitness coordinator (HFC)
High-rise pack carry, 177
Hindsight bias
    defined, 269
    minimization of, 254
    reporting
        cost savings of timely reporting, 254
        hindsight bias, 254
        indemnity, 255
Hose advance test, 177
Hose drag test, 178
Hose hoist, 177
HRFP. *See* Health-Related Fitness Program (HRFP)
HUD. *See* Heads-up display (HUD)
Hypothesis
    creation of, 216
    defined, 220
    research, 194, 195–196

I

IAFC. *See* International Association of Fire Chiefs (IAFC)
IAFF. *See* International Association of Fire Fighters (IAFF)
IAP. *See* Incident action plan (IAP)
IC. *See* Incident commander (IC)
ICC. *See* International Code Council (ICC)
ICS. *See* Incident Command System (ICS)
Identified impacts of grants, 281–282
IDLH. *See* Immediately dangerous to life and health (IDLH)
IED. *See* Improvised explosive device (IED)
IFSAC. *See* International Fire Service Accreditation Congress (IFSAC)
Immediately dangerous to life and health (IDLH), 73, 94
Impact attenuators, 237, 245
Implementation of fire life safety initiatives, 262–268
    information distribution, 264–268
    leverage points, 262–264
        contributing factors, 263–264
        emergency response plan, 264
        future occurrence, reduce change of, 262–263
        provide a barrier, 263
Improvised explosive device (IED), 448

Incident action plan (IAP)
    civil disturbance, 338
    following to prevent unsafe acts, 119
    incident safety office responsibilities, 69
    response standards, 305
Incident Command System (ICS), 69
Incident commander (IC), 26, 305
Incident management, 69–76
    emergency traffic, 72–73
    FIRESCOPE, 69
    ICS, 69
    NIMS, 69. *See also* National Incident Management System (NIMS)
    rapid intervention, 73–76
    responder rehabilitation, 76
    risk management, 202–203
    safety officer, 69–70
    scene accountability, 70–72
        barcode scanning software, 71–72
        passport system, 71
        span of control, 72
        tag system, 71
Incident safety officer (ISO), 69, 318
Incident with potential (IWP), 253, 269
Individual therapy, 361–363
Initiative 1 (cultural change)
    defining cultural change, 198–199
    environment, 114–115
    purpose of, 8
    text of, 2
Initiative 2 (personal accountability)
    enhancing accountability, 199–200
    investigations, 261
    text of, 33
Initiative 3 (risk management techniques)
    application of techniques, 200–203
    standard policies and procedures, 301
    text of, 60
Initiative 4 (eliminating unsafe acts), 98, 203, 252
Initiative 5 (training and certification standards)
    certifications, 264
    computer-based training and education, 206
    emergencies handled by trained professionals, 139
    emergency drills, 100
    implementation of standards, 204–206
    instructor development, 204–205
    live fire training, 205
    professional development, 205–206
    recruit-level training, 204
    text of, 135
    training simulators, 205
Initiative 6 (medical and fitness standards)
    development of standards, 207–208
    responders' physical health, 348
    tests for fitness capabilities, 150
    text of, 160
Initiative 7 (research agenda)
    creation of research agenda, 208–210
    investigative data collection and research, 261–262
    national research agenda, 193. *See also* National research agenda
    program seed grants, 283
    research improvements, 197
    text of, 192

Initiative 8 (technology)
  engineering, 113–114
  technology utilization, 210–211, 226
  text of, 224
Initiative 9 (enforcement)
  eliminating unsafe acts, 115
  investigations and near-misses, 183, 194, 211–212, 250, 261
  learning from accidents, 253
  standard policies and procedures, 301
  text of, 250
Initiative 10 (grants for safety), 212, 274
Initiative 11 (response standards)
  common policies and procedures, 264
  enforcement, 115
  establishment of standards, 212–213, 309
  standard responses to emergencies, 139
  text of, 301
Initiative 12 (response to violent incidents)
  apparatus proximity dangers, 314
  examining response to violent incidents, 213–214
  text of, 327
Initiative 13 (emotional support)
  providing emotional support, 348
  for PTSD, 214
  safety, 118
  for substance abuse, 214
  text of, 348
Initiative 14 (public education)
  education, 113
  enabling public education, 214–215
  fire prevention, 283
  text of, 370
Initiative 15 (residential fire sprinklers), 215, 397
Initiative 16 (engineering safety into equipment)
  building safety into equipment, 215
  text of, 424
Injury patterns, 252–253
In-kind contributions, 294, 297
In-service training requirements, 145–148
  common skills, 146
  funding issues, 147–148
  target hazards, 146–147
Institute for Safe Medication Practices (ISMP), 255, 269
Instructor development, 204–205
Insurance Services Office (ISO)
  defined, 57, 131, 157
  disguised unsafe acts, 110
  fire service ratings, 52
  in-service requirements, 146
Integrated rescue device (IRD), 236, 245
Internal Revenue Service (IRS), 250
International Association of Fire Chiefs (IAFC)
  communication technology, 244
  crew resource management, 122
  defined, 131, 188, 246, 270, 297
  health and fitness standards, 162
  Heritage Program funding, 278
  near-miss reporting, 255
International Association of Fire Fighters (IAFF), 162, 188
International Code Council (ICC), 401
International Fire Service Accreditation Congress (IFSAC), 139, 148–149
International Organization for Standardization (ISO), 53, 57, 302, 323

International Residential Code (IRC), 401
Intravenous line (IV), 105, 131
Investigations
  components, 250
  contributing factor, 250
  defined, 30, 270
  of fatalities, 251–253
  fear of, 250
  fire inspections, 411–412
  Initiative 9, 211–212
  motor vehicle crashes, 233
  near-miss reporting, 253–259
    advantages of, 253–255
    components of, 255–257
    process of, 257–259
  reducing injuries, 261
  resistance to change, 11
  systems for, 259–268
    data collection and research, 261–262
    the event, 260
    implementation, 262–268
    investigations, 261
    near-misses, 261
    personal accountability, 261
IRC. See International Residential Code (IRC)
IRD. See Integrated rescue device (IRD)
IRS. See Internal Revenue Service (IRS)
ISMP. See Institute for Safe Medication Practices (ISMP)
ISO. See Incident safety officer (ISO); Insurance Services Office (ISO); International Organization for Standardization (ISO)
ISO 9001, 302
IV line. See intravenous line (IV)
IWP. See Incident with potential (IWP)

J
Job, safety of, 21
Job description, 330–332
  extracting the news/extricating patients, 331–332
  investigating the news/assisting law enforcement, 331
  overview, 330
  reporting the news/responding to emergencies, 330
Job performance requirements (JPRs), 142, 430
Jones, Tony, 336–337
JPRs. See Job performance requirements (JPRs)
Juvenile firesetter intervention specialist, 373, 395

K
Kinetic energy, 103, 131
King, Rodney, 327
Klein, Gary, 67
Knowledge training programs, 142
Knowledge-based problem solving, 129–130

L
Ladder raise and extension test, 178
Large-diameter hose (LDH), 20, 146, 233
Leadership during change, 7–15
  economic impacts, 8–9
  managing change, 8–9
  opportunities for change, 13–15
  overview, 7–8
  political impacts, 9
  resistance to change, 9–13

  social impacts, 9
  technological impacts, 9
Learn Not to Burn®, 373, 395
LEDs. See Light-emitting diodes (LEDs)
Level II staging, 340, 345
Life safety
  education for firefighter safety, 391–394
    first responder actions, 393–394
    preventing emergencies, 391–392
    reporting emergencies, 392–393
  residential sprinklers and, 409
Light-emitting diodes (LEDs), 232
Line-of-duty deaths (LODDs)
  being caught, trapped, or lost, 242–244
    collapse, 242–243
    robots, 243–244
  being struck
    by debris, 238–239
    by equipment, 237–238
    by vehicles, 236–237
  caused by terrorism, 327
  communication, 244
  defined, 30, 57, 94, 188, 220, 246, 345, 367, 450
  emergency services change, 2
  falls, 233–236
    from apparatus, 233–234
    from buildings, 235–236
    from ladders and aerial devices, 234–235
  flashover, 239–242
    simulators, 239–242
    warning devices on gear, 239
  Initiative 9 and, 194
  maintenance vs. repairs, 183
  motor vehicle crashes, 231–233
    crash investigation, 233
    emergency response, 232–233
    vehicle safety, 231–232
  NIOSH reports of, 36
  responder rehabilitation, 76
  safe design award program for prevention of, 428
  statistics, 228
  stress and overexertion, 228–231
    failure to cope with stress, 352
    incident integration, 230–231
    medical monitoring, 229
Live fire training, 205
LODDs. See Line-of-duty deaths (LODDs)

M
Magazines for spreading knowledge, 266–267
Management
  leadership during change, 8–9
  no-fault management, 22–24
  supervisor responsibilities, 25–27
    barriers, 25–27
    problem filtering, 25
Mannion, Paul, 89–90
Manual on Uniform Traffic Control Device (MUTCD), 430
Manufacturers of equipment, 431–432
Maslow's hierarchy of needs, 199, 220
Mass casualty incident (MCI), 342
Match funding, 279, 297
Maximum heart rate (MHR), 175, 188
Mayday, 63, 94

McCoy, Thomas J., 101
MCI. *See* Mass casualty incident (MCI)
McPhee, John, 328–329
Measured impacts of grants, 282
Medical and fitness standards
   body metabolism, 164–176
      aerobic vs. anaerobic, 174–176
      fuel, 172–174
      overview, 164–165
      oxygen, 165–172
   efficiency factors of firefighting, 184–187
      body composition, 186
      extrinsic factors, 185–187
      fitness level, 186
      movement speed, 185
      muscle fiber composition, 186
      technique, 186–187
      work rate, 185
   firefighter statistics, 163–164
   Initiative 6, 207–208
   maintenance vs. repairs, 182–184
      annual pump testing, 184
      preventive maintenance, 182–183
      truck checks, 183–184
   NFPA 1582 standards, 181–182
   NFPA 1583 standards, 180–181
   overview, 160–161
   physical agility testing, 176–180
      Candidate Physical Ability Test, 177–179
      combat challenge, 176–177, 187
      fit for duty vs. wellness, 179–180
      overview, 176
   push for health and fitness, 162–164
Mental health support, 361
Mentoring, 27, 30
Metabolic equivalent (MET), 172, 188
Metabolism. *See* Body metabolism
MHR. *See* Maximum heart rate (MHR)
Military applications, 227–228
Mine Safety and Health Administration (MSHA), 251
Mini grants, 282–283
Mission statement, 277, 297
Model Procedures Guide for Structural Firefighting, 69
Modeling policies, 321–322
Motor vehicle crashes, 231–233
   being struck by vehicles, 236–237
   crash investigation, 233
   emergency response, 232–233
   vehicle safety, 231–232
Movement speed efficiency factor, 185
MSHA. *See* Mine Safety and Health Administration (MSHA)
Muscle fiber composition, 186
MUTCD. *See* Manual on Uniform Traffic Control Device (MUTCD)
Myocardial infarction, 166, 188

N

N-95 filter mask, 86, 94
NAHB. *See* National Association of Home Builders (NAHB)
Narrative for grants, 289–294
   objectives (tactical), 289–290
   procedures (task), 290–294
   significance (strategic level), 289
National Aeronautics and Space Administration (NASA)
   military applications, 227
   near-miss reporting, 255

National Association of Home Builders (NAHB), 408
National Board on Fire Service Professional Qualifications (NBFSPQ, or ProBoard), 139, 149
National Fallen Firefighters Foundation (NFFF)
   defined, 270, 297–298
   functions, 30
   Heritage Program funding, 278
   opportunities for change, 14
   website, 267
National Fire Academy (NFA), 8, 198
National Fire Fighter Near-Miss Reporting System, 255, 270, 278, 298
National Fire Incident Reporting System (NFIRS), 138
National Fire Protection Association (NFPA). *See also specific NFPA*
   checklists, 123
   committees for equipment safety, 428–430
      apparatus standards, 429
      equipment standards, 429
      personnel standards, 430
   educational programs, 373
   military applications, 227
   purpose of, 5
   residential fire sprinklers, 397
   safety officer, 70
   SCBA standards, 275
   standard policies and procedures, 302
   uniforms, 333
National Firefighter Near-Miss Reporting System, 211, 220
National Highway Traffic Safety Administration (NHTSA), 259
National Highway Transportation Safety Administration (NHTSA), 144
National Incident Management System (NIMS)
   CERT programs, 391
   checklists, 123
   fire service ratings, 52
   grant recipient eligibility, 285
   NIMS 700, 315
   personnel standards, 430
   purpose of, 69
   response standards, 314–316
   standardized communications terminology, 88
   training and certification standards, 138
National Institute for Occupational Safety and Health (NIOSH)
   built-in protection, 435
   fatality reports, 5
   investigations, 250
   LODD reports, 36
   NIOSH Alerts, 264–265, 270
   response standards, 306
   situational awareness, 202
   unsafe acts, 111
National Institute for Standards and Testing (NIST), 99
National Institute of Child Health and Human Development (NICHD), 283
National Institute of Standards and Technology (NIST)
   defined, 220, 246, 270, 450
   equipment safety, 430
   flashover tests, 201
   functions, 131
   military applications, 227
   research projects, 265

National Registry of Emergency Medical Technicians, 23, 30
National research agenda, 197–215
   accountability, 199–200
   creating a research agenda, 208–210
   cultural change, 198–199
   defined, 220
   emotional support, 214
   equipment safety, 215
   grants and safety, 212
   investigations, 211–212
   medical and fitness standards, 207–208
   public education, 214–215
   purpose of, 193
   residential fire sprinklers, 215
   response standards, 212–213
   risk management, 200–203
   technology, 210–211
   training and certification standards, 204–206
   unsafe acts, 203
   violent incidents, 213–214
National Transportation Safety Board (NTSB), 224–225, 445
National Volunteer Fire Council (NVFC)
   defined, 188, 298
   health and fitness standards, 162, 163
   Heart-Healthy Firefighter Program, 278
NBFSPQ. *See* National Board on Fire Service Professional Qualifications (NBFSPQ, or ProBoard)
NCSU. *See* North Carolina State University (NCSU)
Near-miss reporting, 253–259
   advantages of, 253–255
   components, 255–257
      confidential reporting, 256–257
      nonpunitive reporting, 257
      voluntary reporting, 255–256
   defined, 270
   investigations, 250, 261
   process of, 257–259
      contact information, 259
      event description, 258
      incident information, 258
      lessons learned, 259
      reporter information, 257
Newton, Isaac, 9
NFA. *See* National Fire Academy (NFA)
NFFF. *See* National Fallen Firefighters Foundation (NFFF)
NFIRS. *See* National Fire Incident Reporting System (NFIRS)
NFM. *See* No-fault management (NFM)
NFPA. *See* National Fire Protection Association (NFPA)
NFPA 13, Standard for the Installation of Sprinkler Systems, 398, 405
NFPA 13D, Standard for the Installation of Sprinkler Systems in One- and Two-Family Dwellings and Manufactured Homes, 398–401, 402, 405
NFPA 72, National Fire Alarm and Signaling Code, 401
NFPA 96, Standard for Ventilation Control and Fire Protection of Commercial Cooking Operations, 398

494   INDEX

NFPA 450, Guide for Emergency Medical Services and Systems, 322
NFPA 471, Recommended Practice for Responding to Hazardous Materials Incidents, 322
NFPA 1001, Standard on Firefighter Professional Qualifications, Firefighter I, and Firefighter II, 141, 142, 143, 148
NFPA 1021, Standard for Fire Officer Professional Qualifications, 142, 206
NFPA 1026, Standard for Incident Management Personnel Professional Qualifications, 430
NFPA 1035, Standard for Professional Qualifications, 373
NFPA 1201, Standard for Providing Emergency Services to the Public, 322
NFPA 1221, 51
NFPA 1403, Standard on Life Fire Training Evolutions, 50, 99, 127, 205
NFPA 1407, Standard for Fire Service Rapid Intervention Crews, 322
NFPA 1500, Standard on Fire Department Occupational Safety and Health Program
   addressing unsafe acts, 116
   health and safety standard, 50
   long-term planning, 51
   modeling, 322
   no effort, 50–51
   protective equipment, 334
   recertification requirements, 151
   research hypothesis, 195
   some effort, 51
NFPA 1521, 70
NFPA 1561, 51
NFPA 1582, Standard on Comprehensive Occupational Medical Program for Fire Departments, 181–182
NFPA 1583, Standard on Health-Related Fitness Programs for Fire Fighters, 164, 180–181
NFPA 1710, Standard for the Deployment of Fire Suppression Operations, Emergency Medical Operations, and Special Operations to the Public by Career Fire Departments
   health and fitness standards, 162
   research hypothesis, 195
   response standards, 322
   SAFER grants, 284
   staffing reductions, 26
NFPA 1720, Standard for the Organization and Deployment of Fire Suppression, Emergency Medical Operations, and Special Operations to the Public by Volunteer Fire Departments
   health and fitness standards, 162
   research hypothesis, 195
   response standards, 322
   SAFER grants, 284
   staffing reductions, 26
   water tenders, 313
NFPA 1901, Standard for Automotive Fire Apparatus, 429
NFPA 1911, 184
NFPA 1971, Standard on Protective Ensembles for Structural Fire Fighting and Proximity Fire Fighting, 429
NFPA 1981, Standard on Open-Circuit Self-Contained Breathing Apparatus (SCBA) for Emergency Services, 275

NFPA 1982, 242
NHTSA. *See* National Highway Traffic Safety Administration (NHTSA); National Highway Transportation Safety Administration (NHTSA)
NICHD. *See* National Institute of Child Health and Human Development (NICHD)
NIMS. *See* National Incident Management System (NIMS)
NIOSH. *See* National Institute for Occupational Safety and Health (NIOSH)
NIST. *See* National Institute for Standards and Testing (NIST)
No Higher Honor, 67
No-fault management (NFM), 22–24, 47
North Carolina State University (NCSU), 439
NTSB. *See* National Transportation Safety Board (NTSB)
Nuclear weapons, 328–329
NVFC. *See* National Volunteer Fire Council (NVFC)

O

Objective analysis
   defined, 298, 421
   in-kind contributions, 295
   sprinkler plan review, 411
Objectives, 289–290, 298
Occupational Safety and Health Administration (OSHA)
   checklists, 123
   rapid intervention, 73
   response standards, 316–317
   situational awareness, 202
Officer in charge (OIC), 11, 120, 338
Officer promotions, 319–320
OIC. *See* Officer in charge (OIC)
OJT. *See* On-the-job training (OJT)
Oklahoma City bombing, 329, 352
On-the-job training (OJT), 144–145
Open consensus-based teaching method, 142
Operating grants, 280
Operational Risk Management (ORM), 63, 94
Operations level, 28, 154
Opportunities for change, 13–15
Organization
   no-fault management, 47
   personal accountability to the, 47
   responsibility for, 47–49
Organizational accountability, 47–56
   defined, 48, 57, 220
   health and safety, 49–56
   implementation, 54–56
   Initiative 2, 199
   NFPA 1500, 50–51
   no-fault management, 47
   rating systems, 52–54
ORM. *See* Operational Risk Management (ORM)
OSHA. *See* Occupational Safety and Health Administration (OSHA)
Overexertion, 163, 188, 228–231
Oxidizer, 86, 94
Oxygen, 165–172
   oxygen uptake, 171–172, 189
   the pipes, 170–171
   the pump, 165–170

P

Paeno, Lynne Marie, 285
Page, Steven B., 306
Panoramic detection system (PDS), 233, 246

PAR. *See* Personnel accountability report (PAR)
Parasympathetic nervous system, 166, 189
PASS. *See* Personal alert safety system (PASS)
Passive fire protection, 215, 220
Passport system, 71
PDA. *See* Personal digital assistant (PDA)
PDS. *See* Panoramic detection system (PDS)
Peer fitness trainer (PFT), 180, 189
Performance measures for response, 213
Periodicals for spreading knowledge, 266–267
Peripheral vascular resistance (PVR), 171, 189
Personal accountability
   ability, 38
   defense mechanism, 38
   defined, 38, 57, 220
   improving of, 41–47
      become active member, 42–43
      decision to be accountable, 41–42
      speak up, 44
      take responsibility for outcome, 44–45
      take responsibility for safety, 45–47
   Initiative 2, 199
   postincident critique, 39–40
   in professional sports, 40–41
   reducing injuries, 261
   responsibility for outcome of calls, 44–45
      controllable components, 45
      semicontrollable components, 45
      uncontrollable components, 45
   responsibility for safety, 45–47
      equipment, 46
      organization, 47
      services, 46
      training, 46
   taking responsibility, 39
Personal alert safety system (PASS), 88, 128, 227
Personal digital assistant (PDA), 244
Personal identification number (PIN), 155
Personal identification verification (PIV), 155, 157
Personal protective equipment (PPE)
   culture research, 199
   defined, 94, 131, 189, 220, 246, 323, 345, 450
   equipment safety, 426
   HANS device, 105
   limitations, 106
   NFPA standards, 403
   perceptions, 105–106
   for rapid intervention groups, 75
   response standards, 317
   shootings and, 342
   sweating while using, 169–170
   task level risk management, 83
   technology trends, 227
   unsafe practices, 99
Personally owned vehicles (POVs), 208, 313
Personnel
   education of, 390–391
   qualification standards, 430
   responding, 313–316
      apparatus positioning, 314

INDEX **495**

command system, 314–316
risk management, 316
Personnel accountability report (PAR)
communication, 127
defined, 94–95, 131, 270
Saving Your Own curriculum, 266
scene accountability, 71
PFA. *See* Psychological first aid (PFA)
PFT. *See* Peer fitness trainer (PFT)
PHCC. *See* Plumbing-Heating-Cooling Contractors Association (PHCC)
Physical ability testing, 150
Physical agility testing, 176–180
Candidate Physical Ability Test, 177–179
combat challenge, 176–177, 187
fit for duty vs. wellness, 179–180
overview, 176
Physiological response to emergencies, 208
PIN. *See* Personal identification number (PIN)
PIO. *See* Public information officer (PIO)
PIV. *See* Personal identification verification (PIV)
Plumbing-Heating-Cooling Contractors Association (PHCC), 403
PM. *See* Preventive maintenance (PM)
Point of sale, 417, 421
Policies and procedures
collaboration, 321–322
standards, 303–305
Policy
defined, 323
prevention through design, 445
standard policies and procedures, 301
Political impacts of change, 9
Popov, Boris, 431
Positioning for emergency response, 336
Positive securing device (PSD), 307–308, 323
Positive train control (PTC), 446, 450
Positive-pressure ventilation (PPV)
defined, 30, 270
emergency services change, 3
Saving Your Own curriculum, 267
Postincident critique
after unsafe acts, 120
defined, 57
personal accountability, 39–40
Post-incident stress management, 214
Post-traumatic stress disorder (PTSD), 352–363
defined, 367
emotional support for, 214
group therapy, 354–360
individual therapy, 361–363
overview, 352–354
Potential energy, 103, 131
Pounds per square inch (psi), 169
POVs. *See* Personally owned vehicles (POVs)
PPC. *See* Public Protection Classification (PPC)
PPE. *See* Personal protective equipment (PPE)
PPV. *See* Positive-pressure ventilation (PPV)
Practice, prevention through design, 442–445
Preaction system, 400, 421
Presidential Directive 12, 155
Prevention through design (PTD), 435
Prevention vs. preparation, 370–371
Preventive maintenance (PM), 182
Price of projects, 292–293
Proactive change, 4–5, 30

Problem filtering, 25, 30
Problem solving
history of, 372–373
knowledge-based, 129–130
rules-based, 129
skills-based, 129
ProBoard, 139, 149
Procedure. *See also* Policies and procedures
defined, 323
for grants, 290–294
response standards, 301
Procrastination, 10
Professional appearance, 291
Professional development, 205–206
Professional Grant Writer: The Definitive Guide to Grant Writing Success, 285
Professional qualifications, 142, 157
Program seed grants, 283–284
Promotional exams, 55
Proselytize, 363, 367
Protective equipment, standard procedures, 334–335
Proteins for metabolism, 173–174
Proximity dangers, 314
PSD. *See* Positive securing device (PSD)
Psi. *See* Pounds per square inch (psi)
Psychological first aid (PFA), 360
Psychomotor domain, 140, 157
Psychotherapy, 354, 367
PTC. *See* Positive train control (PTC)
PTD. *See* Prevention through design (PTD)
PTSD. *See* Post-traumatic stress disorder (PTSD)
PTT. *See* Push to talk (PTT)
Public education
life safety education for firefighter safety, 390–394
first responder actions, 393–394
preventing emergencies, 391–392
reporting emergencies, 392–393
overview, 370
partnerships, 386–391
funding, 389–390
personnel, 390–391
project management, 386–389
prevention vs. preparation, 370–371
program development, 375–386
atmosphere, 378–386
audience, 377–378
community groups, 380
educational facilities, 379
hot topics, 376–377
media, 385–386
message, 375–377
recreational locations, 380–384
workplace locations, 379–380
progressive message, 372–374
diversification of duties, 372
problem solving, 372–373
proven success rate, 373
project management, 386–389
camaraderie, 388
careful planning, 386–387
champion, 386
coalition, 386
collaboration, 387–388
commitment, 389
compelling case, 387
continuity, 388
creativity, 388
credentials, 387
Public information officer (PIO), 373, 395
Public Protection Classification (PPC), 52, 57
Public shootings, 340–342

Pull station, 311, 323
Pulmonary edema, 169, 189
Purveyor, 415, 421
Push to talk (PTT), 244
PVR. *See* Peripheral vascular resistance (PVR)

Q
Qualifications for grants, 291–292
Qualitative data, 194, 221
Qualitative research, 288, 298
Quantitative data, 194, 221
Quantitative research, 288, 298
Queens College Step Test, 172, 189
Quick-response grants, 282–283

R
RACE, 393, 395
Radio communications, 86–92
citizen band (CB), 212, 220
cognitive radio, 244, 245
effective communications, 89–90
acknowledge and repeat, 90
hail first, 90
politeness, 90
sandwiched transmission, 90
short and to the point, 90
ineffective communications, 87–89
background noises, 88
excitement, 87
multitasking, 89
terminology, 88–89
software defined radio, 244, 246
stressful communications, 91–92
emergency traffic, 91–92
fire is still burning, 92
multiple requests, 91
say it now, say it later, 92
stay here with me, 92
Radiological dispersal device (RDD), 328
Rapid intervention, 73–76
IDLH standards, 73–74
RIC/RIT/FAST, 74
RIG, 74–76
Rapid intervention crew (RIC), 74
Rapid intervention group (RIG), 74–76
best location for, 74–75
equipment, 75
functions, 75–76
special assignment, 74
Rapid intervention team (RIT), 74
Rating systems, 52–54
FLSI 16, 53–54
Insurance Services Office, 52
International Organization for Standardization, 53
RAVE II^{TM}, 241, 246
RDD. *See* Radiological dispersal device (RDD)
Reactive change, 4–5, 30
Reason, James, 107
Recognition-primed decision making (RPD), 67–69, 95
Recruit schools, 54
Recruit-level training, 204
Rehabilitation (rehab), 76, 95
Relative anonymity, 363–364, 367
Remembering learning process, 140
Remote patient monitoring (RPM), 229, 246
Repetitive motion injuries, 439–441
Reports for grants, 296
Reproductive health issues, 207–208

496 INDEX

Rescue, Alarm, Confine, Extinguish (RACE), 393, 395
Rescue drag test, 178
Research
  costs of safety, 447
  data collection, 215–219
    analyzing the data, 217–218
    hypothesis, 216
    during investigations, 261–262
    plan for, 216–217
    utilizing the data, 218–219
  designing safety, 436–445
    proper design, 438–439
    proper equipment, 437
    proper positioning, 437
    proper technique, 437–438
    repetitive motion injuries, 439–441
    specific injuries, 436–439
  due regard for safety, 195
  equipment safety, 430
  for grants, 288
    qualitative research, 288
    quantitative research, 288
  hypothesis, 195–196
  national research agenda. *See* National research agenda
  qualitative data, 194–195
  quantitative data, 194
  research and development, 196–197
  research grants, 280–282
  school resource officers, 192
  as a tool, 194–197
  types of, 194
Residential fire sprinklers
  active vs. passive fire protection, 215
  code enforcement, 410–412
  design factors, 399–400
  dry systems, 400
  future approaches, 415–416
  mandates for, 413
  NFPA 13D, 398–401
  opposition to, 401–410
    cost, 401–404
    freedom of choice, 407–408
    need, 408–410
    resale value, 407
    water availability, 404–405
    water damage, 405–407
  preaction systems, 400
  public awareness of need for, 215
  retrofitting, 416–420
  sprinkler advocacy, 413–415
  wet systems, 400
Resistance to change, 9–13
Resources for training programs, 142–143
Respiratory protection
  Initiative 8, 210–211
  SCBA. *See* Self-contained breathing apparatus (SCBA)
  SCUBA, 86, 95
Responder rehabilitation, 76, 95
Response standards, 301–323
  communications, 212–213
  implementation, 321–322
  levels of standard policies and procedures, 309–320
    Series One (S-1: National), 309–316, 324
    Series Three (S-3: Specialized), 319–320, 324
    Series Two (S-2: Template), 316–319, 324
  overview, 301–302
  performance measures, 213

  policy collaboration, 321–322
  strategic response to risk, 212
  terminology, 302–308
    policies and procedures, 303–305
    SOPs and SOGs, 305–306
    writing style, 306–308
Responsibility vs. accountability, 33–35, 57
RIC. *See* Rapid intervention crew (RIC)
RIC/RIT/FAST, 74
RIG. *See* Rapid intervention group (RIG)
Rinn, P. X., 66–67
Ripkin, Cal, Jr., 40–41
Risk management
  culture research, 199
  decision making, 64–69
    recognition-primed, 67–69
    training-based, 65–67
  defined, 30–31, 57, 95, 131, 221, 323
  emergency services change, 2
  fireground safety, 93
  incident management, 69–76
    emergency traffic, 72–73
    FIRESCOPE, 69
    ICS, 69
    NIMS, 69
    rapid intervention, 73–76
    responder rehabilitation, 76
    safety officer, 69–70
    scene accountability, 70–72
  Initiative 3, 200–203
    collapse prediction, 201
    flashover prediction, 201
    incident management, 202–203
    scientific community risk assessment, 200–201
    situational awareness, 201–202
    wildland firefighting operations, 202
  Operational Risk Management, 63, 94
  overview, 63–64
  personal accountability, 43
  response standards, 316
  risk-benefit analysis, 60–63
  SPE model, 63–64
  standard policies and procedures, 302
  strategic level, 77–81
    rural risk management, 78–81
    urban risk management, 77–78
  tactical level, 81–83
  task level, 83–92
    safe search, 83–84
    safe tools, 84
    situational awareness, 84–92
  unsafe acts, 110–111
Risk Watch®, 275–276, 298, 373–374, 395
Risk-benefit analysis, 60–63
  civil disturbance, 338
  defined, 60, 95, 131, 345
  empowerment and, 102
RIT. *See* Rapid intervention team (RIT)
Roadside memorial, 353, 367
Robots, 243–244
Roentgens, 88, 95
Root cause
  defined, 25, 31, 270, 395
  investigations of, 252
  prevention vs. preparation, 371
  problem filtering, 25
RPD. *See* Recognition-primed decision making (RPD)
RPM. *See* Remote patient monitoring (RPM)
Rules-based problem solving, 129
Rural strategic risk management, 78–81

S
Safe search, 83–84
  accountability, 83
  building aspects, 83
  emergency procedures, 84
  methodology, 84
  personal protective equipment, 83
SAFER. *See* Staffing for Adequate Fire and Emergency Response (SAFER)
Safety. *See also* Health and safety accountability
  changes for, 5–6
  engineering safety into equipment, 215
  equipment, 104–106. *See also* Personal protective equipment (PPE)
  fireground, 93
  personal accountability for, 45–47
Safety culture
  certification level, 27–28
    awareness level, 28
    operations level, 28
    technician level, 28
  changes, 15–28
    certification level, 27–28
    components, 22–27
    creation of, 17–19
    cultural changes for safety, 19–22
    emergency services, 2–3
    evolution to, 16–17
    overview, 15–16
  components, 22–27
    employee responsibilities, 27
    no-fault management, 22–24
    supervisor responsibilities, 25–27
  defined, 31, 57
  personal accountability, 42–43
Safety curve, 425–428
  Improvement Phase 1, 426–427
  Improvement Phase 2, 427
  Improvement Phase 3, 427–428
Safety officer, 69–70
Safety targeted assessment via gateway evaluation (STAGE), 442
SAM. *See* Staging area manager (SAM)
Santayana, George, 250
SARE. *See* Sustainability Agriculture Research and Education (SARE)
Saving Your Own curriculum, 266
SCA. *See* Sudden cardiac arrest (SCA)
Scalar references, 217, 221
SCBA. *See* Self-contained breathing apparatus (SCBA)
Scene accountability, 70–72
  barcode scanning software, 71–72
  passport system, 71
  span of control, 72
  tag system, 71
Scene protection, 314
SCG. *See* Fire Station Construction Grants (SCG)
School resource officer (SRO), 192, 221, 280, 298
School shootings, 340–342
Scientific community risk assessment, 200–201
SCUBA. *See* Self-contained underwater breathing apparatus (SCUBA)
SDO. *See* Standards developing organization (SDO)
SDR. *See* Software defined radio (SDR)
Search test, 178

INDEX

Secondary audience, 377, 395
Secondary device, 329–330, 345
Self-contained breathing apparatus (SCBA)
    common skills requirements, 146
    for crowd control, 338
    instructor use for training, 204–205
    limitations, 106
    medical monitoring, 229
    NFPA standards, 275
    repair responsibilities, 36
    safety standards, 429
    supervisor responsibilities, 26–27
    training-based decision making, 66
Self-contained underwater breathing apparatus (SCUBA), 86, 95
Self-defense, 336
September 11, 2001, 329, 330, 370
Series One (S-1: National) standard policies and procedures, 309–316, 324
Series Three (S-3: Specialized) standard policies and procedures, 319–320, 324
Series Two (S-2: Template) standard policies and procedures, 316–319, 324
Services, personal accountability to, 46
Shielding from debris, 238–239
Shootings, 340–342
SIDS. See Sudden infant death syndrome (SIDS)
Significance-level information, 289
Simulators, 239–242
    realistic environment, 241
    realistic gear, 240–241
    realistic movement, 241
    realistic response, 241–242
Single share, 356–357, 367
Situational awareness
    air management, 86
    crew resource management, 124–125
        mistakes, 124–125
        task saturation, 124
    defined, 95, 131, 270
    history of concept, 84–85
    mismatching expectations with realizations, 85–86
    near-miss reporting, 256
    radio communications, 86–92
    risk management techniques, 201–202
    strategic risk management, 80
Size-up, 335–336
Skills training programs, 142
Skills-based problem solving, 129
Slow twitch (Type I) muscle, 186, 189
Smith, Stew, 163
Social impacts of change, 9
Software defined radio (SDR), 244, 246
SOGs. See Standard operating guidelines (SOGs)
SOPs. See Standard operating procedures (SOPs)
SOTER. See Supervisor of tracking employee resources (SOTER)
Span of control, 72, 95
SPE (severity, probability, exposure) model, 63–64, 95, 447, 451
Speaking up and personal accountability, 44
Special events, 320
Specialized operations teams, 319
Specific gravity, 172, 189
Spirometry, 172, 189
Sports utility vehicle (SUV), 231
Sprinklers. See Residential fire sprinklers
SRO. See School resource officer (SRO)

SRS. See Supplemental restraint system (SRS)
S-T elevated myocardial infarcation (STEMI), 136
Staffing for Adequate Fire and Emergency Response (SAFER), 284–285, 298
STAGE. See Safety targeted assessment via gateway evaluation (STAGE)
Staging, 340–341
Staging area manager (SAM), 430, 451
Stair climb test, 178
Standard of care, 136–137, 157
Standard operating guidelines (SOGs)
    empowerment, 101
    response standards, 302–303, 305–306
    triage and treatment, 342
    writing style, 306–308
Standard operating procedures (SOPs)
    emergency services change, 3
    empowerment, 101
    personal accountability, 43
    response standards, 302–303, 305–306
    triage and treatment, 342
    writing style, 306–308
Standards developing organization (SDO), 302, 324
State Farm® insurance safety grants, 279
Static, defined, 31
Statistics
    evaluation of, 228
    firefighter, 163–164
    public education, 375–376
        local statistics, 376
        national and state statistics, 375–376
Stefano, Michael, 175
STEMI. See S-T elevated myocardial infarction (STEMI)
Strategic level risk management, 77–81
    rural risk management, 78–81
    urban risk management, 77–78
Strategic weapons attack team (SWAT), 213, 306
Stress, 348–352
    burnout, 350
    coping with, 349–351
        alcohol abuse for coping, 350–351
        drug abuse for coping, 351
    failure to cope, 352
    overexertion and, 228–231
    overview, 348–349
    post-incident stress management, 214
    PTSD. See Post-traumatic stress disorder (PTSD)
"Struck by" injuries, 236–239
    by debris, 238–239
        distance, 238
        shielding, 238–239
        time, 238
    by equipment, 237–238
    by vehicles, 236–237
Subjective analysis
    defined, 298, 421
    in-kind contributions, 295
    sprinkler plan review, 411
Substance abuse, 214
Sudden cardiac arrest (SCA), 164
Sudden infant death syndrome (SIDS), 283, 375
Supervisor of tracking employee resources (SOTER), 230, 246
Supervisor responsibilities, 25–27
    barriers, 25–27
    problem filtering, 25
Supplemental restraint system (SRS), 211, 376–377, 430

Support system, 366
Sustainability Agriculture Research and Education (SARE), 280, 298
Sustainability of programs for grants, 287
SUV. See Sports utility vehicle (SUV)
SWAT. See Strategic weapons attack team (SWAT)
SWAT Leadership and Tactical Planning, 336–337
Sweating, 169–170
Sympathetic nervous system, 166, 189
Systole, 166, 189

T

Tactical EMS (TEMS), 331
Tactical level risk management, 81–83
Tag system, 71
Taratogens, 208, 221
Target hazards, 146–147
Task level risk management, 83–92
    safe search, 83–84
    safe tools, 84
    situational awareness, 84–92
Task saturation, 124
Teachable moment, 412, 421
Technician level
    of certification, 28
    of training, 154
Technique for efficiency, 186–187
Technology
    change, impacts of, 9
    fire detection and extinguishing agents, 211
    line-of-duty deaths, types of, 228–244
        being caught, trapped, or lost, 242–244
        being struck, 236–239
        communication, 244
        falls, 233–236
        flashover, 239–242
        motor vehicle crashes, 231–233
        stress and overexertion, 228–231
    overreliance on, 224–226
    respiratory protection, 210
    thermal imaging camera, 224
    utilization of, 226–228
        current trends, 227
        military applications, 227–228
        present needs, 226–227
        statistics, 228
TEMS. See Tactical EMS (TEMS)
Tentative interim amendment (TIA), 429, 451
Terrorism, 327–330
    IEDs, 448
    LODDs, 327
    Oklahoma City bombing, 329, 352
    overview, 327–328
    radiological dispersal device, 328
    secondary devices, 329–330, 345
    September 11, 2001, 329, 330, 370
    weapons of mass destruction, 328–329
Tests
    American Society for Testing and Materials, 302
    annual pump testing, 184
    CPAT. See Candidate Physical Ability Test (CPAT)
    criterion task test, 176, 188
    fitness capabilities, 150
    flashover tests, 201
    National Institute for Standards and Testing, 99
    physical ability, 150

physical agility testing, 176–180
  ceiling pull, 179
  combat challenge, 176–177, 187
  criterion task, 179, 188
  equipment carry, 178
  fit for duty vs. wellness, 179–180
  forcible entry, 177, 178
  high-rise pack carry, 177
  hose advance, 177
  hose drag, 178
  hose hoist, 177
  ladder raise and extension, 178
  rescue drag, 178
  search test, 178
  stair climb, 178
  victim rescue, 177
Queens College Step Test, 172, 189
special situations, 182
Textbooks, 268
Thermal imaging camera (TIC), 224, 381
Thermodynamics, 169, 189
Throttle (heart rate), 166–169
TIA. *See* Tentative interim amendment (TIA)
TIC. *See* Thermal imaging camera (TIC)
Tiered training, 153–154
Timelines of grants, 292
Time-out, 119–120
Tools, safe, 84
TRADE. *See* Training Resources and Data Exchange (TRADE)
Trade-offs, 414, 421
Traffic preemption systems, 225, 246
Training. *See also* Education
  accreditation, 148–149
  business associations, 150
  career firefighter basic training, 143
  certification requirements, 143–145
  certification transfers to other locations, 135–137
    chain of survival, 137
    standard of care, 136–137
  common skills, 146
  computer-based, 206
  credentialing, 154–156
  crew resource management, 123–124
  deficiencies in, 139
  defined, 157
  evasive maneuvering, 336–337
  exercise training program, 180
  FESHE model, 152–153
  health and safety implementation, 55
  higher-educational requirements, 149–150
  history of standards, 138–139
  initial requirements, 139–143
  Initiative 5. *See* Initiative 5 (training and certification standards)
  in-service requirements, 145–148
    common skills, 146
    funding issues, 147–148
    target hazards, 146–147
  knowledge training programs, 142
  live fire training, 205
  on-the-job training, 144–145
  personal accountability to, 46
  physical ability testing, 150
  positioning, 336
  recertification requirements, 151–152
  recruit-level, 204
  resources for programs, 142–143
  self-defense, 336

  seminars, 268
  simulators, 205
  size-up, 335–336
  skills, 142
  tiered training, 153–154
  TRADE, 321, 324
  training-based decision making, 65–67, 95
  volunteer firefighter basic training, 143–145
Training Resources and Data Exchange (TRADE), 321, 324
Transfer valve (volume or pressure), 169
Transportation Security Administration (TSA), 139
Trending, 181–182, 189
Truck checks, 183–184
TSA. *See* Transportation Security Administration (TSA)
Two-in, two-out rule, 73

U
Understanding learning process, 140
Uniforms
  apparel for EMS worker safety, 7
  battle dress uniform, 333
  safety of, 21
  standard procedures, 332–333
United States Coast Guard (USCG), 63
United States Fire Administration (USFA), 153, 275
Unsafe acts
  accidents, 103–108
  crew resource management, 121–130
    communication, 125–129
    policies and procedures, 123–124
    problem solving, 129–130
    situational awareness, 124–125
  defined, 270
  fatalities linked to, 252
  firefighter empowerment, 100–103
    defined, 100
    education, 101–102
    empowerment, 102–103
    enablement, 102
  foundation of, 98–100
  Initiative 4, 98, 203, 252
  prevention of, 112–115
    education, 113
    enforcement, 115
    engineering, 113–114
    environment, 114–115
  recognition of, 108–112
    disguised acts, 109–110
    distraction events, 111–112
    poor risk management decisions, 110–111
    spotlight on unsafe acts, 108–109
  stopping acts in progress, 115–120
    call a time-out, 119–120
    plan for the problem, 115–119
    postincident critique or debriefing, 120
    request task completion, 119
    speaking to your peers, 118–119
    speaking to your supervisor, 117–118
    working with unsafe cts, 120
Urban strategic risk management, 77–78
USCG. *See* United States Coast Guard (USCG)
USFA. *See* United States Fire Administration (USFA)

V
Vanadium oxide (VOx), 227, 246
Vapor density, 172, 189

Vasoconstriction, 171, 189
Vehicle crashes. *See* Motor vehicle crashes
Vent, enter, search (VES), 127, 131
Victim rescue test, 177
Victim syndrome, 36, 57
Videos, 268
Violent incident response
  animals, 343
  bomb threats and possible bombs, 342–343
  civil disturbance, 337–340
  examining responses to, 213–214
  job description, 330–332
    extracting the news/extricating patients, 331–332
    investigating the news/assisting law enforcement, 331
    overview, 330
    reporting the news/responding to emergencies, 330
  school or public shootings, 340–342
  standard procedures, 332–337
    protective equipment, 334–335
    training, 335–337
    uniforms, 332–333
  terrorism, 327–330. *See also* Terrorism
    overview, 327–328
    secondary devices, 329–330, 345
    weapons of mass destruction, 328–329
Viscosity of blood, 171, 189
VO2 max, 171, 189
Volunteer firefighters
  basic training, 143–145
  NFPA standards. *See* NFPA 1720, Standard for the Organization and Deployment of Fire Suppression, Emergency Medical Operations, and Special Operations to the Public by Volunteer Fire Departments
  NVFC. *See* National Volunteer Fire Council (NVFC)
von Clausewitz, Carl, 84–85
VOx. *See* Vanadium oxide (VOx)

W
On War, 84–85
Waves of change, 13
Weapon of mass destruction (WMD), 328–329, 345
Websites, 267–268
Wellness, 179–180, 189
Wellness-Fitness Initiative (WFI), 162, 189
Wet system, 400, 421
WFI. *See* Wellness-Fitness Initiative (WFI)
Wildland Fire Lessons Learned Center, 268, 270
Wildland firefighting operations, 202
WMD. *See* Weapon of mass destruction (WMD)
Worcester Polytechnic Institute (WPI), 228, 430
Work rate efficiency factor, 185
World Trade Center (WTC), 106
WPI. *See* Worcester Polytechnic Institute (WPI)
WTC. *See* World Trade Center (WTC)